INTRODUCTION to DISPLAY ENGINEERING

JunSin Yi | DoYoung Kim | YeonGon Mo | ByungSeong Bae | SungEun Lee
JangKun Song | ByoungDeog Choi | SungJei Hong | YounJung Lee

HONGREUNG

INTRODUCTION to
DISPLAY ENGINEERING

Ph. D. JunSin Yi
Professor, Sungkyunkwan University

Ph. D. DoYoung Kim
Professor, Ulsan College of Science

Ph. D. YeonGon Mo
Director, QCT Engineering/Qualcomm Korea

Ph. D. ByungSeong Bae
Professor, Hoseo University

Ph. D. JangKun Song
Professor, Sungkyunkwan University

Ph. D. SungEun Lee
LG Electronics Institute

Ph. D. ByoungDeog Choi
Professor, Sungkyunkwan University

Ph. D. SungJei Hong
Senior Researcher, Korea Electronics Tech. Institute

Ph. D. YounJung Lee
Research Professor, Sungkyunkwan University

Publisher
HONGREUNG PUBLISHING COMPANY(August 31, 2020)

ISBN-13
979-11-5600-637-4

e-mail
hongpub@hongpub.co.kr

Web
http://www.hongpub.co.kr

Adress
10, Insubong-ro 50-gil, Gangbuk-gu, Seoul, Korea
TEL : (02) 999-2274~5, 903-7037 / **FAX** : (02) 905-6729

₩ 33,000

CORRECTED EDITION 2020

Preface

Display technology is based on the state of the art semiconductor technology and it involves various applications of physical, electrical, electronics, mechanical, materials, chemistry, etc. For display, the industry is activated earlier than the research and development of schools, and manpower of related field is required rapidly. Taking the demand of industry into account, "the introduction to display engineering" was published to be used as a textbook for undergraduate students in Sungkyunkwon University, Korea. The book published about 7 years ago was in Korean. Since globalization is the trend in these days. The textbook is reviewed and is published in English after 3 years of long and lengthy work efforts.

Chapter 1 is the overview of a flat display. Chapter 2 will focus on FET and semiconductors which are basic ideas of LCDs, so that readers can understand thin film transistors (TFT) better to be dealt with in Chapter 2. In Chapter 3, liquid crystal displays (LCD) will be described in detail, especially focusing on an AM-LCD which is the most common mode in relation to LCDs. Liquid crystal types and their movement dynamics are discussed in Chapters 4. Chapter 5 covers driving method for a display panel. Chapter 6 will deal with the overall concept of organic light-emitting diodes (OLED). Chapters 7 probes new types of flexible displays. Chapter 8 will cover the principle and practice of microdisplays. In Chapter 9, plasma discharge and plasma display panel (PDP) related technology will be covered. Various display system now employs finger-tip touch screen. Chapter 10 focus on the various touch screen related concepts. Chapter 11 will focus on CRT, field emission display (FED). These types of displays are invented based on how electrons migrate under an electric field and a magnetic field, Chapter 11 will start from both theoretical electron movement theoretical perspectives. Chapter 12 will cover three-dimensional displays that were developed recently. In the last chapter 13, the current status of the whole display systems and the prospect for the various types of display systems will be discussed.

We hope that this book is useful for those who are related with display industry as well as for students studying display engineering. We thank everyone who helped us to publish this book.

January 2019

Authors

Contents

Chapter 01 | Flat Panel Display

Chapter 02 | Field Effect Transistor

Chapter 03 | Liquid Crystal Display, LCD

Chapter 04 | Liquid Crystal

Chapter 05 | Display system drive

Chapter 06 | Organic Light-Emitting-Diode OLED

Chapter 09 | Plasma Display

Chapter 10 | Touch Screen Panel

Chapter 12 | 3D Display

Chapter 13 | Display Application System

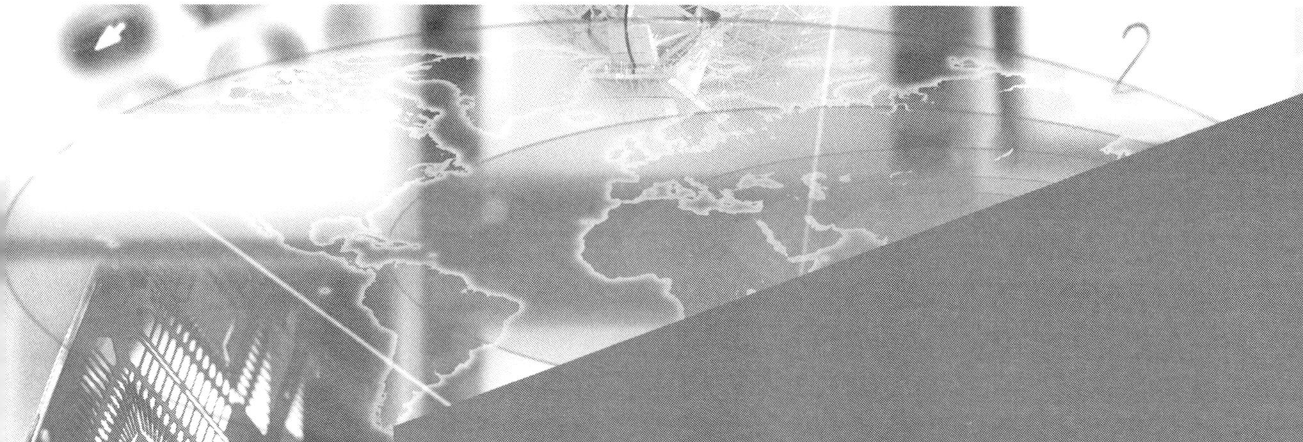

Flat Panel Display

Chapter 01

D i s p l a y E n g i n e e r i n g

*1 What is Display?

A display device is basically an equipment that shows data visually in the form of characters or figures on a screen. The data can be roughly categorized into the following three types: ① Characters such as numbers and alphabets, ② Chinese characters and ③ Figures indicating the result of computer processing (including visuals and patterns). From a functional perspective, a display device is a sort of human-machine interface, as it converts electrical signals from TV, computers, communication devices and/or other industrial machines to human readable information. This is done by converting the electrical signal into a certain pattern that is visually recognizable to human, as shown in Figure 1.1 In other words, it is an electronic device exchanging information with humans through electronic apparatus and visuals. It is also a component device connecting humans and electronic apparatus. Therefore, the role of electronic display devices is getting more important in our information-oriented society, and in recent years they are broadly used and applied in various fields including industry, culture, entertainment, medical science and so forth. Also, the display industry is one of the leading industries in Korea, following the semiconductor industry, with an increasing annual sales. To keep up with the diversifying needs of the information-oriented society, new and advanced electronic display devices continue to be developed.

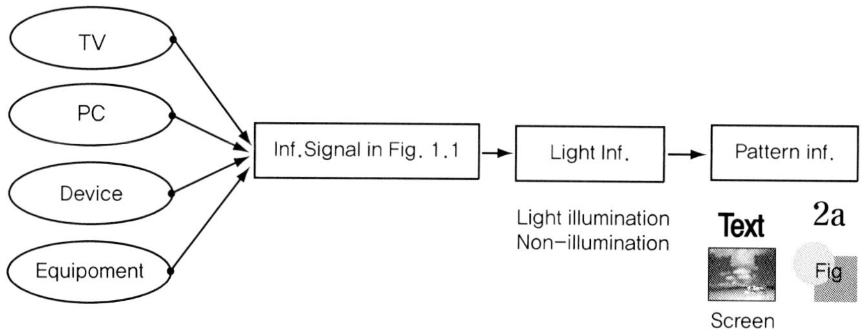

Figure 1.1
Information transfer method of display device

Displays are classified into two types based on how light signals are displayed. One is an emissive display where light signals are displayed by their own light emission. The other one is a non-emissive display in that the light signals are displayed by light modulation, influenced by reflection, scattering and/or interference of the surrounding lights. Figure 1.2 shows different categories of our most common display devices. Among those devices a CRT (Cathode Ray Tube) has the longest history, but has been replaced by other displays such as quid crystal display (LCD) and plasma display panel (PDP), and its market share has sharply decreased. Meanwhile, the LCD has gradually expanded its market share recently, from 9.4% in 2001 to 38.8% in 2007 with over a 40% average increase. Along with such a rapid growth, low power consumption, high-definition, flat, light-weight and touch-screen LCDs are now available to various portable information devices such as electronic note devices, tablet, personal digital assistant (PDA) with a stylus, where CRTs cannot be used. Also, the flat panel of LCDs has the merit of being used as wall-mounted TVs and/or large-screen HDTVs. Large-scale investment in organic light emitting diode (OLED) displays have continued since 2011. OLED uses organic electronic materials, so it has many advantages such as light weight, ultra-thin, low cost, and flexible. OLED is being applied to not only devices that emphasize mobility such as smart phones and tablet PCs, but also to large-screen TVs.

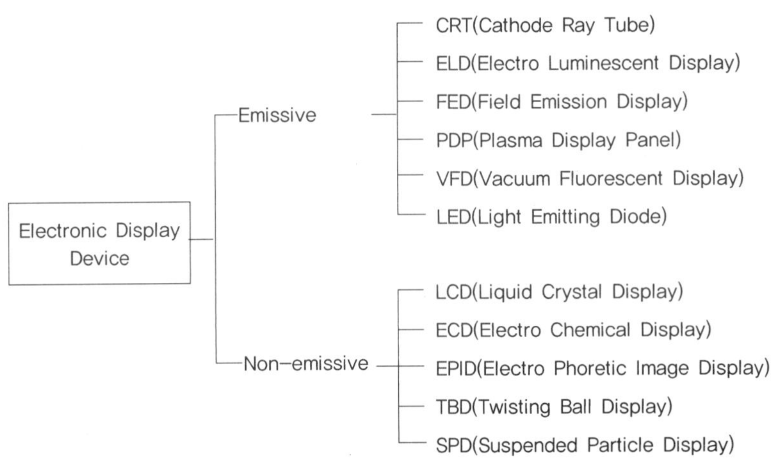

Figure 1.2

The categorization of major display devices based on light emission capability

In accordance with such trend, this book particularly deals with LCDs throughout three chapters. Chapter 2 describes FET and semiconductors which are basic units of LCDs so that readers can understand thin film transistors (TFT). In Chapter 3 and 4, LCD will be described in detail, especially focusing on an AM-LCD (active matrix – LCD) which is the most commonly used. Liquid crystal types and their operational dynamics are discussed. Chapter 5 covers driving method for a display panel. Chapter 6 will deal with the overall concept of organic light-emitting diodes (OLED). Chapters 7 probes new types of flexible displays. Chapter 8 will cover the principle and practice of micro displays. In Chapter 9, plasma discharge and plasma display panel (PDP) related technology will be covered. Various display systems now employ fingertip touch screen. Chapter 10 focus on the various touch screen related concepts. Chapter 11 will focus on CRT and field emission display (FED). These types of displays are invented based on how electrons migrate under an electric field and a magnetic field. Chapter 12 will cover three-dimensional displays that were developed recently. In the last chapter 13, the current status of the whole display systems and the prospect for the various types of display systems will be discussed. This book is intended for students and the general public who may have a personal interest in display systems. This book may also be helpful to the production line engineers in the displays industry with a basic understanding of the industry across the board.

* 2 History of Flat Panel Display

The major history of flat panel displays is explained in Table 1.1. The cathode ray tube (CRT) that was developed by Braun in 1897, is considered the first display device in history although Reinitzer found liquid crystal earlier in 1888. After CRT invention, seventy four years had passed until the first AM-LCD (Active Matrix Liquid Crystal Display) was invented by Brody in 1971. This chapter only summarizes a brief history of flat panel display technologies, with the details followed by next chapters.

Table 1.1 History of Major Technological Advance of Flat Panel Displays

Liquid Crystal Display (LCD)	Electro Luminescence Display (ELD)
• 1888 Reinitzer (Austria), found liquid Crystal • 1968 Heilmeier et al. (US), published a research paper on Technology for DS-LCD and GH-LCD • 1971 Schadt et al.(Switzerland), developed TN LCD • 1972 Microma, Rockwell (US), applied LCDs to clocks and electronic calculators • 1980 Dundee group (UK), developed a-Si TFT-LCDs • 1982 Seiko & Casio (Japan), developed black and white LCD TVs • 1984 Seiko (Japan), developed STN(SBE)-LCD technology	• 1936 Destiau (France), established a theory of Electro Luminescence on ZnS:Cu fluorescence substrates • 1950 Sylvania (US) developed dispersed AC (Alternating Current) driven EL devices • 1968 Bell Lab (US) developed AC driven ELDs with vapor deposited thin-film • 1978 Sharp (Japan) developed thin-film AC driven ELDs with the double-insulating layers • 1980 Lohja (Finland) developed AC driven ELDs with atomic layer epitaxy (ALE) thin-film • 1987 Tang et al (US) published organic thin-film

• 1991 Sharp (Japan), developed LCD projectors • 1996 Hitachi(Japen), Developed IPS (IN-Plane Switching) • 1998-2006 Hitachi(Japen), Evolution from S-IPS to H-IPS technology	displays • 1988 Planar System (US) developed full color ELDs • ELDs have almost disappeared because of LCDs.
Plasma Display (PDP)	**Field Emission Display (FED)**
• 1954 Skellet (US), published DC (Direct Current) driven PDPs • 1956 Burroughs (US), made a prototype of NIXIE tube • 1966 Illinois Univ (US), developed AC driven memory type PDPs • 1968 Philips (Netherland), conducted an experiment to display TV images using DC driven PDPs • 1969 Burroughs (US), developed DC driven PDPs named Self-Scan™ • 1978 NHK (Japan), developed 16-inch DC driven color PDP TVs • 1985 NHK (Japan), developed pulse memory type DC driven color PDP TVs • 1993 Fujitsu (Japan), developed 21-inch full color AC driven PDPs • PDPs have almost disappeared because of OLEDs and low-cost LCDs.	• 1897 discovered an electron emission from metal or semiconductors in a high electric field under high-degree vacuum (10^9 V/cm) • 1926 a field emission was explained through quantum tunneling • 1928 Basic Fowler-Nordheim equations were established, J(E) • 1968 SRI, developed the Spindt arrays • 1989 the first monochrome FED was developed • 1993 LETI, developed 6-inch full color FEDs • 1997 Motorola, developed 5.6-inch high voltage full color FEDs • 1999 Candescent-Sony, developed 13.2-inch full color FEDs • 2001 Samsung, developed 7-inch CNT-FEDs • PEDs have almost disappeared because of low-cost LCDs.
Cathode Ray Tube (CRT)	**3D Display (3D)**
• 1897 Braun (Germany), developed a Braun tube • 1926 Germany, Japan, published an experiment on CRT -based transmission/reception for displaying characters • 1935 Germany, started TV broadcast for the first time in the world • 1950 US, developed circular CRTs displaying color images • 1961 Korea, started TV broadcast in black and white • 1970 CRTs for color displays were put into practical use • 1972 US, commercialized large CRTs for projection display • 1980 Korea, started TV broadcast in color • 1982 RCA (US), Siemens (Germany), developed beam guide-type, plasma excited flat CRTs • 1991 Japan, commercialized High vision TVs • 1994 started an experimental broadcast for commercializing High vision • 1996 Sony(Japan, developed first complete flat CRT • 2005 LG-Pilips Display, developed FTM, Cyber Tube • CRTs have almost disappeared because of LCDs and PDPs.	• mid-1800s conducted various researches including anaglyph • 1839 Daguerre, invented silver-halide photography • 1947 Dennis Gabor (UK), discovered optical holograms • 1990 NHK, Sanyo (Japan) published 3D displays using lenticular and beam splitter methods • 1994 disposable stereo cameras with three lenses came to the market • 1995 Sanyo (Japan) developed 2D/3D converting technology which transforms 2D images to 3D • 2000 Sanyo (Japan), DTI (US), Dresden (Germany), developed 15-inch and 18-inch 3D displays by combining LCDs and stereoscopic optics technologies • 2002-2004 ATTEST(Advanced Three-dimensional Television System Technologies) Project. IST Program

※ Main Terms[1)]

- LCD(Liquid Crystal Display): A display device based on the contrast effect that occurs when the optical characteristic of liquid crystal's initial molecular structure is transformed by external fields such as an electric field or a magnetic field.
- TN LCD(Twisted Nematic LCD): An LCD based on the principle that the transmittance of light varies depending on its polarization state when the long axes of liquid crystal molecules are arranged to be twisted 90 degrees between the upper and lower substrates.
- STN(Super Twisted Nematic): An LCD mode that has the same structure as Twisted Nematic, with the twisted angle larger than 90 degrees.

3 Types of Flat Panel Displays

This chapter describes the features of flat panel displays and future promising displays developed so far. As shown in Figure 1.2, flat panel displays are divided into emissive displays and non-emissive displays, depending on whether the display emits light by itself or not, but both share the same basic structure: the layers of display materials in solid, liquid or gas forms and the electrode layers on the substrates are laminated together.

(1) Liquid Crystal Display (LCD)

LCDs can be classified by the form and the driven method. In terms of the form, LCDs can be divided into projection LCDs and direct view LCDs. Projection LCDs are mainly used to project large images ranging from forty inches to hundreds of inches, by enlarging high-intensity images or data with optical devices and projecting them on a screen. These projection LCDs are again classified into rear projection and front projection, depending on the position of the screen and the light source. Projection LCDs consist of a light valve, optical system, circuit system and an instrument system. The direct view types of LCDs are divided into transmissive LCDs and reflective LCDs. In the case of transmissive LCDs, the light comes from the backlight, and the intensity of light is adjusted by the LCD panel. In the case of reflective LCDs, images are displayed as natural light or the surrounding lights are reflected on the LCD panel.

1) The name of high-definition televisions (HDTV) was first used in Japan. As a first step of new media, there was an experimental broadcast using the MUSE method, which broadcasts High visions through satellite channels. High vision TVs feature 1,125 scanning lines and the wide 16:9 aspect ratio. Today, High vision has been replaced by ISDB, which includes full modern digital technology, and broadcasting has been discontinued.

LCDs can also be categorized by the driven method into two types, i.e, electrically addressed ones and optically addressed ones. Optically addressed LCDs are controlled by optical signals. A spatial light modulator is a good example of the optically addressed LCDs. Meanwhile, electrically addressed LCDs are again divided into passive matrix LCDs and active matrix LCDs, depending on whether an active element arises when pixel electrodes operate. Figure 1.3 describes the structure of direct view TFT-LCDs.

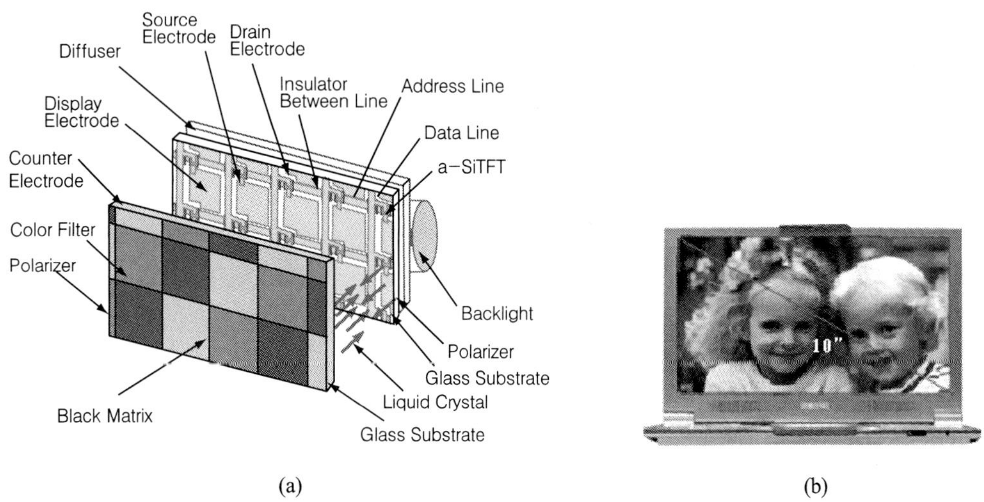

Figure 1.3
The structure of Active Matrix LCD(AM-LCD) panel (a) and an example of 10-inch wide LCD TV (b)

A direct view TFT-LCD is manufactured in the following steps: 1) Arrange liquid crystal molecules to be vertically twisted 90 degrees or 270 degrees between two glass substrates with transparent electrodes. 2) Attach polarizer films to the front and back of the substrates. The polarization direction of these polarizers are aligned 90 degrees to each other. 3) Attach the light source to one side and a color filter to the other side where the light should emit. The LCD controls light using dielectric anisotropy and double refraction.

More specifically, the light from the source is first polarized by the back polarizer and then proceeds in the twisted direction of the liquid crystals. The light then passes through the front polarizer. But if a voltage is applied to the electrodes on the substrates at this moment, the liquid crystal molecules rise vertically against the substrates due to the dielectric anisotropy, by which the light goes straight through the liquid crystals without being twisted. In such case, the light that has passed through the liquid crystals will be completely blocked by the front polarizer. More specifically, depending on whether or not a voltage is applied to the liquid crystals, the light is switched as shown in Figure 1.4.

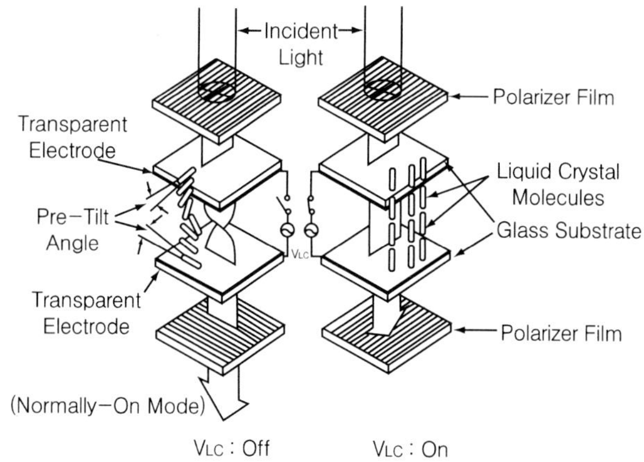

Figure 1.4
The structure of LCD and its operating principle

LCDs display images by controlling light in this way. There are Simple Matrix type and Active Matrix type in LCDs, depending on how the LCD operates. Simple Matrix type LCDs operate by arranging the electrodes on the upper and lower substrates to be perpendicular to each other, as explained above. There is some disadvantage in such LCDs. That is, the signal voltage affects other pixels and thus, when the number of electrodes increases, the contrast ratio drastically gets reduced, which makes the quality of displayed images worse. On the other hand, Active Matrix type LCDs operate by arranging the electrodes vertically on one substrate and then placing an active element such as a TFT at the cross point of the electrodes (a pixel). Controlling pixels with the active elements can eliminate the interference between pixels. Thus, Active Matrix type LCDs can display images of good quality. For such active elements, TFTs made of hydrogenated amorphous silicon are used.

Meanwhile, Passive Matrix type was used for the LCD products (watches, calculators, etc.) developed in the early stage. Passive Matrix type LCDs include Twisted Nematic (TN)-LCDs, Super Twisted Nematic (STN)-LCDs, Ferroelectric (F)-LCDs and Polymer Dispersed (PD)-LCDs, which are classified by the type of the liquid crystals used. TN-LCDs are used in LCD products because of their low price and high response speed, but they have a very weak view angle. An STN-LCD was often used for the early LCD-TVs and/or laptops since the transmittance changed rapidly due to large twisted angle of 240 or 270 degrees. An F-LCD using ferroelectric liquid crystals shows the fastest response of all. A PD-LCD was invented to improve the low light efficiency of TN-LCDs that require polarizers absorbing a half amount of the incident light. A PD-LCD is made by employing the liquidity of liquid crystals and the structure and double refraction of the polymer. More specifically, when mixing polymer materials with liquid crystals, due to the liquidity, the liquid crystal drops have their molecular arrangement influenced by the border conditions and molecular forces. Eventually, each drop shows random molecular arrangement. When the light comes in and passes through the liquid crystal drops with the random molecule arrangement, if an electric field is applied at that moment, the light is refracted in different ways due to the double refraction of liquid crystals.

Active Matrix LCDs, which control each pixel with an active element, are classified into two-terminal LCDs and three-terminal LCDs, depending on the number of the terminals. For two-terminal LCDs, Metal-Insulator-Metal (MIM) method and Diode method are used. Three terminal LCDs, which are used for most of the current LCD products, utilize thin film transistors. Three-terminal LCDs are again categorized into amorphous silicon (a-Si:H) TFT-LCDs, polycrystalline silicon (poly-Si) TFT-LCDs, CdSe TFT-LCDs and so forth, depending on the active layer type of the thin film transistor.

(2) Organic Light Emitting Diode (OLED)

OLED is the next-generation flat panel display made by using self-luminous phenomenon that emits light when current flows in a fluorescent organic compound. The image quality reaction rate is 1,000 times faster than that of an ultra-thin liquid crystal display (TFT-LCD). In terms of the operating characteristic, OLEDs belong to emissive display type, which do not require a backlight, have high brightness and efficiency, excellent contrast ratio and wide viewing angle. Also, OLEDs run fast with a low operating voltage and thus, help consume less electricity. In terms of the manufacturing process, flexible substrates, for instance, made of plastic, can be used for OLEDs due to the low processing temperature, and they can be manufactured in large scale. It is flexible, which helps designing display products in various shapes and connect to other relevant devices. Moreover, OLEDs are slimmer and lighter than other displays and thus, are suitable for portable devices. The basic structure of OLED is as shown in Figure 1.5 (a): there are glass or plastic substrates with the upper and lower electrodes (the positive and negative electrodes), and in between them organic layers are inserted. When it operates, electrons from the negative electrode and holes from the positive electrode are injected into the organic layers. The electrons and holes are recombined in the organic layers, by which light is produced.

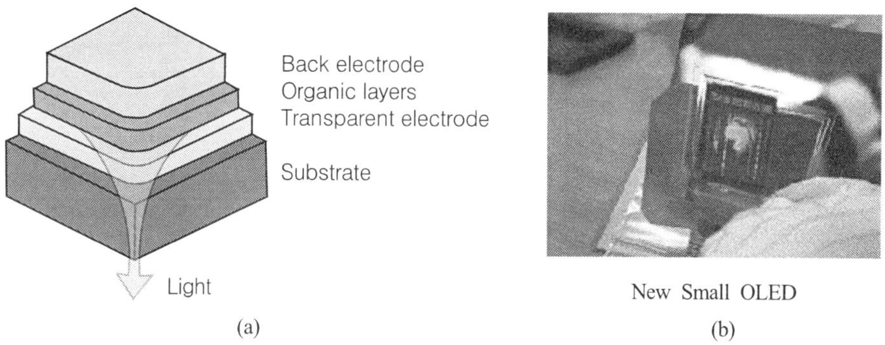

Figure 1.5

The structure of OLED (a) and an example of the OLED-applied products (b), made by ELiA Tech. (Korea)

(3) Plasma Display Panel (PDP)

A plasma display panel is a display device based on how a fluorescent lamp works, consisting of two glass substrates with embedded electrodes and a mixture of gas at a half of atmospheric pressure filled in between the substrates. Each of the glass substrates is less than three millimeters thick and thus, the whole panel is at most six millimeters thick. In the space, 2~10% of xenon gas produces ultraviolet rays, and other gasses such as helium and neon are also mixed. A high resolution color PDP consists of barrier ribs separating each of the pixel, and in between the barrier ribs, phosphors are formed. When ultraviolet rays enter in the barrier ribs, they hit the phosphors and then emit visible light.

The most important characteristic of PDPs is emitting light on their own just like CRTs. Also, they have viewing angle larger than 160 degrees and thus, it is easy to make a large, thin and light-weight screen using a PDP. However, miniaturization is difficult and the power consumption is large due to the plasma. Thus, PDP has been replaced by LCD.

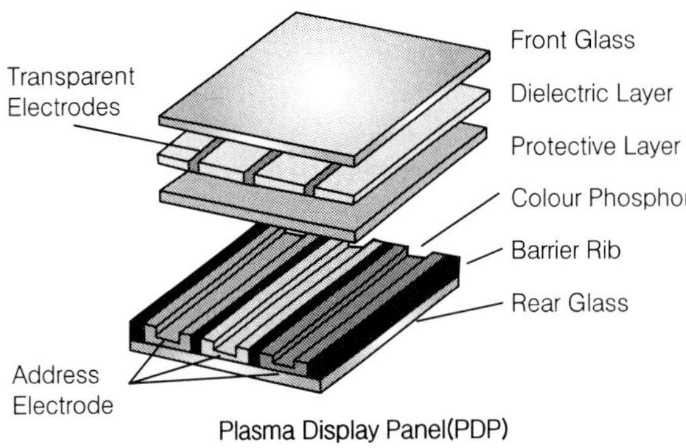

(a)

(b)

Figure 1.6

The structure of PDPs (a) and an example of 61-inch PDP applied products (b), made by Pioneer (Japan)

(4) Field Emission Display (FED)

A Field Emission Display (FED) consists of a field emitter made of metal or semiconductors of ultrafine structure. When an electric field is applied to the emitter, electrons are released into the vacuum tube and hit RGB phosphors, by which images are displayed. FEDs are also called Thin CRT since they can be made light and thin, displaying good quality images as those of CRTs, theoretically. In fact, FEDs have been considered an ideal display device, as they have all the merits of flat panel displays as well as CRTs, such as high brightness, low power consumption, prompt responsiveness, wide viewing angle, high resolution, excellent color display and a broad range of the workable temperature.

However, the FED is unfavorable in comparison to the LCD in terms of price and difficulty to enlarge the size of a display. As a result, the FED slowly disappeared due to the development of LCD.

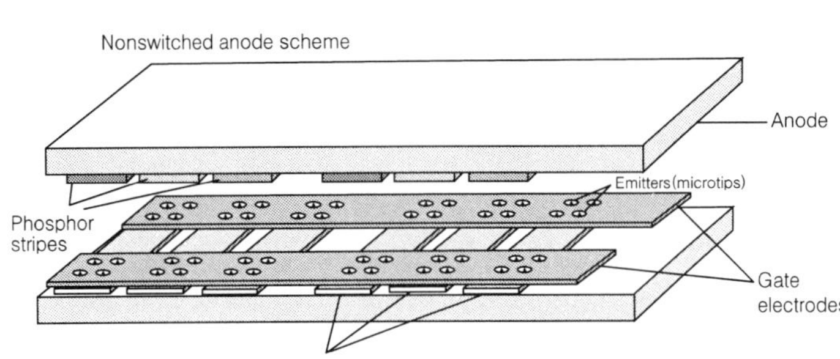

(a)

(b)

Figure 1.7

The pixel structure of FEDs (a) and an example of 1/4-VGA mode-applied products (b)

(5) Cathode Ray Tube (CRT)

So-called Braun tube, Cathode Ray Tube (CRT) is a display device converting electric signals to light signals. When cathode electrons (thermoelectron) are released by the operating electric gun (a thermoelectron generating device) installed in the cathode of a Braun tube[2], the cathode electrons are drawn toward strong positive potential, being scanned along with the tube. The electrons are deflected up and down or right and left by Deflection York, which enables two-dimensional display. These electrons then hit the phosphor screen, located at the back of the screen, resulting in the emission of light by the phosphors and visuals such as characters and figures are displayed on the screen. Although it dominated the early display markets, the popularity of the CRT gradually disappeared.

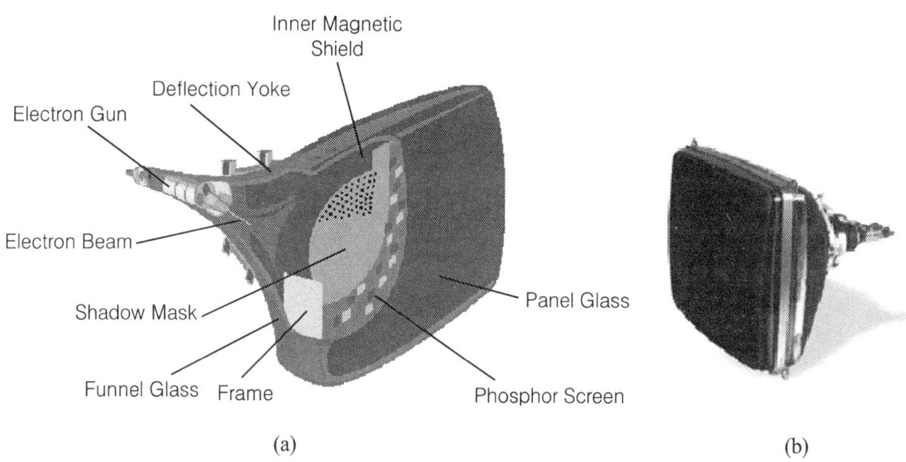

Figure 1.8
The internal structure of a CRT (a) and an example of the actual CRT products (b)

(6) 3D Display

3D stereoscopic images are more realistic images, showing the shape, depth and the space of an object, unlike 2D flat images. This is based on our perception spatial cognition; that is, recognizing images with volume around us by each of our both eyes (right and left eyes) and then combining them in the process of organizing information in our brain. Thus, 3D stereoscopic images give us a vivid feeling as if we are really there in the scene, just like virtual reality. To summarize, a 3D display technology provides a three-dimensional effect to viewers by showing the depth of

2) Braun, Karl Ferdinand (1850~1918); a physicist born in Germany. He studied fundamental physics at first, but while studying electromagnetism afterwards, he invented the Braun electroscope and a cathode ray oscillograph (Braun tube, 1897) for experimental purposes.

images, which is impossible in two-dimensional flat displays and thus makes them feel like they see real objects instead of just images. Humans experience three-dimensional effect as the images that each of the right and left eye sees are actually different from each other, but they are eventually combined by the brain process. This is because of the different position of the two eyes–there is a gap of about sixty five millimeters between the two eyes and thus, each eye sees images in a slightly different direction, causing binocular disparity. Conventional CRTs or flat panel displays, show only two-dimensional images (it can not display the front, back nor the left-right side of an object). The 3D display technology shows an object in various directions, by which viewers can feel a sense of the depth of the object as well as a sense of reality.

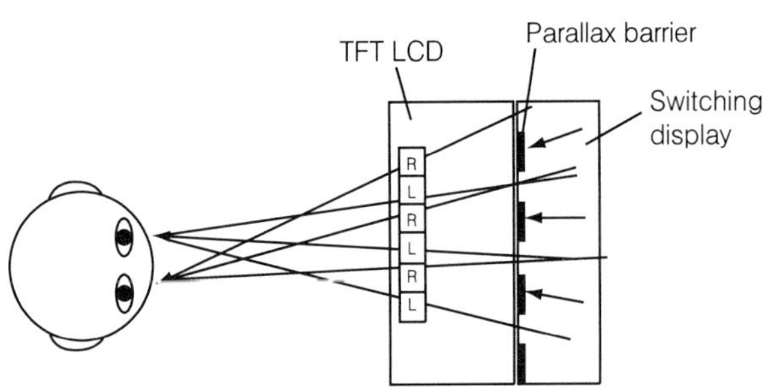

(a)

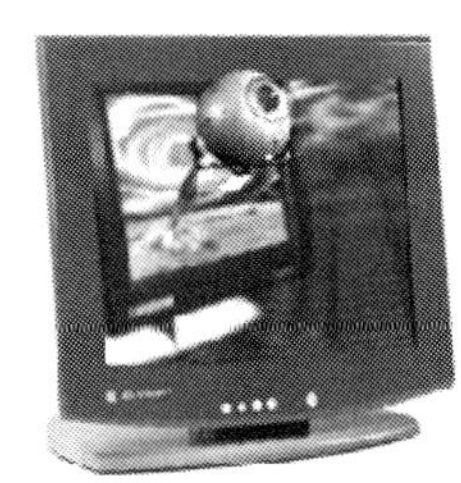

(b)

Figure 1.9

The principle of binocular disparity of 3D displays (a) and an example of 15.1-inch 3D display products (b), made by Duenrkohp & Radicke (Germany)

(7) Future state of the art Displays

1) Nanocell Display

The nanocell display device is a display that increases the color wideness (diversity) and greatly reduces the reflectance by mixing the light absorbing material with the polarizing plate of the conventional LCD. It has a wide viewing angle and high image quality, making it a good candidate for the next-generation display. Although the nanocell process is added to the LCD process, the existing design does not need to be changed.

2) Quantum Dot Display (QLED)

Quantum dot is a semiconductor nano-crystal that emits different colors when it receives light. QLED is made by attaching the quantum dots to a film or by depositing on vacuum glass tubes. Quantum dot TV is evaluated to be as good as OLED TV because its color reproduction is close to 100%. It is basically an LCD–based TV with LED backlight. Thus, it has the advantage of utilizing the existing LCD production line.

(8) Other Displays

1) Electro Chemical Display (ECD)

An ECD is a non–emissive display based on reversible color changes caused by the electrochemical oxidization and deoxidization of solid or liquid substances. Among the main materials for ECDs are transition metal oxides (such as amorphous tungsten oxides (a–WO_3), molybdenum oxides (MoO_3) and iridium oxides (IrO_2)), viologen derivatives, organic compounds (cobalt bipyridine), diphtahalocyanins (rare–earth metal), silver iodides, various redox compounds and so forth.

2) Electro Phoresis Display (EPD)

In a colloidal solution, which is liquid with dispersed particles, electric double layers with positive and negative electrodes are formed, and each of the dispersed particles has either a positive or negative electric charge. When a DC voltage is applied to such a colloidal solution, the dispersed particles in the solution migrate towards the opposite charge by coulomb force. EPDs utilize such electrophoresis phenomenon[3].

3) Vacuum Fluorescent Display (VFD)

A VFD is a display device that utilizes the light emitted by phosphors for the excitation of slow electrons. VFDs are similar to CRTs in that they also use a cathode ray luminescent phenomenon. However, VFDs can choose to apply a voltage to the grid electrode or the anode electrode, whereas CRTs display images by scanning electron beams.

3) Electrophoresis is a technique to move particles by applying a voltage; this is possible as colloid particles can be electrically charged. This has been used for the refinement and analysis of biochemical extracts in genetic engineering.

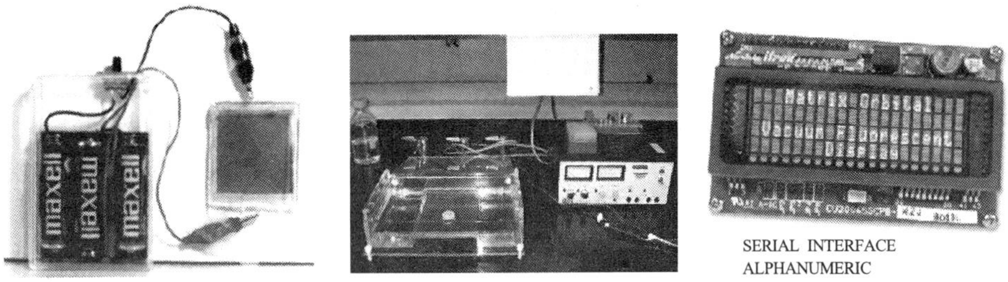

(a) (b) (c)

Figure 1.10

Blue ECD, (b) The experimental device for EPDs, (c) a commercialized VFD product made by Embedded EMJ System (US)

D i s p l a y E n g i n e e r i n g

4 Image Quality of Display

As explained so far, there are various display devices such as OLEDs, LCDs, PDP, CRTs and so on, but each device has different ranges of color reproduction and displays color differently. When we compare different display devices side by side, we notice a difference, even though the same images are being displayed, the colors of the images differ depending on the display device. It is natural for the display devices to exhibit color differently since each device was developed to meet different needs at each stage and also has its unique shortcomings as needs and applicability vary day by day. Thus, it will be important to understand the color gamut and color management/adjustment of each device. Researchers work hard for the purpose of displaying optimal colors for each device. In consideration of colors, both the physical quantity measurable by a sensor (Figure 1.11) and the psychophysical factors should be taken into account. In other words, it will be different to analyze colors based on physical factors such as a distribution of light energy in the visible wavelength range (wavelength 380 nm through 780 nm) or based on psychological factors such as the emotions, experience and culture of an individual. For instance, depending on the state of mind of the viewer, two colors in the same physical device may be seen differently. A professional colorist can distinguish a minor difference in colors. People working at a printing company or at a paint company are very sensible to colors, so they can manage a better quality of colors. Most of the display devices show an excellent color gamut. Displays, which show images by emitting light, are great at displaying bright colors in detail, but weak in displaying dark colors. On the contrary, printing is good at reproducing dark colors as it shows images by reflecting light. LCDs have a blue tint, whereas CRTs have a red tint. Due to this, cosmetic companies that are supposed to show good skin tone tend to use CRT monitors. Of course, the color reproduction of the recent displays has improved very much, but it still remains a big issue to show consistent skin tone of humans. There are differences in the color reproduction rate and the reproduction of color between LED and LCD. The LED have a higher color reproduction rate than that of the LCD. For good output of color, the peak of the graph for each color should be clearly distinguished so that the color reproduction rate becomes high and colors can be expressed more clearly.

Even though the light source of the LCD has all the R, G, and B wavelengths, the wavelengths are not clearly separated. LCD has the blue color spectrum that is clearly distinguished from the other colors in LCD color spectrum, but the difference between the green wavelength and the red wavelength is small. Thus, the blue can be expressed a color close to pure one, but the red and green are not enough to be expressed as pure colors. Therefore, the red and green form small triangles on the color coordinate system.

On the other hand, the color spectrum of OLED shows that the R, G, and B are clearly distinguishable. This means that it can express pure colors. The color near this pure color is located at the edge of the color coordinate system, which means that a larger triangle can be drawn. That is, all the color of the light on the area of the triangle can be expressed in the display so the color radius that can be expressed on the display is wide.

To increase the color reproduction rate, there is a method of narrowing the wavelength–range of the light passing through the color filter and completely prevent it from passing through the other areas. Efforts to improve these characteristics of color filter are continuing. Although the color filter works well, the amount of light passing through the filter is too small and the power consumption becomes a problem if it does not match the light source. It means that matching the light source to improve the characteristics of the color filter is also important. Alternatively, there are efforts on improving the characteristics of the light source. This ensures that the R, G, and B output wavelengths are clearly separated without mixing in the white light–source so that it is possible to print out high–quality images with hight color–purity. Since the R, G, and B wavelengths are more clearly distinguished than the conventional light sources, the output light can be clearly distinguished by color even if the same filter is used.

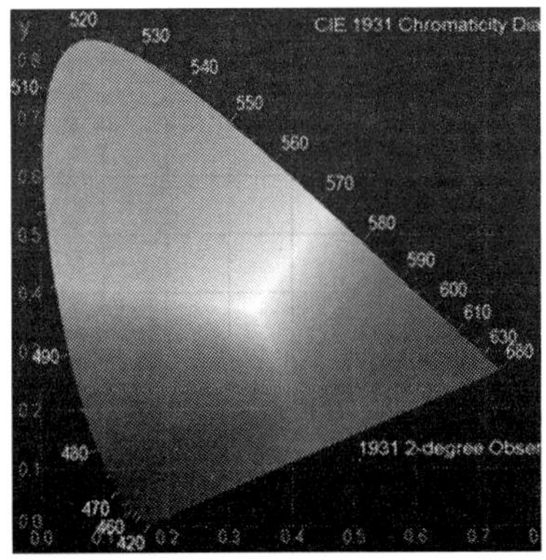

Figure 1.11
CIE Coordinates

5 Prospect of Next-Generation Displays

Development of flexible substrates is mainly led by Japanese and American companies. DUPONT–TORAY, Kaneka, and SKC–Kolon–PI are developing and commercializing PI (polyimide) substrates for flexible displays. Currently, organic EL is mainly used in flexible display devices.

For the operating elements, amorphous silicon, low–temperature polycrystalline silicon transistors, organic TFTs made with organic semiconductor materials, nano–TFTs using nanowires, transparent TFTs using transparent inorganic semiconductors, etc. are popular. Especially low–temperature polycrystalline silicon transistors (LTPS) are mainly used for flexible display.

The flexible display market is mainly led by Samsung Display and LG Display, both of which are producing CURVED–type OLEDs. It is used in smartphones such as Samsung's Galaxy S8, wearable devices, and large OLED TVs. In the early stage of mass production, a display with a curvature radius of 400 mm was first applied to a smartphone, and the curved products with a curvature radius of 7 mm or less are being developed and used. JDI in Japan and AUO in Taiwan seem to have secured technology for pilot–level production. However they have not yet started mass production.

Future display devices can be modified from its curved version to rollable and foldable displays. Prototype products of 10–inch and 5.5–inch foldable displays were exhibited in 2014 by SEL in Japan, they exhibited an 8K–class high–resolution foldable display device. Samsung Electronics is expected to release a smartphone with a foldable display in 2017~2018. In case of rollable display, Samsung Display has exhibited 5.7–inch prototypes rollable displays in SID2016 assuring that the original technology was secured.

QLED (Quantum dot LED) display is attracting attention as a next–generation display after OLED. QLED is a display device that displays color by emitting electroluminescence from a quantum dot device made of a nanocrystal semiconductor material with a three–dimensional quantum dot structure. The quantum dots are used to increase the color reproducibility of LCD. The research has been focused on developing a QLED device that is capable of self–luminescence.

E–paper displays, which can sustain images using EPD (Electro Phoretic Display) and have less eye fatigue, have been adopted in e–books, and development is continuing in E Ink and SiPix. The development of these displays is helping to develop future technologies such as VR, AR, UHD–broadcasting.

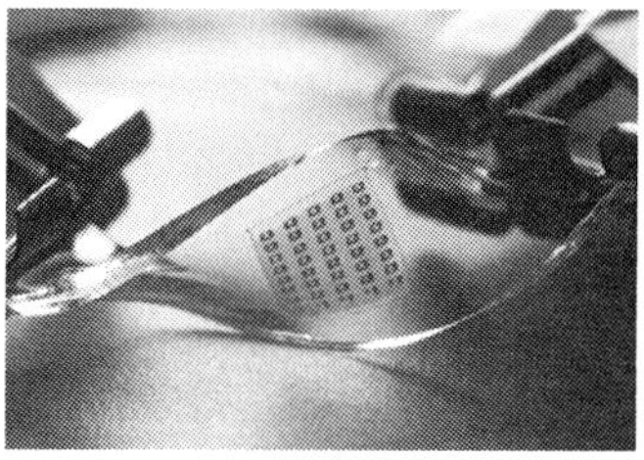

(a)

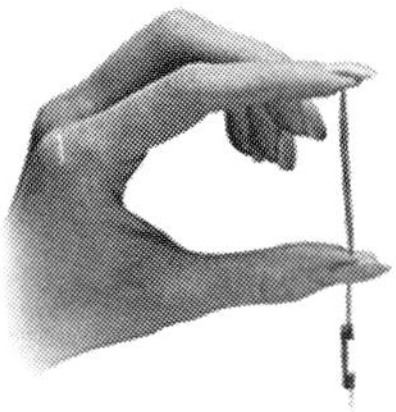

(b)

Figure 1.12

(a) A freely bendable and expandable electronic circuit, (b) A thin and light display

Exercise

1. Classify display devices into Emissive type and Non–Emissive type. And explain the characteristics of each device.

2. List up the major flat panel displays and explain the characteristics of each device.

3. Investigate why CRTs, which are excellent in terms of price, brightness and life span, need to be replaced by the technologies of flat panel displays.

4. Investigate the history of each display device and draw a chronological table.

5. Investigate a new concept of displaying method that is not mentioned in this textbook.

References

[1] Shouichi Matsumoto, Isao Oota, Hisashi Okada, Yoshimasa One, Takao Kishino, Tatsuhiko Niina, Hidel Mori and Satoshi Watanabe: *Electronic Displays*, Ohmsha, Ltd. Tokyo, 1995.

[2] Byung Gwon Joo, Jae Gyeong Kim, Yoon Hee Lee: "Organic EL: Next-Generation Display dreaming of Infinite Application," *Electronic Industry Information*, 94-98 (March 2001).

[3] Eui Sun Jung: "Study on PDP Life Characteristics against Operation and Poeration Frequency,“ Doctoral Dissertation, SungKyunKwan University (2001).

[4] Yoon Ho Song, Jin Ho Lee, Sang Jik Gwon: "Development Trend and Prospect of Field Emission Display," *Electrical and Electronic Materials*, vol. 15-12: pp.11-18 (Dec 2002).

[5] Eun Soo Kim: "Domestic and Abroad Technological Development Trend and Market Prospect of 3D Display System," *Journal of Korea Information Display*, vol.3-2 (2002).

[6] Cheol Joo Kim: "Amorphous Semiconductors," *Academy Research* (1998).

[7] T. Tsukada: *TFT and LCD*, Bookshill. Seoul, 2000.

[8] Bong Gyu Roh, et al.: *LCD Engineering*, Sung An Dang. Paju, 2000.

[9] Pierret: *Semiconductor Device Fundamentals*, Addison Wesley. Boston, 1997.

[10] Sang Woo Kim: *Display Engineering*, Chungbum Publishing Co. Seoul, 2001.

[11] Choong Hoon Lee, Cheol Ha Kim: "Application of Hydrogenated Amorphous Si,“ *Electrical and Electronic Materials*, vol.12-2: pp.7-13 (Feb 1999).

[12] Jin Jang: "Development Trend of Low Temperature Polycrystalline Silicon Thin Film Transistors", *Electrical and Electronic Materials*, vol.12-2: pp.1-6 (Feb 1999).

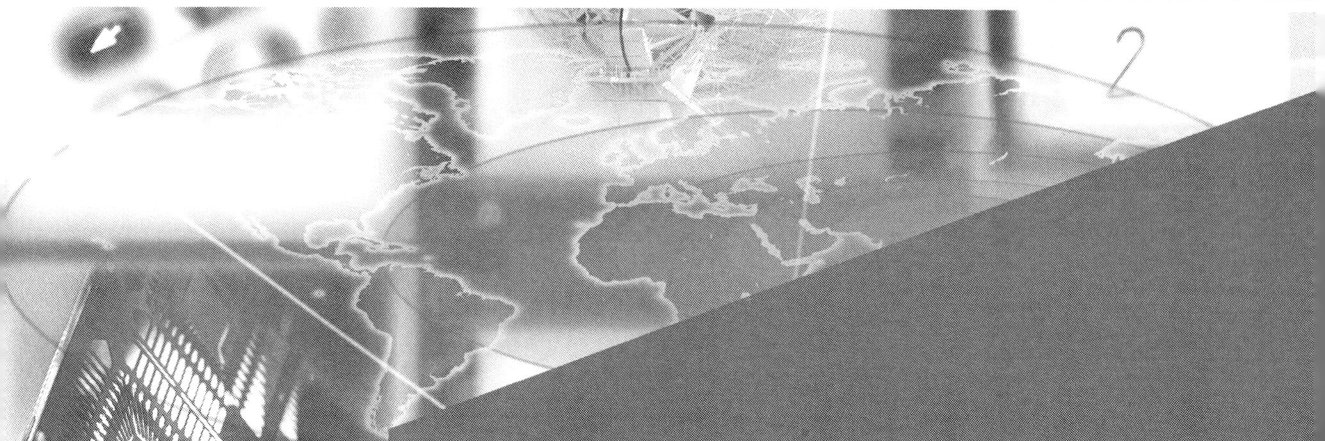

Field Effect Transistor

Chapter 01 → 02

1 Summary of Characteristics of MOSFET

1.1 Why we need to learn MOSFET?

A pixel is a word coming from "picture element", and is identified by X– and Y–coordinates in display devices. A display device contains a large number of pixels. In order to combine these pixels in a matrix form, and to control each pixel, a three–terminal switching element is required. In other words, a transistor with a three–terminal switching element is used to address a pixel. The addressed matrix element or pixel displays the ON/OFF state, while the collective response of all the pixels in a display device results in a desired information. A thin film transistor, similar to MOSFET (Metal Oxide Semiconductor Field Effect Transistor), is used as a switching element. Therefore we will learn the basics iof MOSFET in this chapter. We need to understand two facts in studying MOSFET. First, metal–SiO_2–Semiconductor or MOS (metal oxide semiconductor) controls the ON/OFF state of a transistor according to the voltage applied to the gate electrode. Second, a pixel is a p–n junction semiconductor element, like light emitting diode (LED), externally connected between the drain and the source of the MOSFET. Figure 2.1 shows the structure of a MOSFET transistor and a conceptual diagram of the composition of a display matrix.

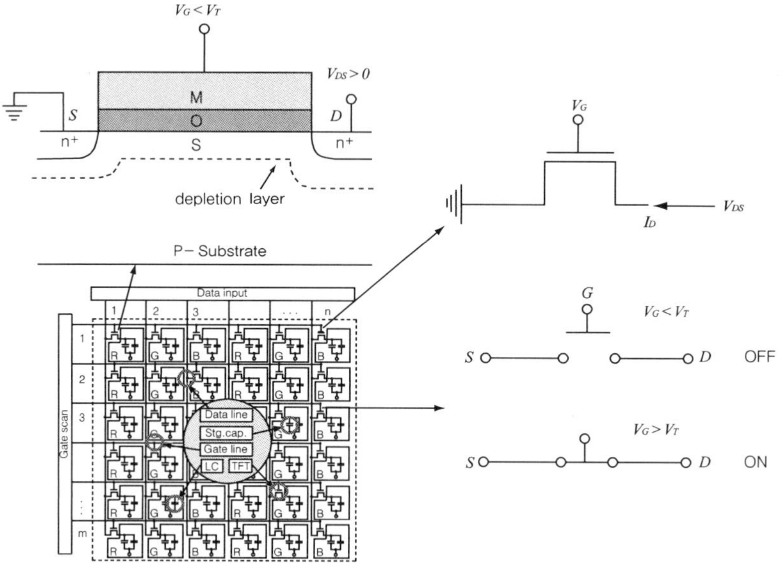

Figure 2.1

The structure of a transistor and a conceptual diagram of the composition of display matrix

In Chapter 2.1, we will briefly introduce MOSFET and deal with thin film transistors used in the display devices. In the following chapters, LCDs driven by such transistors will be covered.

1.2 Structure and Operation of MOS Capacitor

The core of MOSFET is an MOS capacitor, which is vertically placed at the gate and is known as Metal–Oxide–Semiconductor. A controlled electric field in the MOS structure enables or disables the electric current flow from the source to the drain, by increasing or decreasing the number of electrons in the surface layer of the silicon semiconductor, depending on the way of applying a voltage and the level of the applied voltage. The structure of Metal–Oxide–Semiconductor and its energy band in an ideal case are described in Figure 2.2. C_{OX} refers to the equivalent capacitance of the oxide film, and C_S refers to the capacitance of the silicon part which can be changed by the gate voltage. E_{fm} and E_f are the Fermi level of metal and silicon, respectively. Also, E_c, E_i, E_v represent the energy levels of the conduction band, intrinsic Fermi level and the valence band, respectively. $q\phi_M$, $q\phi_{si}$ are the work functions of metal and silicon, and $q\phi_F$ is the Fermi potential. The ideal MOS capacitor has the following characteristics: (1) there is no disparity in the work function between silicon electrodes and metal electrodes, (2) an electric charge exists only in the metal electrodes and silicon and (3) the Oxide film is an effective insulator where the electric charge from the external voltage source cannot pass through. However, in real devices there is disparity in the work function between silicon and metal, and electric charges could exist even in the oxide film.

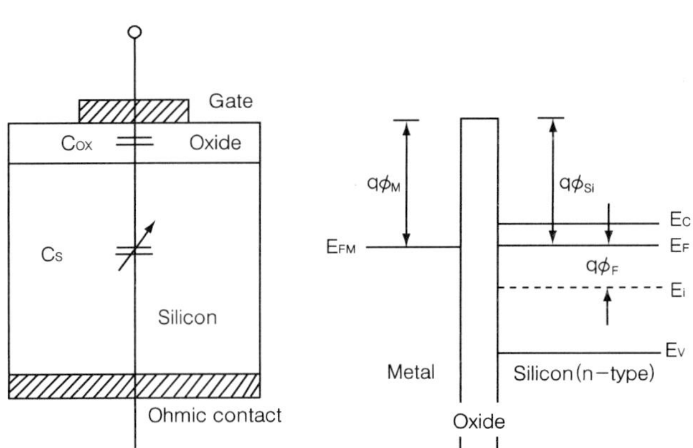

Figure 2.2
The structure of MOS capacitor and the energy band of silicon

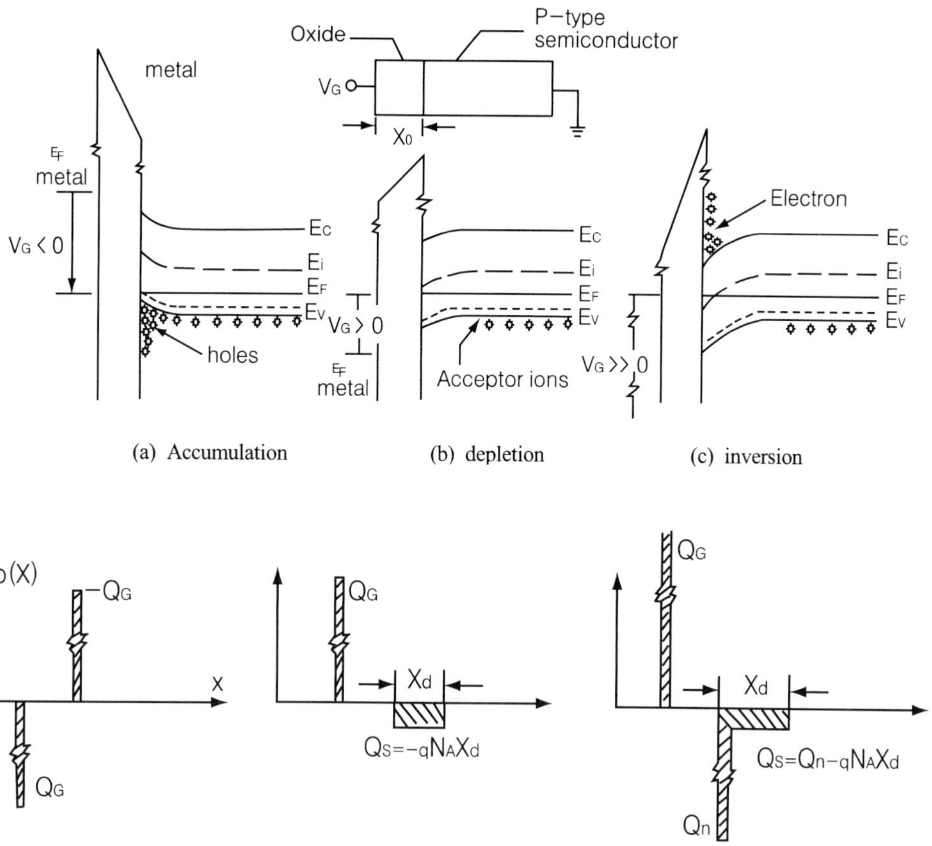

Figure 2.3

Changes of the energy band depending on the change of the external voltages applied to a p–type silicon MOS capacitor

Figure 2.3 shows the changes of the energy band when the external voltages applied to a p–type silicon MOS capacitor are changed. As described in the Figure, the energy band around the edges of the conduction band and the valence band gets bent due to the difference in the electric potential of each of the surfaces. The energy band also changes depending on the changes of the external voltages. The width of the induced space charge is indicated as X_d. When a negative voltage is applied to the gate, a voltage drop occurs in the insulating Oxide film, and positive holes are accumulated on the surface of the semiconductor (i.e., Accumulation mode). When a positive voltage is applied to the gate, positive holes are pushed away from the surface of the semiconductor, causing depletion mode. Assume that a higher positive voltage is applied to the upper metal gate. We may expect that the size of the induced electric field increases, and so does the number of the corresponding negative charges (electrons) on the MOS capacitor. With a sufficiently high gate voltage is applied, the surface of the semiconductor is inverted from p–type (dominated by holes) to n–type (dominated by electrons), and at the Oxide film/semiconductor interface, the inversion layer of electrons is induced (i.e., Inversion mode).

A threshold voltage (V_T) is defined as the minimum gate voltage required to reach the inversion point, and the threshold inversion point is defined as the state when the surface potential is $\phi_s = 2\phi_F$ in the p-type semiconductor. From the time the inversion layer starts to be formed by the applied gate voltage, a channel is formed and current flows from the source to the drain. In other words, MOSFET switches the current flows between the source and the drain due to a change in the applied gate voltages.

1.3 Types of MOSFET

What we have covered thus far is a MOSFET using the p-type substrates (bulk) and the n+ type source and drain layers, which needs a positive gate voltage to operate MOSFET by forming channels of electrons between the source and the gate. As this MOSFET type operates by n-type channels (electrons), it is called an n-channel MOSFET. In another type of MOSFET, the n-type substrates (bulk) and the p+ type source and drain layers are used. Here, the channels connecting the source with the drain are formed by positive holes (p-type carriers) and thus, such MOSFET is called a p-channel MOSFET. The n-channel MOSFET and p-channel MOSFET are often abbreviated as NMOS and PMOS.

The characteristic of the n-channel MOSFET and p-channel MOSFET share in common is that they both are in the OFF state with no gate bias applied. As there is no channel formed between the source and the drain at this moment, the drain current is zero. Accordingly, such MOSFET is considered to be a normally OFF MOSFET. The transfer characteristic of the n-channel MOSFET is that the drain current turns the transistor to the ON state with a high positive gate voltage, whereas the drain current of the p-channel MOSFET does the same with a negative gate voltage. This indicates that a certain gate voltage is required to create the conducting channels of either electrons in NMOS or holes from each of the n-channel and the p-channel. Generally, each of these MOSFETs are called an enhancement MOSFET.

On the other hand, MOSFETs can also be made in the form where channels exist without externally applied gate voltage, if the thickness of the silicon layer and the Oxide film, the doping concentration, metal materials, etc. are properly adjusted. Such MOSFET does not need any gate voltage to turn into the ON state and thus, it is calso called an ON MOSFET. It is also called a depletion-mode MOSFET, as the electrons or holes should be removed from the channels by enlarging the depletion region. Its OFF state is achieved by applying cut-off voltage to its gate. An n-channel depletion-mode MOSFETs needs a negative voltage to reduce the drain current, whereas a p-channel depletion-mode MOSFET needs a positive voltage to turn itself to the OFF state.

1.4 Comparison between TFT and MOSFET

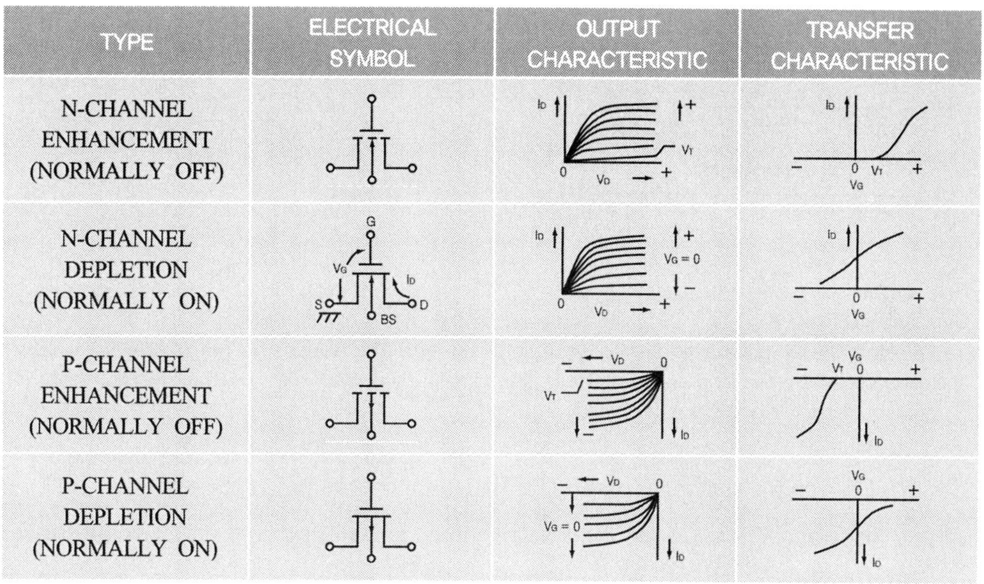

Figure 2.4

The characteristic of the current-voltage in each MOSFET type

All thin film transistors used for MOSFETs and displays are the elements with the MOS gate structure. However, there is a critical difference between TFTs used in MOSFETs and in displays, which is their mode of operation. TFTs operate in the Accumulation mode. In other words, when a gate voltage is applied, the density of certain carriers increases in the channel. In order to operate these TFTs it is critical to use thin films with the adjustable conductivity. For example, the p+pp+ junction device using the silicon single-crystal substrates (used for the production of MOSFET) that is produced in the accumulation mode (accumulating electrons or holes, being opposed to the applied gate voltage, between the insulating layer and the silicon surface) does not work. This is explained in Figure 2.5 (a). if a negative gate voltage is applied, positive holes are accumulated on the surface of p-type substrates; however, the holes in the p-type substrates are majority carriers and have a mobility. Therefore, the negative gate voltage can not control the current. In order to operate the insulated–gate field effect transistors as memory devices, the opposite gate voltage (i.e., a positive gate voltage) is needed to cause inversion layer in the p–type substrate. As shown in Figure 2.5 (b), the positive gate voltage forms the channels of electrons on the surface and take a role of the switch turning ON/OFF the current flowing at the n+pn+ junction between the source and the drain.

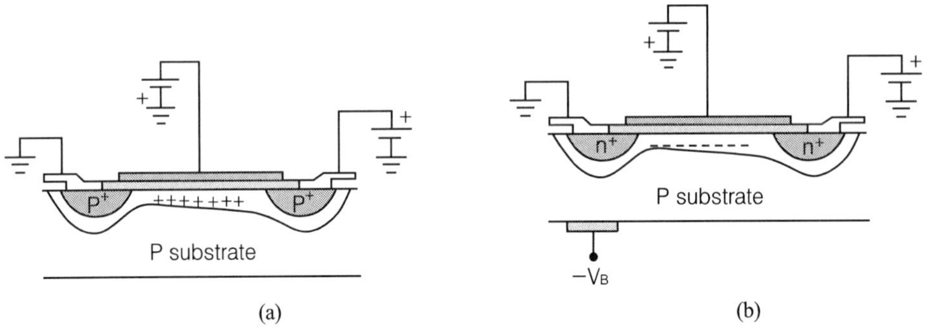

Figure 2.5

Field Effective Transistor with the insulated gate on the p–type substrates
(a) The structure for generating the Accumulation mode, (b) The actual structure of MOSFET, where the opposite electric field applied to the gate comes into contact with n+. The inversion layer is generated by the positive gate voltage.

Figure 2.6 shows the structure of a transistor used in display devices and its current–voltage characteristics. Such a transistor is manufactured on glass substrates with high insulation. The structure of such a TFT is as the following: the gate placed at the lower part (unlike the upper gate structure that we discussed earlier), a silicon nitride insulating thin–film that is grown under low temperature, an intrinsic silicon thin–film grown with continuous high resistance (mostly being amorphous with a very high resistivity of over 10^8 ohm–cm, unless impurities are injected) and silicon thin–films including impurities such as n^+ or p^+ that are formed around the source or the drain. In such a structure, as the electrons leaving the source move to the silicon channel layers formed below the gate electrode and then move towards the drain. This is the ON state current the drain.

As shown in Figure 2.6, when a positive voltage is applied to the gate, electrons are accumulated on the surface of silicon forming channels. Accordingly, it is possible to switch the transistor with the gate voltage.

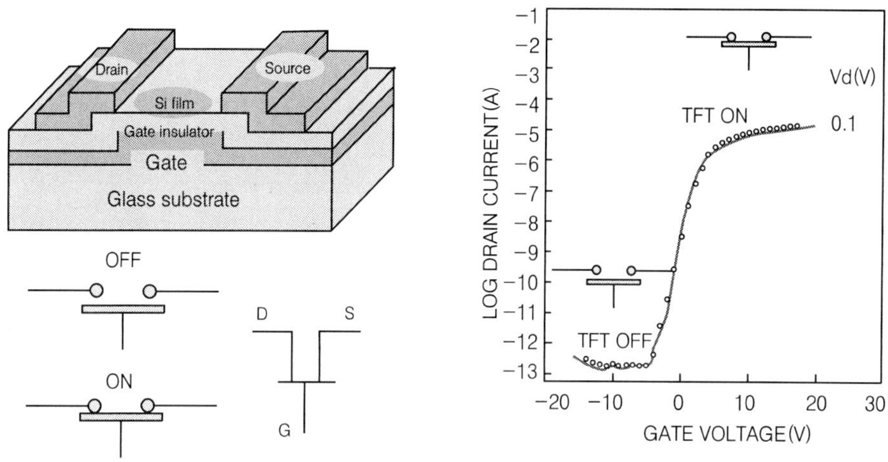

Figure 2.6

The TFT structure form on a glass substrate and the current–voltage curve

Display Engineering

2 Background and Operating Principle of Thin Film Transistors

2.1 Background of TFT

A Thin Film Transistor (hereafter, "TFT") can be called as a Field Effect Transistor (FET), manufactured with semiconductor thin–films on the insulating substrates. Just like FETs, TFTs consist of three terminals, i.e., the gate, drain and the source, and their main function is the switching operation. The operating principle of TFTs is very similar to that of FETs. That is TFTs switch ON or OFF the current flow between the source and the drain. This is achieved by adjusting the gate voltage, here gate is the third electrode. TFTs are used in sensors, memory devices and/or optical devices, but mainly used as a pixel switching element in Active Matrix (AM) type flat panel displays.

In AM–LCDs, the driving voltage to each pixel is adjusted by a TFT, by which the ON/OFF state or the optical transmittance of each pixel is determined.

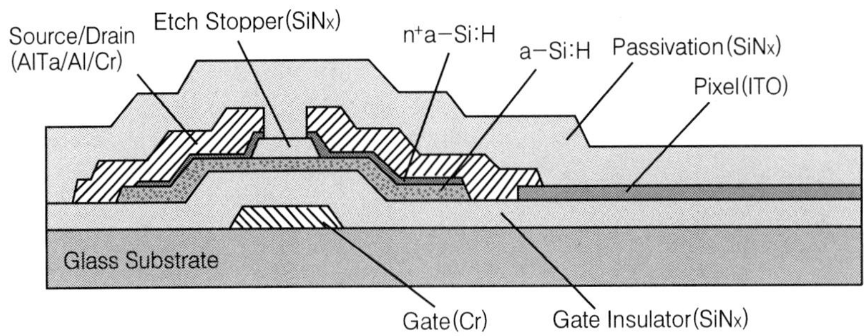

Figure 2.7
The schematic diagram of a cross–section of a–Si TFTs used for LCDs

(source: LG–Philips LCD homepage)

In the TFT amorphous silicon is used as the active layer for the switching element of large AM–LCDs in the industry. This is because the amorphous silicon thin–films can be grown at a substrate temperature lower than 350°C. Therefore, the amorphous silicon can be used with a low cost and large–area glass substrates. Also, amorphous silicon TFTs have a high drain current (enough to supply electric charges even in the displays with more than 1000 scan lines) and a low OFF current (discharging little electric charges in one frame). Thus, it has an advantage of being used for a large display with the contrast ratio higher than 100:1.

The display industry is continuously developing to control smaller pixels. Switching technology is developed to precisely control the pixels as diodes. The pixel diodes are nonlinear elements. Further more, displaying of a pixel requires a transistor, just like AM–LCDs that compose the whole screen by connecting a switching element (TFT) with each pixel. AM–LCDs are the result of the combination of display technologies, semiconductor technologies and liquid crystal technologies.

2.2 Operating Principle of TFTs

The operating region of TFTs are broadly divided into the linear region and the saturation region, just as in MOSFET (Metal Oxide Semiconductor Field Effect Transistor). When the drain voltage is low, the characteristic between the drain and the source appears to be ohmic, resulting in the drain current proportional to the drain voltage. On the other hand, in the saturation region the drain voltage is high, here the drain current remains nearly constant regardless of the increase of the drain voltage.

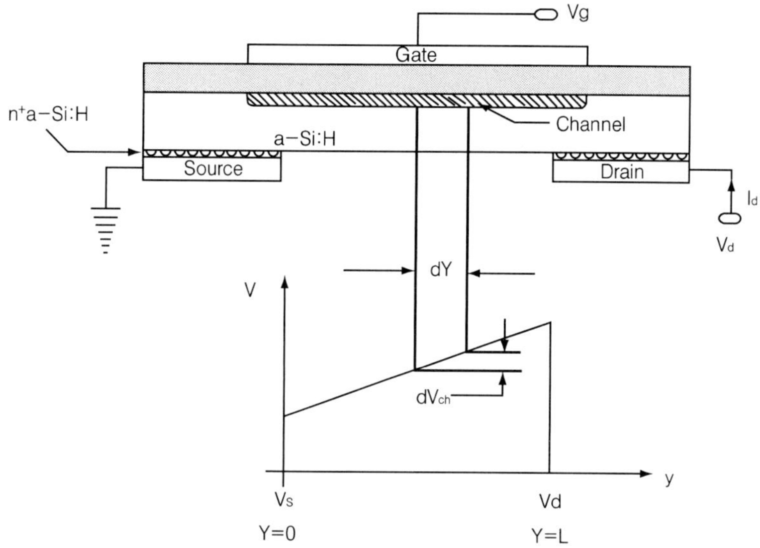

Figure 2.8

The cross-sectional diagram of amorphous silicon thin-film transistors

2.3 Linear region

When analyzing the electrical characteristic of the TFT shown in Figure 2.8, the gradual channel approximation method is used. That is, the electric field on the x-axis (vertical axis) forms a channel, and the electric field on the y-axis (horizontal axis) has the drain current flow through amorphous silicon, with the width of the channel gradually moving towards the drain from the source. This assumes that the width of the channel is affected only by the gate voltage on the x-axis, not the y-axis.

Let's say the gate voltage is higher than the threshold voltage (V_T); that is, electric charges are induced into the channel. In such cases, the relationship between the induced charges (Q_I) and the gate voltage (V_G) are expressed as follows:

$$Q_I = -C_{SiN_x}(V_G - V_{TH}) \quad \text{(Equation 2-1)}$$

C_{SiNx} refers to the capacitance of the gate insulator at the channel. (Equation 2-1) assumes that the channel voltage is zero V, but in practice, the voltage V is formed at the y-axis by the bias between the drain and the source. Thus, the charges induced in the channel should be amended as follows:

$$Q_I = -C_{SiN_x}(V_G - V_T - V) \quad \text{(Equation 2-2)}$$

Meanwhile, the channel current is defined by:

$$I_D = W\mu_n Q_I E_y$$ (Equation 2-3)

Where W is the width of the channel, μ_n is the electron mobility, Q_I is the induced charge amount and E_y is the electric field in the y-direction.

Now if we substitute $E_y = -dV/dy$ and (Equation 2-2) for (Equation 2-3),

$$I_D dy = W\mu_n C_{SiN_x}(V_G - V_T - V)dV$$ (Equation 2-4)

And then if we integrate this from y=0 to L, and from V=0 to V_D (the drain voltage), it becomes:

$$I_D = C_{SiNx}\mu_n \frac{W}{L}[(V_G - V_{TH})V_D - \frac{1}{2}V_D^2]$$ (Equation 2-5)

If V_D is in the very small linear region ($V_D < 1V$), (Equation 2-5) is given by:

$$I_D = C_{SiN_x}\mu_n \frac{W}{L}(V_G - V_T)V_D$$ (Equation 2-6)

Thus, from (Equation 2-6) we see that the factors affecting the current-voltage characteristic of TFTs are the thin film insulator capacitance, electron mobility, the length and width of the transistor channel, the gate voltage, the threshold voltage and so forth. These factors are summarized in Table 2.1 below.

Table 2.1 The characteristics of TFTs and their influential factors

Characteristic parameter	Factor
On current	• W/L • Mobility • Interface • Ohmic contact • Gap states • Back surface(or interface)
Off current	• W/L • Fermi level(a-Si:H) • Interface • Back surface(or interface) • Ohmic contact • Band gap
Mobility	• Width of band tails • Interface states
Gate voltage swing	• Gap states(defect states) • Interface states

2.4 Saturation Region

If the drain voltage gets higher, causing the gate voltage to be neutralized, the channel disappears from the drain part (pinch–off phenomenon), resulting in no further increase of the drain current. Accordingly, the equations to calculate the drain current in the linear region, as described above, are no longer valid. To solve this, Q_I should be replaced by zero. Therefore, if we substitute $V=V_G-V_T=V_D$ in (Equation 2–5), it can be expressed by:

$$I_D = \frac{C_{SiN_x}\mu_n W}{2L}(V_G - V_T)^2 \quad \text{(Equation 2–7)}$$

It is possible to calculate the field–effect mobility in the saturation region based on (Equation 2–7). The characteristics of amorphous silicon transistors are influenced by the characteristic of amorphous silicon itself (e.g., the density of states, band mobility, etc.), the characteristic of the interface between amorphous silicon and the insulator, the contact between amorphous silicon and metal, the geometry of the device and so on. These factors are summarized in detail in Table 2–1.

Figure 2.9 and 2.10 show the output transfer and output characteristics of amorphous silicon TFTs, respectively. The gate voltage swing (S) is the gate voltage (V_g) required to multiply the drain current (I_D) by ten times. This is defined as (Equation 2–8).

$$S = \frac{dV_g}{d(\log I_d)} \quad \text{(Equation 2–8)}$$

The gate voltage swing calculated from the transfer characteristics in Figure 2.9 is 0.27 V/decade.

The method of calculating field–effect mobility by using transconductance (g_m) is commonly used in MOSFETs or polycrystalline silicon TFTs and defined as follows:

$$g_m \equiv \frac{\partial I_d}{\partial V_g} \mid V_{d=const.} = \frac{W}{L} C_i \mu_n V_d \quad (V_d < V_{dsat}) \quad \text{(Equation 2–9)}$$

I_d is the drain current, V_g is the gate voltage and V_d is the drain voltage. W and L are the width and length of the TFT channel, respectively. C_i is the capacitance per unit area of the insulator and μ_n is field–effect mobility. V_{dsat} is the drain voltage when the drain current is saturated.

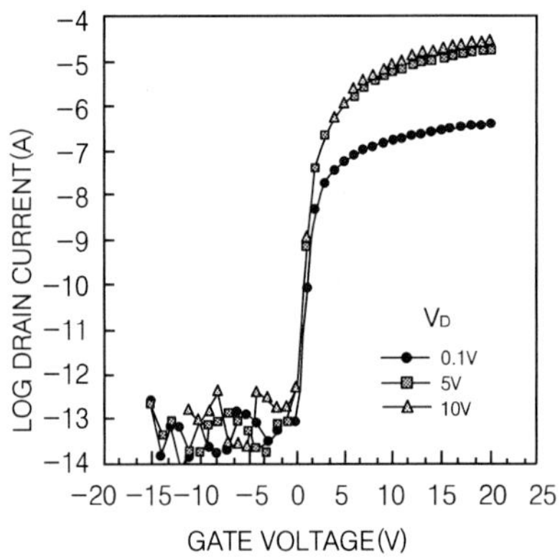

Figure 2.9

The transfer characteristics of amorphous silicon TFTs (W/L=96 μm/10 μm)

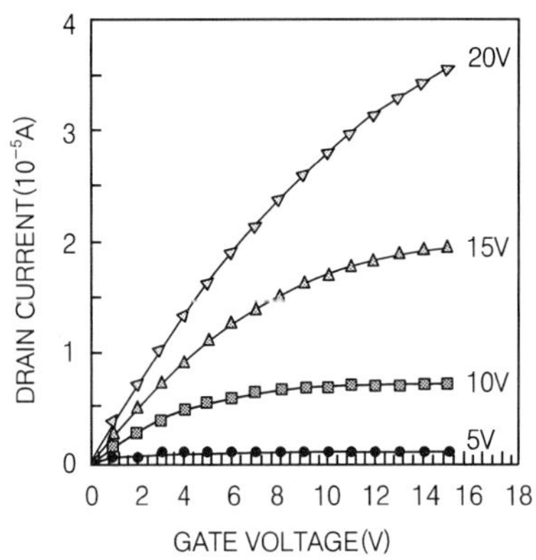

Figure 2.10

The output characteristics of amorphous silicon TFTs (W/L=96 μm/10 μm)

2.5 Gain

Gain K is defined as $K=(C_{SiN_x}\mu_n/2)(W/L)$ and thus, determined by a process parameter, i.e., mobility (μ_n), the thickness of the oxide film ($T_{OX}=\epsilon_r\epsilon_0/C_{SiN_x}$) and the width ($W$) and length ($L$) of the gate (a design parameter). Due to the undesirable lateral diffusion (X_j), the channel length reduces from the mask length (L) to effective channel length (L_{eff}), where $L_{eff} = L - 2X_j$. Such changes in the channel length are described in Figure 2.11 (a). A similar reduction in channel width takes place; the effective channel width, reduces from W (mask width) to, W_{eff}. In addition, as shown in Figure 2.11 (b), it should be considered that the effective length is further reduced in the saturation state, resulting from the channel pinch-off. To calculate the gain (K) and the intrinsic transconductance (K) in the saturation state, (Equation 2–7) is used. Also, to get K (a function of I_D and V_G) without the influence of V_T, the device should be measured by two other I_D values.

(1) Measurement of Gain in the Saturation Region

In Figure 2–12, I_{D1} and I_{D2} are expressed as follows:

$$I_{D1} = K(V_{G1} - V_T)^2 \quad \text{(Equation 2–10)}$$

$$I_{D2} = K(V_{G2} - V_T)^2 \quad \text{(Equation 2–11)}$$

And if (Equation 2–10) and (Equation 2–11) are denoted by K, it is expressed as:

$$K = \left(\frac{\sqrt{I_{D2}} - \sqrt{I_{D1}}}{V_{G2} - V_{G1}} \right)^2 \quad \text{(Equation 2–12)}$$

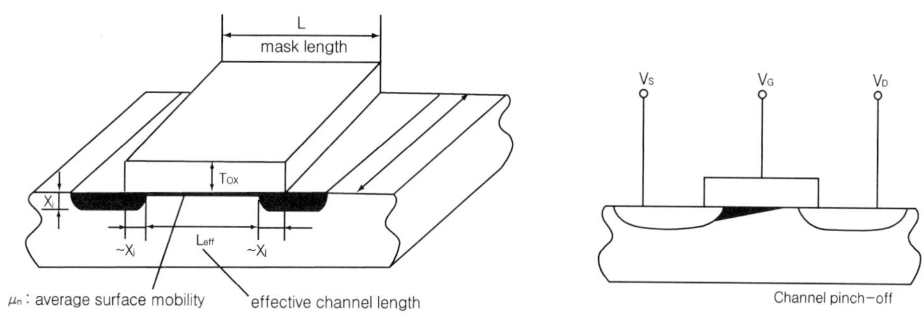

(a) (b)

Figure 2.11

(a) The structure diagram of the channel of transistors (b) Channel pinch–off

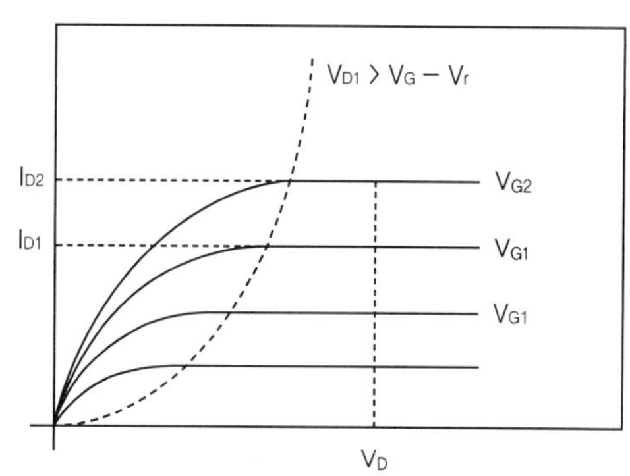

Figure 2.12

The measurement of K in the saturation region

As $K=K'W/L$, K is a function of W/L, which depends on the geometrical form of the device. Once we get the channel width (W) and length (L), it is also possible to get K' from K.

The K' can be expressed as $K'=K(L/W)=(\mu_n/2)(\epsilon_{OX}/T_{OX})$, by which we know that the K' depends on a process parameter. However, we should also consider that the effective channel length L_{eff} is influenced by V_D in the saturation state. And lateral diffusion further lessens the effective channel length. Due to this, the effective data for calculating K' can only be obtained from the device with a relatively longer gate (〉 25 μm), where the influence from V_D can be ignored. In addition, if we subtract $2X_j$ (the effect by lateral diffusion) from L, we get a more accurate value for L_{eff}. $L_{eff}=L-2X_j$, where X_j is the depth of side contact.

To place a side view device in the saturation region, the drain and the gate should be tied together to make V_G equal to V_D ($V_G=V_D$), as shown in Figure 2.13. The voltage–current relationship in such a case is described in Figure 2–14. By doing this, with any drain voltage larger than zero (V_D〉0), the device will operate in the saturation region.

K can be measured much conveniently by designating the value for I_D between two measurement points to be a square of a certain number. For instance, let's say the two measured values for I_D are 9 μA and 4 μA. In such case, after measuring V_G, K can be simply calculated by:

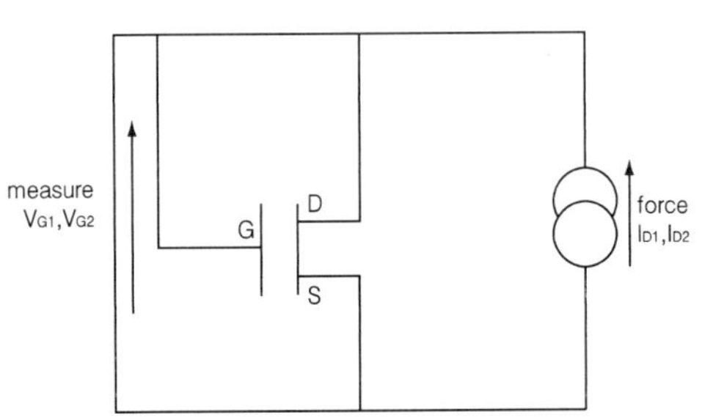

Figure 2.13 Characteristics in the saturation region

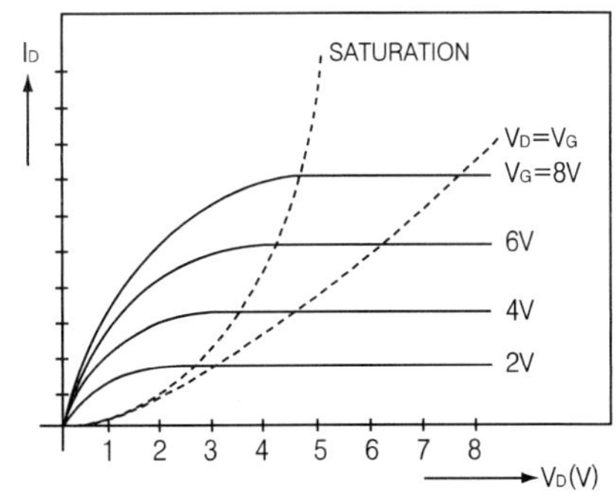

Figure 2.14 Current curve when $V_D=V_G$

$$K=\left(\frac{\triangle\sqrt{I_D}}{\triangle V_D}\right)^2=\left(\frac{\sqrt{9}-\sqrt{4}}{\triangle V_D}\right)^2=\frac{10^{-6}}{(V_D)^2}\Bigg\} \quad \text{(Equation 2–13)}$$

(2) Measurement of Gain in the non-saturation region

In order to eliminate the channel length modulation in the saturation region, it is more accurate to measure K in the non-saturation region where continuous channels are formed between the source and the drain. For this purpose, the condition $V_D < V_G - V_T$ should be satisfied.

K, which is a function of I_D, V_D and V_G, should be measured by switching V_G into two different states with the fixed V_D, based on (Equation 2-5).

In Figure 2.15, I_{D1} and I_{D2} are expressed respectively, as:

$$I_{D1} = K'\frac{W}{L}(2(V_{G1} - V_T)V_D - V_D^2) \quad \text{(Equation 2-14)}$$

$$I_{D2} = K'\frac{W}{L}(2(V_{G2} - V_T)V_D - V_D^2) \quad \text{(Equation 2-15)}$$

When subtracting (Equation 2-14) from (Equation 2-15), the equation is given by $I_{D2} - I_{D1} = K'(W/L)[2(V_{G2} - V_{G1})V_D]$. If we express this by K', it is given by:

$$K'\frac{W}{L} = \frac{I_{D2} - I_{D1}}{2V_D(V_{G2} - V_{G1})} \quad \text{(Equation 2-16)}$$

If we measure K under the condition described in Figure 2.15, K is as follows:

$$K = K'\frac{W}{L} = \frac{I_{D2} - I_{D1}}{2 \times 0.2 \times (6-4)} = \frac{I_{D2} - I_{D1}}{0.8} \quad \text{(Equation 2-17)}$$

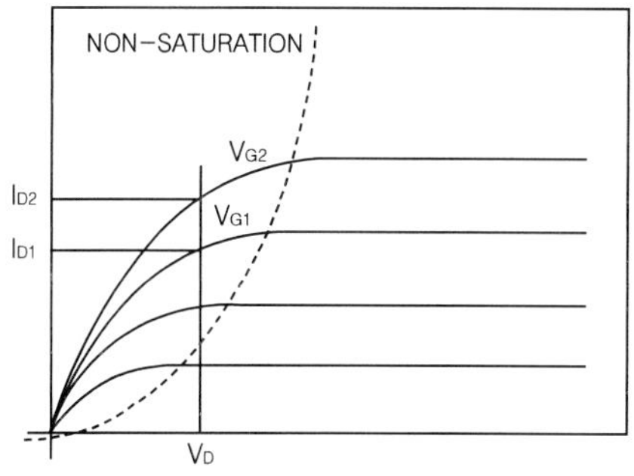

Figure 2.15

The measurement of K in the non-saturation region

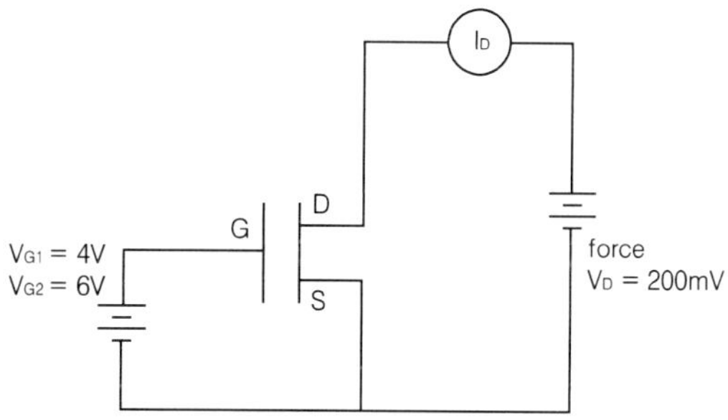

Figure 2.16

The circuit of K measurement in the non-saturation region

Eventually, the intrinsic transconductance K' is measured by:

$$K' = \frac{\Delta I_D}{0.8} \cdot \frac{L}{W} \quad \text{(Equation 2-18)}$$

2.6 Threshold Voltage

A threshold voltage is defined as a gate voltage that causes to form the continuous inversion layer within the channel by a strong inversion occurring close to the gate electrode. This is expressed as (Equation 2-7) in the saturation region. To eliminate a design parameter K, the following equations are given after experimenting with two I_D values.

$$I_{D1} = K(V_{G1} - V_T)^2 \qquad \text{(Equation 2-19)}$$

$$I_{D2} = K(V_{G2} - V_T)^2 \qquad \text{(Equation 2-20)}$$

If we divide (Equation 2-20) by (Equation 2-19) and denote it by V_T, it becomes:

$$V_T = \frac{(\sqrt{I_{D2}}/\sqrt{I_{D1}})V_{G1} - V_{G2}}{(\sqrt{I_{D2}}/\sqrt{I_{D1}}) - 1} \qquad \text{(Equation 2-21)}$$

For example, if I_{D2} equal to $4I_{D1}$ ($I_{D2}=4I_{D1}$) as shown in Figure 2.16, for a simple testing, then V_T can be expressed by:

$$V_T = 2V_{G1} - V_{G2} \qquad \text{(식 2-22)}$$

And then the bias should be adjusted so that the I_{D2} value is put into the saturation region. I_{D1}, then, it is automatically put into the saturation region. Figure 2.17 describes such a test circuit.

As a way of measuring a more accurate threshold voltage, V_T can be calculated from the cross point of the extended straight line of the I_D-V_D characteristic curve and the V_D-axis, as described in Figure 2.19.

The slope of the straight line is expressed as:

$$\left.\frac{\partial I_D}{\partial V_D}\right|_{V_D\to 0} = \frac{\partial}{\partial V_D}\left[\mu\frac{W}{L}C_{SiN_x}(V_G - V_T - \frac{V_D}{2})V_D\right] = \mu_n C_{SiN_x}\frac{W}{L}(V_G - V_T) \qquad \text{(Equation 2-23)}$$

In such case, $K=\mu_n C_{SiNx} W/L$ should be obtained in the non-saturation region. In the same way, we can calculate V_T when reverse bias (V_B) occurs.

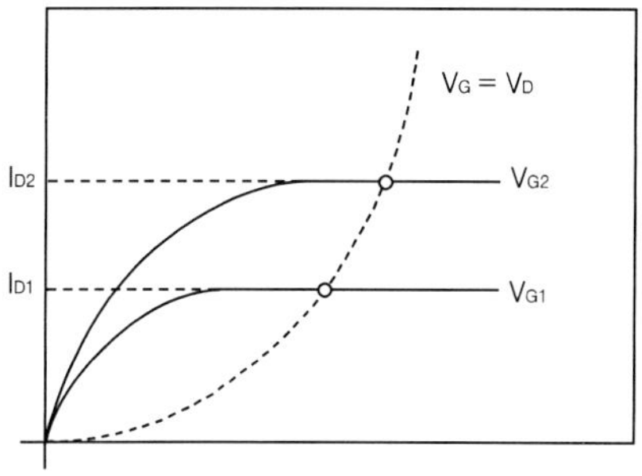

Figure 2.17

I_D against V_D in the saturation region

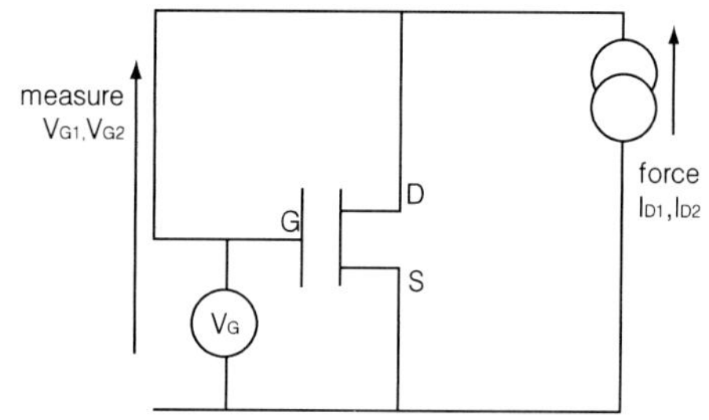

Figure 2.18

A test circuit of channel enhancement transistors

2.7 Mutual Transconductance g_m

A small signal parameter g_m represents the change of I_D, in accordance with a certain change from V_D to V_G. As the gap between gate voltage steps varies, the g_m value changes depending on the operating point to be chosen. The I–V characteristic depending on the gate voltage is described in Figure 2.19. The dependency of g_m on I_D can be obtained by:

$$g_m = \left.\frac{\Delta I_D}{\Delta V_G}\right|_{V_D} = \mu_n C_{\mathrm{SiN_x}} \frac{W}{L} V_D \qquad (V_D < V_{sat}) \qquad \text{(Equation 2-24)}$$

$$g_m = 2K' \frac{W}{L}(V_G - V_T) \qquad (V_D > V_{sat}) \qquad \text{(Equation 2-25)}$$

The typical device for measuring g_m is described in Figure 2.20. The equation is given by:

$$g_m = \frac{I_{D2} - I_{D1}}{V_{G2} - V_{G1}} \qquad \text{(Equation 2-26)}$$

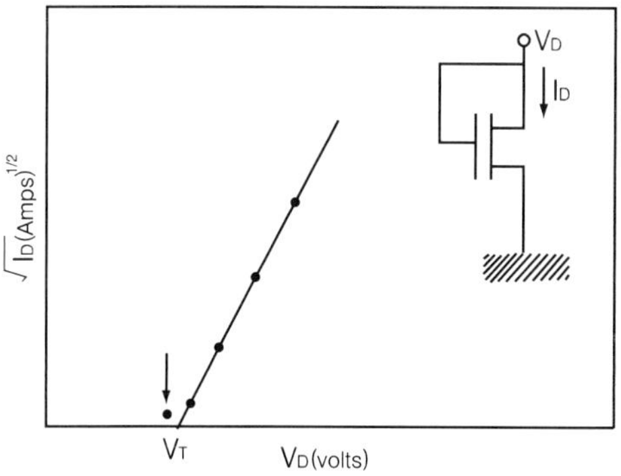

Figure 2.19

Measurement of a threshold voltage

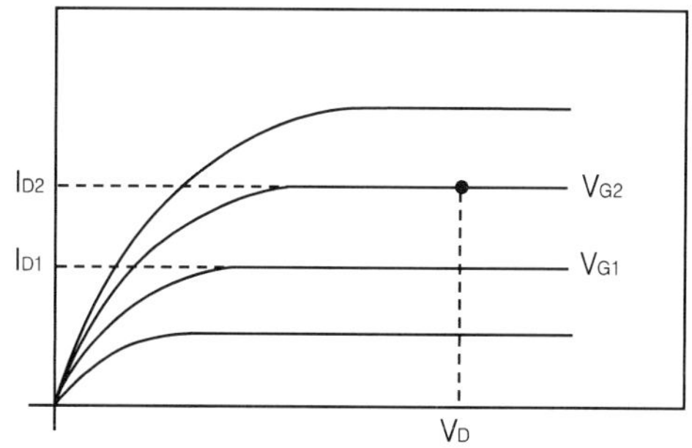

Figure 2.20

I–V characteristic depending on the gate voltage

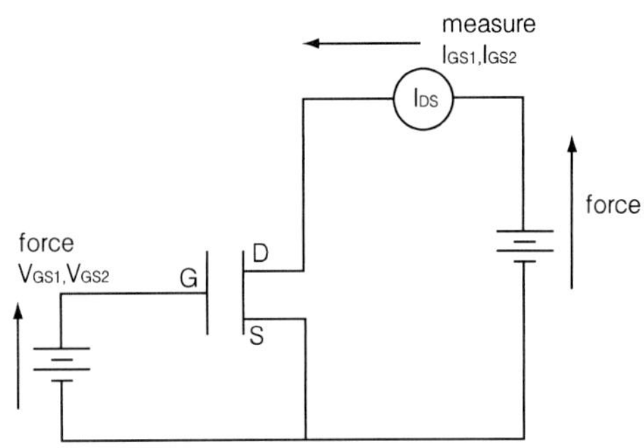

Figure 2.21
A circuit diagram for g_m measurement

With this simple circuit, it is possible to measure g_m between the designated V_G and V_D. There are also some cases where it is better to measure g_m between the designated I_D and V_G. V_G can be adjusted for this purpose, using a manual device. When using an automatic system, V_G can be adjusted by programming until I_D meets the required condition.

*3 Display Engineering

Amorphous Silicon Thin Film Transistors

3.1 Amorphous Silicon

Various Electronic devices use amorphous silicon in various fields recently. The "amorphous" refers to a material having irregularly arranged atoms that have short-range order (but no long-range order at all). Short-range order means that the number of the nearest neighbor atoms, the bond distance and the bond angle are approximately ordered.

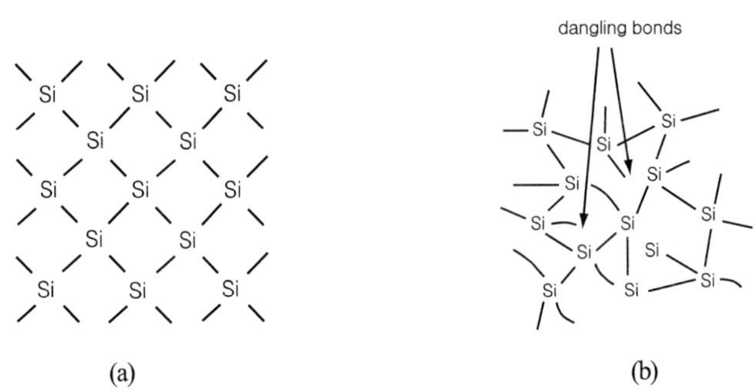

Figure 2.22
The bonding structure of crystalline silicon (a) and amorphous silicon (b)

In crystalline silicon, each silicon atom has a covalent bond with four other neighboring silicon atoms, featuring three-dimensional periodicity. Amorphous silicon also shares similar coordination number, bond angle and bond distance with crystalline silicon, but there is no regularity (periodicity) in grouping atoms in amorphous silicon. That is, regularity exists in short-range only, not in long-range.

Figure 2.23 schematically describes the bonding of atom and energy band diagram of amorphous silicon having no hydrogen. The defect density of such amorphous silicon is approximately 10^{20} $cm^{-3}eV^{-1}$, which is much larger than that of the hydrogenated amorphous silicon. In other words, the hydrogenated amorphous silicon, as shown in Figure 2.23, has less dangling bonds and localized defect states than that in the amorphous silicon without hydrogenation. The hydrogen in amorphous silicon significantly reduces the dangling bond density and reduces localized states. Also, the forbidden gap exists differently from that of crystalline silicon, since there is no long-range regularity in amorphous silicon. Due to such a disorder, the band tail state exists in amorphous silicon.

In amorphous silicon, hopping conduction occurs through the localized states, because of which the mobility of charge carriers are very low. Also, the defect states of amorphous silicon generated by dangling bonds exist in the forbidden gap. Unlike single crystal, it is not possible to define the accurate band gap of amorphous silicon, due to the localized states and band tails. Instead, for amorphous silicon, the level where the mobility of electrons and positive holes gets extremely low should first be determined based on the relationship between energy and mobility; this should then be considered to be the mobility gap, corresponding to the band gap of a single crystal.

The main reason for using single crystal silicon in IC or LSI technology is that it is possible to change the intrinsic silicon to p-type or n-type material by adding a few impurities such as B or P atoms respectively. A transistor in IC or LSI is produced by suitably combining such p-type, n-type materials. On the contrary, even if impurities are added to amorphous silicon, it is not possible to precisely alter the materials into p-type or n-type due to the presence of dangling bonds or band tail. Furthermore, the conductivity of the amorphous silicon formed by decomposing silane gas vastly fluctuates with the addition of B or P. Such silicon materials are produced by the decomposition of silane contains about 10at.% of hydrogen. The lattice invasion of hydrogen relaxes the deformation of the lattice and reduces the defect density due to the bonding of dangling bonds.

The amorphous silicon without hydrogen produced by sputtering makes localized states with high density in the band gap. Due to this, the fermi level of semiconductors is fixed or pinned near the center of the band gap. In such semiconductors with many defects, localized states of approximately 10^{20} cm^{-3} exist and thus, it is not possible to form donor level or acceptor level even if the semiconductors are doped with dopant impurities of 10^{18}~10^{19} cm^{-3}. On the other hand, the hydrogenated amorphous silicon has a lower density of localized states and thus, it is possible to form donor level or acceptor level by doping. The defect density of the hydrogenated amorphous silicon is approximately 10^{15}~10^{17} cm^{-3}, and it is possible to control valence electrons with the addition of impurities of 10^{18}~10^{19} cm^{-3}.

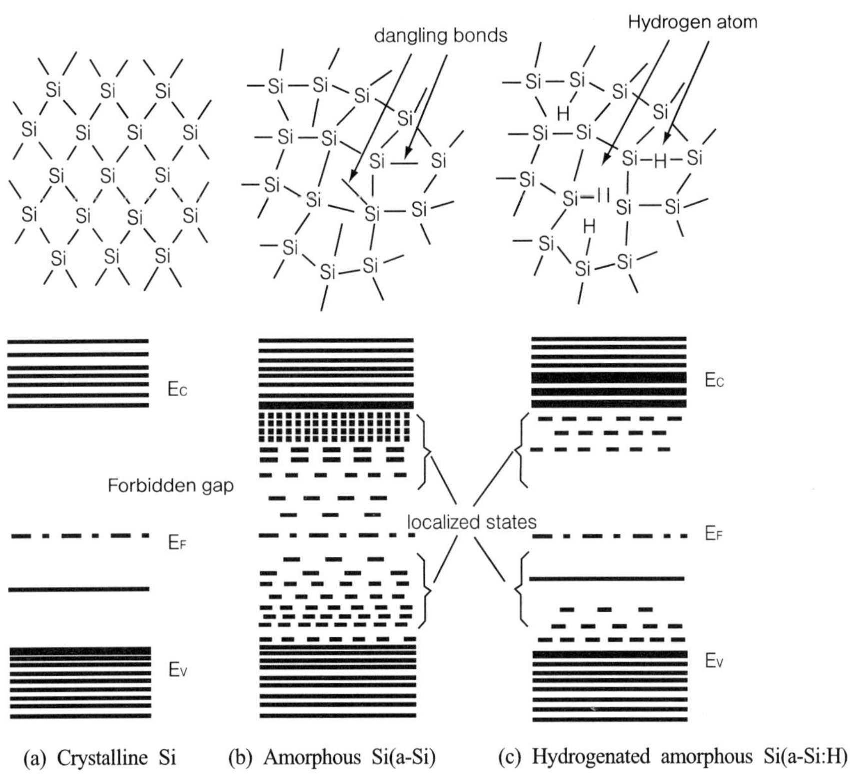

Figure 2.23
The atom bond and energy band of crystalline silicon and amorphous silicon

3.1.1 Hydrogenated Amorphous Silicon

In crystalline silicon, each silicon atom has four covalent bonds with four neighboring silicon atoms. The lattice constant of silicon is 5.431 Å with three-dimensional regularity. Amorphous silicon also shares similar coordination number, bond angle and bond distance with crystalline silicon, but there is no regular periodicity in longer length scale. In amorphous silicon, many of the silicon atoms exist with dangling bonds, which does not form bonds with the necessary four atoms. However, a hydrogen atom is combined with the dangling bonds in the hydrogenated amorphous silicon, eliminating dangling bonds, by which the density of localized states reduces in the band gap. The below figure shows the atom bond of the hydrogenated amorphous silicon (a–Si:H).

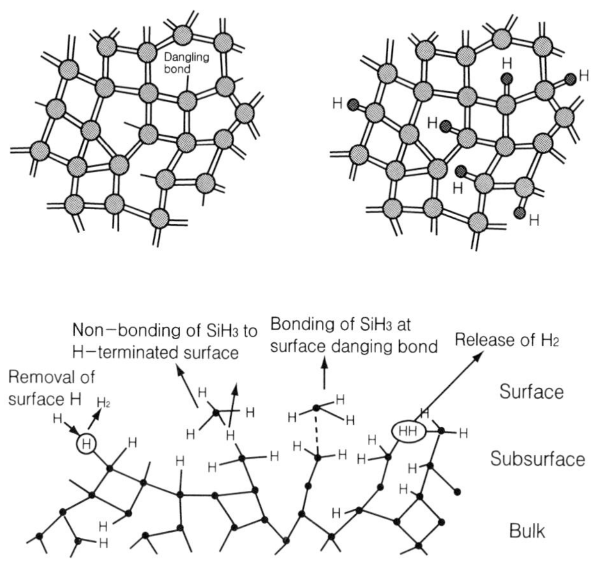

Figure 2.24

The apparatus for growing the hydrogenated amorphous silicon (a–Si:H) thin–film

As shown in Figure 2.24, the combined shape of silicon and hydrogen (Si–H) is important in the growth of a–Si:H thin films. Among various bonds of silicon and hydrogen, Si–H bond with relatively higher bond energy (3.2 eV) is much more stable than other bonds such as Si–H_2 bonds. At the silicon deposition temperature of 200~300°C, the decomposition of Si–H_2 by thermal energy decreases, resulting in an increase in the Si–H_2 bonds in amorphous silicon. If the substrate temperature is higher than 300°C, the chemical bond type is strongly influenced by the deposition method, rather than the temperature. It is known that the Si–H_2 bond weakens the general characteristics of semiconductors in a–Si:H thin films. In the non–hydrogenated amorphous silicon, the defect density is very high, approximately 10^{20} $cm^{-3}eV^{-1}$. As the defect density of the amorphous silicon thin films grown by RF sputtering or e–beam is too high (higher than 10^{20} $cm^{-3}eV^{-1}$), causing doping to be not effective, such amorphous silicon thin films were not used in practice for electric/electronic devices. However, W.E. Spear, P.G. LeComber and S. R. Ovshinsky at Dundee university prepared hydrogenated amorphous silicon by glow discharge hydrogenation or fluorination, by which the internal density of the forbidden gap has greatly reduced, causing doping to be effective in amorphous silicon. The defect density of these hydrogenated a–Si:H thin films was approximately $10^{15}cm^{-3}eV^{-1}$. This is because hydrogen eliminates the broken bonds of silicon (dangling bonds) and reduces localized states.

3.1.2 Characteristics of a–Si:H Thin Films

The general characteristics of a–Si:H are listed in Table 2.2. The a–Si:H is a semi–direct–gap semiconductor with the energy band of 1.72 eV, and has a larger optical absorption coefficient than crystalline silicon (c–Si). The changes in the optical absorbance of silicon thin films are described in Figure 2.25.

Table 2.2 The general characteristics of a–Si:H

Deep Localized States Density at $E=EF_o$	10^{15}~10^{16} [cm^{-3}/eV]
Deep States Characteristics Energy	86 [meV]
Tail Localized States Density at $E=E_C$	~2×10^{21} [cm^{-3}/eV]
Tail States Characteristics Energy	23 [meV]
Width of Tail States Band	~150 [meV]
Dielectric Constant	~11
Diffusion Constant at 300 K	[cm^2/s]
- electrons in conduction band	0.26 ~ 0.52
- holes in valence band	0.13 ~ 0.26
Effective Density of States in Conduction Band at 300 K	~10^{19} [cm^{-3}]
Effective Density of States in Valence Band at 300 K	~10^{19} [cm^{-3}]
Energy Gap at 300 K	1.72 [eV]
Index of Refraction	3.32
Mobility at 300 K	[cm^2/V · s]
- electrons in conduction band	10~20
- holes in valence band	5~10

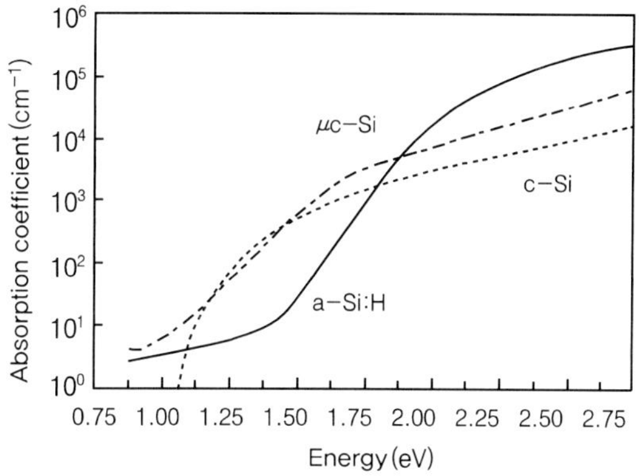

Figure 2.25

The optical absorbance with photon energy

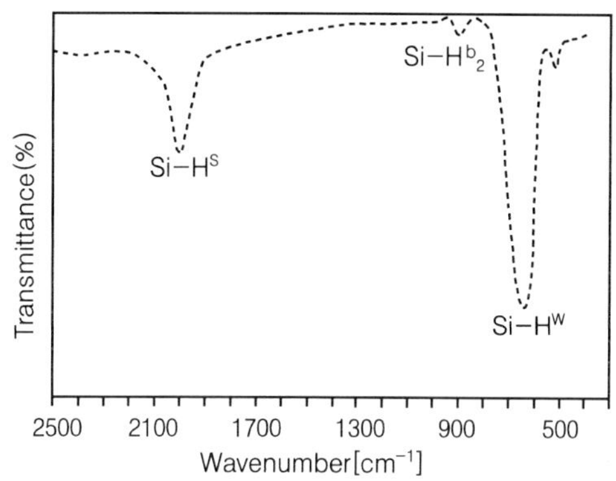

Figure 2.26

A typical result of FTIR analysis of a–Si:H thin film

Figure 2.26 shows different Si–H bonds in hydrogenated amorphous silicon films. Figure 2.27 shows the diffusive–movement of hydrogen between Si–H bonds. Hydrogen moving between Si–Si bonds, with repetition of trapping and de–trapping process in loosely combined localized states. If Si–H bond exists right next to strained (weak) Si–Si bond under a stable structure, it is possible to break the strained (weak) Si–Si bond with the bond energy of non–radiative electron–holes that occurs when the light is applied. Then a Si–Si bond is broken into two Si atoms, one of which bonds with hydrogen and the other remains a dangling bond. Such a–Si:H thin film is sensitive to light, and its initial physical properties could change after UV light is applied.

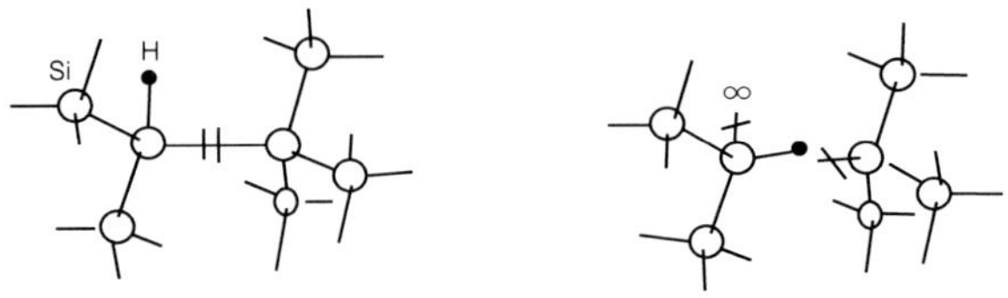

Figure 2.27

The micro model of metastable dangling bond formation with light applied

The electrons in amorphous silicon can exist in the extended and localized states, especially in the energy gap. The energies E_c and E_v that distinguish the extended and localized states are called mobility edges. Carriers have a band mobility of dozens of $cm^2/V{\cdot}s$ in the conduction band and valence band. The characteristic of electron transport in amorphous silicon is determined by the interaction between charge carriers and localized states. If an electric charge that is trapped in localized states is transported in the conduction band by thermal excitation, this is called de-trapping; transporting a charge carrier by direct tunneling in localized states is called hopping.

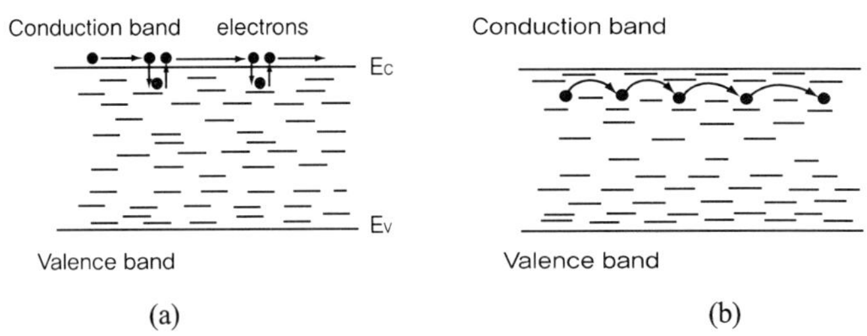

Figure 2.28

The carrier transport model of electrons by trapping(a) and hopping(b) in hydrogenated amorphous silicon

The hydrogenated amorphous silicon is a material that have characteristics to be used in electronic devices. The advantages of using the amorphous silicon are as follows:

① Large area deposition is effectively done.

② Deposition is possible under low temperature.

③ As a-Si:H is an amorphous material, it is easy to form a dimorphous boundary with other materials, still maintaining good boundary characteristics. Thus, it is possible to deposit amorphous silicon on various types of substrates such as insulators, metals and/or semiconductors. The formation of such good boundary with any materials helps to flexibly invent and employ various TFT processes in a process line.

④ Amorphous silicon is resistant to scratches (Vickers hardness HV=1,5002,000 kg/mm^2). And it is possible to form fine patterns with photolithography technology.

⑤ The hydrogenated amorphous silicon has a high electric resistance when it is not doped, but once it is doped, it gets higher electrical conductivity.

⑥ It is non-toxic.

3.1.3 Structure of α-Si TFT

The structure of a-Si TFT, consisting of three electrodes such as the gate, drain and source, can be classified into a Staggered structure and a Coplanar structure, as shown in Figure 2.29, due to the positions of the three electrodes in relation to the a-Si layer.

- Coplanar
- Staggered
  - Normal Staggered (top gate)
  - Inverted Staggered (bottom gate)
    - BCE(Back Channel Etched)
    - ES(Etch Stopper)
      - Semi-self aligned
      - Fully-self aligned

Figure 2.29
The classification of TFT based on the structure

Coplanar structure refers to the structure that the three electrodes are placed on the same side, whereas Staggered structure means that the electrodes are divided by the a-Si layer. The Staggered structure is again divided by the position of the gate into normal Staggered and inverted Staggered, where the gate is placed at the lower part. The mimetic diagram of each structure is described in Figure 2.30.

The inverted Staggered structure is generally used in a-Si TFT. For this structure, the layers are deposited in the order of the insulator, the a-Si layer and then n^+ a-Si. Thus, the processing time, as well as a number of impurities, is reduced, and the interfacial properties between the a-Si layer and the gate insulator becomes excellent.

The inverted Staggered structures are BCE (Back Channel Etched) and ES (Etch Stopper). BCE structure deposits the a-Si layer and subsequently the n^+ a-Si layer (on top of the a-Si layer), and then forms the source/drain electrodes. Once the source/drain electrodes are formed, the n^+ a-Si layer is etched; the a-Si layer is also etched at the same time and thus, the a–Si layer has to be thick enough. The issue is that the thicker the a–Si layer, the higher its sensitivity to light, causing a leakage current to increase.

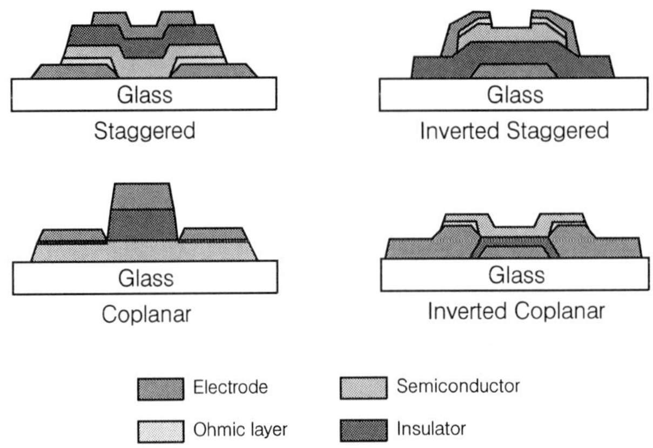

Figure 2.30
The mimetic diagram of the basic structure of TFT

ES structure deposits the n^+ a–Si layer above the patterned Etch Stopper layer and forms the source/drain electrodes. Once the source/drain electrodes are formed, the n^+ a–Si is etched; In this case, the Etch Stopper layer on the channel side prevents the a–Si layer from being etched. However, for the patterning of the Etch Stopper layer, the device is to be taken out of the vacuum state.

Figure 2.31 shows the cross–sectional diagrams of BCE structure (a) and ES structure (b) of the inverted a–Si TFT.

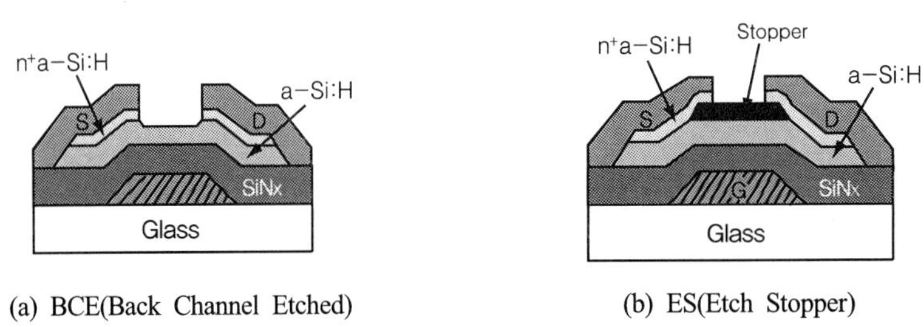

Figure 2.31
Two structures of the inverted staggered TFT

BCE structure is the most widely used for the inverted staggered a–Si TFT, due to its simple manufacturing process. However, when etching the n^+ a–Si layer, the a–Si layer should also be over–etched. For this reason, the a–Si layer should be thicker than that of the ES structure. As mentioned above, if the a–Si layer gets thicker, the photo–induced leakage current increases when the light is applied. The increase of the leakage current leads to lower quality of the images. There have been many reported research results, aiming to decrease the leakage current.

3.2 The manufacturing process of a-Si TFT

3.2.1 Gate metal formation process (gate step)

This is a process to grow thin film by sputtering[4] gate metal and form a storage electrode at the same time. The thin-film is grown mostly by Al or Cr electrodes, and electrodes are defined using one gate mask through a photolithography process. Such a series of the unit processes are described in Figure 2.32. New equipments and materials for this purpose continue to be developed; for instance, there are several other methods for metal deposition than the traditional sputtering method, such as Spin On Metal, Metal printing by ink-jet and so forth.

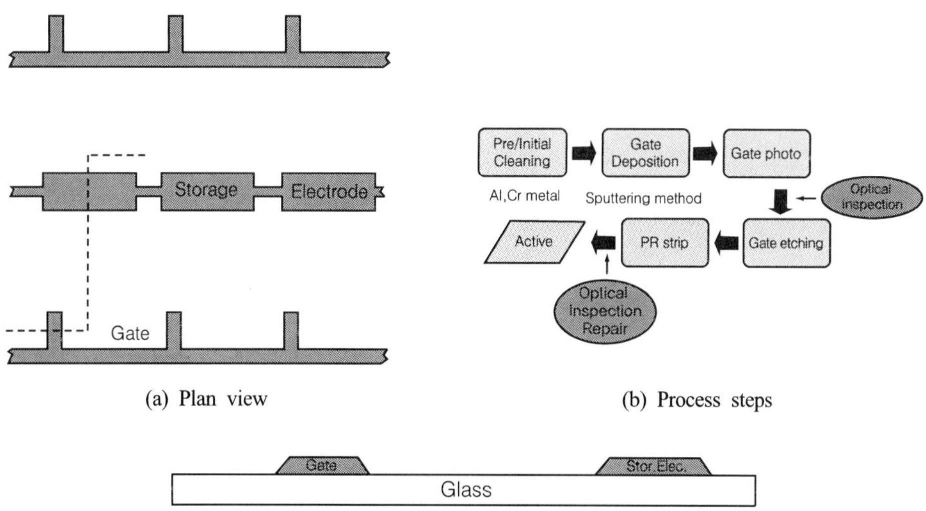

(c) Sectional view

Figure 2.32
The gate metal formation process of TFT

In recent years, low-resistant and mechanically strong Al alloys tend to be used. A gate material should be not only low resistant but also strongly adherent to a glass substrate as they are deposited first on the substrate. Further, various

4) Sputtering is a physical phenomenon that positive ions with high energy, generated from plasma gas, eject target materials with a negative pole, by which the target materials are deposited on the substrate.

thin films continue to be compiled on the gate in the subsequent processes and thus, the surface of a gate material should be flat and easy to etch and taper. Also, the material should be strongly resistant for various chemical solutions to be used in the subsequent processes. The examples of gate materials used in Japan in 1990's were Cr chosen by NEC, Ta by Sharp and MoTa alloy by Toshiba. However, as the size of the panel is getting larger, the resistance of the gate materials has reached its limit and thus new metal structures continue to be developed.

3.2.2 Active layer growth process (Active step)

On the upper part of the gate metal, the hydrogenated a-Si thin-film should be grown by Plasma Enhanced Chemical Vapor Deposition (PECVD) method, and at the same time, n^+ a-Si and SiN_x should also be laminated and grown. After this growing process, all the parts except for the gate electrode and the SiN_x insulator should be etched with an optical mask, to obtain the structure shown in Figure 2.33 (a). As a method of growing the active layer, Catalytic CVD (or Hot wire CVD[5)]) and other methods have been introduced in addition to PECVD. These are new technologies that came out from the doubt on the possibility of growing thin-film with plasma even for glasses larger than 1,000mm× 1,000mm, and from the need for an economizing process to deal with the high-priced equipment. Such issues actually occur with PECVD; for the purpose of depositing a large area by PECVD, it is inevitable to enlarge the size of the flat panel electrode, which entails the increase of the volume of plasma. Also, in order to maintain the deposition rate, the rf power should be incremental along with the volume of plasma, causing a higher possibility of abnormal discharge. To lessen such possibility and maintain the consistency of the film, the density of plasma should be as uniform as possible at all the time regardless of the environment. More specifically, there should be no distortion or bumps on the surface of the electrodes, and the distance between the electrodes should be consistent. Nevertheless, it is quite difficult to technically realize the required flatness and parallelization of the electrodes for the panels larger than 1 meter.

5) Hot wire CVD is known as catalytic thermal CVD or evaporative surface decomposition.

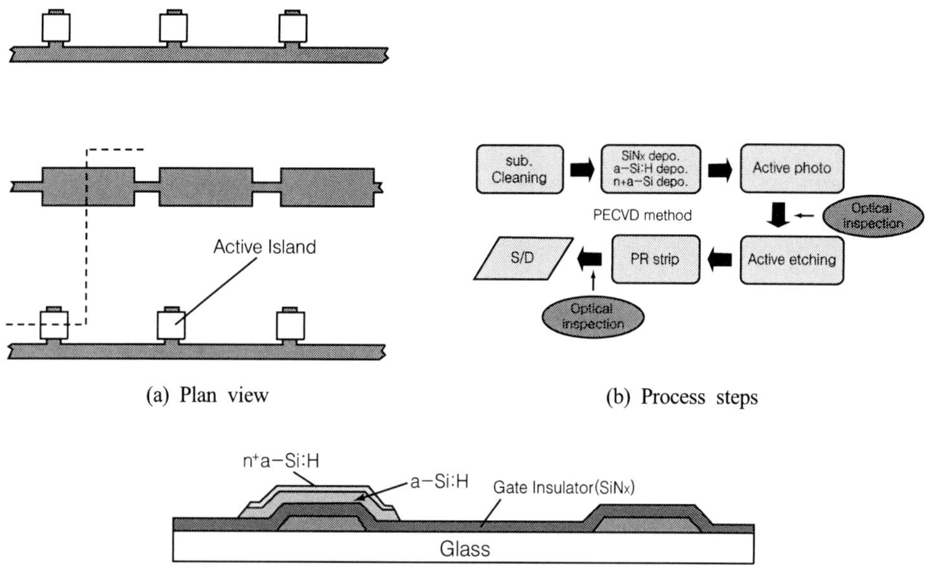

(a) Plan view

(b) Process steps

(c) Sectional view

Figure 2.33
The growing process of the active layer and the insulator of TFTs

(1) Gate Insulator Technology

An oxide film that takes a key role in the operation of a semiconductor device is used for field oxide, protective layers, insulators between layers and gate insulators. Due to such importance, researches on the production and assessment of insulators have been actively undertaken for decades since MOSFET was invented. The formation of insulators is made with various methods; among them, the insulator made by thermal oxidation has been mostly used for semiconductor devices due to its excellent interfacial and insulating properties. However, thermal oxide film needs high-temperature process with the silicon substrate temperature of over 800°C, and thus cannot be used in low-temperature process. Therefore, the production of a-Si:H TFTs inevitably requires a gate oxide film made by low-temperature process.

There are several ways to produce insulators under low temperature, among which PECVD (Plasma Enhanced CVD) is the most commonly used. The silicon oxide film produced by PECVD, among various insulators used for gate dielectrics, used to have weak electric characteristics and thus could only be applied as the second protective layer or delamination material. However, there have been studies on good quality silicon oxide film recently. Currently, the gate insulator generally used for polycrystalline silicon TFTs is PECVD SiO_2. For the deposition of SiO_2, SiH_4+N_2O+He or TEOS+O_2 gasses are mainly used as the raw material. PECVD SiO_2 has advantages in that its deposition temperature is low and it can be made large, but the problem is that impurities are also generated. In this regard, there was a recent report that PECVD SiO_2 by TEOS shows improved uniformity and step coverage. Meanwhile, it is known that SiO_2 by ECR–CVD can be processed under low temperature and shows excellent insulating properties such as high breakdown voltage and low interfacial state density; however, it requires a second deposition of another insulator, which makes it difficult to apply to large devices.

For a gate insulator of TFTs, silicon nitride film (SiN_x) or silicon oxide film (SiO_2) can be used. Silicon nitride film has good interfacial properties between SiN_x and poly–Si, but there are also some disadvantages; that is, the bulk capture of the positive holes and electrons (contributing to electric conduction) in the sample are huge, and the driving voltage is unstable. On the other hand, silicon oxide film does not have as good interfacial properties (between SiO_2 and poly–Si) as those of silicon nitride film, but the density of bulk capture in the sample is low with the stable driving voltage.

Table 2.3 Comparison of the characteristics of low–temperature SiO_2 depending on the deposition method

Method	T_s	Large area	Step coverage	Electrical characteristic
Sputtering	excellent	excellent	bad	good
LPCVD	bad	bad	good	bad
APCVD	bad	good	good	bad
ECR-CVD	excellent	bad	bad	excellent
PECVD(SiH_4, N_2O)	excellent	excellent	good	good
PECVD(TEOS, O_2)	excellent	excellent	excellent	good

In recent years, silicon oxide film of good quality was produced by PECVD method with the electric properties and low bulk capture density similar to those of silicon oxide film produced by thermal oxidation. The gate insulator in polycrystalline TFTs has a huge influence on the characteristics and stability of the device. Table 2.3 shows the electric characteristics of polycrystalline silicon TFTs depending on the deposition method.

(2) Ion doping technology for the growth of n+ a-Si:H

Ion doping is employed to adjust the threshold voltage of TFTs as well as to form ohmic contacts in the source/drain area of TFTs. The devices mainly required for ion doping are an ion implanter and ion shower. Many devices produced with the ion shower, which is simple to compose and drive, are being reported recently. The ion shower is easy to use for large-area processing and thus is mostly used for doping process in the production of low-temperature polycrystalline silicon TFTs (Figure 2.34).

Polycrystalline silicon with a crystalline structure is amorphized when ions are injected and thus, to activate the injected impurities after the ion doping process, heat treatment process with high temperature is required. For the heat treatment methods are furnace method, RTA method, ELA method and so forth. The advantages and disadvantages of each method follow: first, furnace method is good for mass production, but it requires high-temperature process, meaning that inexpensive glass substrates cannot be used for this purpose and that the activation efficiency gets lower. The merits of RTA method are short processing time and good activation efficiency, but the method also requires high-temperature process which is improper for the processing of glass substrates. Meanwhile, ELA method can employ low-temperature process and its activation efficiency is also excellent. However, the processing time is long, and high-priced laser equipment has to be used.

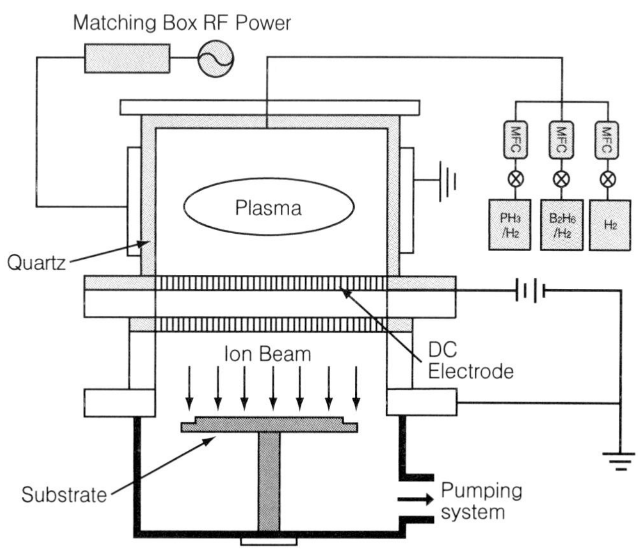

Figure 2.34

The equipment drawing of ion shower and the distribution chart of impurities in polycrystalline silicon after the ion shower

Figure 2.35 indicates the sheet resistance of polycrystalline silicon with thermal treatment after the doping process. Normally the ion doping and activation processes are conducted independently in the production of polycrystalline silicon TFTs, but it is good to conduct both processes simultaneously, considering the processing time and efficiency. For instance, GILD (Gas−Immersion Laser Doping) is a doping method that flows impurities in the gas form and at the same time it re−crystallizes the area that is doped, with a laser. This method helps get an excellent doping effect without any separate activation process.

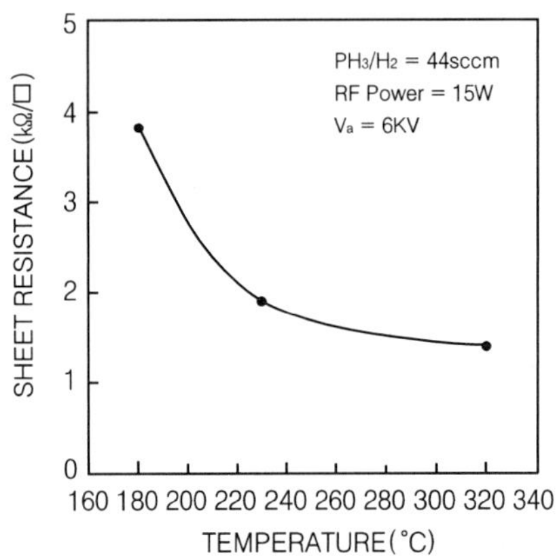

Figure 2.35
Sheet resistance of doped polycrystalline silicon depending on the heat treatment temperature

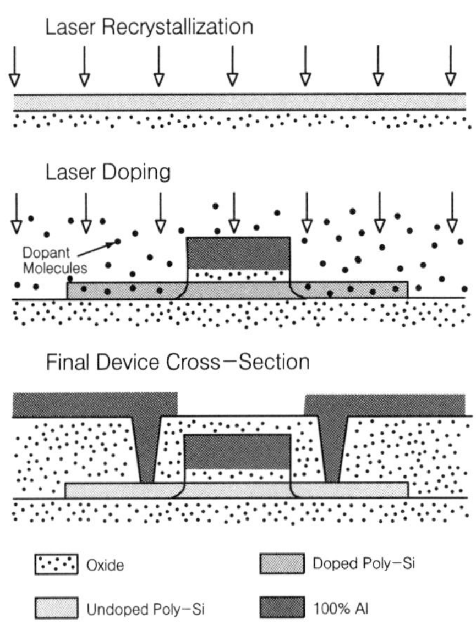

Figure 2.36
The processing steps of GILD

3.2.3 Source/drain metal formation process (Data step)

To form the source and drain electrodes, metal thin–film should be deposited generally by sputtering and treated with photolithography process, and then the source and drain are defined simultaneously. The drain and source defined in this way are to be used as self–aligned masks in the dry etch process to define the channel area.

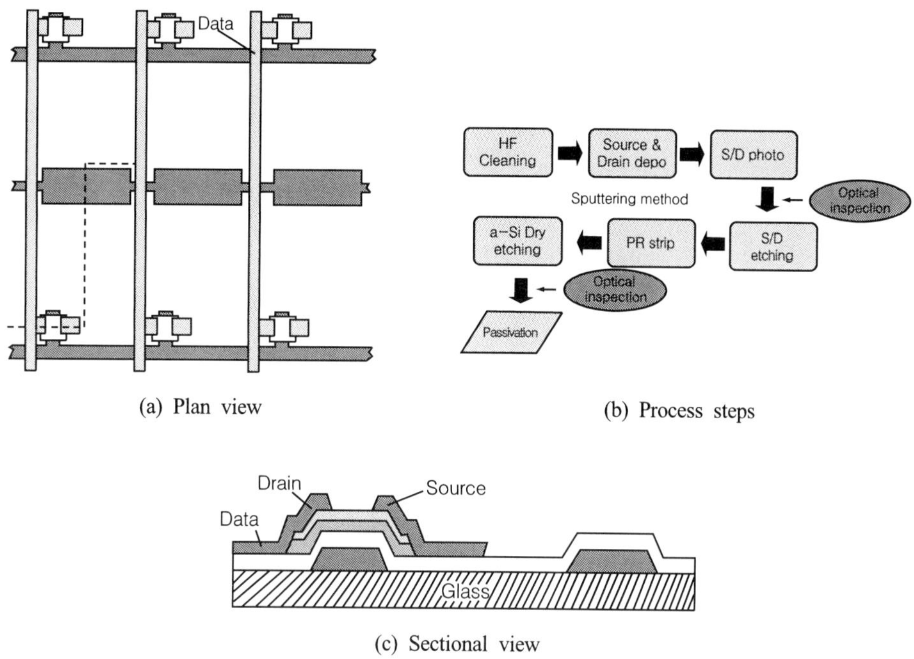

Figure 2.37
The formation process of the source and drain electrodes of TFTs

The key conditions for the source/drain (S/D) materials are (1) the ohmic contact with a–Si:H, (2) no cracks throughout all the step areas, (3) no ITO and battery response under Top ITO Transistor structure and (4) no erosion in the chemigraphic solution in the processes following the S/D process. Condition (1) is related to TFT characteristics and more important than low resistance condition. The ohmic contact with a–Si:H will only be possible if the material can form silicide together with a–Si. The examples of such materials to be stably used in production are Cr, Mo and Ta. Cr has been used by Hitachi and NEC for a long time. Mo and Ta are used by Toshiba and Sharp, respectively. Cr is a stable S/D material used by many companies, satisfying Conditions (1), (3) and (4) and showing stable transistor characteristics. Also, its mechanical solidity is excellent and thus, the Cr/ITO structure is used as TAB IC Bonding Pad for the gate pad and data pad. As Cr is also used for the black matrix of

C/F (color filter), using Cr as the S/D material helps reduce the number of types of sputtering targets to manage.

Figure 2.38 is a graph showing the changes in the size of TFT LCDs. As the size gets bigger, the injection and maintenance of liquid crystals and faster responsiveness become essential, and high mobility of TFTs should be realized. The enlargement of TFTs brought out RC delay issue, which should be resolved soon.

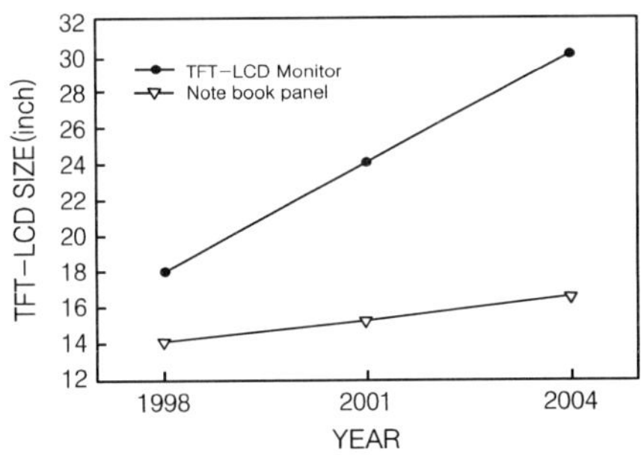

Figure 2.38

The changes in the size of TFT–LCDs using single substrate

Table 2.4 Low resistance metal for large TFT–LCDs

Metal	Resistivity (mWcm)	Melting Point(°C)	CTE(ppm/°C)	Issue
Cu	2.7	1083	16.5	adhesion, unavailability of dry etch, acid-resistance
Al	3.1	660	23.6	acid-resistance, surface morphology
Mo	11.5	2610	4.9	acid-resistance
Cr	21.0	1875	6.2	Detached during thickening, unsuitable for taper etch
aTa	25	2996	6.5	TaN/Ta structure is required for low resistance
bTa	173.2	2996	8.4	high resistance
Ti	86.5	1668	4.6	high resistance
W	18.1	3410	13.3	delamination
Ni	50	1453	–	adhesion, unavailability of DC sputter
ITO	200	–	–	dependancy on the deposition conditions

For a 17–inch XGA LCD, approximately 350 meters of the gate line and 1 kilometer of data line are used, which clearly shows how the use of the low resistant material will impact on the RC delay. To solve the RC delay issue requires metal with low resistance, even lower than that of Cr or Mo currently used. The studies on such metal have been conducted actively since 1990's; the researches on some low resistance metal such as Al, Cu, Al-alloy have already reached a pretty high level. The characteristics of each metal are described in Table 2.4. As it is possible to produce 40-inch TFT LCDs using pure Al, further researches to use Al for gate/data lines are going on.

3.2.4 Channel Formation Etch Process (Channel step)

The channel formation process is to etch n^+ a-Si:H by reactive ion dry etching method, with the drain/source electrodes being a mask themselves; at the same time a-Si:H thin film is overly etched, for there is no proper selectivity between n^+ a-Si:H and a-Si:H. The channel formation is an important process which determines the characteristics of thin film transistors. And which metal is used as S/D has a huge impact on the subsequent processes.

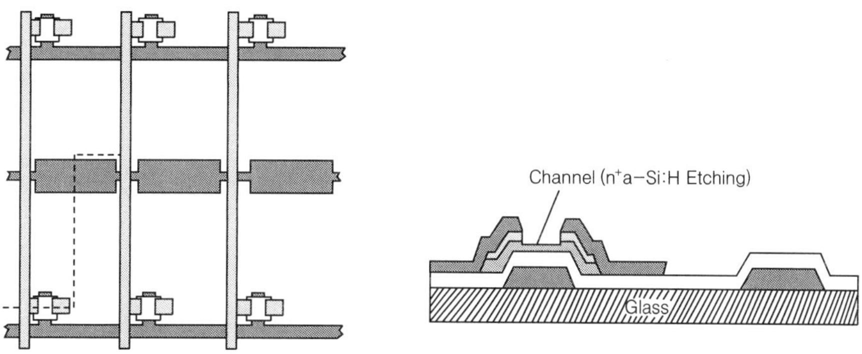

(a) Plan view (b) Process steps

Figure 2.39
The channel formation steps of TFTs

3.2.5 Passivation step

The formation of SiN_x (nitride silicon) passivation using PECVD is a process to protect the thin film transistor, especially the gate and data line. Thus, the ITO electrode and the pad part of the gate line are open by the PR process as they are to be connected with electrodes subsequently.

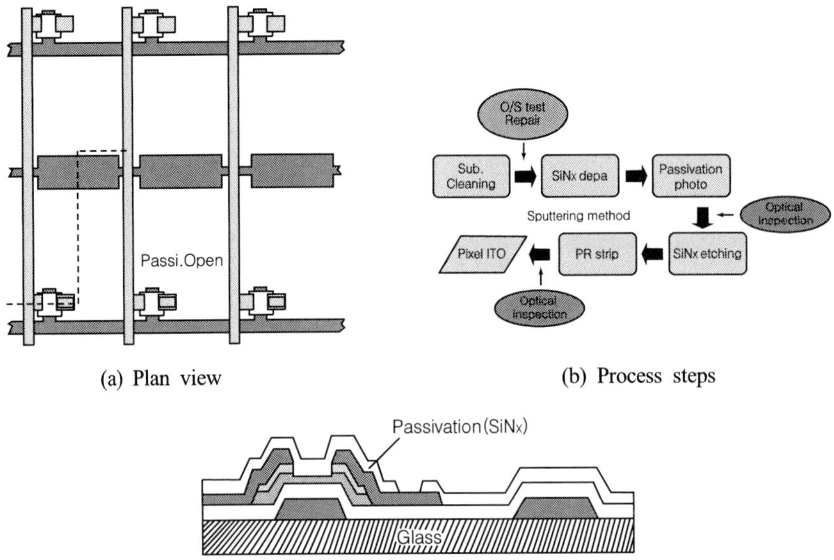

(a) Plan view (b) Process steps

(c) Sectional view

Figure 2.40
The passivation step of TFTs

3.2.6 ITO step

The ITO thin film that grew by the sputtering method is patterned as pixel. Masks and formed as the opposite electrode of the color filter ITO. For such ITO electrode, step coverage needs to be secured. Also, this requires a special attention since there could be defects in pixels if the electrode gets disconnected due to its contact failure with the electrode at the bottom of the protective layer. Commercialized ITO targets are produced by forming and sintering mixed powder of In_2O_3:90wt.% and SnO_2:10wt.%. The level and price of an ITO target are determined by the sintered density. The lower the sintered density is, the more vacancies the target has; as the target is consumed, its vacancies are exposed, making many nodules. It is reported by analyses on the nodule that it is an insulating material, the core of which is produced by an arc discharge that occurs with the exposed vacancies. What is more important in relation to ITO is to improve the processes to obtain low resistant film and better transmissivity, rather than developing new processes.

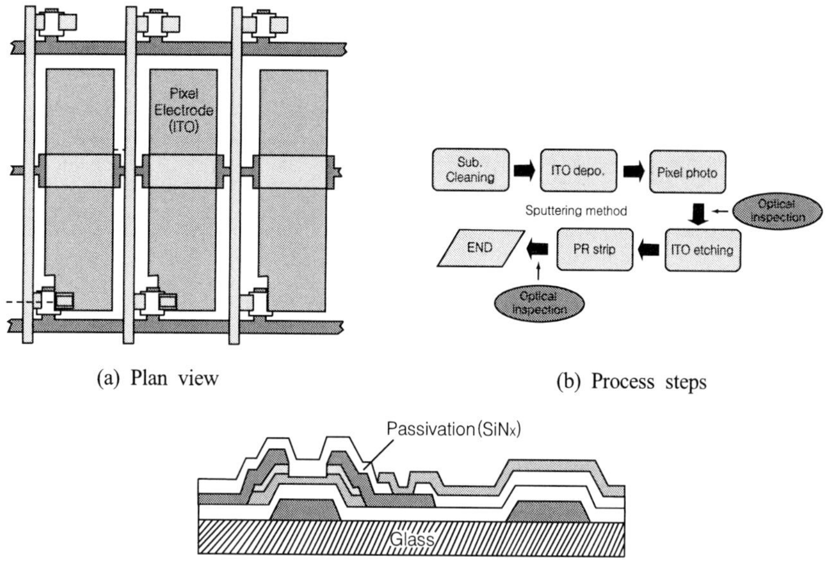

(a) Plan view

(b) Process steps

(c) Sectional view

Figure 2.41
ITO Step of TFTs

D i s p l a y E n g i n e e r i n g

4 Polycrystalline Silicon Thin Film Transistors

4.1 high-temperature Poly-Silicon TFTs'(HTPS TFTs')

4.1.1 Characteristics of HTPS TFT

The main characteristic of HTPS TFT is that the same manufacturing processes as the existing semiconductor wafer processes are adopted and thus, it is possible to produce a more stable TFT device. More specifically, using furnace treatment and an ion implantation equipment under high temperature helps to produce polycrystalline silicon with good crystallinity and the gate insulator with stable insulating nature. However, in order to endure the high growth temperature of approximately 1000°C, expensive quartz substrates have to be used for HTPS TFT. All of the products that have already commercialized or are being developed are used for high-definition TFT-LCDs in less than two inches. The reasons that HTPS TFTs are used for such small displays are to reduce the expenses required for the display panel itself, by producing many displaying pixels compared to the size of the substrates, and to embed the driving circuit above the substrates, which will make HTPS TFTs competitive and differentiated from a–Si TFTs in the market, which require external driving circuits.

An HTPS TFT has much higher electron mobility than that of a–Si TFT and thus, it is possible to embed the driving circuit on the substrates where the HTPS TFT is laminated. One of the benefits of the high electron mobility of a transistor is that the channel of an HTPS TFT can be made smaller than that of an a–Si TFT, by which a high aperture ratio of the screen can be obtained. The comparisons for the characteristics of HTPS/LTPS TFTs and a–Si TFTs are listed in Table 2.5.

Table 2.5 Comparisons for the characteristics of a–Si TFTs and poly–Si TFTs

Items		a–Si TFT	Poly–Si TFT (Low Temperature)	Poly–Si TFT (High Temperature)
Manufacturing Process	Temperature	Max. 350°C	< 450°C	over 1000°C
	Substrate	Glass	Glass	Quartz
Device Structure		Bottom gate-Key align	Top gate-Self-align	Top gate-Self-align
Transistor Properties	Electron Mobility	0.5~1 $cm^2/V \cdot s$	10~500 $cm^2/V \cdot s$	30~100 $cm^2/V \cdot s$
	Embedded Driver	×	○	○
Stability/Uniformity of TFT		○	△	○
Enlargement		○ ≤50 inch	△ ≤15 inch	× ≤2 inch
Aperture Ratio		small	large(small TFT)	large(small TFT)

4.2 Low-Temperature Polycrystalline Silicon TFT

4.2.1 Comparison of A-Si/LTPS/HTPS TFT-LCDs

Table 2.6 shows the comparison of TFT-LCDs made of a-Si, LTPS and HTPS.

Table 2.6 Comparison of TFT-LCDs made of a-Si, LTPS and HTPS

	With LTPS TFT, it is possible to realize System Display, which integrates image processing functions and data processing functions on glass, by combining microfabrication technology (such as LSI) and CMOS circuit technology with LCD technology.		
	High-Temperature Polycrystalline Si TFT (HTPS)	Low-Temperature Polycrystalline Si TFT(LTPS)	a-Si TFT
Substrate Temperature	> 600°C	< 600°C	< 350°C
Transistor	CMOS (nMOS) (pMOS)	CMOS (nMOS) (pMOS)	nMOS
Driver	- Transistor DAC - Embedded Video Memory - Chemical conversion treatment circuit	- Transistor DAC - Embedded Video Memory - Chemical conversion treatment circuit - First step for system display	Attached to the external part
Precision	20μm pitch(400ppi)	40μm pitch(200ppi)	80μm pitch(100ppi)
Substrate	quartz glass	alkali-free glass	alkali-free glass

4.2.2 Manufacturing Process of Low-Temperature Poly-Si TFTs

There are basically two types in LTPS, i.e., Top gate and Bottom gate. The typical structures of these types are shown in Figure 2.42.

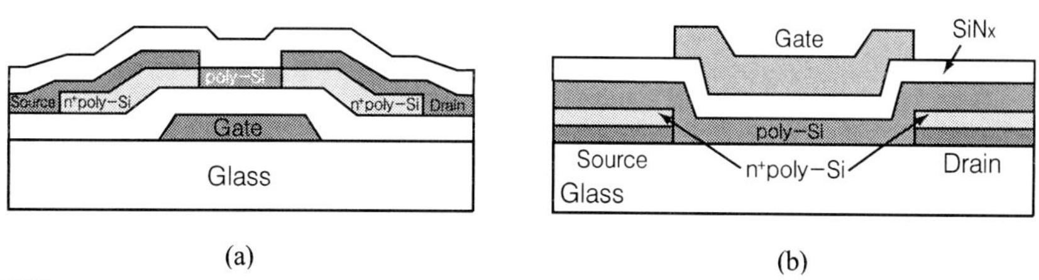

Figure 2.42
The structure of Bottom Gate (a) and Top Gate TFT (b)

(1) Top gate type LTPS TFT Process

Figure 2.43 describes an example of the process of manufacturing LTPS TFTs with off-set structure, using the top gate.

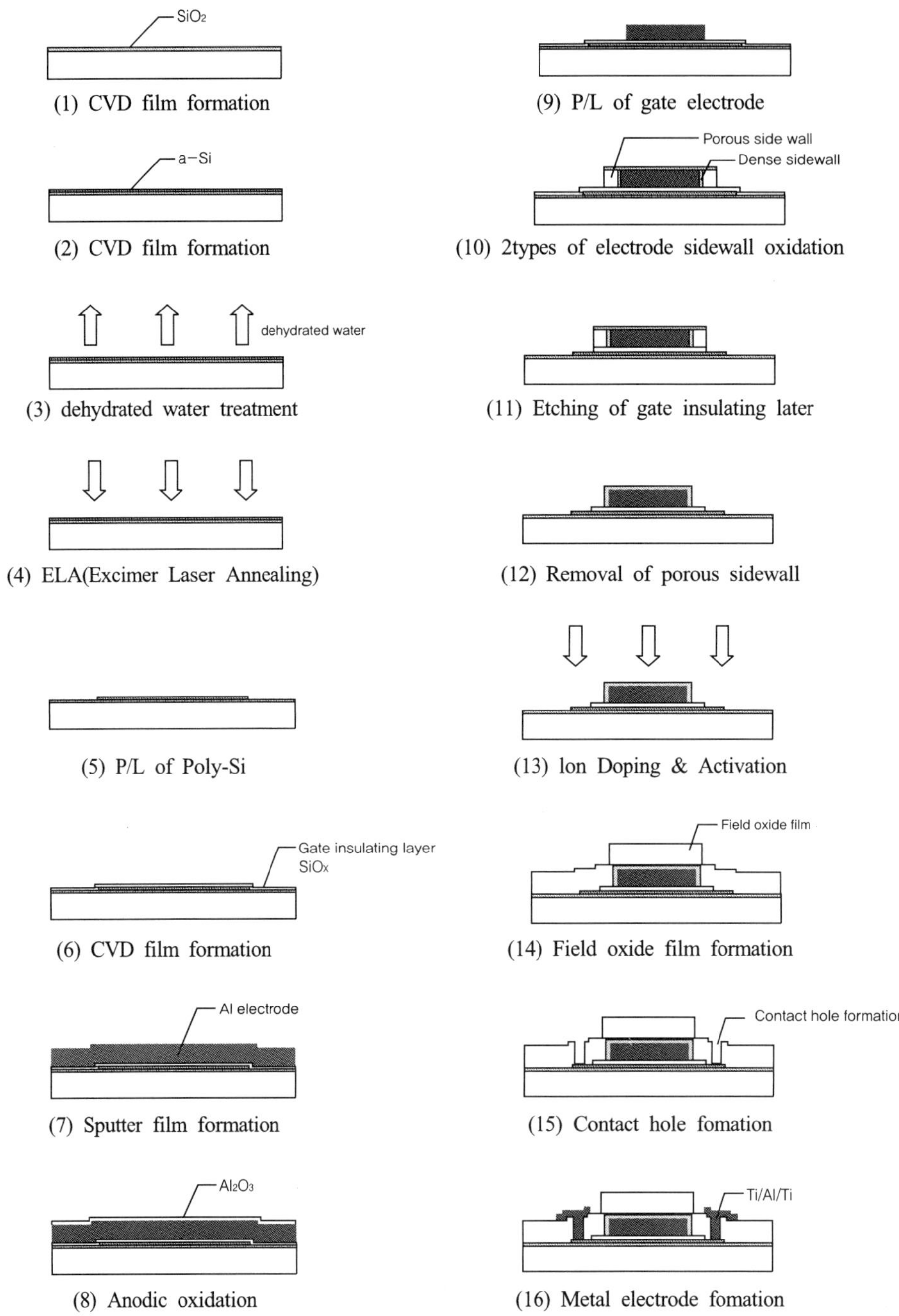

Figure 2.43 The example of the manufacturing process of Top gate type LTPS TFTs

4.2.3 LTPS TFT Manufacturing Process using Bottom Gate

An example of the manufacturing process of LTPS using Bottom gate is described in Figure 2.44. The properties of the bottom gate type LTPS TFT are the same as those of a–Si TFTs currently produced, and it can be produced by converting the a–Si TFT production line. In addition, the silicon layer is formed at the back of the gate insulator and thus, there is no possibility of defect creation with impurities from the glass substrate. No layer is needed since external light is blocked by the gate electrode. The example refers to the process published by HITACHI. This is an old method, but it is still interesting as it does not use ion implantation method. It is based on limited researches since not every detail of the process is published, like the Top gate type above. Some conventional processes such as the existing process of fabricating a–Si TFT could be omitted here.

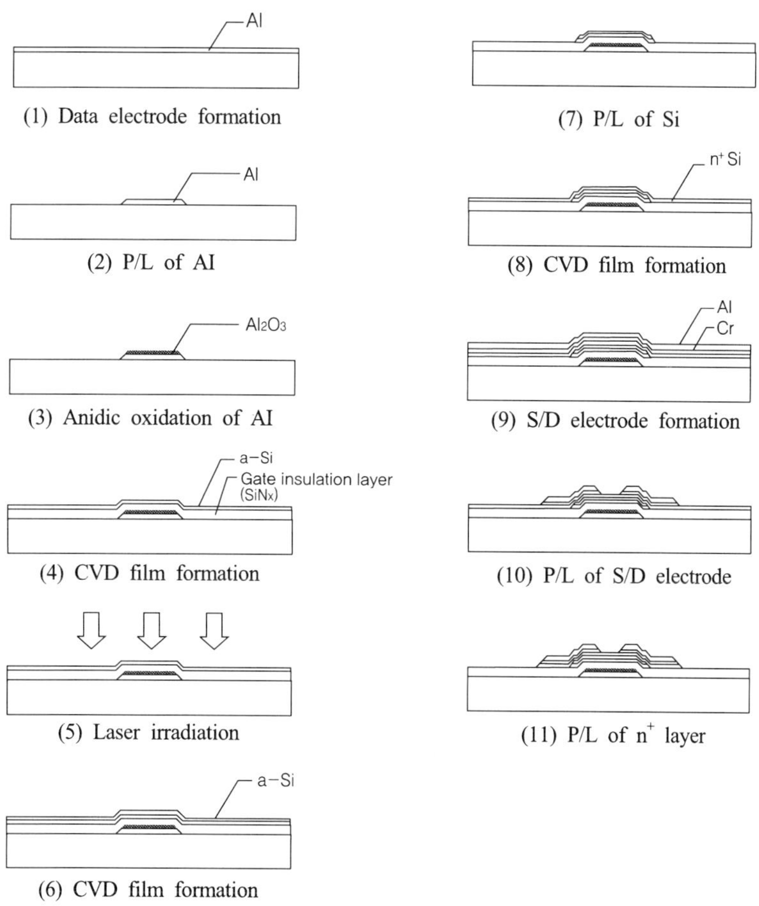

Figure 2.44
The example of the manufacturing process of Bottom gate type LTPS TFTs

4.2.4 Structure of LTPS TFT Device

The device structure of LPTS TFT devices is well established for both top gate and bottom gate type; there has not been much change to the structure thus far. The top gate type are considered to be good in that it is stable and easy to produce, but the bottom gate type is also used much nowadays; it is not affected by the external light and has accumulated many records of being used in practice. Thus, it is hard to say which one is better than the other at this moment. For instance, the bottom gate structure is used by Sanyo Electric. a–Si TFT line is easy to adopt, with a merit that poly–Si can be formed simply by crystallizing a–Si with a laser. However, it may be hard to make a good gate insulating layer and good interface with poly–Si, by the method that forms the gate insulating layer first, creating the interface with a–Si, and then transforms a–Si to poly–Si. Also, if the gate electrode and gate insulating layer already exist at the bottom, laser radiation conditions such as the material and thickness for laser crystallization need to be fine–tuned, which could be tough to be done.

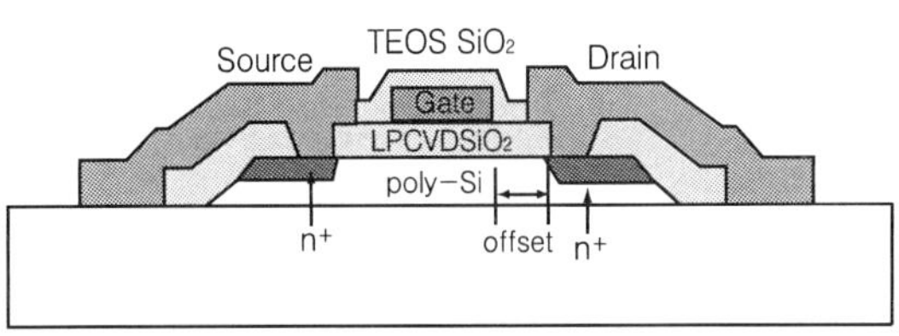

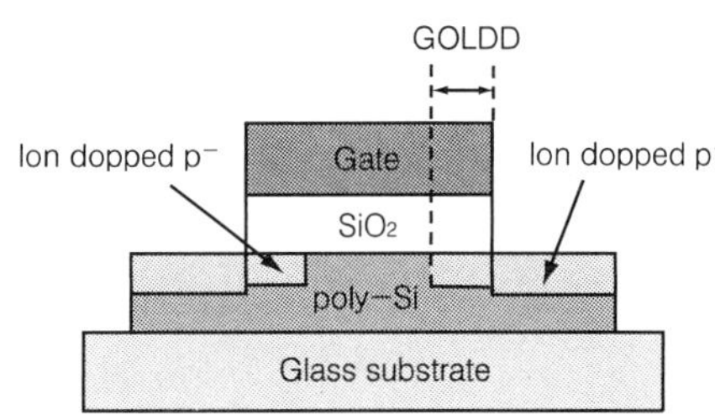

Figure 2.45 Cross–sectional Diagrams of Offset Structure and GOLDD (gate overlapped lightly doped drain) Structure

To prevent leakage currents, offset or LDD structure can be advantageous. However, production of the offset structure is too complicated, and the fall of the drain current is unavoidable. The offset distance has to follow sub–micron order (less than 0.4 μm) in order to secure the drain current, which could possibly obstruct the high production yield from being achieved.

4.2.5 Materials for LTPS TFTs

The core technology of manufacturing LPTS TFTs consists of two major steps, i.e., producing a–Si from semiconductor materials with a PECVD device and transforming it to poly–Si by laser crystallization under the temperature less than that of producing a–Si. Although some research results, were published using the

Si–Ge, it is not necessary to use such composite material. Also, the possibility that said major steps will be changed seems low.

For the top gate structure, the n^+ Si and p^+ Si layers are made by injecting phosphorus and boron ions after producing poly–Si. For the bottom gate structure of Sanyo Electric, ion implantation is followed by laser crystallization. A high–priced ion implantation equipment is required for this purpose, but the method directly depositing a–Si mixed with impurities by PECVD is no longer used for manufacturing LTPS TFTs recently. As the semiconductor layer of LTPS TFT is very thin, less than 500 nm, it is hard to secure proper selectivity when etching the n^+ Si and p^+ Si layers.

Thus, it is unlikely that the material for this method will be changed. Although SiN_x is often used for a–Si TFT, the SiO_2 layer or the double layer of SiN_x and SiO_2 are mostly used for the gate insulator of Low–Temperature poly–Si TFT. Both SiN_x and SiO_2 are generally formed by PECVD.

The source/drain electrodes of LTPS TFTs are the same as those of a–Si TFT; that is, silicide metal is used for the contact with the semiconductor layer, and the double–layered wiring of Al is used.

4.2.6 Various technological issues of LTPS TFT

There are two types of LTPS TFTs among the LTPS TFTs with embedded driving circuits. One is used for the operation of liquid crystal pixels, and the other is used as a driving circuit. As the properties required for each of these types are different, the structure and manufacturing process of LTPS TFTs are getting more complicated. Meanwhile, LTPS TFTs for the operation of liquid crystal pixels can be produced with NMOS–TFT (just like the structure of a–Si TFTs), but to produce LTPS TFTs for the driving circuit with CMOS, two types of TFTs, both NMOS and PMOS, have to be made first. It is still possible to produce LTPS TFTs for the driving circuit with NMOS only, but CMOS is more stable in terms of the signal margin and power consumption.

The technological issues for the formation of poly–Si improve the properties of LTPS TFTs and control their uniformity. The technologies can be summarized as the following five items: 1) a–Si formation, 2) de–hydrogenation, 3) poly –crystallization, 4) reduction measure for I_{off} currents, 5) threshold voltage (V_{th}) control

(1) a-Si formation

This is classified as an HTPS formation method, as it requires a high temperature over 900°C, although it can also be possible to produce the poly-Si layer directly by LPCVD (Low-Pressure CVD). The major formation method of the LTPS layer is forming a-Si and then generating the poly-Si layer from the crystallized a-Si. The most commonly used material for preparing Si is silane (SiH_4). There are some other methods using the Ge-Si layer, but such composite material does not have to be used.

Because Te is easy to crystallize, researches on the semiconductors based on Te were actively conducted in Europe. However, this did not draw any attention in Japan due to various problems; PMOS could not be produced with this method, it was difficult to control etching and the material was toxic. For the formation of the a-Si layer, a plasma CVD device is used. The formation temperature is around 250°C, and the thin film is 40~50 nm thick. The crystal grain size of poly-Si strongly depends on the thickness of the CVD thin film. Sanyo Electric is making an effort to meet the uniformity of the thin-film thickness within ±2%, in order to maintain the uniformity of the properties of poly-Si TFT; however, the plasma CVD devices currently sold have the thin-film thickness uniformity of only 5~10%, which needs to be improved by the device manufacturers. Such inconsistent crystal grain size directly affects the field mobility of electrons and consequently has a huge influence on the circuit design margin. More specifically, the TFT size has to be enlarged for stable driving, or the transference number could drop if the properties of TFT are not good due to the non-uniformity of crystal grain size.

(2) Dehydrogenation

a-Si produced with the plasma CVD conditions that are the same as a-Si TFT generally contains about 10% of hydrogen. If the a-Si layer is laser-crystallized without preconditioning process, the hydrogen will be evaporated too much, which causes the a-Si layer to be detached and shattered. In order to avoid this, the concentration of hydrogen should be reduced before beginning the full crystallization process. The target is less than a 2% of hydrogen. For this purpose, several dehydrogenation methods are being suggested, which can be broadly classified into two types: one is to dehydrogenate hydrogen gas mixed by PECVD before crystallization, and the other is to form the a-Si layer containing less hydrogen in the first place. Explanations on four specific methods in this regard follow:

1) Dehydrogenation by Batch process

The most common Dehydrogenation method is a batch process applying heat to a–Si at a temperature of 430°C for two hours. By this process, the concentration of hydrogen becomes less than 2%, which will not cause any problems during the crystallization.

2) Dehydrogenation by excimer Laser

This method consists of two steps: first eliminating hydrogen from the thin–film by crystallizing it with an excimer laser, whose optical output power is less than 100 W, and then crystallizing it to poly–Si with an increased output power of the excimer laser. This does not require many equipments, but the disadvantage is that the excimer laser should be used twice. This leads to low productivity.

3) Formation of the a–Si layer with less than 2% of hydrogen in the first place

The a–Si layer can be formed with less than 2% of hydrogen by changing the production conditions of a–Si. Japan Vacuum Technology (currently running a business as "ULVAC Co., Ltd.") is actively carrying this method forward, which is promising. As published in [SID 97], the detailed production method is the reduction of the concentration of hydrogen by decreasing rf Power.

4) Formation of a–Si by evaporation

Unlike a–Si produced by PECVD, the a–Si layer produced by evaporation does not contain hydrogen. Such a–Si layer has many dangling bonds and thus, even if TFT is made, it is not possible to control the drain current by the gate potential, which acquires much higher leakage current. That is, such TFT cannot be used as a–Si TFT. Nevertheless, according to some cases reported by NTT and others, this could result in the production of LTPS TFTs with high mobility when they are poly–crystallized by an excimer laser. But this requires an ultra–high vacuum state for the evaporation, which is at the level of 10^{-7} torr and thus, there is a low possibility that this method can be commercialized.

(3) Poly–crystallization

There are laser and lamp crystallization in poly–crystallization. SPG (Solid

Phase Growth) could be used for HTPS, but for LTPS the substrate cannot endure the method and thus, it should be treated in a short time. The most common poly–crystallization measure is to use an excimer laser with an output power of 200 W.

However, there are also many weaknesses in the method; the operational runtime is low, the uniformity of the excimer laser output is about ±6% (which does not meet the required uniformity of ±2%), and a very sophisticated skill is required for crystallization (as it needs fine–tuning). Further, if the driver circuit area only needs to be poly–crystallized, power YAG lasers could be used instead of excimer lasers. YAG lasers are continuous–wave and easy to form a uniform layer by scanning the substrates only.

Meanwhile, there are still many things to be improved by TFT LCD manufacturers in the poly–crystallization process. For example, when an excimer laser continues to be radiated on spots, the spots overlap with each other. But even in the overlapping area, the face and the end should be separately radiated. Otherwise, the nested irradiation in the overlapping becomes noticeable like a border between them, causing TFT to lack uniformity. Such overlaps should be minimized by optimizing the arrangement of TFT. Sanyo Electric and other companies are also considering using Rapid thermal anneal (RTA) based on a lamp, but it is being pointed out that if the thickness of a–Si gets thinner, the distribution of energy on the surface gets unstable. Thus, this method will unlikely be a threat to the superiority of the excimer lasers. The field mobility of electrons and positive holes of the LTPS produced by laser anneal varies depending on the manufacturer. Even in terms of the non–uniformity of data, the electric properties of the poly–Si layer have a high dependency on the manufacturing conditions. According to the publication of Sanyo Electric, there were some cases where the field mobility of electrons can reach 420 $cm^2/V \cdot s$ to the maximum. In this regard, the field mobility of single–crystal Si electrons is approximately 600 $cm^2/V \cdot s$.

(4) Reduction of I_{off} current

I_{off} current is one of the important characteristics of TFT. NMOS is used for TFT in driving liquid crystal pixels and the leakage current (I_{off} current) should be less than 1 pA. However, the leakage current of poly–Si is high and reducing the leakage current is one of the remaining tasks. The major five solutions published for this purpose so far are as follows:

1) Partial crystallization of the driver circuit (to LTPS TFTs)

At the current technology level, an excimer laser is radiated on a spot–by–spot basis. Thus, in order to transform a large size of a–Si to poly–Si, we have to carefully radiate the laser several times by overlapping the spots. This can be used the other way around. That is, the driver circuit can be transformed into poly–Si, with the pixel part still remaining a–Si. The leakage current of a–Si is low in the first place, and high speed is not needed as well. If this method is used, there is no need to radiate the laser on the display part, which dominates the most area of LTPS TFT LCDs and thus, the productivity can be improved.

2) Multi–transistor structure

This method is to reduce the I_{off} current by arranging more than two NMOS in series. TFTs for driving pixels commonly have this structure. In LTPS TFT, it is not that all of the TFT leakage currents become high; the leakage current could get high in accordance with a certain probability, depending on the size of the grain boundary shape. Thus, stochastically, the leakage current can be reduced by connecting multiple transistors in series. There is a report that this method could reduce the leakage current by two orders of magnitude, but when two resistors (the leakage currents can be considered resistance) are simply connected to each other in series, the leakage current decreases by half and thus, it is obvious that such huge two–order reduction effect will not happen.

Some TFTs have the leakage current two order of magnitude as high as that of general TFTs, in which case the leakage current is restrained by arranging the TFTs in series. Such multi–transistor structure can be used together with the following two methods.

3) offset structure

This method was mentioned above in the LTPS TFT manufacturing process; that is, installing a high resistance area between the source/drain electrodes and the channel area right below the gate electrode. This method has a disadvantage that the drain current has to get lower, but it is said that Fujitsu method resolved this issue by controlling the width of the offset area at the sub–micron level. If we leave the silicon oxide layer outside the gate electrode as it is, and then inject ions on that part, the ions can only be injected into the area where the silicon layer is exposed. The principle sounds simple, but the production method is actually quite complicated, as previously explained.

4) LDD structure

Matsushita Electric Industrial Company reduces the I_{off} current applying LDD (Lightly Doped Drain) structure to NMOS. The LDD structure is based on the phenomenon that phosphorous ions pass through TaOx less than 60 nm thick; the structure can be obtained only by one-time doping. The only weakness it has is that the poly-Si TFT based on this structure has low field mobility of electrons, which is considered to be due to the formation process of poly-Si. To form the LDD structure, a thin TaOx layer needs to be installed (as much as a number of ions) outside the gate electrode, and ions should be injected. The LDD part will be formed in this way.

5) Wing structure

Wing structure has already been applied to HTPS TFTs, but can also be used for LTPS TFTs as well. This structure employs the offset and LDD structure simultaneously. This wing structure comes from the idea to electrically control the device, which are originally impossible to control electrically, once they are formed. The wing structure was suggested by NTT in 1990. NTT uses a sub gate electrode between the gate electrode and insulator. As the sub gate contains the source/drain part, the area between the channel and the source/drain can also be controlled by the sub gate potential. This makes it possible to flow a higher drain current when the gate electrode is on, and restrain the leakage current when the gate electrode is off.

There is another method: producing the main gate electrode with continuous two times of metal formation at the last step. The NTT method has to go through two separate metal formation processes, whereas said metal formation process is done at one time. A thick insulator is laminated on the thin gate insulator (silicon oxide film), and the thick layer should be partly removed. However, it is still possible to form an insulator with high etch selectivity, depending on the metal oxide layer used.

(5) Threshold Voltage (V_{th}) control

V_{th} can be controlled when a-Si is doped with boron. Normally the V_{th} of NMOS gets higher if boron is mixed in the channel of NMOS. But at the same time, a pn junction is formed between the source/drain area and the channel area of TFTs (as it is in single–crystal transistors), which surely reduces the leakage current. V_{th} affects the silane flux during the formation of a–Si and thus, needs to be delicately controlled.

D i s p l a y E n g i n e e r i n g

5 Technical Trend and Application of Thin Film Transistors

5.1 Application of a–Si TFT

(1) a–Si:H TFT for Monitors

The recent trends of TFT–LCD development can be summarized as enlargement, pixelization and wide viewing angle. One of the issues that arises along with the enlargement of TFT–LCD is RC delay, which occurs due to the resistance of the gate line. To resolve this issue, Samsung has used a dual scan segment driving method for 30–inch TFT–LCD, recently. Further, in order to develop LCD products to be used for monitors, Japanese companies have been actively putting efforts into developing core technologies for viewing angle improvement, large screens, ΔVp reduction, flicker reduction, low–cost products and so forth.

(2) TFT for CNS(Car Navigation System)

The size of LCDs used for CNS is mainly five to seven inches. NTSC LCDs should be used for CNS as it works as a portable TV as well. The LCDs used for CNS should be working under a broad range of temperature and display clear images even in the bright place; thus, a backlight with low reflexibility and high–intensity is required. Accordingly, FEDs (Field Emission Display) with the merit of broad working temperature and high intensity may possibly create the market for CNS LCDs. As large–sized or high–definition LCDs are not needed for this purpose, most of TFT–LCD manufacturers, such as Hitachi, NEC and Sharp, supply NTSC TV in five to seven inches, which can be installed on CNS. Sharp has been selling 5.8–inch and 7–inch wide screen TFT–LCD TV·s since the summer of 1997; the number of products they manufacture monthly is two thousand products of each type, and the unit price is about one hundred thousand yen. Also, these products are equipped with another CNS kit model the company sells, which is about two hundreds seventy thousand yen.

Figure 2.46
7-inch LCD TV of SHARP

(3) Application to Sensor

1) Image Sensor

The image sensors (or optical sensors) based on amorphous silicon are catching attention as a new field worth to develop in the 21st century. The common optical sensors made based on the principle of a-Si image sensors are photodiodes and phototransistors. The characteristics of photodiodes are ① good linearity of the output current compared to the amount of the induced light, ② low sensitivity to the temperature changes, ③ fast responsiveness and ④ low measurement errors. The image sensor is a sensor that constitutes an array in which the optical sensor is composed of pixels. There are three types of image sensors: CCD (Charge Coupled Device), CMOS (Complementary Metal Oxide Semiconductor) and a-Si image sensors.

CCD is a semiconductor device, published by Boyle and Smith in Bell Lab in 1969, functioning as an electronic eye that converts visual signals to electric signals. This is applied to a vast range of fields such as household, education, medical, industry, space industry and so on, being boosted by the dramatic advance of semiconductors. CCD takes an important role in the fields requiring visual inputs, working as the eyes of a human or the film of a camera.

CMOS image sensors (or contact image sensors) began to be developed by Fairchild, RCA, etc. in 1967. Hitachi in Japan also focused on developing the sensors from 1979, but abandoned the development of MOS-type image sensors and turned to CCD in 1989: as the MOS sensors had worse image quality, more complicated circuit and lower packing density compared to CCD. Also, in terms of the price, there is no difference between CMOS image sensors and CCD. The chip size of the sensors was way too large. Afterward, amplified MOS imager (AMI) was published by NHK/Olympus in 1990. In 1993, the first CMOS-type camera chip was published by University of Edinburgh, and in the same year, CMOS-type Active Pixel Sensor

(APS) was published by JPL. Then since 1995, CMOS image sensors have been actively developed by U.S. colleges and institutes.

Meanwhile, CCD image sensors have relatively difficult manufacturing process compared to CMOS, and random access to CCD is not possible, which is available for CMOS. The disadvantages CMOS image sensors previously had begun to be overcome in the late 1990's along with the development of the CMOS processing technology and the improvement of signal processing algorithm; further, CCD processes started to be applied to CMOS as well, by which the product quality of CMOS has been much improved and consequently, CMOS has grown to share the image sensor market with CCD. However, these products still have some weaknesses; to obtain images the products require an optical system. Thus, the volume of the devices has to be large and an additional optical device has to be installed.

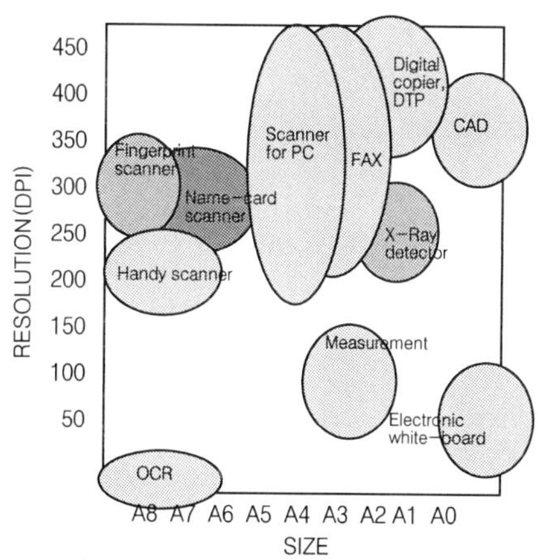

Figure 2.47
An example of application of image sensors depending on the resolution and the size of detection area

2) X-ray Detecting Sensor

The diagnostic X-ray testing method widely used for medical purposes covers the shooting process using an X-ray detecting film, the development of films and the printing process. There are some problems with this method, i.e., it takes some time to get the testing result with such method, a certain place to store the films needs to be prepared and there is a possibility of loss or damage of the films. To cope with such problems, digital X-ray image sensors using amorphous silicon have been researched and developed recently. A digital X-ray image sensor uses a-Si TFT as the switching element, and its optic acquisition part consists of a-Si TFT or a PIN diode. With the use of this sensor, it is possible to build digital databases, provide real–time medical consultation from remote sites and dramatically reduce the time taken from the shooting to the testing. Figure 2.48 is a mimetic diagram of the structure of a digital X–ray image sensor. X–rays coming from the outside are transformed to visual rays by the fluorescent substance on the upper part of the image sensor, which is detected and digitalized into images by a–Si PIN at the optic acquisition part. Figure 2.49 shows an example of the images obtained by digital X–ray image sensors.

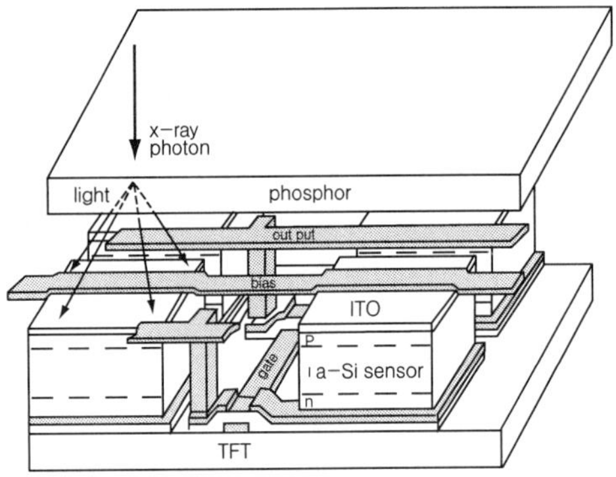

Figure 2.48

A mimetic diagram of the structure of a digital X–ray image sensor

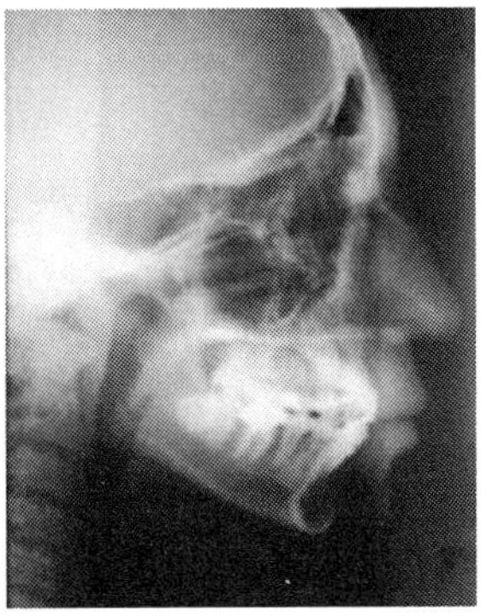

Figure 2.49

An example of image obtained by digital X–ray image sensors

Also, fingerprint detectors and document input devices are important examples where a–Si image sensors can be applied to. In accordance with the development of online transactions through the internet and the increasing necessity of personal security, personal identification devices are getting more important, with a growing demand of portable document input devices. In this situation, the image sensors using amorphous silicon are considered a proper alternative and thus, they are being actively studied and developed.

5.2 Application of Poly–Crystalline TFT

(1) Development of Next–Generation TFT–LCD

a–Si technologies have been used in the TFT–LCD industry for a long time, and mass–production based on the technologies has been already realized. The main field where a–Si is applied to is still large–sized display devices such as PC monitors, which has not changed much from the past. Although a–Si is used for some small–medium sized panels such as GPS system for vehicles, LCD TV·s, digital cameras, etc., it has not yet fully satisfied the demand for high performance of terminal-applied systems, due to its low resolution and slow response. For instance, 3G mobile phones require colorful moving images consuming low electricity, but a-Si cannot meet the requirement. Also, the display panels used for DVD players should have fast response, high color saturation and power-saving features, to which a-Si cannot conform. Meanwhile, LTPS was considered next-generation LCD technology in the industry from the earlier stage. But the manufacturers that started working on the development at that time were not so successful, as the market size was small and the growth of the small-medium sized panel market slowed down. Nevertheless, the industry has been focusing on the technology since 2000, which has brought much more opportunities to the mass-production of LTPS.

(2) Focus of Next-Generation LTPS Research

The researches on LTPS TFTs are focusing on the following four subjects: ① the enhancement of mobility and the reduction of cost, ② the enhancement of reliability and mass-production, ③ the development of devices with high resolution to keep up with the trend of scaling down semiconductors and ④ the design of a high performing circuit. The industry has been constantly suggesting possible solutions for the four main subjects since late 2000; for instances, there is a method called MILC (Metal Induced Lateral Crystallization) that enlarges the crystals by stimulating the growth of the crystals in the glass substrate with a metal catalyst. Using this method can enhance the mobility up to 300 $cm^2/V \cdot s$ and reduce the production cost. CG Silicon suggested by Sharp is an example of MILC. Further, Philips developed a method to resolve the defects discovered in the growth of crystals by MILC.

Even though LTPS has a potential to dominate the market and to be applied to a vast range of fields in the future, we cannot surely say that the technical advances will happen in the way the industry is expecting. In fact, the system panels produced using LTPS are similar to a blueprint of the semiconductor technologies made ten years ago, in that PCB substrates should be replaced by glass substrates. The difference between them appears to be whether they are to be integrated with TFT or not. That being said, the biggest difference between the two actually is the temperature of the manufacturing process; that is, the temperature of forming Si wafers for semiconductors reaches more than 1,000°C, whereas glass substrates can only bear 400~500°C. This poses a huge obstacle against the enhancement of mobility and the micronization of wiring intervals, all of which remain the tasks to be solved in the industry.

Meanwhile, in terms of the market condition, LTPS has gradually been accumulating good results in the market. In terms of the size, Matsushita, Toshiba and Sony have already released mini-sized laptops using LTPS panels in six to ten inches. Also, LTPS LCD panels are being used for the color PDAs made by Olympus and Compaq. According to the forecast of ST-LCD, which is a joint venture of Sony and Toyota, about an 80% of the small-medium sized TFT-LCDs will be based on LTPS and for this reason, the development of LTPS will be inevitably accelerated.

References

[1] David W. Greve: *Field Effect Devices and Application: Devices for Portable, Low-Power, and Imaging Systems*, Prentice Hall. New Jersey, 1998.

[2] Ben G. Streetman, Sanjay Banerjee: *Solid State Electronic Devices 5th ed.*, Prentice Hall. New Jersey, 2000.

[3] Neamen Donald A.: *Semiconductor Physics and Devices (Korean edition)*, Bando Publishing Company. Seoul, 2001.

[4] Yoon Ki Shin: *Electronic Circuit Fundamentals*, Intervision. Seoul, 1997.

[5] Chang Yeop Park, et al.: *Semiconductor Device Engineering*, Kyobo Book. Seoul, 1998.

[6] Gyu Ho Yeon: Electronic Circuits, Woo Yong Publishing Company. Seoul, 1999.

[7] Jong Deok Lee: *Silicon Integrated Circuit Process Technology*, Dae Young Publishing Company. Seoul, 2000.

[8] Hong Joon Park: *Design of CMOS Analogue Integrated Circuit I*, Sigma Press. Seoul, 1999.

[9] Hang Geun Jeong, et al.: *Semiconductor Device & Process for Integrated Circuit Design,* Hong Leung Science Publishing Company. Seoul, 1997.

[10] P.P.Ko: "HOT-Electron Effect in MOSFETs," Ph.D. Thesis, U.C.Berkeley (1983).

[11] Pierret: *Semiconductor Device Fundamentals*, Addison Wesley. Boston, 1997.

[12] Won Chan Kim: *Understanding of Semiconductor Device*, Dae Young Publishing Company. Seoul, 1999.

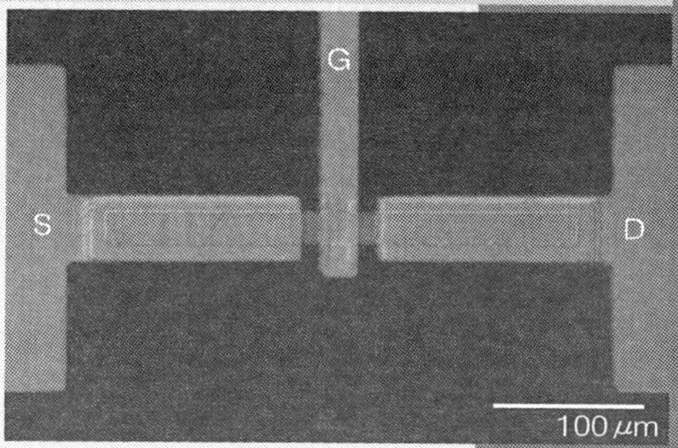

Liquid Crystal Display, LCD

Chapter 02 → 03

1 Summary of Liquid Crystal Displays

Since lyotropic liquid crystal was discovered by Virchow in 1854, there has been much progress in the LCD industry, thanks to the efforts of many scholars and corporations. Even Korea has been leading the worldwide TFT-LCD market since last few years. In this chapter, it will be discussed what an LCD is, what kinds of types and characteristics it has and finally what a color TFT is, which is the aggregation of the LCD technologies that have been developed so far.

LCs (liquid crystals) in the shape of thin and long stick let an incident light pass through or block it, depending on the externally applied voltage. They are in the liquid form under the certain temperature range and show the optical double refraction. Other materials normally turn into the transparent liquid from the solid form at their melting point, whereas liquid crystals turn into a untransparent and turbid liquid at the melting point and then become transparent if the temperature further goes up. The term "liquid crystal" is used in two ways: one is to describe the state in between the solid form and liquid form. The other is to refer to the materials under such in-between state.

1.1 Definition and General Characteristics of LC (Liquid Crystal)

(1) Definition of LC (Liquid Crystal)

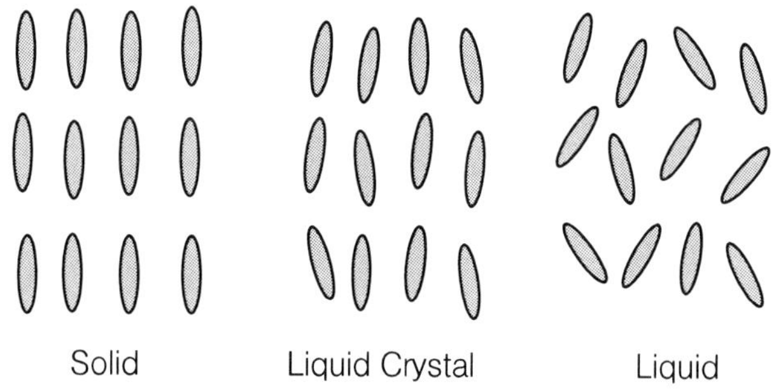

1) What is Liquid Crystal?

Liquid crystals are the materials whose molecules and the direction of molecular axis are in the middle of the two states, in other words, they can have a complete regular pattern like solid materials and they can also have no regular pattern like isotropic liquid. Such state of materials is called mesomorphic phase or mesophase. That is, the materials in such in–between state appear to be fluidic and sticky liquid, but externally appear to be anisotropic crystals.

The temperature for liquid crystals is in the middle of the temperature when the material is solid and when it is liquid. The change of state depending on the temperature of liquid crystals is described in Figure 3.1. The proportion of the materials with the liquid crystalline characteristics to the currently existing organic materials is 1/200.

The molecule structure of liquid crystals is thin and long. For example, the molecule of MBBA (N–(P–methoxy benzylidene) – p′butylaniline), which shows the mesophase under the temperature from about 20℃ to 47℃, is a few angstrom (Å) long and wide. Such molecules change its arrangement and movement by a very minimal stimulus like an electric field, magnetic field and heat from the outside. Using this characteristic, the optical properties of liquid crystals can be easily changed, which is the primary reason for liquid crystals to be used as display materials.

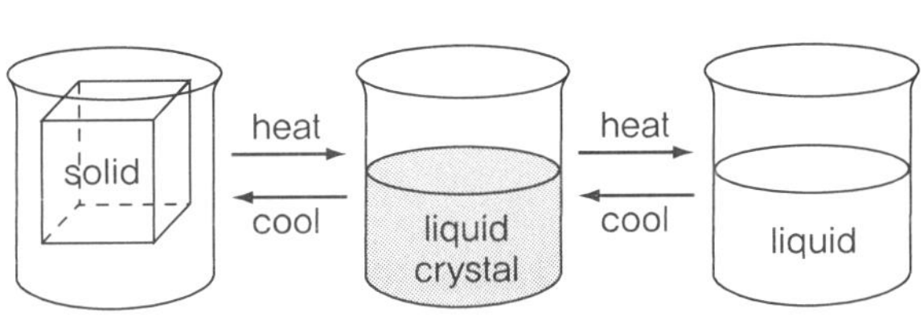

Figure 3.1
The state change of liquid crystals by the temperature change

The most noticeable characteristic of the liquid crystalline state is that the long molecules tend to orient towards the common axis called a director, which is different from the liquid molecules with no certain pattern. In the solid state, molecules are arranged in good order and thus cannot be moved. Thus, the directional order of liquid crystalline state is in between the solid and liquid state.

To describe the degree of order in materials, an order parameter is defined. As a formula to define an order parameter, the following is used conventionally.

$$ds = \frac{1}{2} < 3\cos\theta^2 - 1 >$$

Here θ means the angle between the director and the long axis of each molecule. In isotropic liquid solution, the average value of cosine is zero and thus the order parameter is also zero. The order parameter of the complete crystalline state is one. The same of liquid crystals ranges from 0.3 to 0.9, which depends on the temperature due to the movement of molecules.

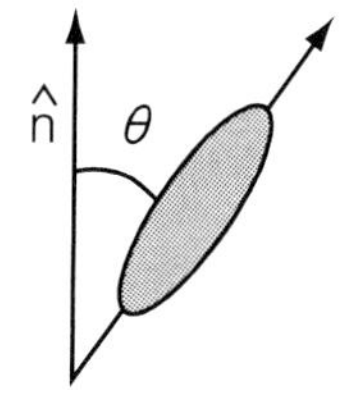

Figure 3.2
The long axis direction of molecules towards the director

(2) Types of LC (Liquid Crystal)

Liquid crystals can be found in the materials under a line-shaped or flat board-shaped molecule structure. They can be classified into three types by how molecules are arranged, such as smectic, nematic and cholesteric.

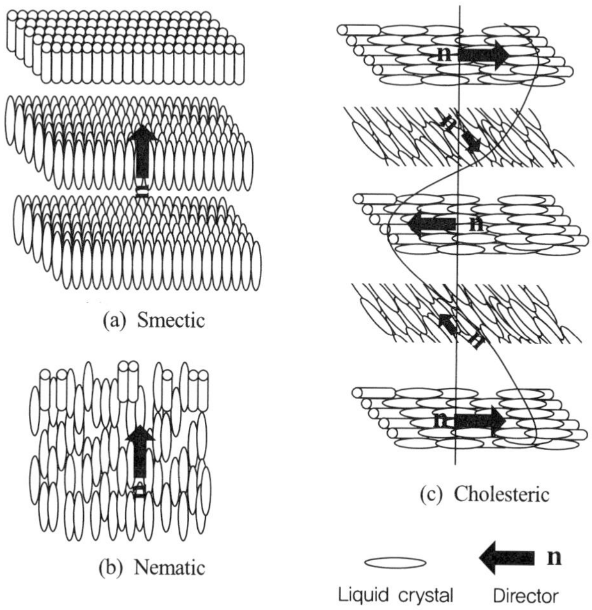

Figure 3.3
The molecule arrangement structure of liquid crystalline state

1) Nematic LC

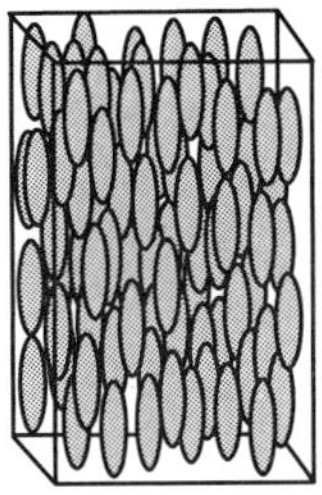

A theoretical representation of the nematic phase (left) and a photo of a nematic liquid crystal (above).

photo countesy Kent State University

Nematic LCs have no regularity in the molecular position but have the orientation order towards the molecule axis. As the direction of molecules at the top and bottom of the liquid crystal are almost equal to each other, the polarization is cancelled off, resulting in no ferroelectricity.

2) Cholesteric LC

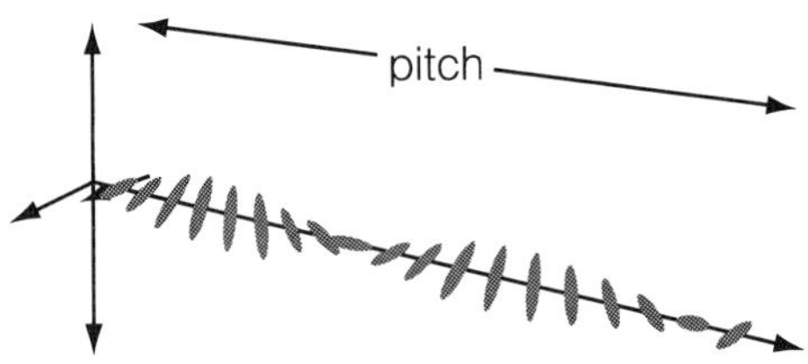

When the molecule axis is twisted or chiral molecules are mixed with nematic LCs, the director **n** of the nematic LC gets twisted. The liquid crystals with such arrangement are called cholesteric LCs. The molecular layer of the cholesteric LCs is very thin, and the molecules are arranged towards the long axis of the layer, whose surface is flat. As shown in the above figure, the direction of the long axis of molecules in each layer is misaligned with that of the adjacent layer, forming a spiral structure. Such structure is named 'cholesteric' since it can be often found in cholesteric derivatives, although there is no liquid crystalline characteristic in cholesteric itself.

3) Smectic LC

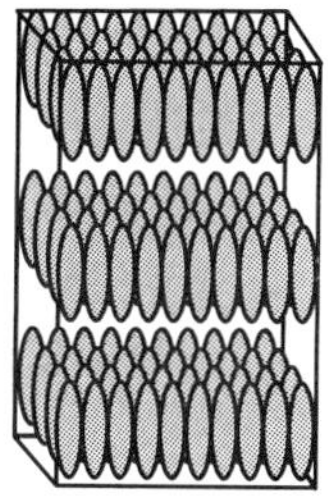

theoretical

A theoretical picture of the smectic phase (left) and a photo of the same phase (above).

photo countesy Kent State University

The molecule arrangement layer of Smectic LCs is more regular than nematic or cholesteric LCs. In smectic LCs, there is no regularity in the positional order of molecules in the direction parallel to the surface of the layer, but the molecules do have regularity in the perpendicular direction of the layer. In other words, the molecules keep positional order in one direction, and the molecule axis as a whole is oriented towards one direction.

(3) Basic Characteristics of LC (liquid crystal)

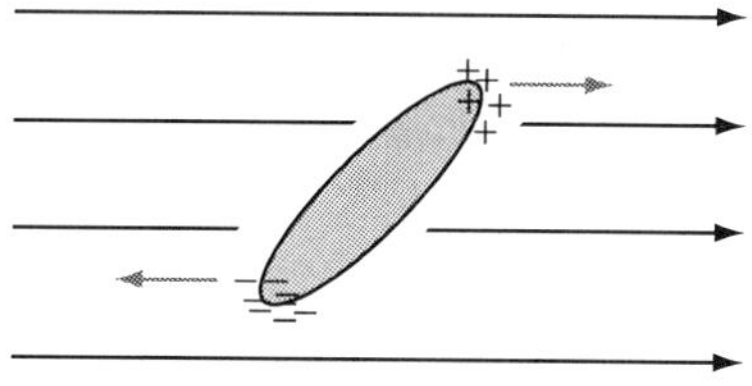

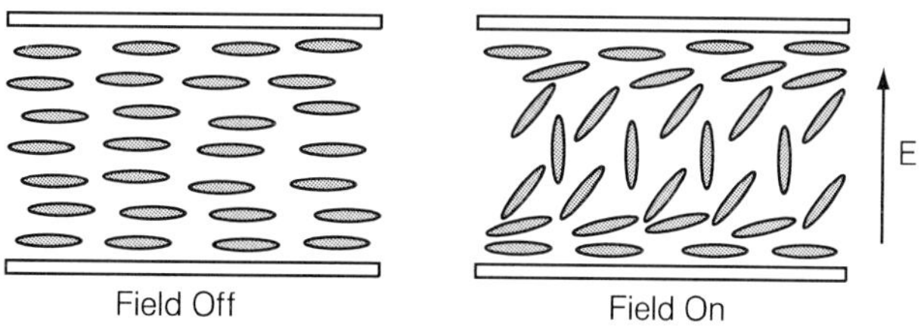

Figure 3.4

Alignment by electric field

The reaction of the liquid crystal molecules to the external electric field stems from the electromagnetic properties of the molecules. If one end of the molecule has a positive charge and the other end has a negative charge, the molecule becomes a permanent electric dipole. When an external electric field is applied to liquid crystals, the dipoles(liquid crystal molecules) are aligned towards the direction of the electric field. Even if the molecules are not permanent dipoles, they can still be affected by the external electric field. In some cases, the electric field can re-arrange the charge of the molecules to generate induced electric dipoles. Such induced dipoles are not as strong as permanent dipoles, but the induced dipoles also tend to arrange towards the electric field.

(4) Molecule Arrangement of LC (Liquid Crystal)

Homogeneous Arrangement	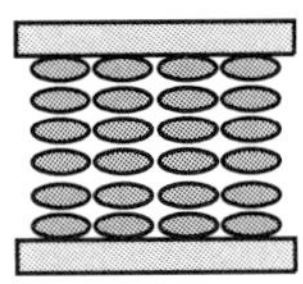	All LC molecules are parallel to the surface of both substrates of the liquid crystal cell, being arranged in the same direction as the substrates.
Homeotropic Arrangement	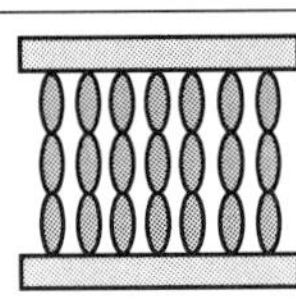	All LC molecules are perpendicular to the surface of both substrates of the liquid crystal cell.
Pre-tilted Arrangement	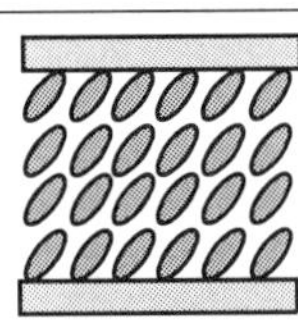	All LC molecules are tilted at the same angle to both substrates of the liquid crystal cell, being arranged in the same direction as the substrates.
Hybrid Arrangement	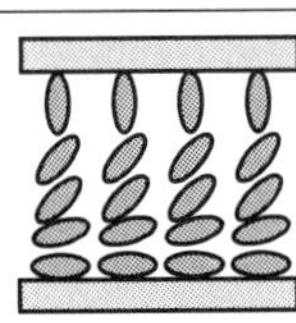	LC molecules are parallel to one substrate and perpendicular to the other substrate. Also, the LC molecule array is arranged to be continuously rotated 90 degrees between the two substrates.
Twisted Arrangement	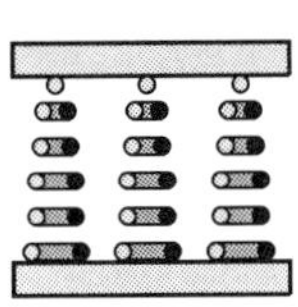	All LC molecules are arranged parallel to the surface of both substrates, but the arranged direction is twisted by 90 degrees with respect to both substrate surfaces. Thus, the direction of LC molecules is continuously twisted by 90 degrees between the two substrates.
Planar Arrangement	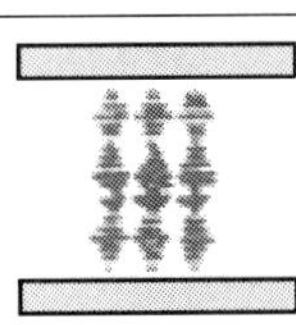	LC molecules are arranged to have the Helical axis of LCs perpendicular to the surface of both substrates. This is also called Grandijean arrangement.

Focalconic Arrangement	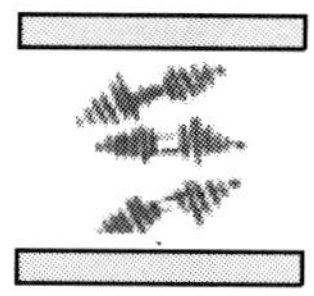	LC molecules are arranged to have the Helical axis of LCs parallel to the surfaces of both substrates. However, the direction of the Helical axis is not consistent. The molecules can be controlled to be oriented towards one certain direction on the electrode surface by applying various physical or chemical treatments to the transparent surface of the electrode.

The characteristics of LC have been briefly covered so far. Table 3.1 lists the advantages and disadvantages of making display devices using the characteristics of LC.

Table 3.1 Advantages and Disadvantages of LCD

Advantage	1. LCD is an energy-saving device which can drive for a long time with low power consumption(less than 100 μW/cm^2). 2. LCD can operate at low voltage(less than 10V), which enables direct IC operation as well as the miniaturization and simplification of the driving electronic circuit. 3. The element is thin(less than 10 mm) and can be used for large-screen displays(around 80 ~ 90 cm) to small-screen displays(less than 10 mm). LCD is especially suitable for portable devices. 4. LCD is a light receiving display and thus can display clear images even in a bright place. 5. It is easy for LCDs to display images in color, so the enlargement and diversification of display function can be done.
Disadvantage	1. LCD is a non-light emitting device. Reflective displays show less clear images in a dark place. 2. LCD needs a backlight to display very clear images or color images. 3. There is a limitation on the viewing angle, as the contrast displays tend to depend on the viewing direction. 4. The response time varies depending on the ambient temperature and thus, there are some difficulties in driving at low temperature (-30 ~ -40°C).

1.2 History of LCD

CRTs started to be developed when Braun invented an electron tube (cathode ray tube or Braun tube) in 1887. The development of LCDs followed one year later after F. Reinitzer discovered liquid crystals (thermotropic LC) in 1888 for the first time. Liquid crystals appear to be liquid at a glance, but they are anisotropic like crystals in terms of optics and become liquid under certain temperature ranges. Thus, they are called 'Thermotropic Liquid Crystal'. However, it took some time for such invention or discovery to be applied to useful products. A CRT was applied to a vacuum tube TV in 1934 by a British, Schonberg, which was in a relatively short time compared to the time that took for the first use of LCs: they were first applied to displays in 1968 by a U.S. company, RCA(Radio Company of America). However, LCDs have been actively used for electronic calculators and clocks since 1973. LCDs then went through several stages of development, resulting in the commercialization of STN LCDs and small TFT-LCDs in 1986. In the 1990s, with the mass production of 10-inch TFT-LCDs, LCDs became the iconic display of laptops. Nowadays, LCDs are getting the spotlight as a display to replace CRTs in the industry.

Table 3.2 History of LC

Year	R&D	Remark
1854	Discovery of lyotropic LC	R. C. Virchow (Germany)
1888	Discovery of thermotropic LC	F. Reinitzer (Austria)
1930	Discovery of Freedericksz transition	V. Freedericksz
1962	Patent application for DSM-LCD	Wiliams (RCA, US)
1971	TN-LCD	Schadt, Helfrich (Switzerland)
1972	Suggestion of Active Matrix	P.Brody
1980~1983	Ferroelectric LCD	K. Yoshino
1984	STN-LCD	Scheffer, Nehring (Switzerland)

Table 3.3 History of early stage of TFT-LCD

Year	R&D and Product	Remark
1978	MOS-FET 1.75" 30625 pixel LCD Cdse-TFT active matrix image displaying LCD	Lipton et al. /Brody and Luo
1979	Suggested applying a-Si TFT to LCD	LeComber et al.
1982	Development of a-Si TFT 3" (125 ×160) TV Development of a-Si TFT 4" (220 ×240) TV	Sanyo
1983	Development of poly-Si TFT 2. 14" (240×240) TN-LCD color TV	Seiko-Epson
1984	Commercialization of poly-Si TFT 2" (240×220) LCD color TV Commercialization of CdSe-TFT (128 ×192) LCD	Seiko-Epson /Panelvision
1985	Development of a-Si TFT 10" (640×480) Color LCD	Toshiba
1987	Commercialization of a-Si TFT 6" LCD TV Development of a-Si TFT 4" (640 ×480) LCD color TV	Hitachi/Sharp
1988	Commercialization of a-Si TFT 4" LCD color TV	Sharp

1.3 Types and Characteristics of LCD

LCDs can be classified by the form and the driven method. In terms of the form, LCDs can be divided into projection LCDs and direct view LCDs. Projection LCDs are mainly used to project large images ranging from forty inches to hundreds of inches, by enlarging high–intensity images or data with optical devices and projecting them on a screen. Such projection LCDs are again classified into rear projection and front projection, depending on the position of the screen and the light source. Projection LCDs consist of a light valve, optical system, circuit system and an instrument system. Meanwhile, direct view LCDs, which displays the light directly from LCD, are divided into transmissive LCDs and reflective LCDs.

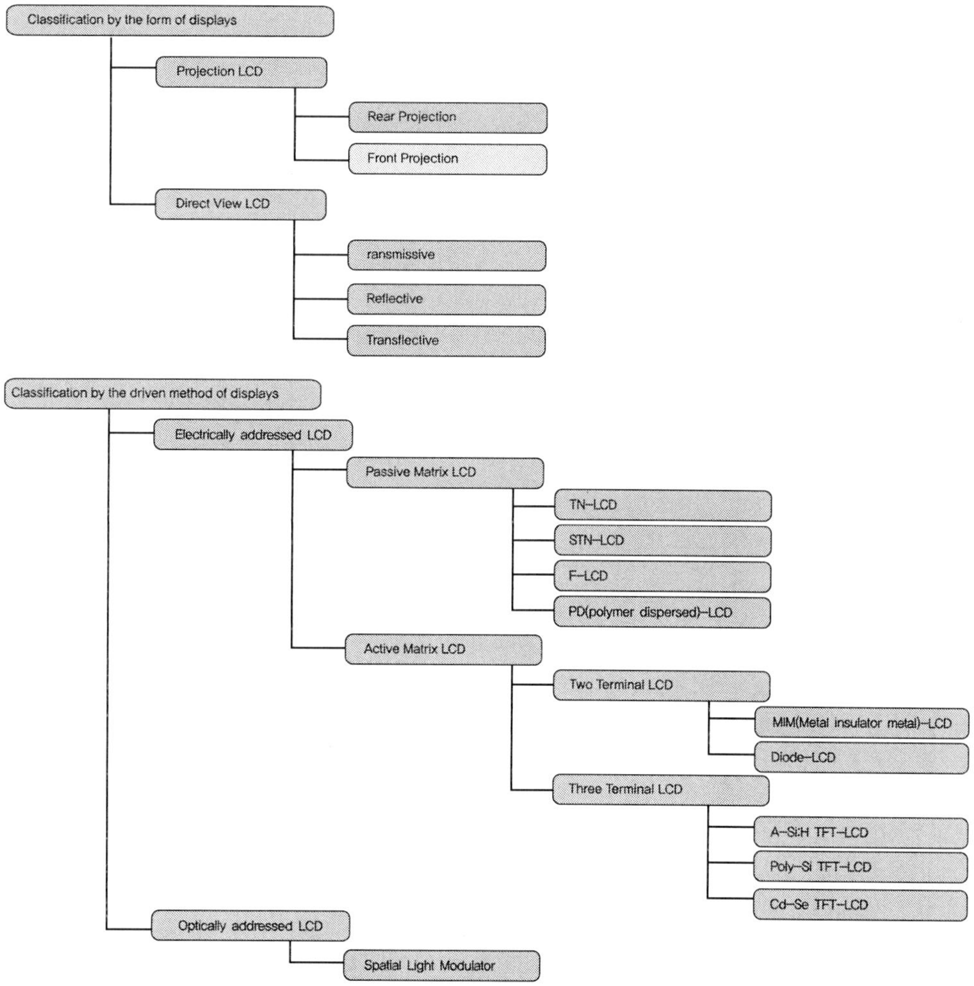

Figure 3.5
Classifications of LCDs

In the case of transmissive LCDs, the light comes from the backlight, and the intensity of light is adjusted by the LCD panel. In the case of reflective LCDs, natural light and ambient light are reflected from the LCD panel to form an image.

LCDs can also be categorized by the driven method into two types, i.e, electrically addressed ones and optically addressed ones. Optically addressed LCDs are controlled by optical signals. A spatial light modulator is a good example of the optically addressed LCDs. The electrically addressed method can be classified into a passive matrix LCD and an active matrix LCD depending on the presence or absence of an active element when the pixel electrode is driven. Passive Matrix type was used for the early LCD products (watches, calculators, etc.). Passive Matrix type LCDs include Twisted Nematic (TN)–LCDs, Super Twisted Nematic (STN)–LCDs, Ferroelectric (F)–LCDs and Polymer Dispersed (PD)–LCDs, which are classified by the type of the liquid crystals used. A TN–LCD is the most commonly used one in LCD products nowadays, but it has several problems in terms of responsiveness, viewing angle, the working temperature range and so forth. STN–LCD was used in early D–TVs and notebooks because of the fact that the twist angle varies from 240° to 270° according to the applied voltage. An F–LCD using ferroelectric liquid crystals shows the most prompt responsiveness of all. A PD–LCD was invented to improve the low light efficiency of TN–LCDs which require polarizers absorbing a half amount of the incident light and consuming it as heat. A PD–LCD is made by employing the liquidity of liquid crystals and the structure and double refraction of the polymer. PDLC utilizes fluidity of liquid crystal, structure of polymer material and birefringence. When a polymer material and a liquid crystal are mixed, due to liquid crystal droplets are aligned at molecular boundaries due to molecular forces and boundary conditions. Macroscopically, each of the droplets has an arbitrary molecular arrangement. When parallel light is incident in this state, the birefringence of the liquid crystal causes the liquid crystal droplets to pass through the liquid crystal layer, and the refraction direction changes according to the application of an electric field.

An example of TN–LCDs, which are the most commonly used LCDs around the world, is the black and white liquid crystal of household appliances. To get a better understanding, the first thing to know is polarized light. Polarized light is the light that has passed through polarizers, which are thin plates allowing the light vibrating in only a certain direction to pass through. The axis equal to the direction of light when it passes through polarizers is called a transmission axis. There are a number of narrow cracks (transmission axes) on the polarizers. Figure 3.6 describes horizontal and vertical polarized lights, with the vertical waves and horizontal waves removed, respectively, when the light was passing through polarizers.

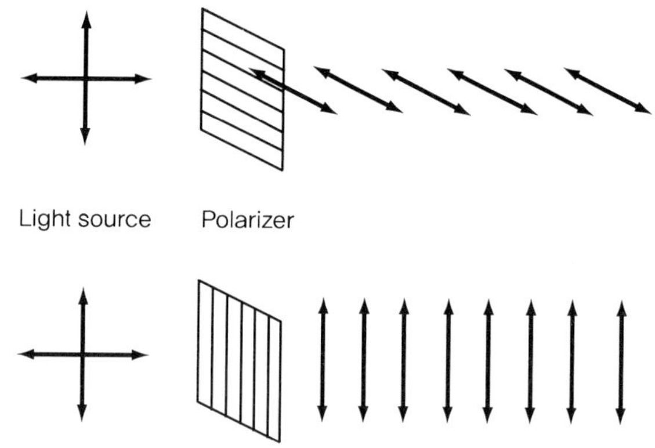

Figure 3.6
Polarization of light by Polarizers

Polarized light can be passed or blocked depending on whether a voltage is applied to the liquid crystal layer inserted in between the two parallel polarizers, as shown below.

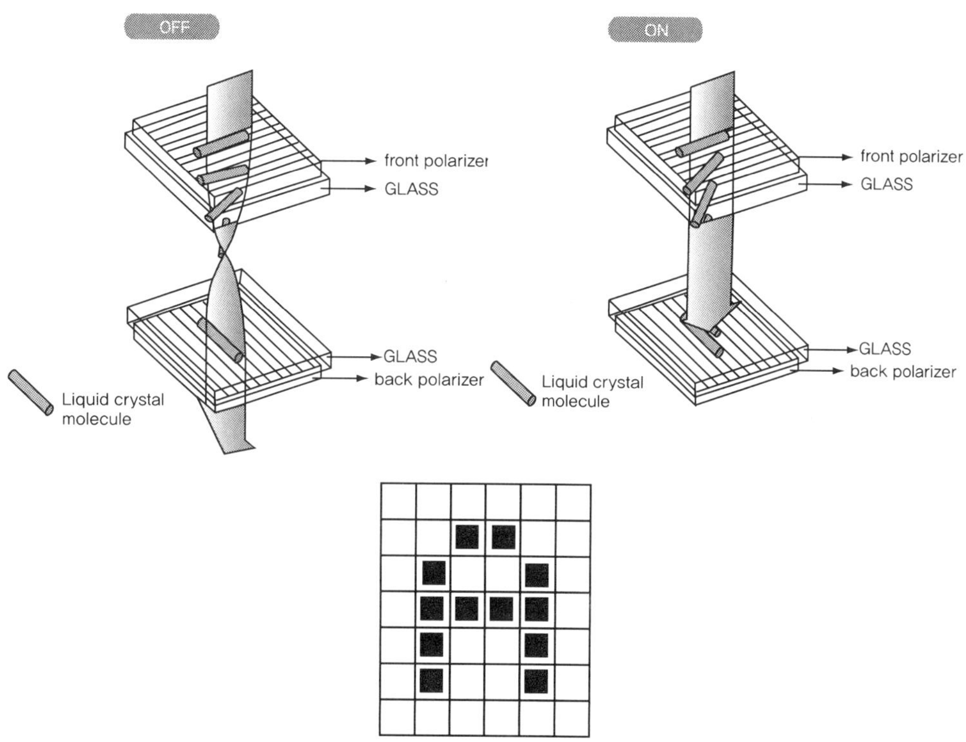

Figure 3.7
he operating principle of Twisted Nematic LC

The 'OFF' state above means that there is no electric voltage applied, allowing the light that has passed through one polarizer to be twisted along with the arrangement of the liquid crystal molecules and to pass through the other polarizer, which is called white state. In other words, when no voltage is applied, light passes through the polarizers. The 'ON' state means the state when a voltage is applied to the device. Then the liquid crystal molecules rise up along with the direction of the electric field, the light which has passed through one polarizer, is directly transmitted to the other polarizer and blocked by the polarizer. That is, when a voltage is applied, light is blocked, by which characters are displayed. In this way, it is possible to display shapes and characters by applying voltage to the top and bottom electrodes.

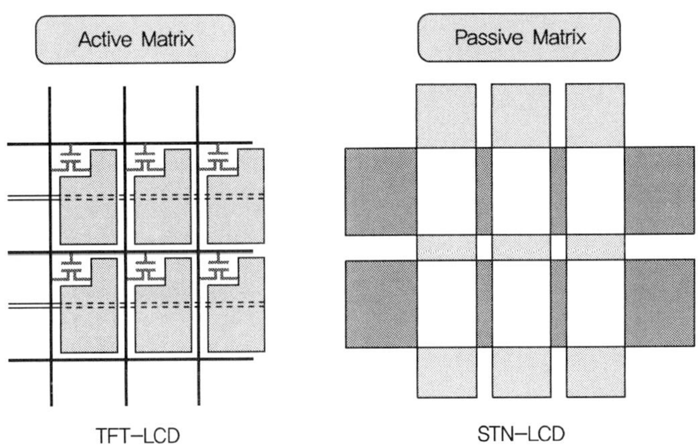

(a) AM-LCD and PM-LCD

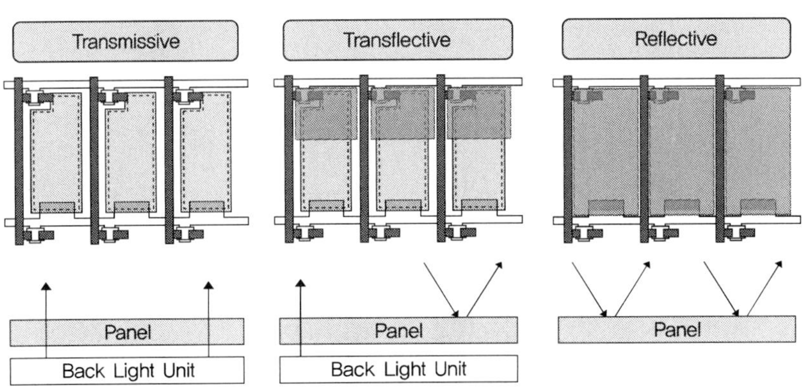

(b) Structure of AM-LCD (Transmissive, Transflective, Reflective)

Figure 3.8

The structure of AM–LCD and Comparison with PM–LCD

TFT–LCDs can be divided into Passive Matrix (PM) LCDs and Active Matrix (AM) LCDs, as shown in Figure 3.8 (a). Unlike PM–LCDs, the performance of

AM−LCDs has improved with the use of a transistor, which is an active element. This enabled AM−LCDs to display colors and contrasts as freely as possible in line with the input signals of TFT, and accordingly, AM−LCD method has become the mainstream method. AM−LCDs can be manufactured in three types, Transmissive, Transreflective and Reflective: among these the Transmissive type is the majority.

* 2 Structure and Operating Principle of AM−LCDs

Display Engineering

2.1 Structure of Color TFT−LCD

TFT−LCDs can be divided by the display method as following:

Table 3.4 TFT−LCD Classification based on the structure

Direct view type	Notebook PC, LCD monitor, Small TV
Projection type	LCD projector, OHP, projection TV
Reflective type	HHP, PDA
Trans-Reflective type	

Among these, Direct view type and Projection type are the most commonly used. Projection type is again divided into front projection and rear projection. Front projection is based on the operating principle of the projectors used for presentation or home−theaters. Rear projection is used for projection TVs easily found in any household. The operating principle of front/rear projection is described in Figure 3.9.

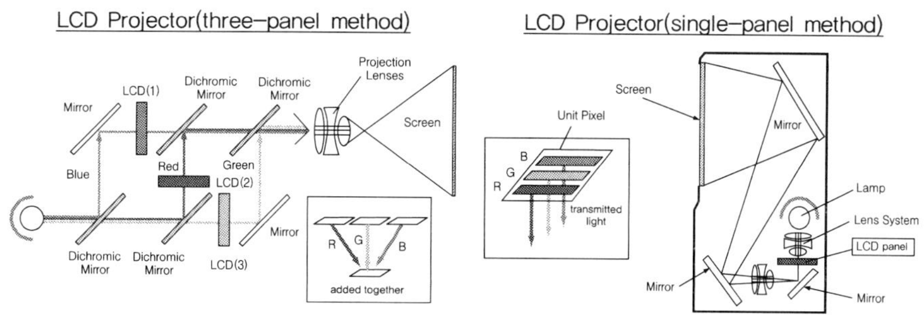

(a) Front projection　　(b) Rear projection

Figure 3.9
Two different Projection type LCDs

Meanwhile, Direct view type LCDs are used for popular LCD products including LCD TVs and laptop computers. The operating principle and structure of Direct view type LCDs are shown in Figure 3-10.

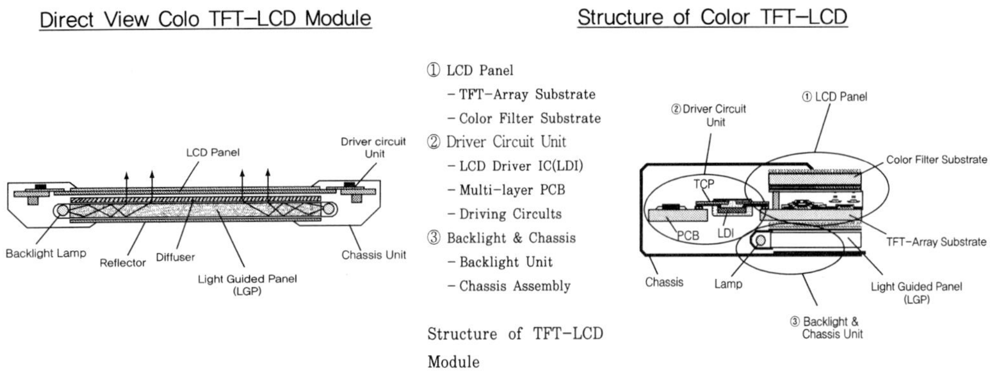

Figure 3.10
The structure of Direct View type TFT–LCD Module

(1) Structure of TFT Panel

The structure of a color TFT–LCD panel is as follows.

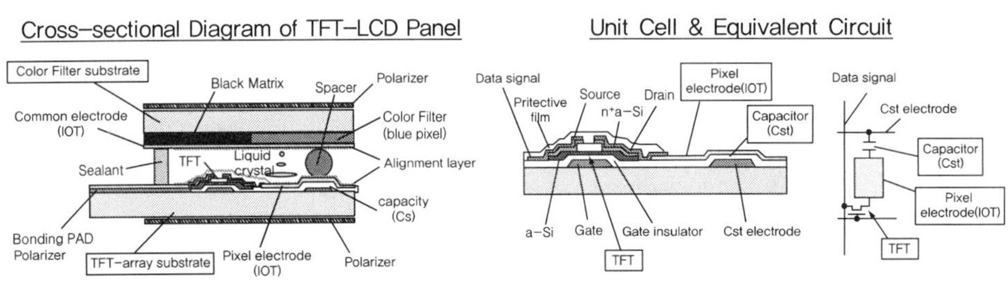

Figure 3.11
The structure of a TFT–LCD Panel

The above diagram on the left consists of the following: TFT–Array substrates are glass substrates on which more than two different types of metal films, an insulating film, an amorphous silicon layer and ITO (indium tin oxide) are evaporated in the form of thin films. By processing the substrates, a TFT is produced. With the TFT, storage capacitance and pixel electrodes form a unit element, making an individual pixel. Such pixels are connected with each other by metal signal lines as a matrix, and at the end of each signal line is a bonding pad applying the driving signal of LDI (LCD Driver IC). The figure on the right shows a unit cell in detail and the design of an equivalent circuit. The electric field of ITO controls the rotation of liquid crystals.

(2) Structure of Color Filter

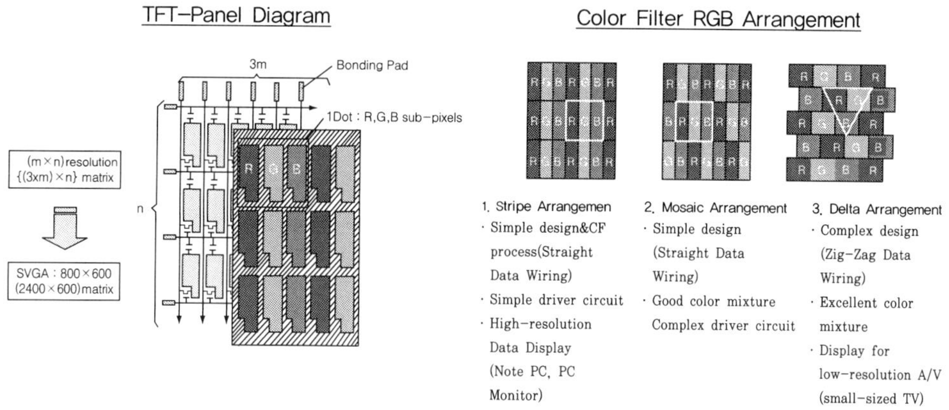

Figure 3.12
Structure of Color filter

As shown in the diagram on the left, when the resolution of a TFT is $m \times n$, the number of columns gets tripled as three pixels (RGB) are used for one dot. The color filter (CF) substrates are produced by a separate manufacturing process, in which black matrix (BM) blocking the penetration of light between pixels is formed with Cr thin film on the same glass substrates as used for the TFT-Array, and the RGB color filter is installed aligned with each pixel area. The ITO thin film, which is used as the common electrode, is evaporated on the whole surface of the CF substrates regardless of pixels.

(3) Backlight and Mounting Unit

Backlight, which is used as the light source of LCDs, forms a surface light source with consistent brightness, using a fluorescent lamp. For the fluorescent lamp, a small but super-luminescent cold cathode-ray tube fluorescent lamp (CCFL) is used. There are two types of the installation of CCFL, i.e., Direct-light type and Edge-light type, as shown in Figure 3.13.

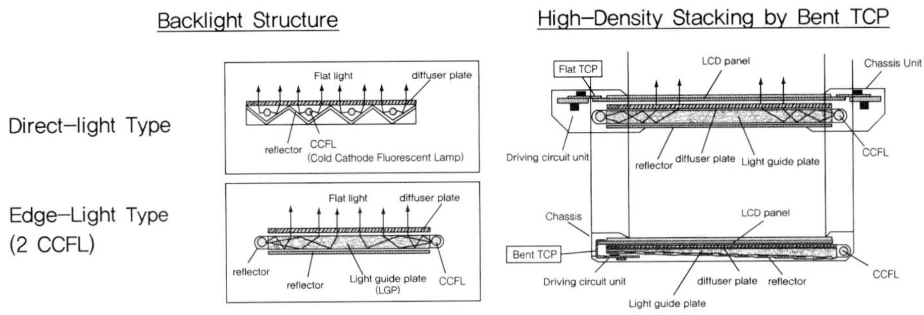

Figure 3.13
The structure of Backlight Assembly

As shown in the left figure above, the lamp can be placed right under LGP (Light guide panel) or on the side of LGP, which are called Direct–light type and Edge–light type, respectively. For small–sized LCDs and LCDs for laptops, single edge light type is used as shown in the right figure above. This is because the single edge light type has relatively better luminous efficiency than the direct–light type and it is possible to produce LCD products that are lighter and thinner with the single edge light type. Thus, the single edge light type is mainly used for the products with high portability.

2.2 Operating Principle of Color TFT–LCD

The characteristics of liquid crystals required for LCDs are 1) broad working temperature range, 2) fast response, 3) high contrast ratio, 4) driven by low voltage, 5) the least transmittance of high voltage, 6) reduction of reaction time, 7) wide viewing angle and 8) reliability. Such characteristics continue to demand a change of paradigm in the production of TFT LCDs. Accordingly, new and various attempts for the driving methods that differ from those used in the earlier production of TFT LCDs have been made. The operating principles we will cover from now on are the most commercialized ones at the moment.

(1) Gray Scale Displaying Method

When a voltage less than V_{th} is applied to the TN (Twisted Nematic) mode liquid crystal cells used in most of the TFT–LCDs, the liquid crystal molecules are arranged to be twisted 90 degrees in the form of spiral, which is the same state as the early state, i.e., the state right after the liquid crystals are injected, as shown in the left figure in Figure 3.14.

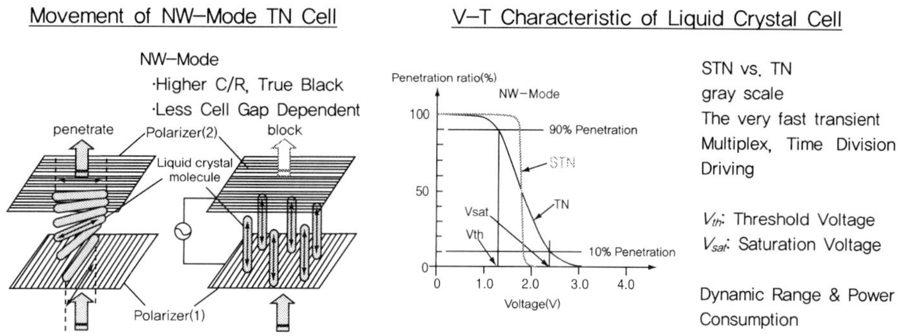

Figure 3.14

The movement of NW mode TN liquid crystal cells and the V–T Characteristic curve

When polarizers are placed on the top and bottom of the liquid crystal cells, the incident light, which is linearly polarized, proceeds along with the spiral arrangement of the liquid crystal molecules with the polarizing axis rotated 90 degrees, causing the light to pass through the polarizer on the opposite side. If the applied voltage is increased higher than V_{th}, the molecules get rearranged in the direction of the electric field, and the linearly polarized incident ray passes through the liquid crystal layer, rotating less than 90 degrees, by which the amount of light that penetrates and comes out of the other polarizer starts to decrease. When a sufficient voltage is applied, the molecules are arranged in the same direction as the electric field, causing the linearly polarized light neither to rotate nor to pass through the other polarizer.

The relationship between the applied voltage and luminous transmittance of NW (normally white) mode TN cells and STN cells can be described as the V–T characteristic curve on the right in Figure 3.14. If the applied voltage gradually increases from the point where the luminous transmittance is maximum (white state), the transmittance gradually decreases. When the applied voltage is modest, the transmittance changes quite largely, along with the change of the voltage. If a higher voltage is applied again, the transmittance reaches the lowest point (black state), where the transmittance rarely changes even if the applied voltage changes. Thus, not only the gray scale but also medium brightness can be displayed by the modest amount of voltage. However, the V–T characteristic drastically changes in the STN mode and thus, it is not possible to display medium brightness by the amount of voltage applied to the liquid crystal cells. The STN mode is based on the passive driving method without storage capacitance to maintain the applied voltage for some time. Thus, gray scale in the STN mode is displayed by multiplexing or time division driving. Meanwhile, the voltage causing the TN cell to be completely black when it is applied is defined as dynamic voltage. If the dynamic voltage is high, the power consumption of the driver unit, including the driver IC, increases. Thus, the LCDs with low dynamic voltage is required to achieve the low power consumption of TFT-LCDs.

As shown on the right in Figure 3.14, there is a non-linear region in the V-T characteristics. In order to display gray scale with linear characteristic, the data voltage should properly be adjusted to control the TN liquid crystal cells so that each gray scale has the linear characteristic.

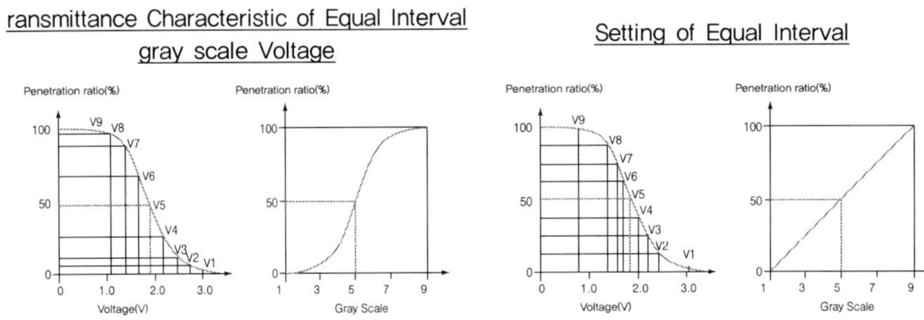

Figure 3.15

The equal interval gray scale voltage and the gray scale characteristics after dividing the voltage by the equal interval transmittance

As shown in the left of Figure 3.15, if we divide the voltage based on the equal intervals, the transmittance characteristic of the gray scale voltage gets non-linear. On the contrary, as shown in the right of Figure 3.15, if the voltage is divided into equal intervals based on transmittance, the transmittance characteristic becomes linear.

However, when it comes to real application, it should be considered that human eyes are not so good at discerning the degree of brightness in a bright screen. Considering this, the linearity of the transmittance per gray scale should also be adjusted. Such adjustment process is called γ-correction, and Figure 3.16 shows the relationship between the gray scale and transmittance in γ-correction.

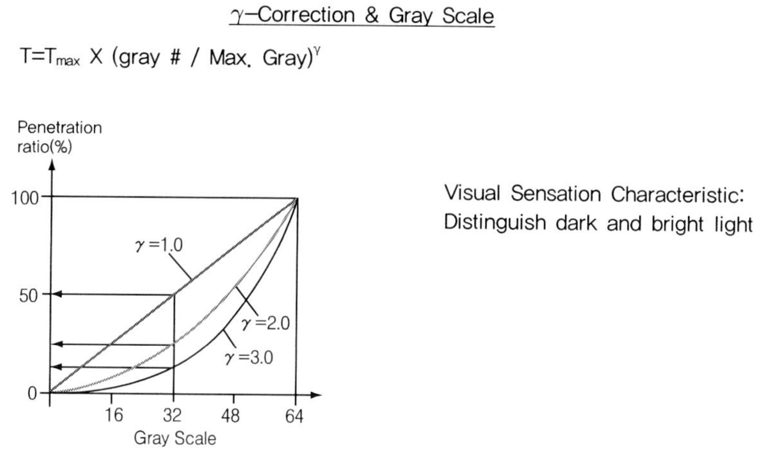

Figure 3.16

The relationship between γ–correction and gray scale

Such γ–correction can be conducted by adjusting the gray scale standard voltage in the LCD driver circuit unit and providing the same to the data driver IC or by processing the visual data with software.

(2) Color Display Method

As mentioned earlier, each dot consists of three pixels, i.e., RGB. Then let's find out how many colors can be displayed in connection with gray scale.

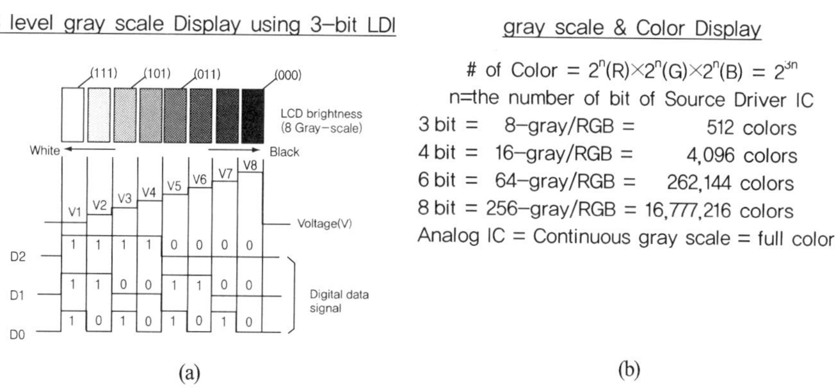

Figure 3.17

8 level gray scale Display & the numbers of the relevant displayable colors

Assuming that the penetration of light can be either completely blocked or fully allowed in liquid crystal cells, i.e., only either one of the OFF/ON state is possible, the number of colors that each sub–pixel can display are calculated by combining the possible RGB states which are expressed as $2\times2\times2=8$. If there are only two available options to control the sub-pixels, i.e., ON or OFF, each of RGB can be controlled by the one-bit digital data signal. Figure 3.17 (b) shows that a set of three-bit can display the gray scale of eight levels ($=2^3$). Thus, if each of RGB can display a set of three-bit, 512 combinations of colors can be made ($8\times8\times8=512$). In sum, the number of displayable colors is determined by the number of the bits of the digital actions in the driver IC. This can be expressed as follows:

$$N=2^n(R)\times2^n(G)\times2^n(B)=2^{3n} \qquad \text{(Equation 3-63)}$$

Theoretically, if 6bit data driver IC is used, 262,144 colors (2^{18}) can be displayed. However, in order to realize this theory, the gray scale distinction in each sub-pixel of RGB needs to be clear. More specifically, in order to display full colors, i.e., 16.7M (2^{24}), using 6bit IC, the gray scale should be discernable as many as 256 levels (2^8).

To maximize the numbers of the gray scale levels, the contrast ratio of LCD (the difference between the brightest and the least bright states) should be relatively high, i.e., at least 100:1 of contrast ratio is required for excellent color display. The maximum contrast ratio of STN is about 50:1, but TFT-LCDs show better color display with the contrast ratio higher than 300:1. The reason TFT-LCDs use NW mode is that the mode has better contrast ratio than that of NB mode. Also, the reasons TN liquid crystal cells are used for TFT-LCDs are that TN cells have absolutely higher contrast ratio and faster response time than STN cells, and that it is easy for TN cells to display gray scale with the liquid crystal voltage since the V-T characteristic of TN cells has a gentle slope, as shown in Figure 3.14.

(3) Display Performance of LCD

The resolution of LCD screens is determined by the number of bits of the source driver IC. Currently, most of the TFT-LCDs use 6bit LDI, and 8bit LDI is also used. The video formats of AM-LCDs are standardized VGA, SVGA, etc. Table 3.5 lists the resolutions that have already been commercialized.

Table 3.5 The resolutions of TFT-LCD products and their pixels, display ratio and standardized names

No.	Resolution	Dot #	pixel #	Display ratio	Standardize
1	320×240	76,800	230,400	4:3	Quarter VGA
2	640×400	256,000	768,000	16:10	EGA
3	640×480	307,200	921,600	4:3	VGA
4	800×480	384,000	1,152,000	15:9	Wide VGA
5	800×600	480,000	1,440,000	4:3	SVGA
6	1024×600	614,400	1,843,200	~17:10	Wide SVGA
7	1024×768	786,432	2,359,296	4:3	XGA
8	1280×1024	1,310,720	3,923,160	5:4	SXGA
9	1400×1050	1,470,000	4,410,000	4:3	SXGA+
10	1600×1200	1,920,000	5,760,000	4:3	UXGA
11	1920×1080	2,073,600	6,220,800	16:9	HDTV
12	1920×1200	2,304,000	6,912,000	16:10	Wide UXGA
13	2048×1536	3,145,728	9,427,184	4:3	QXGA
14	3200×2400	7,680,000	23,040,000	4:3	QUXGA

When the size of an LCD gets bigger, the length of the gate lines extends. Then the time constant of the signal lines also increases, causing the signal delays. Also, if the resolution goes up causing the number of the gate lines to increase, each signal line would be selected less frequently than before and the data voltage cannot be fully written on the liquid crystal cells through TFT. To resolve this problem, there have been many studies going on, with many advances in the technology. It is expected to see the development of TFT-LCDs bigger than 60 inches in the near future.

(4) Operating Principle of Active Matrix

We learned what Passive Matrix and Active Matrix are in Chapter 1. Now we will learn the operating principles of Active Matrix used in TFT-LCDs.

The schematic diagram of the whole system operating TFT-LCD is shown in Figure 3.18 (a). More specifically, digital image data coming out of the graphic controller of a computer are input into the LCD timer control ASIC. If the input data are analog, an AD converter, transforming the analog signals to digital signals, needs to be used together with the LCD timer control ASIC, since TFT-LCD operates digitally. The time control ASIC creates control signals, using the input data, in the form of digital signals that can be processed by the source driver IC. The analog circuit generates a standardized voltage needed to display multi–level gray scale and provides the same to the data driver IC. The voltage that turns on or off the TFT is also produced in the analog circuit and supplied to the gate driver IC. The inverter supplies power to CCFL (Cold Cathode Florescent Lamp) used as the backlight lamp and adjusts the brightness of the LCD by adjusting the lamp current. Meanwhile, in a high–resolution LCD, the frequency of the digital data signals and clock signals increases, causing some problems with the interface between the computer system and the LCD module, e.g., signal distortions and EMI (Electro Magnetic Interference). To address this issue, a new technology that converts the signals to the LVDS (Low Voltage Differential Signaling) format and then transmits has been developed and applied.

Schematic Diagram of the Operating System of TFT–LCD Module | Gate Signal Line Selection & Scan Pulse Supply

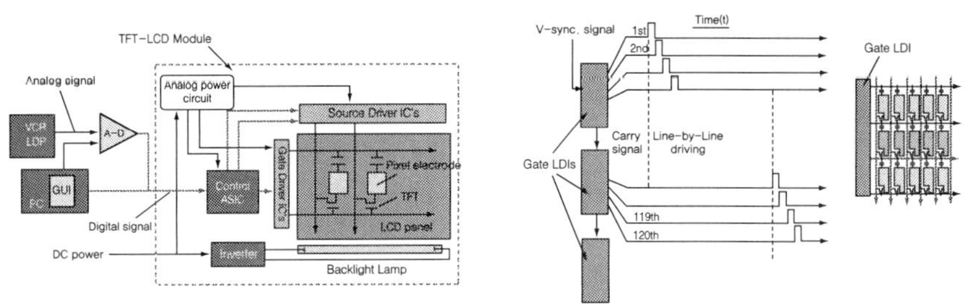

(a) (b)

Figure 3.18

The operating system of TFT–LCD Module / Selection of the gate signal lines & Scan pulse supply

Two types of driver IC are used for TFT–LCDs: a row driver IC (gate driver IC) selects the gate signal lines of the TFT–array in order and creates scan signals. A column driver IC (data driver IC or source driver IC) converts the digital image data to the pixel voltage and it is applied to the signal lines. When the gate driver IC selects a scan line and applies a scan pulse, turning on the TFT, the source driver IC applies the signal voltage to the liquid crystal cells through the signal line of each data. The operating principle of the gate driver IC with multi outputs are described in Figure 3.18. The gate driver here is a sort of shift register that creates the ON–OFF signal voltage of TFT stepwise. In the case of high–resolution AM–LCDs, normally multiple gate driver ICs are used, but the output signal should come out of only one line among the whole gate lines in one frame.

Now let's take a look at the operating principle of the source drive IC with 6bit digital type, which is the most commonly used for high-resolution TFT-LCDs. The input signals include 1) the digital image data in 18bit (6bit for each of RGB), 2) the standard voltage for display gray scale, i.e., V1~V9 and 3) various timer signals that control the digital actions of the driver IC. When the shift register receives the clock and carries signals, informing of the start of operation, and starts to operate, each sub-register transports pulses stepwise in accordance with the shift clock and at the same time, stores the input digital data one-by-one in the data register. These processes repeat until data in one horizontal line are all stored. Once one line of data is stored, the data are loaded onto pulses and sent down to the latch at one time. The image data entered in the latch selectively output sixty-four gray scale voltages (assuming that the nine gray scale standard voltages are externally applied by the level shifter and the DA converter), the current of which is amplified by the voltage follower. After that, the image data come out of all of the channels through the data signal lines. As such, the data voltage is applied to all the pixels connected to one gate signal line at one time, TFT-LCD operates stepwise on a line-by-line basis.

(5) Viewing angle

Viewing angle means the angle at which we can see the display. Figure 3.19 shows a more accurate definition. When an LCD screen is viewed right in front of it (six o'clock viewing direction), which is at the best viewing angle, a clear image is seen; however, when the viewing point is moved to the left or right, the screen looks darker. This is the weakness of LCD displays that does not exist with CRT displays. Let's figure out how to improve the viewing angle of LCDs.

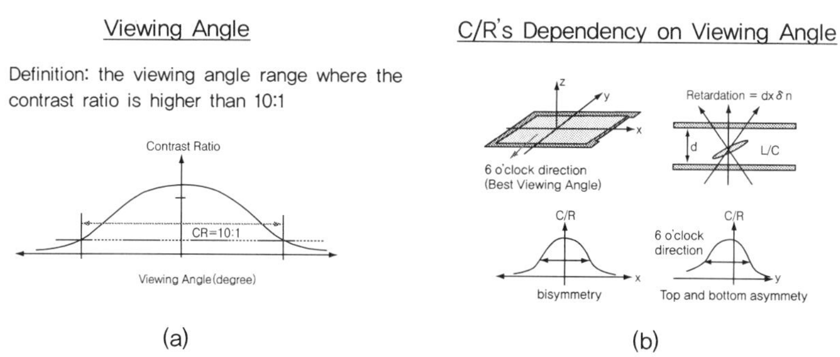

(a) (b)

Figure 3.19

Viewing angle & Contrast Ratio(C/R)'s dependency on viewing angle

The first answer is the alignment technology of TN mode. The viewing angle can be improved by changing the rubbing direction of LC in the TN mode, which is described in Figure 3.20. However, it is difficult to get a much broader viewing angle with this technology only, therefore, a wide viewing angle technology that changes the rotating direction of LC was introduced.

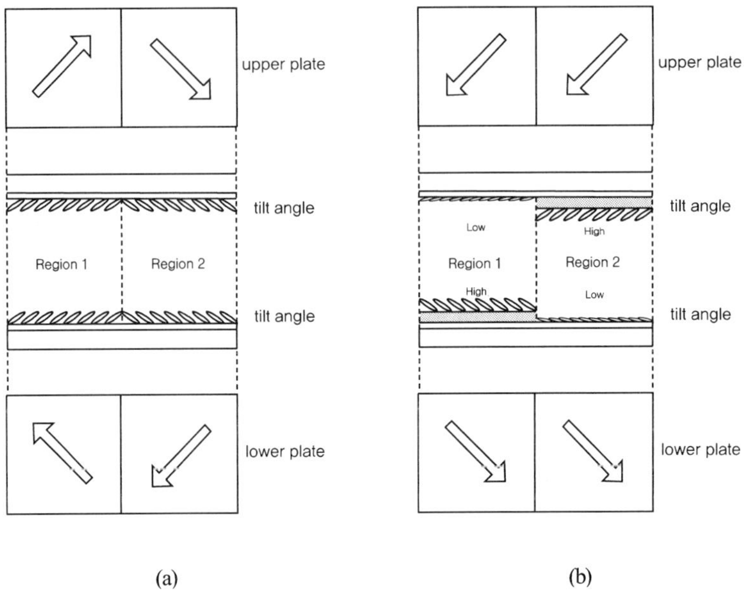

Figure 3.20
The of Two Domain TN cell with different rubbing direction and pre−tilt angles

The structure described in Figure 3.20 (a) improves the viewing angle characteristic, whereas the structure in Figure 3.20 (b) obtains the wide viewing angle using high and low pre−tilt angles. The alignment film is composed of organic (the diagonal−lined part) and inorganic matters. There are IPS (In−Plane−Switch) and VA (Vertically Aligned) modes in the wide viewing technology, as shown in Figure 3.21. The two modes are compared to the most basic TN mode, where LCs are arranged horizontally when power is OFF, and if a voltage of the electrodes on the top and bottom is applied, LCs are arranged vertically, which is the ON state. As mentioned earlier, this is called NW mode.

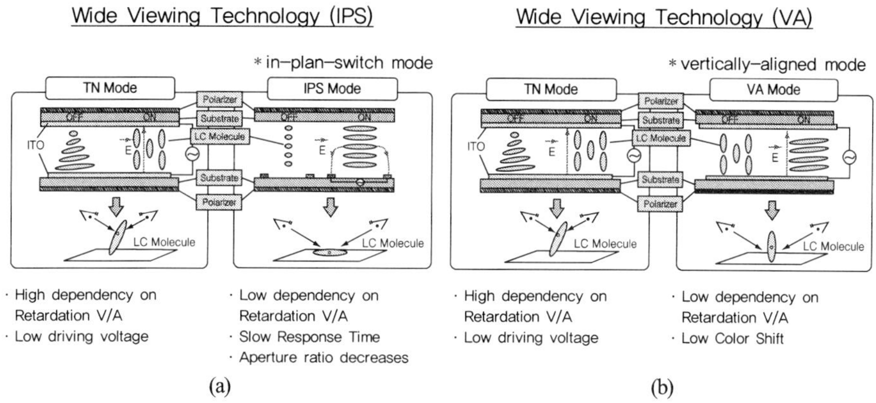

Figure 3.21
Wide Viewing Technology of IPS and VA

Unlike the TN mode, the LC arrangement changes horizontally when power is on in the IPS mode. Also, the electrodes are placed on the left and right, not on the top and bottom, and thus LCs rotate in a horizontal direction as well when a voltage is applied. There is no retardation in the IPS mode that may be caused when LCs are arranged vertically. Nevertheless, as the LC arrangement changes horizontally with the horizontally generated electric field, the response time gets slower and the aperture ratio decreases, which causes the viewing angle to get smaller.

The VA mode changes the LC arrangement vertically like the TN mode, but the difference is that LCs are vertically arranged when power is OFF. The electrodes are also placed on the top and bottom and thus, the electric field is generated vertically. When power is ON, LCs are arranged horizontally, thus the VA mode has a low dependency on the retardation caused by the vertical arrangement of LCs, and even the color shift occurs less frequently. Overall, the VA mode is the most commonly used nowadays.

(6) Operating Principle of Reflective type TFT-LCD

In the case of direct-view-type TFT-LCDs, a 70%~80% of the whole power is consumed by the backlight unit. To reduce this, reflective-type TFT-LCDs are being developed. However, the reflective-type has a strong dependency on the surrounding environment and thus needs to be improved much. At any rate, the reflective-type can be innovative if the operating principle can be fully understood and the advantages of its merits are taken. The operating principle and structure are described in Figure 3.22 as below.

Circular Polarize

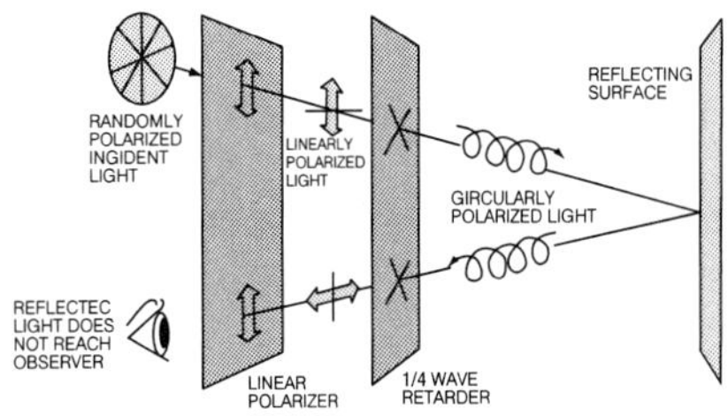

(a) Structure & Penetration Characteristic of Circular Polarizers

DMGH (Double Metal Guest Host) Reflective Type Color

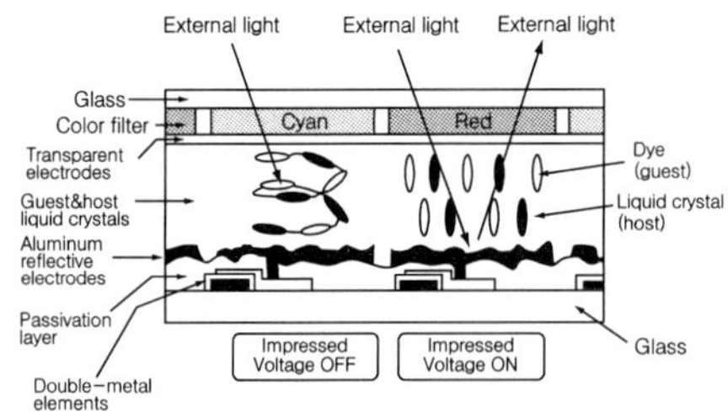

(b) Mimetic Diagram of DMGH Reflective Type Color LCD

Figure 3.22 The operating principle and structure of Reflective type TFT–LCD

Figure 3.22 (a) shows the role of circular polarizers, which are one of the prominent characteristics of the reflective type. The external light polarized in various directions passes through the first linear polarizer and then a circular polarizer, by which the light is heading to the reflecting surface. The reflected light passes through the circular polarizer and subsequently, the linear polarizer (the opposite direction) and then it is seen by human eyes. Figure 3.22 (b), a side view of the reflective type LCD, shows how light passes through in accordance with the LC arrangement. The optical characteristics depending on the reflective display types are listed in Table 3.6, for further reference.

Table 3.6 Optical Characteristics depending on Reflective Display Types

Display Type	PCGH	PDLC	ECB	STN	TN
Optical Characteristic	Absorption	Absorption/Dispersion/Interference	Double Refraction	Double Refraction	Rotation
Reflection - Luminance (Single Color)	Excellent	Fair-Good	Fair	Poor-Fair	Fair
Contrast Ration (Single Color)	Fair-Good	Fair-Good	Fair-Good	Poor-Fair	Excellent
Color Display Performance (with CF BL) (without CF BL)	Good -	Good Good	Good -	- Good	- -
Low Power Consumption	Good	Good-Poor	Excellent	Excellent	Excellent
Response Time	Fair-Good	Good	Good	Fair-Good	Good

Operating Principle	Active-Matrix	Active-Matrix	Active-Matrix Passive-Matrix	Passive-Matrix	Active-Matrix
(PCGH : Phase Charge Guest Host, PDLC : Polymer Dispersed Liquid Crystal, ECB : Electrically Controlled Birefringence, STN : Super-Twisted Nematic, TN : Twisted Nematic)					

2.3 Design of TFT-LCD

Once the screen size and resolution of an LCD product is determined, the size of unit pixels is determined accordingly. Thus, the components of unit pixels, i.e., TFT, pixel electrodes, storage capacitance and the size and arrangement of signal lines, are determined considering various design parameters. More specifically, the design parameters of TFT-arrays are categorized as shown in Figure 3.12.

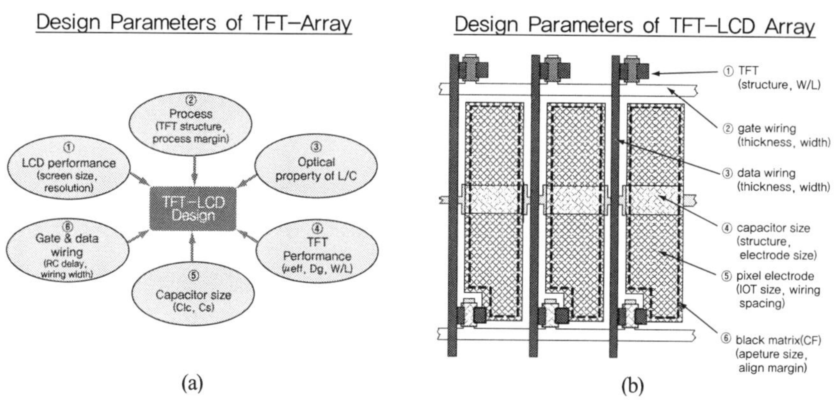

Figure 3.23
Parameters taken into account when designing TFT-LCD Array

The cycle of TFT-LCDs renewing the displayed image is sixty frames per second (60Hz) and thus, if the resolution increases, the time allotted to each gate scan line will decrease. If the resolution of TFT-LCD increases as shown in Figure 3.24, the ON current should flow enough to charge or discharge the liquid cell capacitance (C_{lc}) and storage capacitance (C_s), as much as the data voltage applied from the source driver IC, in a shorter gate signal line selection time. Thus, for the TFTs to be used for high-resolution screens, the performance or the ON current should be enhanced by increasing the W/L ratio of the TFT channels, as to charge enough amount of the load capacitance in a relatively short TFT On time. However, if the length of channels is shortened less than the minimum length determined by the relevant process capability, for the purpose of increasing the ON current of the TFT, problems such as short-circuit will also increase. With this problem set aside, the TFT ON time will still be reduced at any rate, as the TFT becomes larger. If enough ON current does not flow for such TFTs, there will be a switching error which decreases the charging rate of the liquid crystal cells.

Numbers of Gate Signal Lines & Selection Time per Resolution

Resolution	Pixel Arrangement	Gate Signal Line	Selection Time (60frame/s)	Screen Size (Example)
VGA	640×480	480	less than 34 μs	9.4", 10.4"
SVGA	800×600	600	less than 27 μs	10.4", 12.1"
XGA	1024×768	768	less than 21 μs	14.1", 15.0"
SXGA	1270×1024	1,024	less than 16 μs	17.0", 18.1"
UXGA	1600×1200	1,200	less than 13 μs	21.3", 30.0"

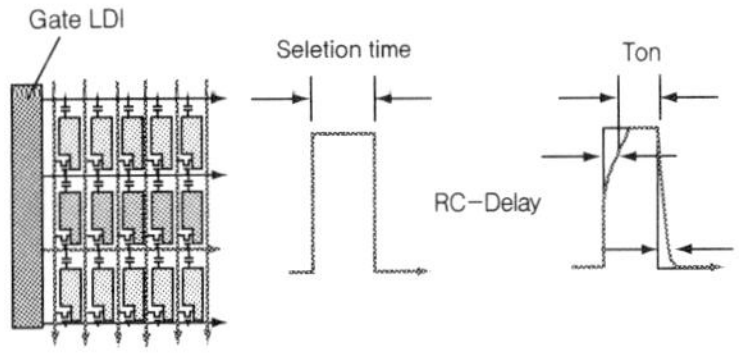

(a)

Limitation of TFT–LCD screen sizes & Resolutions

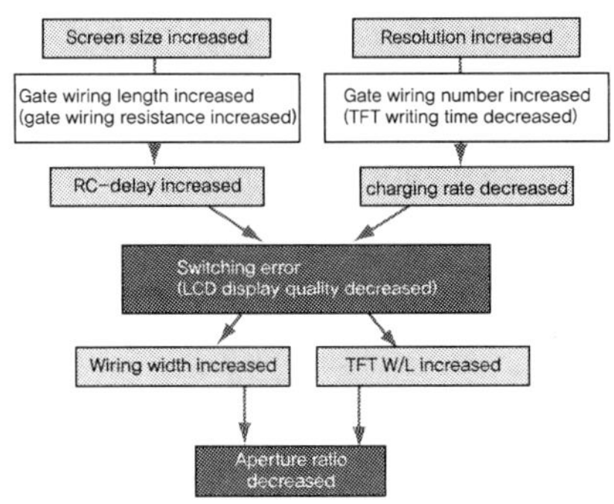

(b)

Figure 3.24

Numbers of Gate Signal Lines & Selection Time per Resolution Limitation of TFT–LCD screen sizes & Resolutions

Therefore, the W/L ratio should be determined in consideration of the process capability of the TFT manufacturing, the minimum width of lines, etc. If the width of the channels is larger than necessary, the parasitic capacitance between the gate electrode and the drain electrode (connected to the pixel electrodes) will increase, and the aperture ratio will be reduced along with the enlargement of the TFT size. Thus, when determining the ideal *W/L* ratio and size of TFT, the following should be considered comprehensively: the performance of the TFT (including the characteristics of the gate insulator and a-Si thin film, field-effect-electron mobility that is determined by the structure of the TFT, and the current gain *K*), the process capability of the TFT-array to be used, the cell capacitance and storage capacitance to load the TFT, etc.

Figure 3.25
The shading effect of a black state caused by writing error of TFT

When the whole LCD screen is displayed in black, the voltages applied from the left and right sides to the liquid crystals displaying black state become different from each other, as shown in Figure 3.25, by which a shading effect occurs. To prevent this error, the time constant of gate signal lines ($\tau = R \times C$) should be the minimum and the performance of the TFT should be fully enhanced so even the TFT connected farthest with the input area could get enough charge.

If the gate signal lines are made too thick, for the purpose of reducing the resistance, the step coverage of the thin film and the data lines to be laminated on the gate signal lines will be reduced, causing defects to increase. Furthermore, increasing the width of the gate signal lines for the purpose of reducing the resistance will increase the capacitance on the common electrode, which causes an increase of the time constant (τ) and a decrease of the aperture ratio. Thus, the thickness and width of the lines should be carefully determined by optimization of the design.

For each unit cell, the *W/L* of the TFT, the width of the gate and data signal lines and the size and arrangement of the pixel electrode, which determines the volume of the storage capacitance, should be decided based on the process design that has taken the process capability of each unit process and the characteristics of the materials used for the gate insulator and lines into consideration. To determine the design, the unit cell operational characteristics and the TFT operational design should precede.

(1) Operational Characteristics of Unit Cell

A TFT perform switch actions in the LC cells, and the operating principle is shown in Figure 3.26.

① A gate signal line is chosen, and V_{on} voltage is applied thereto; in the meantime, the switch is closed, and the TFT ON current (I_{on}) flows and charges C_{lc} and C_s with V_d^+ (write operation).

② Once the selection period ends, V_{off} voltage continues to be applied to the TFT, opening the switch and maintaining the charged V_d^+ voltage for one frame (60 Hz=1/60 sec).

③ In the second frame, when the data voltage polarity is inverted to V_d and the V_{on} voltage is applied again to the TFT gate, the switch is closed, causing C_{lc} and C_s to discharge the V_d^+ voltage (refresh operation) and charge new data voltage V_d (write operation).

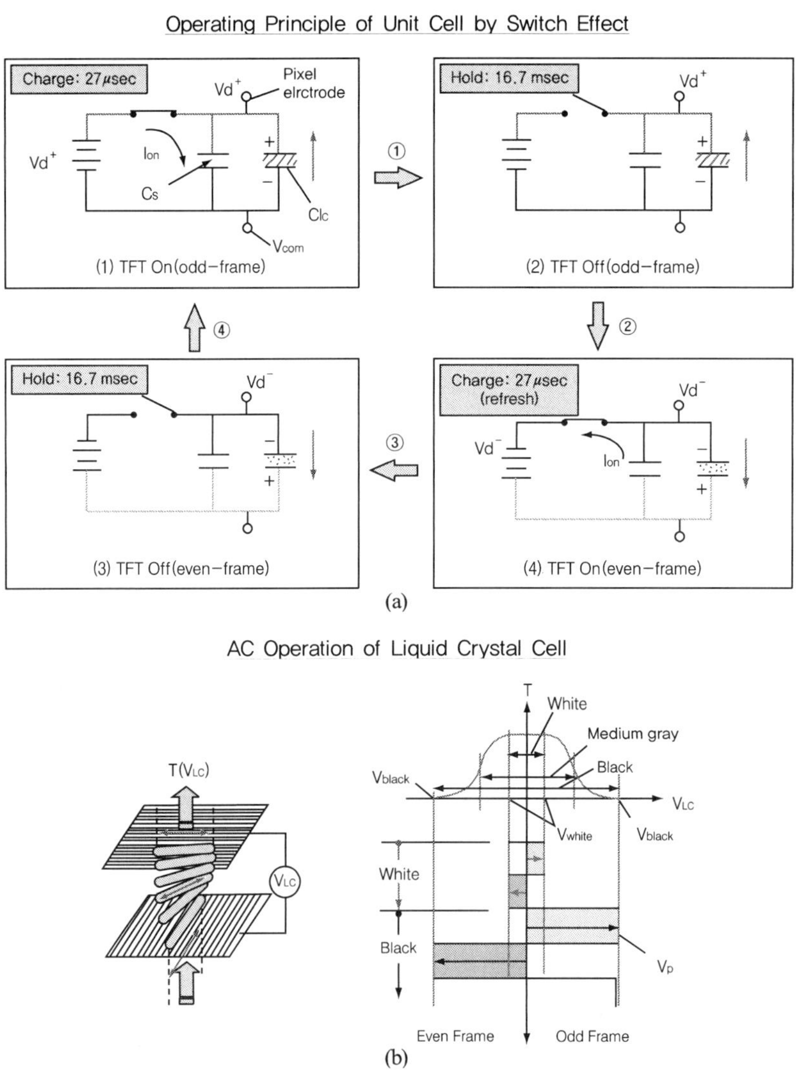

Figure 3.26

Operating Principle of Unit Cell by Switch Effect

④ Again the selection period ends, and when the V_{off} voltage is applied to the TFT gate, the switch is open, and the LC cell maintains a certain amount of voltage V_d^- until new refresh & write operations occur.

To improve the display characteristics of an LCD screen, dot-inversion-driving is used, which is to have each LC cell AC-operated and orient the electric fields of the pixels nearby in different directions. The AC-operation is shown in Figure 3.26 (b) and the driving method is shown in Figure 3-27, which describes the operating principle of the active addressing of 3×3 pixels in the C_s-on-gate method.

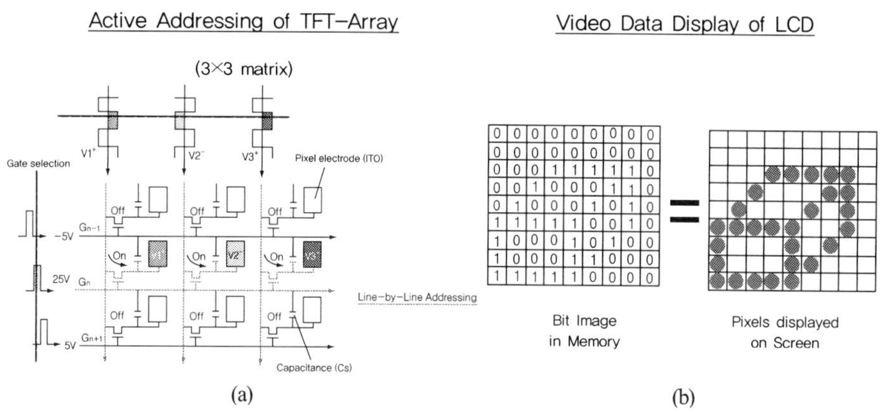

Figure 3.27
Active Addressing of TFT & Video Data Display of LCD

① When the G_n gate signal line is chosen and the TFT ON pulse voltage (V_{on}) is applied, the TFT connected to the G_n gate signal line simultaneously turns the ON state; then the current flows and each of C_{lc} and C_s start to be charged, applying each data signal voltage, V_1^+, V_2 and V_3^+, to the pixel electrodes.

② Once the charging is complete, the G_n gate signal line turns to the TFT Off voltage (V_{off}). Then the G_{n+1} gate signal line is chosen, with the repetition of refresh & write operations to the G_{n+1} line.

③ Although the data signal voltage changes for the write operation of the G_{n+1} line, the voltage applied to the G_n line remains unchanged, as the TFT is OFF, which is electrically blocked from the data signal lines. This state lasts for one frame period ($T_f = 1/60\text{sec}$), i.e., until the G_n line is chosen again in the next frame.

④ Once the polarity of the data signal voltage is inverted as V_1^+, V_2^- and V_3^+ against V_{com} and the G_n line is chosen in the next frame, the pixel voltages are refreshed as V_1^+, V_2^- and V_3^+. With such line-by-line writing method, it is able to operate all the pixels of a whole LCD screen.

(2) I–V Characteristic and Operational Simulation of TFT

The ON–time of TFT decreases as the resolution of the TFT–LCD increases. Accordingly, TFTs should be designed to have the sufficient current driving capability to fully charge the load capacity ($C_t = C_{lc} + C_s$) in a short time.

Measurement of the I–V characteristics of TFT involves the ON and OFF operations. For the ON operation, the ON current (I_{on}) flows by applying a voltage of 20 V to the TFT gate. For the OFF operation, a voltage of −5 V is applied to the gate. The ON cycle (charging cycle) of TFT is usually 20~30 μsec, but the OFF cycle (discharging cycle) is relatively longer, i.e., 16.7 msec(1/60 sec). Thus, for a TFT to perform switch actions, the ON–OFF current ratio (I_{on}/I_{off}) should be greater than 10^6. An a–Si TFT has I_{on} of a dozen of μA and I_{off} of sub–pA and thus, can be used as a switching element with excellent ON–OFF characteristics. The a–Si TFTs are the most commonly used for AM–LCDs.

In general cell operation of TFT–LCD, the gate voltage V_g should be around 20~25 V with the drain voltage (V_d) of about 0~10 V for a write operation. Then the threshold voltage (V_{th}) is less than 5 V, satisfying the condition $V_g > V_d + V_{th}$. The TFT operates in the non–saturation region.

The current driving capability (K) of TFT can easily be measured with the I–V characteristics in the saturation region. The TFT driving condition is always $V_g < V_d + V_{th}$, which means the TFT can maintain the operation state in the saturation region.

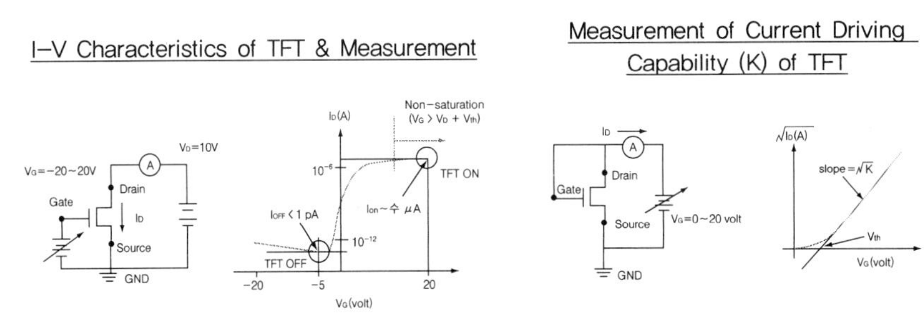

Non–saturation Condition

$$(V_G > V_D + V_{th})$$
$$I_d = K\{2(V_G - V_{th}) - V_D\}V_D$$

(a)

saturation Condition

$$(V_G < V_D + V_{th})$$
$$I_d = K(V_G - V_{th})^2$$
$$K = \frac{1}{2}\mu_{eff} C_g \frac{W}{L}$$
$$\sqrt{I_d} = \sqrt{K}(V_G - V_{th})$$

(b)

Figure 3.28

I–V Characteristics of TFT & Measurement of Current Driving Capability (K) of TFT

As shown in Figure 3.28, if we plot V_g versus the square root of I_d, the slope of the linear region will be the square root of K, and the cross point with x-axis should be V_{th} of TFT. W/L of TFT determines K and thus, the write operation mode is determined by simulation. If we calculate K based on all of the above:

$$K = \frac{C_t}{2(V_g - V_{th} - V_d)t} \times \ln \frac{[2\{V_g - V_{th} - V_d\} + \{V_d - V_p(t)\}] \times [V_d - V_p(0)]}{[2\{V_g - V_{th} - V_d\} + \{V_d - V_p(0)\}] \times [V_d - V_p(t)]} \quad \text{(Equation 3-1)}$$

(where V_g=gate voltage, V_d= drain voltage, V_p= pixel voltage, V_{th}= threshold voltage)

$V_p(0)$ is the pixel electrode voltage at the time of starting the charging (t=0). The charging rate (R_{charge}), which represents the charged degree of pixel electrodes after T seconds, can be expressed as follows:

$$R_{charge} = \frac{V_p(T)}{V_d - V_p(0)} \times 100\% \quad \text{(Equation 3-2)}$$

In case the maximum I_{on} is needed for a write operation in LC cells, the maximum voltage needs to be applied to the LC cells (i.e., black screen display). A TFT should be designed by taking such situation into account. That is, in a practical application of TFT-LCDs, the charging rate of TFT in black screen should be set high enough.

(3) Leakage Current and Optimization of Storage capacitance (C_s)

The structure of LC cell storage C_{lc} is as follows: a flat electrode capacitor filled with dielectric substances, i.e., liquid crystals, between the pixel electrode of TFT-array substrates made of ITO and the common electrode of color filter substrates. LCs are dielectric substances with high resistivity and need to satisfy the following formula to maintain the LC voltage V_{lc} charged to C_{lc} for a certain period of time.

$$R_{lc} \cdot C_{lc} = \frac{\rho_{lc} \cdot d_{lc}}{A} \times \frac{\epsilon_{lc} \cdot A}{d_{lc}} = \rho_{lc} \cdot \epsilon_{lc} \gg T_f \quad \text{(Equation 3-3)}$$

The permittivity of LCs generally ranges from 5ε to 10ε(ε=8.854×10^{-12} F/m) and thus, the resistivity of LCs to be used needs to be higher than 10^{10} ohms-cm in order to satisfy Equation 3-3.

The LC cell voltage V_{lc} applied during the writing operation decreases due to discharging by I_{off} of TFT as well as the leakage current of the LC cell itself. To supplement this characteristic, C_s needs to be installed in parallel with C_{lc}, as shown in Figure 3.29.

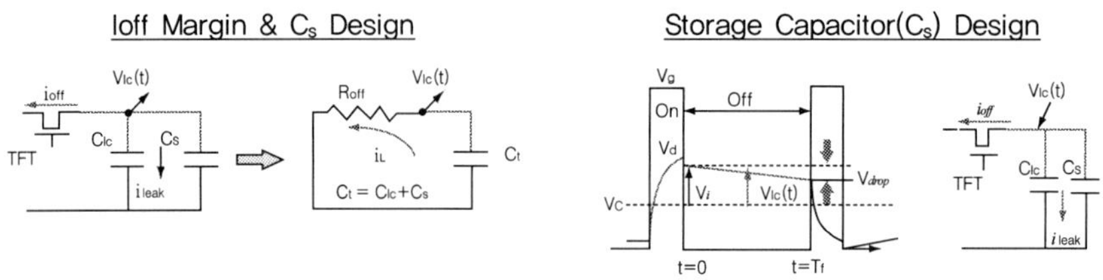

$V_{lc}(t) = V_{lc}(0)[\text{Exp}(-t/R_{off}C_t)]$
$V_{drop} = V_{lc}(0) - V_p(t)$
$V_{drop} = V_{lc}(0)[1-\text{Exp}(-T_f/R_{off}C_t)]$
$R_{off} > (T_f/C_t)/L_n\{V_{lc}(0)/(V_{lc}(0)-V_{drop})\}$
$V_{drop} \ll 1$ gray level(<20 mV for 64-*gray*)

(a)

V_i : LC voltage when the charging is complete (Black displaying voltage)
T_f : Frame cycle (TFT off cycle, 1/60sec)
V_{drop} : Voltage drop by leakage current

(b)

Figure 3.29
I_{off} Margin & C_s Design / Storage Capacitor Design

The equivalent circuit of the LC cell when a TFT is off can be described as Figure 3.29, assuming that the leakage current of the LC cell itself including TFT is I_{off}, and the leakage resistance is R_{off}. The formula to satisfy R_{off} with $V_{lc}(t)$ and the voltage drop V_{drop} after t seconds pass is defined as follows:

$$R_{off} > \frac{T_f}{C_t \times \ln\left[\dfrac{V_{lc}(0)}{\{V_{lc}(0) - V_{drop}\}}\right]} \quad \text{(Equation 3-4)}$$

Meanwhile, the I_{off} characteristics of a-Si TFT are very good, comparable to sub-pA, so a-Si TFT is good enough to be used as a switch element of TFT-LCDs. The voltage holding ratio of LC cell can be expressed:

$$VHR = \frac{V_{lc}(T_f) - V_{com}}{V_{lc}(0) - V_{com}} \times 100\% \quad \text{(Equation 3-5)}$$

When $C_t(=C_{lc}+C_s)$ increases, so does VHR; however, the load capacitance of TFT increases as well. To meet the load capacitance, the current driving capability (K) should be enhanced enough by increasing W/L. However, even in this case, the size of TFT and the storage capacitance electrode ($C_s=\varepsilon A_s/d_s$) gets larger, decreasing the aperture ratio (where A_s: the width of the electrodes, d_s: the thickness of the insulator between the electrodes). Thus, when designing the LC cell, these factors, such as *W/L* of TFT, the size of C_s and so forth, should properly be determined through simulations in consideration of the process parameters. Once V_{drop} acceptable in the unit cell design is set, the C_s value to maintain V_{drop} will also be determined.

(4) Signal Delay and Optimization of Signal Lines

The load capacitance of gate signal lines consists of six factors as shown in Figure 3.30 (a). Thus, the gate signal lines are equivalent to the RC equivalent circuits horizontally connected by resistance to the capacitor as many as the number of the pixels; in this case, RC–Delay can be calculated by the following formula:

$$\Gamma_{RC}=\sum_{i=1}^{3m}(R \cdot C)_i \qquad \text{(Equation 3-6)}$$

As the gate signal voltage is distorted by RC–Delay of the signal lines, the design of signal lines should consider the charging rate of TFT connected farthest from the gate signal lines. Let's say the acceptable RC–delay of gate lines should be less than 4 μsec, i.e., a 20% of the width of the TFT ON voltage pulse. In that case, the current driving capability (*K*) of the TFT should be properly designed in a way that the TFT should still be working when the writing time is 17 μsec.

Load Capacitance of Gate Signal Lines

RC–Delay of Gate Signal Lines

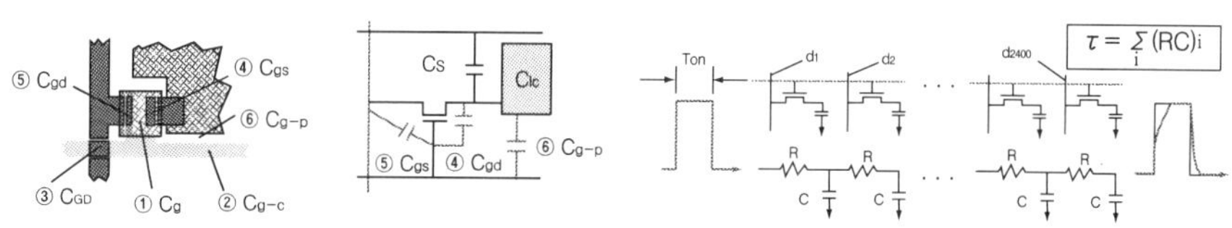

① C_g : Gate capacitance of TFT (SiN_x)
② C_{g-c} : Capacitance between gate signal lines and counter electrodes (L/C)
③ C_{GD} : Capacitance between gate signal lines and data signal lines (SiN_x)
④ C_{gd} : Capacitance between Gate and Drain of TFT (SiN_x)
⑤ C_{gs} : Capacitance between Gate and Source of TFT (SiN_x)
⑥ C_{g-p} : Capacitance between gate signal lines and pixel electrodes (L/C)

(a)

Effective TFT ON time = Gate Pulse Width –RC Delay
Ex) SVGA(800×600)
Ton = 27 μsec
Design Rule : RC〈 20% of Ton → RC–Delay〈 5 μsec

(b)

Figure 3.30
Load Capacitance of Gate Signal Lines & RC–Delay of Gate Signal Lines

Also, to design the gate line with RC–delay less than 4 μsec, the gate line material should first be decided, and the width and thickness of the gate signal lines should then be determined in a way to satisfy the time constant (τ) given in Equation 3–7.

The data signal line load capacitance also consists of four factors as shown in Figure 3.31. Due to the structural characteristics of TFT–LCD, the time constant of data signal lines are smaller than that of gate signal lines and thus, the materials stable to process, such as Cr, Mo and Mo–W, rather than low resistant lines, are mostly used for data signal lines. The width of the lines is generally designed to be as narrow as possible in consideration of the process, for high aperture ratio. However, for large and high–resolution products, the low resistant lining is needed due to RC–delay. Thus, it should be considered to use materials such as Al for data signal lines, like the gate signal lines.

Load Capacitance of Data Signal Lines

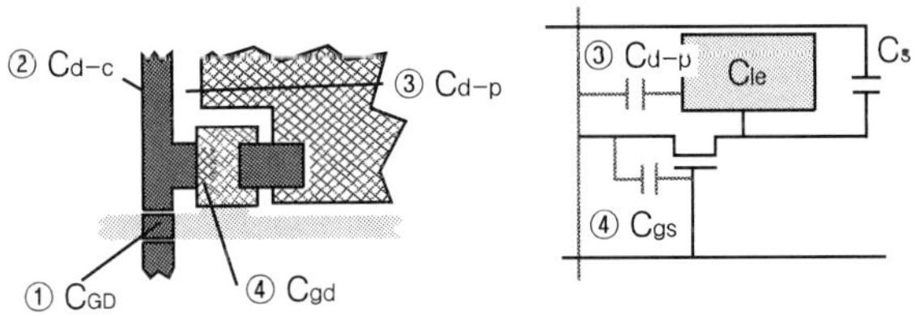

① C_{GD} : Capacitance between Data Signal Lines and Gate Signal Lines (SiN_x)
② $C_{d\text{-}e}$: Capacitance between Data Signal Lines and Common Electrodes (L/C)
③ C_{gs} : Capacitance between Gate and Source of TFT (SiN_x)
④ $C_{d\text{-}p}$: Capacitance between Data Signal Lines and Pixel Electrodes (L/C)

Figure 3.31
Load Capacitance of Data Signal Lines

Production Methods of TFT-LCDs

The manufacturing process of TFT−LCDs mainly consists of four parts, i.e., TFT process, color filter process, cell process and module process.

3.1 TFT Process

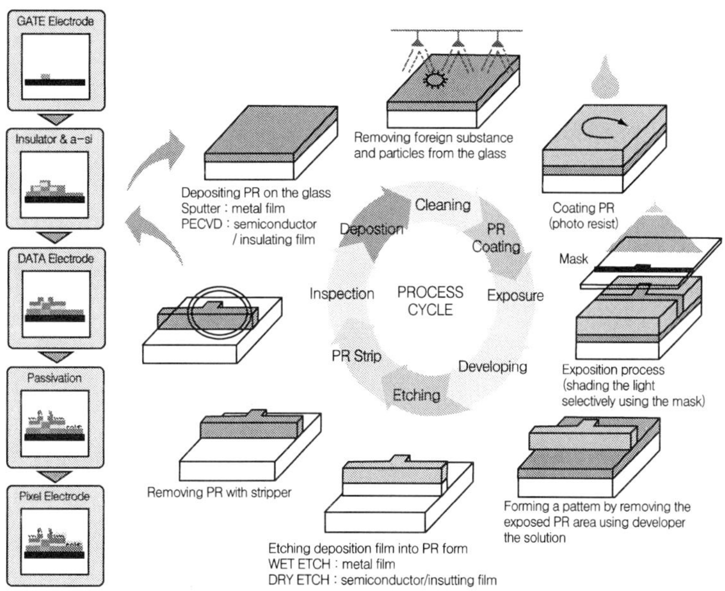

Figure 3.32
TFT Process

The TFT process is very similar to semiconductor manufacturing process, arranging and generating a TFT on the glass substrate by repeating deposition, cleaning, photolithography, etching and inspection. However, glass is used for the TFT process instead of the wafer and thus, the process temperature of 300 ~ 500°C needs to be maintained, whereas the process temperature of the semiconductor manufacturing process is about 1,000°C. Thus, the TFT technology could be more difficult than the semiconductor technology.

(1) Cleaning

Cleaning is basically the process to prevent defects by eliminating any contaminations or particles that could be left on the surface of TFT or the substrate before/while the whole process is undergoing. Further, the cleaning process also aims to enhance the adhesiveness of the thin film to be deposited and its TFT characteristics. In a broad view, this process could also cover the process eliminating a sensitive film followed by the etching process.

Possible forms of contamination during the LCD manufacturing process include ionic contamination such as Na^+, residual metal and semiconductor films generated during the process, and particulate contamination due to handling or equipment. There may be film–like contamination formed on the surface of an existing film such as an oxide film or an organic film. The cleaning methods used for the LCD manufacturing process can be divided into physical cleaning, chemical cleaning and dry cleaning.

(2) Deposition

For the deposition of thin films in the manufacturing process of TFT–arrays, the sputtering method is used for metal films and transparent electrodes, and PECVD (Plasma Enhanced Chemical Vapor Deposition) is used for Si and insulating films.

The basic principle of the sputtering method is that Ar gas is injected between the target made of metal with a high voltage and the anode electrode and then Ar^+ is excited by plasma discharge. If the moving energy of Ar^+ accelerated by the high voltage is stronger than the combined energy of the metal atoms, the atoms can be detached from the surface of the metal. The sputtered atoms consume their energy by inter–collision and interference and combine themselves on the surface of the glass substrate, growing in the form of a thin film.

For depositing metal films, magnetron sputtering using DC plasma is generally used. The magnetic field, which comes out of the diode structure with magnets installed around and the back of the source target, isolates the secondary electrons that keep Ar from being excited. This makes it easy to adjust the plasma, increasing the deposition efficiency. Meanwhile, reactive sputtering is used to deposit compounds or oxidant films in the form of layers, by injecting reactive gas into the reactor while the target is being deposited, which causes the sputtered atoms and the injected gas to react with each other.

The principle of PECVD deposition equipment is as follows: electrons that are excited by plasma decomposes neutral gas compounds by colliding with them, and then re–combine the gas compounds by the interaction of the formed gas ions and the heat energy from the glass substrates, by which thin films are formed and grow. It shows the reactions at each stage, depending on the position of the electrode and substrates in an actual PECVD reactor; the first reaction is decomposition of the gas compounds of plasma flown in from the electrode. The second reaction is the interaction of the decomposed gas ions and excited ions. The third reaction is the formation of a thin film after a nucleus is generated by the interaction of atoms produced by the recombination of the gas ions and excited ions on the glass substrates.

The gas compounds to be introduced vary depending on the type of the film to be formed; normally, SiH_4 and H_2 are used for the deposition of a–Si:H film. For the deposition of an insulating film such as SiN_x, mixed gas of SiH_4, H_2, NH_3 and N_2 is used. PH_3 is added for depositing n^+ a–Si:H films doped with phosphorus (P). The characteristics of a thin film to be formed are determined by rf power, deposition temperature, gas inflow and the distance between the electrode and substrate.

(3) Photolithography

Photolithography is a process of transcribing the patterns on a mask on the glass surface with deposited thin films, which is basically the same as the developing process of general photos. The main processes are Photo Resist (PR) Coating, Align & Exposure and Development. The manufacturing processes of TFT–LCD, the productivity of which should be considered important, are performed by automated in–line facilities.

(4) Etching

Etching is a method of implementing the actual thin film patterns by removing thin films selectively in accordance with the patterns formed by PR on the glass substrate, using physical and chemical reactions. The area of the thin film where the patterns are formed remains, but the area without PR is removed. There are two etching methods, i.e., Dry Etching using gas plasma and Wet Etching using chemical solutions. Once the etching process is done, PR Strip process is carried out by an organic solvent.

(5) Inspection

In TFT–LCD manufacturing process, it is very important to perform an inspection after each unit process is finished to check the completeness of each unit process and the electrical and optical characteristics of the products. Such inspection contributes to finding any defects and their causes and understanding/improving the characteristics of the products, which would reduce the production costs and improve the quality eventually. Inspection processes roughly comprise the assessment of the thickness of deposited thin film and etching, checking on the pattern formation, the assessment of electrical properties and so forth.

(6) TFT Panel Manufacturing Process

Currently, most of the LCD manufacturers use a TFT with the inverted staggered bottom gate, which is relatively easy to produce and does not require formation of the light shield layer of TFT. Such inverted staggered TFTs can be divided into a back channel etch structure and etch stopper structure, depending on the channel formation process.

1) Etch–Back(E/B) Type

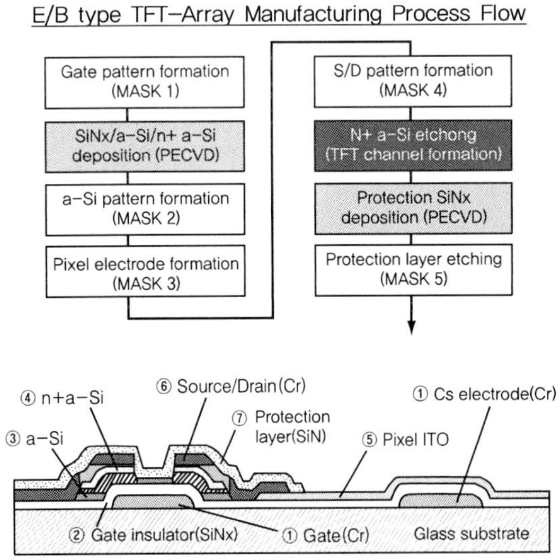

Figure 3.33

Etching in the manufacturing process of E/B type TFT–Array

The n^+ layer is deposited on the active layer of a TFT and source–drain electrodes are formed. The drawback is that amorphous silicon has to be thick since the active layer (made of amorphous silicon) needs to be etched to remove the n^+ layer completely. The thicker amorphous silicon is, the more sensitive to light it gets, which causes leakage current by light.

2) Etch–Stopper(E/S) Type

After depositing the nitride layer on the amorphous silicon, the nitride layer is patterned. The n^+ layer is deposited on the patterned nitride layer and the source–drain electrodes are formed. The nitride layer on the channel side prevents amorphous silicon from being etched when etching the n^+ layer. Unlike the back channel etched type, the thickness of amorphous silicon can be determined while deposition is in the process. However, the patterning of the nitride layer has to be done outside of the reactor. Although the oxidation rate of amorphous silicon is small, the interfacial characteristics are influenced by thinly formed oxide film and impurities.

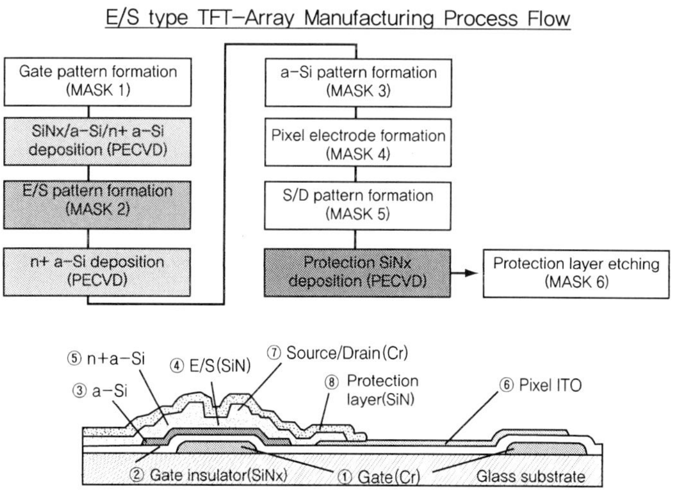

Figure 3.34

Deposition of Protective SiN_x Layer in the manufacturing process of E/S type TFT–Array

3.2 Manufacturing Process of Color Filter

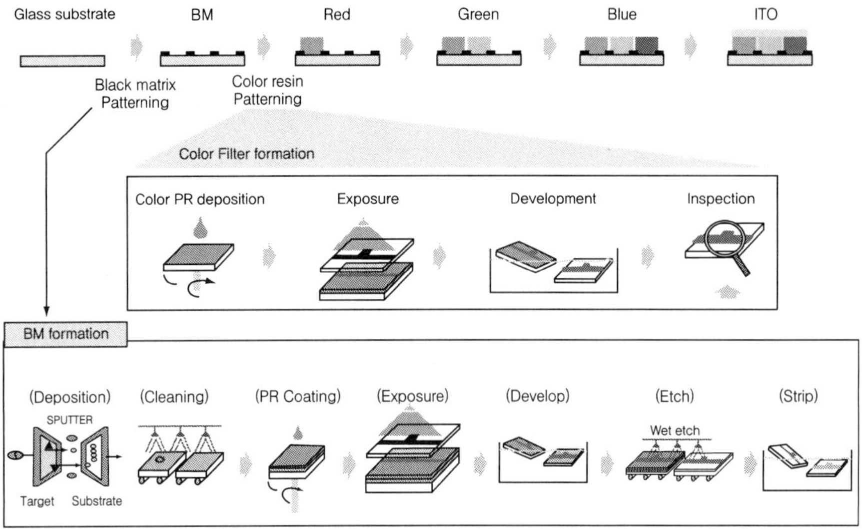

Figure 3.35
Manufacturing Process of Color Filter

The color screen of TFT–LCD is displayed by a TFT adjusting the transmittance of the white light of the backlight, the actions of liquid crystal cells and the additive color mixture of the three primary colors that have penetrated the color filters of Red, Green and Blue. There are two types of color filters, dye type and pigment type, depending on the material of the organic filter used in the manufacturing. Also, color filters can be categorized into dyeing method, dispersion method, electrochemical deposition, a printing method, etc, depending on the production method. The most commonly used method of manufacturing color filters is pigment dispersion method.

The C/F substrate consists of color filter patterns that realize colors, Black Matrix that separates R.G.B cells from each other and blocks light, and the common electrode (ITO) that applies a voltage to liquid crystal cells. Black Matrix is normally placed in between RGB patterns of the color filter for the purpose of shielding Reverse Tilted Domain which is formed around the pixel electrode and the part with no pixel electrode. Also, Black Matrix blocks direct light radiation and subsequently prevent the increase of the leakage current of TFT. In other words, in a TFT with staggered structure, light is blocked at the channel part consisting of the bottom gate electrode and Black Matrix of the color filter.

For the materials of Black Matrix, metal thin films with the optical density of 3.5

or higher, such as Cr, or carbonaceous organic materials are used. The Black Matrix with double layers of Cr/CrO_x is also used for the purpose of low-reflection. Pigment grains generally are untransparent scattering light, but they can transmit light when the size of grains is smaller than the wavelength of light. Thus, the smaller the grains are, the more transparent they are, which enhances dispersal characteristics.

The color filter substrates with low-reflection are made through Black Matrix formation using Cr/CrO_x, color filter formation by pigment dispersion method and the common electrode formation.

Just as in the manufacturing processes of TFT-arrays, once the glass substrates are cleaned, Cr/CrO, Black matrix(B/M) material is deposited by reactive sputtering. Since the formation of B/M patterns is directly related to the aperture ratio, the structural margin to block disclination line by the reverse tilted domains, prevention of leakage current by reflected light and the assembly margin of liquid crystals processes should be considered. The part except TFT portion, gate and data wiring is opened. CrO_x is generally etched by Cr etching so Cr/CrO_x should be etched at the same time.

Once BM patterns are formed, RGB patterns to display colors should be formed with the forementioned photolithography technology, with one difference of using color resists for photoresists. Normally RGB patterns are formed by shifting an identical mask. The application of color resists is done with a spin coating method, which consumes more amounts than printing method does, but is consistent in thickness. For the exposure, proximity method is used. As photopolymerization reaction by the exposure will stop once it reaches a certain level, a post exposure bake should be carried out to maintain the reaction by heat. For development, dipping, puddle, shower or spray method is used. Color resists normally bear negative PR and thus, the part that is not exposed will be removed.

Once RGB patterns are formed, the pixel electrode on the bottom of the TFT substrate and the common electrode for driving liquid crystal cells are formed. For this purpose, ITO, which is a material used for transparent electrodes with good transmittance, electric conductivity and chemical/thermometric stability, is deposited by sputtering. The ITO used in the common electrode does not require pattern formation. In some cases, to protect and flatten the RGB patterns, overcoat films are formed with acrylic or polyimide resins before the common electrode is formed. Among the component of a color filter, the reflectivity of red, green and blue is

1.1%, 1.3% and 0.3%, respectively. These rates are the measurements of specular reflection and have consistent values regardless of the diameter of the light source. The transmittance of a polarizer and glass is 39.2% and 85.2%, respectively.

3.3 Cell Process

The cell process flow is shown in Figure 3.36. The alignment layers are formed on the upper substrate with a color filter and TFT bottom substrate and liquid crystals are properly the alignment layers. Then spacers are distributed and the cell is combined with seal printing. Thereafter, the liquid crystals are injected into the cell by a capillary phenomenon and the inlet is capsuled to complete the LCD process.

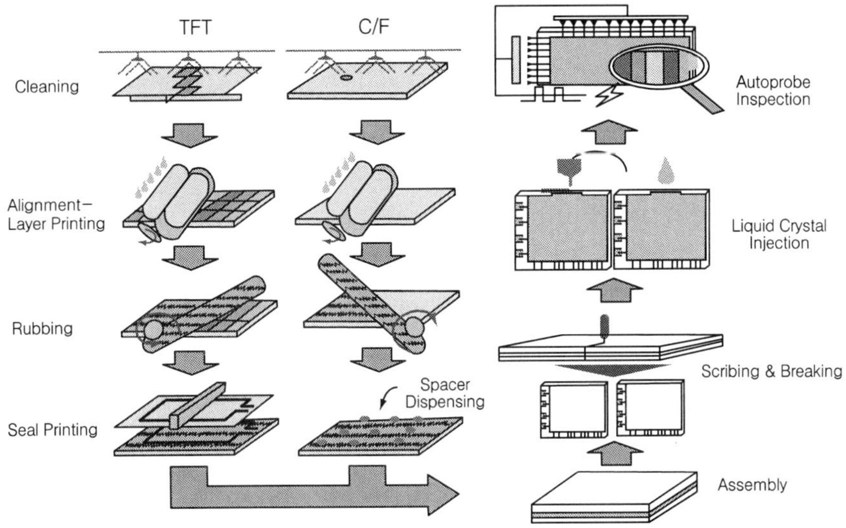

Figure 3.36
Cell Process Flow

(1) Application of Alignment Layer

This process is to apply a thin and consistent alignment layer on the pixel part of the upper and bottom substrates. Once polyimide is applied thinly and consistently on the drum, the polyimide is printed to the patterned rubber plate and then applied to the color filter of the upper substrate and the TFT array of the bottom substrate. Thereafter, the polyimide is hardened in a hardening furnace.

(2) Rubbing

Rubbing is to make straight furrows on the hardened polyimide in a certain direction, so as to have liquid crystals oriented in the direction. Such furrows are made by rubbing the patterns in a certain direction, with the rubbing cloth attached to a rotating drum.

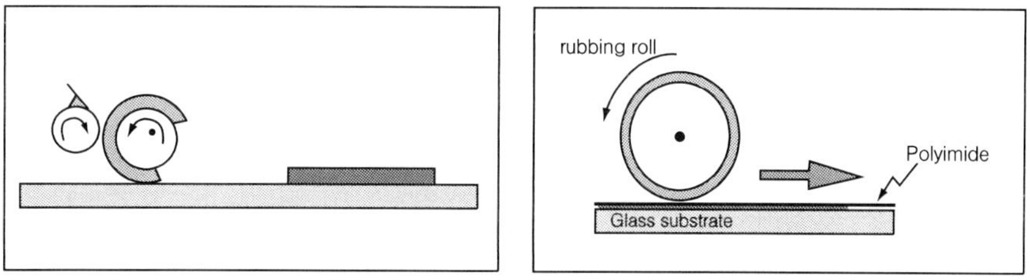

(a) (b)

Figure 3.37
Rubbing Process

(3) Spacer Distribution

This process is to spray spacer evenly, to secure a certain cell gap when sealing the upper and the bottom substrates. The spacer is mixed in a solution, transported through a pump to the distributing nozzle, and sprayed on the substrates from the nozzle at high pressure. The solution is evaporated by heat dry, and the proper density of spacer is about 100~200 spacers/cm^2, depending on the screen size. A spacer is ball-shaped with the diameter of 4~5 μm.

(4) Injection of Liquid Crystals

The next process is injection of liquid crystals between the upper and the bottom substrates, which have been sealed, using a capillary phenomenon and pressure difference. The equipment for this process consists of a vacuum exhauster, vacuum reactor, up-down moving system and a jig to which liquid crystals are soaked and N_2 purge system. When the internal space of the cell gap maintains the vacuum of 10^{-3} Torr by vacuum exhaustion, due to a capillary phenomenon, liquid crystals are sucked up into the cell. Once the cell is filled with about 80% of liquid crystals, slowly purge N_2 into the vacuum chamber, by which liquid crystals will fill the remained space of the cell, due to the pressure difference between the inside and outside of the cell.

3.4 Module Process

This is the process where the quality of the product is eventually determined. A module production is completed by attaching polarizers to the finished panel, populating Driver-IC, assembling PCB (Printed Circuit Board) as well as the backlight and the other components.

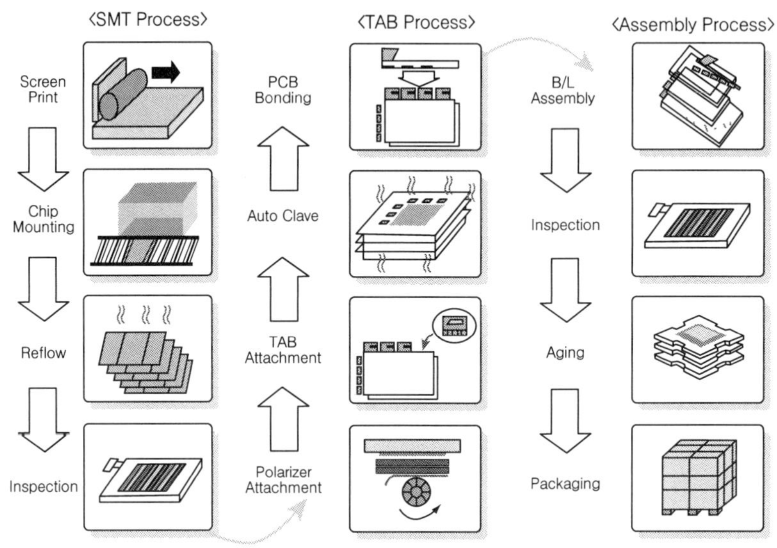

Figure 3.38
Schematic Diagram of Module Process

4 Technical Trend and Perspective of TFT-LCDs

4.1 Securing Price Competitiveness

It is important to secure price competitiveness of TFT-LCDs by simplifying the manufacturing processes, using a single driver circuit, etc. The technology for simplifying the manufacturing processes can be realized by reducing the number of the process masks. Since there are at least 10 to 30 processes linked to one mask, the technology of minimizing the number of masks is very important to save the production costs. Furthermore, if a new liquid crystal injection method called ODF (One Drop Filling) is used, the injection process, which originally takes hours, can be quickly done in minutes: this will help reduce the time for completing the entire processes, contributing to the significant simplification of the processes.

4.2 Demands for Wideview Angle & Fast Response Technologies

PC displays do not require a wide viewing angle since the users mostly see the screen right in front of it. However, the view angle of LCD TVs should be wide, since such products are seen by several people at a time, being in different positions and postures. In addition, fast response technology is needed to display images on larger screens without any pause or cut-off. For this purpose, researches should be continuing on the improvement of the materials used for liquid crystals and the development of a methodology for properly adjusting the circuit voltage, which helps liquid crystals driving faster.

4.3 Development Trend of Glass & Plastic Substrates

There are some tasks to develop in connection with glass substrates: improvement on the formation method, the sophistication and full automation of the manufacturing process, and the glass making technology. More specifically, when a glass substrate is made, it can be overly shrunken or expanded by heat treatment, and it makes a large amount of glass powder while it is being cut. Its polishing process is expensive. Such problems should be prevented or minimized by improving the formation method. Also, in order to prevent the possibility of any specks of dust or scratches on the substrate during the post processes, the manufacturing process should get more sophisticated and fully automated. There will be demands for large glass substrates that are less than 0.5 mm thick for the production of AM-LCDs. Once such issues are solved, we can achieve the goal of producing a thinner, lighter and cheaper AM-LCDs in the future.

Meanwhile, the plastic substrates that have been developed for use until now are polycarbonate (PC), polyimide (PI), polyethersulphone (PES), polyacrylate (PAR), polyethylenenaphthelate (PEN), polyethyleneterephehalate (PET) and so forth. The processing temperature allowed for PI is 275°C, but with colors, PI is not suitable to be used as the substrate of AM-LCDs. PES can be processed under the processing temperature of 200°C, whereas the allowable processing temperature of other plastic substrates is usually below 150°C. However, the processing temperature required for the production of AM–LCDs is higher than 250°C, in order to improve the performance of TFT. Thus, the development of plastic substrates which can endure 250°C will enable the production of AM–LCDs that are light and bendable.

4.4 Lower Power Consumption

The most power–consuming part of LCDs is the backlight lamp unit, i.e., dominating a 60~70% of the whole amount of power consumption. Thus, maximizing the efficiency of the backlight unit is very critical in reducing the power consumption. For this purpose, researches and developments are demanded on the following areas: ① increasing the luminous intensity of the light source, ② developing an inverter that increases the brightness with relatively lower power input and optimized drive frequency for operating a fluorescent lamp, ③ implementing a highly efficient and consistent backlight by developing a diffusing plate with low backscattering and absorbance, and ④ increasing the luminous efficiency by improving the performance of an external aluminum mirror.

4.5 Higher Brightness

Assuming that the total light radiated from the light source is 100%, only 10% of the light eventually reaches our eyes, and the rest of it diminishes while passing through the components inside an LCD, such as a layer–polarizer, color filter, alignment layers, etc. Considering this, the lamp to be used as a light source should be bright enough.

4.6 Thinner/Lighter Product & Lower Production Cost

In addition to lower power consumption and higher brightness, the development of a thinner and lighter LCD will improve its portability. Researches and developments are being accelerated based on the points mentioned.

PRACTICE

1. Explain why most of a–Si:H TFTs are staggered types, whereas MOSFETs and Poly–Si TFTs are coplanar types.

2. List at least more than two advantages and disadvantages of low–temperature polycrystalline TFT–LCDs, in comparison with a–Si TFT–LCDs.

3. Describe advantages and disadvantages of Directive type and Reflective type.

4. Let's say there is a 4bit digital source driver IC. What is the maximum level of gray scale that can be made with this IC? How many combinations of colors would it be possible?

5. Explain the difference between Active matrix and Passive matrix, and describe how to operate Active matrix.

6. Describe the difference between IPS method and VA method. Explain which method is more advantageous.

7. What is the difference between Etch–stopper and Etch–back methods in the manufacturing processes of TFT–LCD panels?

8. List the manufacturing processes of TFTs in a proper order, with a brief explanation of each process.

9. List color filter processes in a proper order, with a brief explanation of each process.

10. Summarize and briefly explain technical trends of TFT-LCDs.

References

[1] Sang Soo Kim, et al. Display Engineering,. Seoul: Chung Chungbum Publishing Co, 2000.

[2] The website of Eng Sung (http://eun-sung.net)

[3] The website of Jun Shin Lee (http://joy.skku.ac.kr)

[4] Strategy Technology Management Institute. *Next Generation Display Market – Technological Analysis*. Seoul: Strategy Technology Management Institute, 2003.

[5] Bob Johnstone. *We Were Burning : Japanese Entrepreneurs and the Forging of the Electronic Age, Vol. 1*. Trans. Park, Jeong Tae. Seoul: Good Morning Books, 2003.

[6] T. Tsukada. *TFT & LCD*, Seoul: Books Hill, 2000. pp.60~63, pp.87~89.

[7] K. H. Lee, J. K. Park and J. Jang, 『High performance polycrystalline silicon thin film transistor using silicon nitride gate insulator』, IEEE Trans. on Elec. Dev. Vol. 45, 2548, 1998.

[8] S. K. Kim, S. I. Cho, Y. J. Choi, K. S. Cho, M. Pietruszko and J. Jang, 『Coplanar amorphouse silicon thin film transistor fabricated by inductively coupled plasma CVD』, Thin Solid Films, Vol. 337, 200, 1999.

[9] S. K. Kim, Y. J. Choi, K. S. Cho and J. Jang, 『Coplanar amorphouse silicon thin film transistor fabricated by inductively coupled plasma CVD』, J. Appl. Phys, Vol. 84, 4006, 1998.

[10] J. H. Choi, C. S. Kim, B. C. Lim, and J. Jang, 『A novel thin film transistor using double amorphous silicon active layer』, IEEE Trans. on Elec. Dev. Vol. 45, 9, 1998.

[11] Chang Soo Kim, et al. 『Boron ion doping effect on the free surface of the back channel etched a-Si:H TFT』, Applied Physics 12(1), 47, 1999.

[12] Jun Shin Lee, et al. *Trends on New Technologies of Thin Film Transistors*. Daejeon: Korean Intellectual Property Offce, 2002.

[13] 『Summary of LCD』, *Monthly Semiconductors*. 1996.

[14] Internet "LC, LCD, TFT" sites

www.eio.com/lcdintro.htm

stefan.www.media.mit.edu/people/stefan/liquid-crystal

tie.kaist.ac.kr/field/organic.html

tftlcd.kyunghee.ac.kr/

[15] Geun Gil Lee and Young Chae Gang. *Physics for Electronic Engineers: Basics & Practice of LCD Electronics*, 1992.

Liquid Crystal

Chapter 03 → 04

1 Liquid Crystal

2 LCD mode

1 Liquid Crystals

In this chapter, we will cover the physical properties and the characteristics of liquid crystal (LC) materials used in displays and their applying principle, at beginner level. Any readers who want to learn the physical properties of LCs in depth, it would be recommendable to read "Handbook of Liquid Crystals (edited by D. Demus et al, Wiley-VCH, Weinheim, Germany)" and "The Physics of Liquid Crystals (P. G. de Gennes and J. Prost, Oxford University Press, New York, USA)." The readers who want to understand the optical principle of LCDs, please refer to "Optics of Liquid Crystal Displays (Pochi Yeh and Claire Gu, John Wiley & Sons, Inc., New York, USA)."

1.1 What is LC?

A LC is a material in mesophase or mesomorphic phase, having both fluidities of liquid and periodicity of solid crystal materials. Since it was first discovered and named "LC" by an Austrian plant physiologist, Friedrich Reinitzer, and a German physicist Otto Lehmann, more than ten thousands of new materials with mesophase have been discovered or compounded until now. LCs in the mesophase, with both the liquid and crystalline nature, have various physical characteristics that are not found in other materials, and a few of these make it possible to use LCs in display applications. Based on such physical characteristics, LCD (liquid crystal display) technology started to be developed in 1970's, and has now become a major technology in the display industry.

A LC mixture consists of stick-shaped organic molecules with the length of 1~2 nm. Such LC is called a calamitic LC, and all the LCs currently used for commercialized LCDs belong to the calamitic LC. Various other types of LCs have been synthesized and studied, including a discotic LC, a bent-core LC, and a bimesogenic or dimer LC, but those have not been used for commercialized LCDs, although discotic LCs are sometimes used as a solidified film for a compensation film of TN (Twist nematic) LCDs. In this book, we will focus on calamitic LCs.

To help readers' understanding of LC materials, one of the well-known materials called 5CB is described in Figure 4.1. A calamitic LC is the combination of the central group, comprising a few benzene linkages or cyclohexane, the terminal group connecting the central group (X and Y in Figure 4.1) and the lateral groups. Each of the chemical structure, flexibility, and electron affinity play an important role in determining the physical characteristics of the LC state.

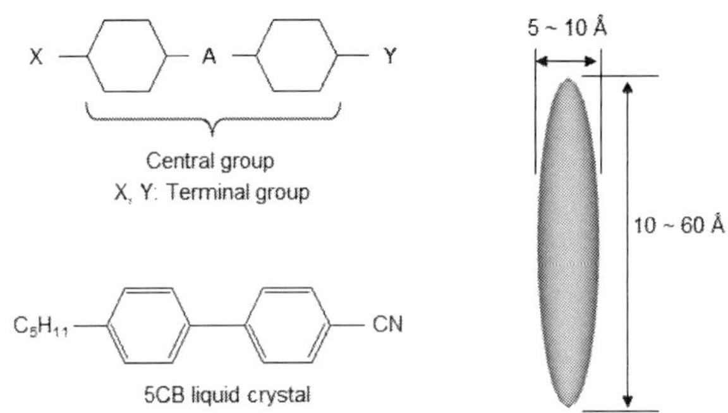

Figure 4.1

General structure of the calamitic (left top) example (left bottom) and size (right)

The arrangement of molecules in LCs have various orderings, depending on the type and temperature of the material. The lowest level of ordering is that molecules only have directional ordering, without spacial ordering. That is, as described in Figure 4.2(a), the calamitic molecules are arranged in one direction, which is called a nematic LC. A higher level of ordering can be found in a smectic LC with a layer structure, as shown in Figure 4.2(b). Due to the layer structure in the smectic LC, molecules are isolated in each layer, which has spacial ordering. Also, the molecules in each layer are arranged in a certain direction, so the smectic LC has directional ordering at the same time. However, the display devices using smectic LCs have rarely been commercialized, with only a few exceptions, and nematic LCs are mostly used for commercialized LCDs.

One of the major reasons to use nematic LCs in displays is the stability of molecular alignment. The initial alignment of nematic LCs is easily recovered even if they are distorted by an external force. On the other hand, smectic LCs which have spacial and directional regularities cannot be easily restored to the original state once their initial alignment is disordered and thus, it is difficult to use them in displays.

Some LCs intrinsically have twisted structure. Such peculiar structure occurs when a chiral dopant is added in nematic or smectic LCs. Nematic LCs with twist structure is called a chiral nematic or cholesteric LC and smectic LCs with twist structure is called a chiral smectic LC.

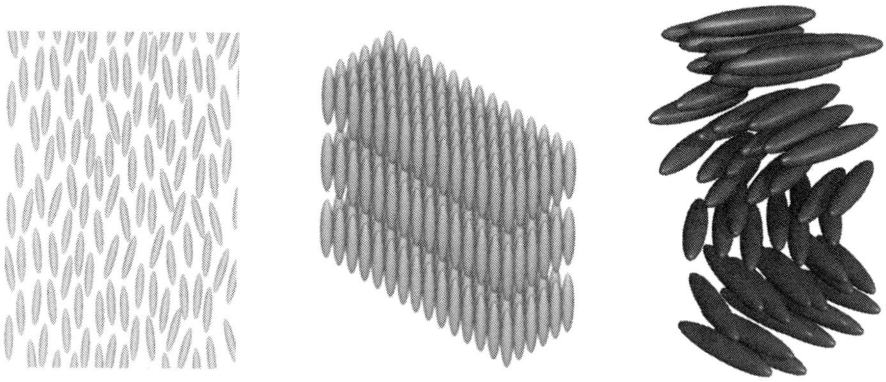

(a) Nematic liquid crystal (b) Smectic liquid crystal (c) Cholesteric liquid crystal

Figure 4.2

Mimetic Diagrams of the Arrangment of Nematic and Smectic LC

To quantify the degree of the LC arrangement, order parameters should be defined. The order parameter (S) of nematic LC can be expressed as a simple formula, as follows:

$$S = \frac{1}{2}\langle 3\cos^2\theta - 1\rangle \quad \text{(Equation 4-1)}$$

Here θ means the angle between the direction of the long axis of LC molecules and the average arrangement direction of LC molecules, and $\langle\ \rangle$ means the average values measured for all of the LCs in the observation area. The order parameter of general nematic LCs ranges from 0.5 to 0.7. The order parameter gradually decreases when the temperature increases, and at a certain temperature, it suddenly turns 0, losing its spontaneous ordering. The temperature is called Phase Transition Temperature. When the temperature is higher than this point, LCs show characteristics of the disordered liquid, i.e., isotropic state. The LCs whose arrangement change depends on the temperature are called thermotropic liquid crystals, which are currently used for all of the display devices.

As explained above, LCs have the characteristics of both liquid and solid crystals, at the same time. That is, LC molecules are relatively free in motion like liquid, have liquidity, and thus LCs easily respond to external stimulation. At the same time, LC molecules are spontaneously arranged with directional ordering like solid crystals, which causes them to be anisotropic.

Thanks to such contrary characteristics, i.e., the spontaneous and regular arrangement of molecules and the facile modulation of the arrangement depending on the external force or the boundary conditions, LCs are widely used in various optical devices including displays. Several important material properties of LCs that play an important role in display applications will be explained from now on.

1.2 Alignment of LCs

When the liquid crystal is injected between two glass substrates and then placed between two polarizers, complex patterns called Schliren texture is observed through a microscope as shown in Figure 4.3(a). Because the LC molecules do not have a consistent alignment direction, despite the spontaneous arrangement, the LCs on the substrates are arranged in all different directions. Figure 4.3(a) describes the alignment direction of the LC molecules. The molecules are aligned along one direction in a narrow area, but when we observe a large area, it shows random alignment. Such LC cell with random molecule arrangement cannot be used as an optical device and thus, all the molecules on the substrates have to be arranged in a certain direction. A process of arranging the molecules in the desired direction is called alignment process.

LCs respond sensitively to external stimulation, as explained above. The external stimulation includes the stimulations to the boundaries of a container of LCs. Thus, if we do treat surface properly on the boundaries contacting LCs, all the LCs in the LC cell can be aligned along a certain direction. In this regard, the average arrangement direction of thousands of LC molecules in the adjacent area is called a director. The tilting angle of the director near the surface from the substrate is called a pretilt angle.

The alignment used in LCDs can be broadly classified into planar alignment (or homogeneous alignment) and vertical alignment (or homeotropic alignment), which are described in Figure 4.3. Planar alignment is to align LCs with a pretilt angle of less than 10 degrees, which will be almost parallel to the surface of the substrates (Technically, planar alignment should refer to the case when the pretilt angle is 0 degree, and when the pretilt angle is larger than 0 degrees, it is called tilted alignment. However, it is common to refer to the pretilt angle of less than 10 degrees as planar alignment in the relevant industry and thus we follow this practice in this chapter). On the other hand, when a pretilt angle is larger than 80 degrees, almost perpendicular to the surface of the substrates, the alignment is called vertical alignment. There is also tilted alignment with a pretilt angle between that of the planar alignment and the vertical alignment, but this is not used for the commercialized LCDs at the moment. When both the two substrates are aligned horizontally but the azimuthal angle direction of the molecule arrangements is vertical to each other, it is called twist alignment. Meanwhile, two substrates could also have different alignment from each other. The case where one substrate has horizontal alignment and the other one has vertical alignment is called hybrid alignment.

In order to obtain uniform alignment of the liquid crystal to the substrate surface, an alignment layer is used. Polyimide type of polymers are commonly used for alignment layers, and the thickness is usually about 700 ~ 1000 nm. The printing method is usually used to coat polyimide or poly–amic acid solutions on the surface of the substrate and applying proper heat treatment to form a solid polyimide layer.

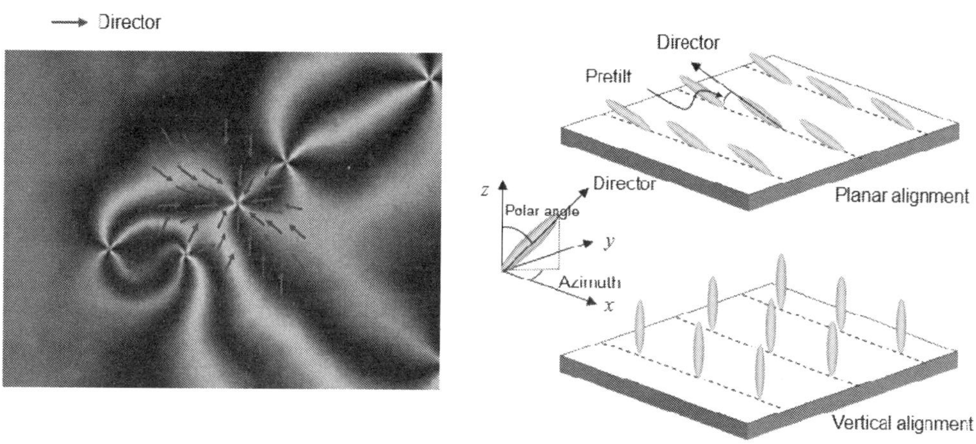

(a) Schliren texture

(b) Planar and vertical alignment

Figure 4.3

(a) Texture by irregular arrangement of LCs, (b) Planar and Vertical Alignment

The alignment of molecules on the surface of the substrates can be indicated as a polar angle and an azimuthal angle, as shown in Figure 4.3. Depending on the value of polar angle, which is the subtraction of the pretilt angle from right angle, the alignment is classified as planar or vertical alignment. The adjustment of a polar angle of the LC alignment is normally determined by the characteristics of the alignment layer material. More specifically, the pretilt angle is determined by the molecule structure, length, and density of the side chain attached to polymers used in the alignment layer. In the case of vertical alignment where the pretilt angle is large, it is likely that a long Alkyl chain is attached densely. For planar alignment with a small pretilt angle, it is likely that a shorter chain is used or its density is low. The strength of the molecule alignment on the substrate surface can be expressed as anchoring force or anchoring energy, where anchoring energy means the energy needed to rotate the molecules on the unit surface area by unit angle. The energy can be determined for the polar angle direction and the azimuthal angle direction, independently, which are called polar anchoring energy and azimuthal anchoring energy, respectively. Normally the alignment layer used in LCDs has strong anchoring energy and thus, for general mathematical calculations and computer simulations, the anchoring energy is assumed to be infinite, which is called strong anchoring assumption.

It is possible to lead an LC cell to vertical alignment by just putting the alignment layer with vertical alignment characteristics. Whereas not only the polar angle but also the azimuthal angle should be controlled for planar alignment. There are many ways of adjusting the azimuthal angle of the LC alignment, among which we will focus on the major two methods in this chapter.

The first one is rubbing alignment, which is to rub the surface of the alignment layer in one direction, with a rubbing cloth (Figure 4.4). As the roller covered with a rubbing cloth spins and rubs the surface of the alignment layer, the layer is aligned in one direction, with consistency. Once the rubbing process is complete, the alignment layer would have fine and long grooves as well as polymers arranged towards the direction where the rubbing cloth passed back and forth. The long axis of the LC bordering the rubbed alignment layer will be aligned along the rubbed direction. Normally, after the rubbing process, the alignment layer for planar alignment will have the pretilt angle of about 1~10 degrees, which is determined by the molecule structure of the alignment layer material. Because the required pretilt angle is different depending on LC mode, proper alignment layer materials should be used.

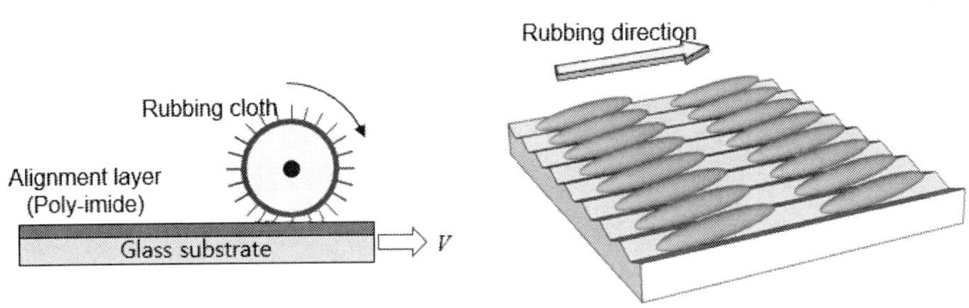

Figure 4.4

Rubbing Alignment Process using a Rubbing Cloth(left) & LC Arrangement on the rubbed layer(right)

The other major alignment method is photo-alignment (Figure 4.5). The alignment layer in the photo-alignment method is usually polyimide type of polymers similar to that of rubbing method. However, the polymers for the photo-alignment method contain light-responsive materials. There are three types of light-responsive materials: Photo-polymerization materials in which photopolymerization occurs by light, photo-isomerization materials in which transition between rod-shaped trans molecule structure and bent cis molecule structure by light, and photo-dissociation material in which polymer chains are decomposed and cut off by light. In general, after applying the polymer, polarized UV light is applied to induce the alignment layer to have anisotropic characteristics. The liquid crystals on the photo-alignment layer are thereafter arranged parallel or perpendicular to the polarized direction.

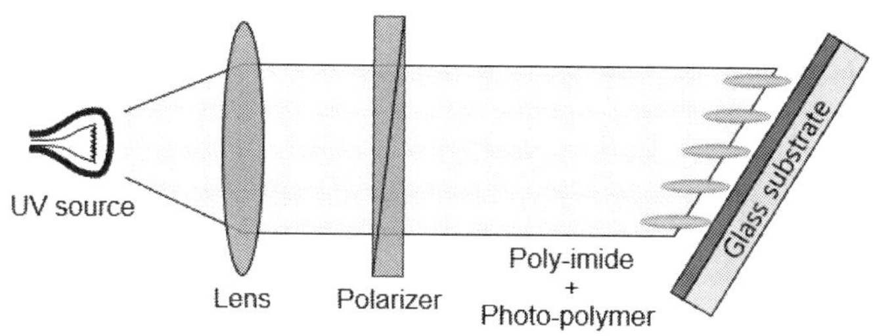

Figure 4.5

Photo-Alignment Method using Radiation of Polarized Ultraviolet Ray

The rubbing alignment method has been used for a long time due to the simple process and the stable material, but as the substrate gets bigger, it is more difficult to perform the rubbing process, and the dust from the process could cause the LCD to have lower image quality. Moreover, because of the various wires formed on the substrates, their surface would be uneven, by which some parts are not properly arranged by the rubbing alignment process. Because of this disadvantage, the photo-alignment method, which is a non-contact technique, has been researched as an alternative technique for a long time; however, the method had not been applied to LCD products, due to the lack of long-term reliability of the LCD panels made with the method. But since 2010, the photo-alignment method has been applied to some products, gradually extending its application range. These days, it is widely used in mobile display such as smart phones.

1.3 Optical anisotropy

Aligned LCs has a birefringence characteristic; that is, the refractive index of the LCs parallel to the alignment direction, n_e, is different from that along the normal direction the alignment direction, n_o. The subscript 'e' and 'o' stand for 'the extraordinary axis' and 'the ordinary axis,' respectively. In rod-shaped LCs, ne is larger than no. The difference in the refractive index of the LCs in the two directions is referred to as anisotropy (Δn) or birefringence.

$$\Delta n = n_e - n_o \qquad \text{(Equation 4-2)}$$

When polarized light passes through a material having birefringence properties, the polarization nature of the light changes, which has become a basis of the optical principle of LCDs.

For those who want to get a deep knowledge of the polarization of light inside LCs, please refer to a book named "Optics of Liquid Crystal Displays (Pochi Yeh and Claire Gu, John Wiley & Sons, Inc.)" as this chapter will treat the basic level of knowledge in that regard.

Light is an electromagnetic wave that oscillates periodically, and its electric field and a magnetic field are vertical to the direction of light propagation. Normally the electric field in natural light is composed of all electric fields vibrating in any direction on the surface vertical to the light direction. Such light is referred to as unpolarized light. However, as described in Figure 4.6, the electric fields of the light that have passed a polarizer vibrate in one direction only; such light is polarized light. Assuming the components of the electric fields towards the two axes x and y on the vertical surface of the light direction (z) are E_x and E_y, respectively, the following equations are given:

$$E_x = A_x e^{i(wt - kz)}$$
$$E_y = A_y e^{i(wt - kz)} \qquad \text{(Equation 4-3)}$$

ω and k represent angular frequency ($2\pi/T$ where T is the period) and wave number ($2\pi/\lambda$ where λ is a wavelength), and ($\omega t - kz$) is called the phase of a wave.

The light vibrating in a diagonal direction, as shown in Figure 4.6, can be decomposed into two electric fields with x and y directions, and each electric field passes through the double refractive LC layer with different refractive indices. After the light passes through the liquid crystal layer with a thickness of d and a refractive index anisotropy of Δn, the electric field components of x and y direction have the phase delay of Δnd. Δnd is called retardation. The two electric fields that have passed the LC layer can be expressed as follows:

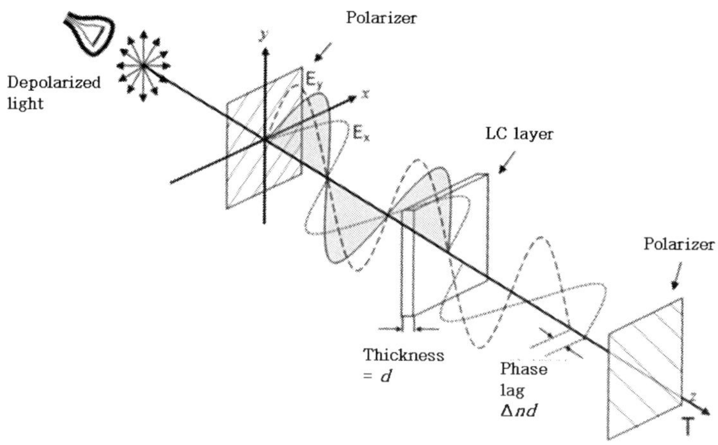

Figure 4.6

Polarizetion of Light & Polarization Changes by Retardation

$$E_x = A_x e^{i(\omega t - kz)}$$
$$E_y = A_y e^{i(\omega t - kz)} \quad \text{(Equation 4-4)}$$

The vector sum of the two electric field components has different forms in response to a phase difference value of the two electric field components. For instances, if Δnd is half of the wavelength (λ), the vector sum will be an electric field 90 degrees rotated from the vector sum calculated before the light has not passed the LC layer. In this way, the polarization of light continuously changes as the light passes through the LC layer.

When the LC layer between two polarizers is aligned uniformly without twist, the transmissivity of light can be defined as follows (the derivation of the formula can be done by Jones Vector or 4×4 matrix calculations, but the details are omitted):

$$T = \frac{1}{2}\sin^2(2\psi)\sin^2\left(\frac{\pi \Delta nd}{\lambda}\right) \quad \text{(Equation 4-5)}$$

where ψ: the angle between the axis of the polarizer and the alignment direction of LCs (Equation 4-5) is a formula that applies only when there is no twist of the liquid crystal, but it is frequently used for the calculation of the transmittance, such as VA LCD, IPS LCD and FFS LCD.

It is possible to adjust the transmissivity of the cell including LCs, by adjusting the alignment direction of the LC layer (ψ) and the retardation value (Δnd); this is generally done by applying an electric field thereon.

1.4 Dielectric Anisotropy

Liquid Crystals are dielectric materials, that is, insulators. When an electric field is applied to a dielectric material, polarization occurs. Polarization can be justified as the vector sum of the unit volume of dipole moments when an atom or a molecule gets to have dipole moments spontaneously or by an external electric field. Normally polarization (P_i) in the axis direction (i) of the dielectric material can be expressed:

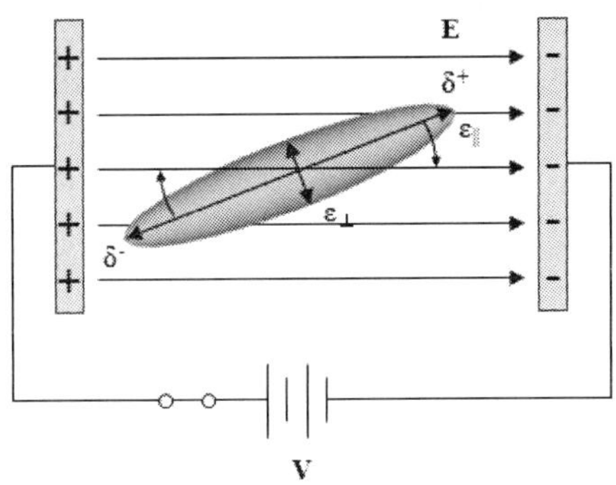

Figure 4.7

Torque of LC(Dielectric Material) by Electric Field

$$P_i = P_{0i} + \epsilon_0 \sum \chi_{ij} E_j + \sum \beta_{ijk} E_j E_k + \ldots\ldots \qquad \text{(Equation 4-6)}$$

The subscriptions represent the axis direction, and E stands for the strength of the electric field of each axis direction. In particular, ϵ_0, χ, and β are permittivity of free space, electric susceptibility, and its higher term, respectively. The first term on the right-hand side of Equation 4-6 represents the spontaneous polarization of the dielectric material. Spontaneous polarization appears under asymmetrical structure such as twisted structure or some part of smectic liquid crystals, but does not occur in general nematic LCs; thus, can be ignored. The higher terms in the electric susceptibility is usually ignorable due to their small values. Thus, the polarization of nematic LCs has the value proportional to the first term of the electric field. Therefore, the polarization (P) of nematic LCs is expressed:

$$P = \epsilon_0 \chi E \qquad \text{(Equation 4-7)}$$

In a material, the displacement field (D) is expressed as the sum of the electric field (E) and the polarization (P), as follows:

$$D = \epsilon_0 E + P \quad \text{(Equation 4-8)}$$

From Equations 4–7 and 4–8, the relationship between the displacement field and the electric field can be expressed:

$$D = \epsilon_0 E + \epsilon_0 \chi E = \epsilon_0 (1+\chi) E = \epsilon_0 \epsilon E \quad \text{(Equation 4-9)}$$

Here ϵ is called a dielectric constant or relative permittivity.

Due to the asymmetrical attribute of LCs, the dielectric constants for the directions parallel and vertical to the alignment axis are different from each other. The dielectric constant for the direction parallel to the alignment axis is generally referred to as $\epsilon_{\parallel}$, and the same for the vertical direction is referred to as $\epsilon_{\perp}$. The difference between them is dielectric anisotropy, which can be defined as follows:

$$\Delta\epsilon = \epsilon_{\parallel} - \epsilon_{\perp} \quad \text{(Equation 4-10)}$$

The LC material with positive dielectric anisotropy is called positive LC, whereas the LC with negative dielectric anisotropy is called negative LC.

In dielectric anisotropic materials, the size of polarization varies depending on the direction of the electric field. As described in Figure 4.7, when the applied electric field does not match the direction of each axis, the strength of the polarization in each direction can be calculated by computing the polarization induced by the electric field for each axis direction, namely, $E_{\parallel}$ and $E_{\perp}$. The electric energy in the dielectric material is expressed as follows:

$$U_{dielectric} = -\frac{1}{2} P \cdot E \quad \text{(Equation 4-11)}$$

Based on the above, the electric energy of the dielectric anisotropic material such as shown in Figure 4.7 can be calculated as follows:

$$\begin{aligned} U_{dielectric} &= U_{\parallel} + U_{\perp} = -\frac{1}{2} P_{\parallel} \cdot E_{\parallel} - \frac{1}{2} P_{\perp} \cdot E_{\perp} \\ &= -\frac{1}{2}\left(\epsilon_0 \chi_{\parallel} E \cos\theta\, \hat{x}\right) \cdot E - \frac{1}{2}\left(\epsilon_0 \chi_{\perp} E \sin\theta\, \hat{y}\right) \cdot E \\ &= -\frac{1}{2}\epsilon_0 \Delta\epsilon E^2 \cos^2\theta + const. \end{aligned} \quad \text{(Equation 4-12)}$$

The electric energy of a dielectric material tends to depend on the angle θ between the dielectric material and the electric field. Therefore, the dielectric material has torque, which is the tendency that the dielectric material rotates toward the lowest point of energy. The torque of the relevant dielectric material can be calculated by differentiating the electric energy by the angle.

$$\tau = \frac{\partial U}{\partial \theta} = \epsilon_0 \Delta\epsilon E^2 \cos\theta \sin\theta \quad \text{(Equation 4-13)}$$

The director of a positive LC gets the torque to rotate parallel to the appled electric field . On the contrary, a negative LC gets the torque to rotate vertical to the electric field; from such characteristic, the principle of electrically modifying the optical characteristics of the LC layer is derived, leading to the operation of LCDs.

1.5 Elasticity and Viscosity

Liquid crystal consisting of numerous molecules is fluidic, despite the considerable interaction force between the molecules. Molecules in LCs have a self-assembly property and are likely to form a assembled structure with the least distortion in alignment. Usually, the least distortion energy is obtained in one directional alignment. Also, if deformation of the arrangement is generated, they tend to restore the initial state with the least distortion. Such restoring force in LC alignment is similar to the restoring force in deformed rubber, which means the storing force of LCs can be described by the mathematical expression used in elastic materials. The theory that considers LCs as continuous elastic materials is referred to as continuum theory.

As explained above, the directors of nematic LCs tend to be arranged uniformly. When an external force is applied to such LCs in the uniform alignment, the LC alignment can be deformed. The alignment state of LCs is described by the director n. The director n is similar to the vector whose size is 1, but the director doens not have positive or negative direction differently from vectors. That is, n will be the same when it is rotated 180 degrees. Occurring of internal deformation of LCs means that the LCs have different arrangement depending on the position, which means that the director is a function of position, that is, n = n(r) (where r is the position vector).

When a spring, a typical elastic material, is extended by a certain length, the restoring force (P) of the spring is proportional to the extended length and determined as follows:

$$P_{spring} = -K\Delta x \quad \text{(Equation 4-14)}$$

K is an elastic coefficient, and χ is displacement. Then the energy accumulated inside the spring is the value obtained by integrating the force required to increase displacement from 0 to χ, which is given as follows:

$$F_{spring} = \frac{1}{2}K(\Delta x)^2 \qquad \text{(Equation 4-15)}$$

Similarly, the internal energy that increases by transformation in the LC arrangement can be expressed to be proportional to the square of transformation. In the case of one dimension (the x axis), the internal energy is defined as:

$$F_{spring} = \frac{1}{2}K\left(\frac{\partial n}{\partial x}\right)^2 \qquad \text{(Equation 4-16)}$$

The restoring force (P) by elasticity can be obtained by partial-differentiating the elastic energy by distance.

$$P_{spring} = -K\frac{\partial^2 n}{\partial x^2} \qquad \text{(Equation 4-17)}$$

In practice, however, elastic deformation occurs in various ways, far beyond the extent that can be described by Equation 4-15. In the case of nematic LCs, the energy by various elastic deformations can be expressed as the combination of the three major elastic deformations, as follows:

$$F = \frac{1}{2}K_1(\nabla \cdot n)^2 + \frac{1}{2}K_2(n \cdot \nabla \times n)^2 + \frac{1}{2}K_3(n \times \nabla \times n)^2 \qquad \text{(Equation 4-18)}$$

This is referred to as Frank's elastic deformation energy density. In equation (18), The equations in the parentheses in the first, second, and the third terms correspond to the splay, twist and bend deformations, respectively, and K_1, K_2, K_3 are their corresponding elastic coefficients (Figure 4.8).

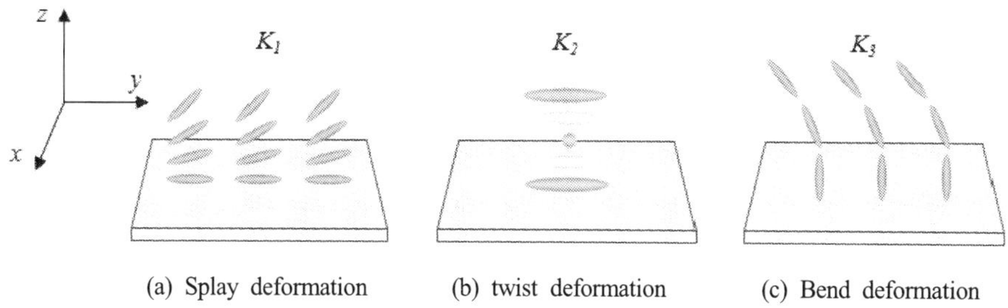

Figure 4.8

Elastic Deformation Types of LCs

The initial elastic deformation in LCDs is determined by the surface confinement conditions; for example, twist nematic (TN) mode has twist deformation initially, and vertical alignmnet (VA) mode and in-plane switching (IPS) mode do not possess any deformation initially. However, most of elastic deformations are induced by an electric field. The elastic deformations consist of the combinations of the three typical deformations shown in Equation 4-18. However, it is very difficult to analytically calculate the elastic deformation due to the complexity of the calculation process and thus, the elastic deformation state is mostly obtained using numerical analysis on the computer.

K_1 and K_3 of general LC materials have similar values to each other, with the K_2 value of about a half of the K_1 and K_3 values, but could vary depending on the molecule structure.

Meanwhile, Equation 4-18 expresses the elastic deformation energy when an external force is applied to LCs, from a statics perspective, i.e., when the system has reached an equilibrium. In order to describe the process in transition to the equilibrium, a theory from a dynamics perspective is needed separately. For this purpose, Leslie and Ericksen suggested a theory to describe the dynamic activities of LCs, i.e., Leslie-Erickson theory. The theory explains various dynamic activities that occur when an external force is applied to LCs, such as flow and rotation of LCs, based on several molecule movement types. In this textbook, however, only part of the theory related to the rotation of the LC molecules, which has a significant influence on LCDs, will be covered.

When external torque (Γ) is applied to LCs, the director of LCs gets to rotate, the angle of which (Φ) changes depending on time. Thus, the differential value ($\dot{\Phi}$) of the angle (Φ) by time is finite, proportional to the external torque.

$$\Gamma = -\gamma_1 \dot{\Phi} \quad \text{(Equation 4-19)}$$

The proportional factor γ_1 is referred to as rotational viscosity.

Rotational viscosity is a force to obstruct the rotation of molecules, slowing down the rotational speed. Thus, LCs with low rotational viscosity are better materials to get the faster speed of response from LCDs.

1.6 Electro-optical Characteristics of Simple LC Cell

By understanding physical properties of the LCs as described above, i.e., (i) LCs orientation by boundary surface in contact with LCs, (ii) the anisotropic refractive index of LCs and polarization modulation of light in LCs, (iii) anisotropic permittivity of LCs and the rotational force of LCs by the electric field, (iv) the elastic deformation and restoration force and rotational viscosity characteristic of the liquid crystal, we can understand all of the basic principle of LCs used in the LCD.

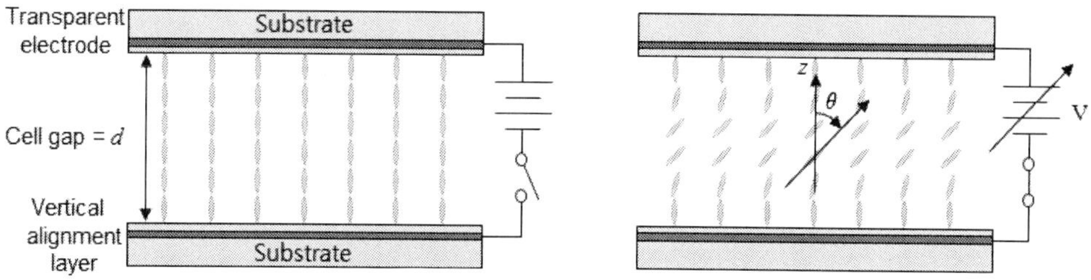

Figure 4.9

Let's say there is a vertically aligned LC cell, as shown in Figure 4.9, which can be made by applying a vertical alignment layer on the surface of two substrates with transparent electrodes, and by stacking the two substrates together with an LC layer in the middle. The LC is a negative LC with negative dielectric anisotropy. Let's assume that the polar angle of the LC is inclined by about one degree in a particular direction from the vertical direction. This is in order to predetermine the molecular tilting direction when the electric field is applied. When a voltage is applied to the two electrodes, the LC layer is experienced the electrical torque by Equations 4–12 and 4–13. However, the LC molecules near alignment layers are fixed strongly to the surfaces. According to the strong surface anchoring energy assumption, the LC molecules on the surface can be assumed to be not moving, unaffected by the electric field applied. However, the LC molecules in the center of the LC cell will be rotated as the electric field increases. Such rotation of LCs by the electric fields causes the molecules in the LC cell to be arranged irregularly, namely, elastic deformation. The elastic deformation, contrary to the stress by the electric field, has the restoring force which tends to restore the original regular alignment. Thus, the director of LCs will stop moving in the equilibrium of the stress by the electric field and the restoring force by the elasticity. This state can be expressed as:

$$F=\frac{1}{2}\int_0^d\left[K\left(\frac{d\theta}{dz}\right)^2-\frac{\epsilon_0\Delta\epsilon}{2}E^2\sin^2\theta\right]dz \qquad \text{(Equation 4–20)}$$

d represents the thickness of the cell, and θ represents the polar angle of the LC. This integral calculus is used, to sum up, all the elastic and electric energies inside the LC cell.

Such equations as Equation 4–20 are generally difficult to calculate and thus, Equation 4–20 can be changed as follows, using the Euler–Lagrange equation:

$$-K\frac{d^2\theta}{dz^2}+\frac{\epsilon_0\Delta\epsilon}{2}E^2\sin\theta\cos\theta=0 \qquad \text{(Equation 4–21)}$$

The first and second terms on the left–hand side show the elasticity and the torque by the electric field, respectively. The director will get arranged at the equilibrium point where the sum of the elasticity and the electric torque equals 0. By finding the value for Equation 4–21, we can calculate how the LC director is distributed inside the cell. However, it is not easy to find an equational value even for such a simple equation. Equation 4–21 would have a complex value called an elliptic equation, which is far beyond the range of the contents to be covered in this chapter. For this reason, in most of the cases, the value is calculated by numerical analysis through computer simulation.

This time, let's take a quick look at how to deal with the dynamic attributes in the same case as Figure 4.9. In order to know the attributes, we have to know the variation by time, for which a term related to its viscosity, i.e., Equation 4–19, needs to be taken into account. By adding the term for viscosity, we can complete the equation for the LC movement by time.

$$-K\frac{d^2\theta}{dz^2}+\frac{\epsilon_0\Delta\epsilon}{2}E^2\sin\theta\cos\theta=\gamma_1\frac{d\theta}{dt} \qquad \text{(Equation 4–22)}$$

By finding a solution to equation (22), it is possible to know the dynamic attributes of LCs but they are also generally obtained via numerical analysis. A solution of equation (22) gives the information about the dynamic attributes with time term which is associated with the liquid crystal response speed.

In order to examine the characteristics of LC cells according to the electric field let's further simplify the cell of Figure 4.9, and let it be graphically visualized as shown in Figure 4.10. Assume that liquid crystal is attached on the substitute by a spring. At this time, the rotational force by the electric field can be derived from equation (10) as follows.

$$\tau_{electric}=\epsilon_0\Delta\epsilon E^2\cos\theta\sin\theta \qquad \text{(Equation 4–23)}$$

On the other hand, the restoring force of the spring, in the simplified structure as shown in Figure 4.10, can be written as follows:

$$\tau_{elastic} = -K\sin\theta \qquad \text{(Equation 4-24)}$$

Considering that the balance of force will come at the point where the sum of Equations 4-23 and 4-24 is 0, $\tau_{electric} + \tau_{elastic} = 0$. From this, we can get:

$$\cos\theta - \frac{K}{\epsilon_0 \Delta\epsilon E^2} \qquad \text{(Equation 4-25)}$$

The cos function on the left-hand side is always less than 1, in order for Equation 4-25 to be valid, the term on the right-hand side should also be smaller than 1. From this, the following result will be drawn:

$$E > \sqrt{\frac{K}{\epsilon_0 \Delta\epsilon}} \qquad \text{(Equation 4-26)}$$

Equation 4-26 means that LCs will not respond to the external electric field at all when the electric field is weaker than a certain degree of strength. The minimum voltage required to drive LCs is called a threshold voltage. From Equation 4-25, we learn that when the electric field increases, θ gradually increases accordingly, from 0 to $\pi/2$, with the voltage higher than the threshold voltage. This can be roughly described in Figure 4.10(b).

From Equation 4-5, we note that the transmissivity of an LC cell increases if the anisotropy of the effective refractive index increases. Thus, if the LC director's angle θ gets larger, the anisotropy of the effective refractive index of the LC increases, causing the increase of transmissivity. Here the anisotropy of the effective refractive index means the difference in the refractive index between the axis with the largest average refractive index and the other axis with the smallest average refractive index when looking down an LC cell from the above. The refractive index becomes large in the direction where the long side of LCs can be seen the best. As such, if the average angle θ gets larger, the anisotropy of the effective refractive index increases.

Now we see that the transmissivity of an LC cell increases when an electric field is applied to the cell. The curve describing the relationship between the applied voltage and the transmissivity is called a V-T curve. In some LCDs, the transmissivity could increase along with the increase of the applied voltage, but in other types of LCDs, it could decrease when the voltage increases, depending on the type of the LC cell. The first is called normally black mode, and the latter is called normally white mode. The major examples of the normally black mode are VA, IPS, and FFS modes, which will be covered in the next chapter, and TN mode is a major example of the normally white mode.

Equations 4–23 to 4–26 expressing the operating principles of LC cells, based on the very simple LC cell shown in Figure 4.10, are somewhat different from the equations practically used in LCD design, due to the oversimplification. The equations in practical use will be explained, following the descriptions of LC mode in the next chapter.

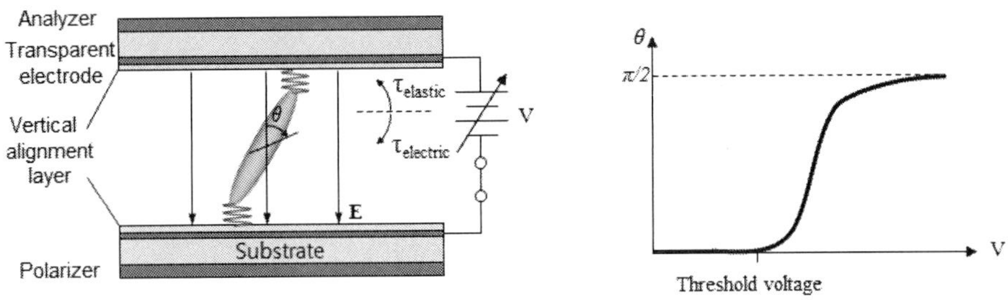

(a) Deformation & Restoring Principle of LC (b) Change in Average Polar Angle by the Applied Voltage

Figure 4.10

Besides the V–T curve, such attributes as response speed and viewing angles are also important in relation to the electrooptical characteristics of LCDs. The response speed of LCs is an index showing how much short time it takes for LCs to change their arrangement when an electric field is applied to the LC cell; normally, when the transmissivity of light passing through an LC cell is transferred from the early transmissivity to the final transmissivity, the time taken from a 10% of transfer to a 90% of transfer is defined as response speed. In connection, the times taken for LCs to rise up and lie down are named the rising time and falling time, respectively. When the applied voltage increases, it is called ON time, whereas when the voltage decreases, it is called OFF time.

The viewing angle characteristic generally refers to the state when the side image is worse quality than the front image. To quantify this the viewing angle characteristics based on contrast ratio is often used. If the contrast ratio, which is the ratio between a dark gray–scale of 0 and a bright gray–scale of 255, is lower than 10 when viewed from side directions, the angle of which is a viewing angle in comparison to the contrast ratio. Normally when the gray-scale of a display increases, the display should get brighter. However, in the case of some LCD displays, when viewed from side directions, bright gray-scales seem darker and dark gray-scales seem brighter, i.e., gray-scale inversion occurs. The side angle at which the gray-scale inversion occurs is called a gray-scale reversal angle, which is sometimes used as a quantitative index for the viewing angle characteristic. However, viewing angles in comparison to the contrast ratio and gray-scale reversal angles may be somewhat too limited to quantitatively show the viewing angle characteristic of the recent high-definition LCDs. Such standard was used for the LCDs with poor image quality in the 1990's; instead, the method of comparing the Gamma curves of the front view and the side view has been in use in recent years to determine the image quality of the front view and the side view.

2 LC mode

2.1 Twisted Nematic (TN) Mode

TN Mode has been used for various commercialized products due to its excellent characteristics to be used for displays since it was suggested by Schadt and Helfrich in Switzerland in 1971, because of its excellent characteristics to be used for displays such as simple structure and manufacturing process, high transmissivity, fast response speed, etc.

Currently, TN mode is used for most of laptops and monitors except for high-end laptops, and electrical appliances as well as small electronic devices. The structure and operating principle of the optical device adopting TN mode are shown in Figure 4.11. On the inner surface of the upper and lower glass substrates, transparent ITO electrodes are placed, on which rubbed planar alignment layers are covered. At this point, the rubbing direction of the upper and lower substrates is vertical to each other; normally the rubbing directions of the upper and lower substrates are applied to +45° and -45°, respectively, based on the horizontal axis from the front view. The LC molecules near the substrate surfaces are arranged parallel to the relevant rubbing direction, being fixed by anchoring energy regardless of the applied voltage. The nematic LC layer between the two substrates has a twisted structure as the optic axis rotates 90 degrees continuously from the lower substrate to the upper substrate. When polarizers are attached on the upside and downside of the TN mode panels, the transmissivity of light can be expressed by the Gooch–Tarry's formula as in Equation 4–27.

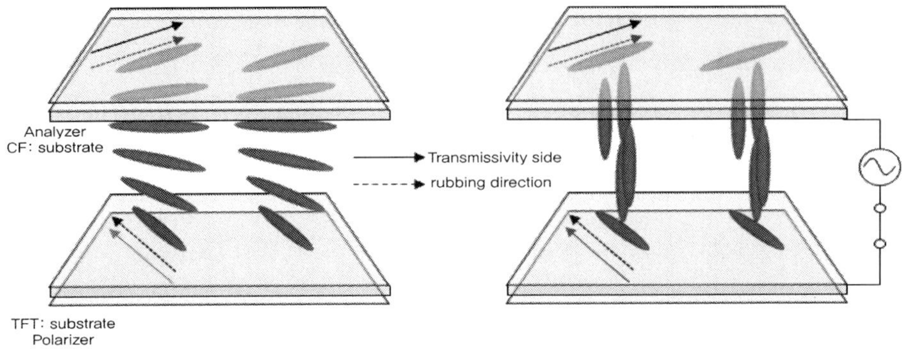

Figure 4.11

$$T = 1 - \frac{\sin^2\left(\frac{\pi}{2}\sqrt{1+u^2}\right)}{1+u^2}, \text{ where } u = 2d\frac{\Delta n}{\lambda} \qquad \text{(Equation 4–27)}$$

Here Δn, d, λ represent the double refraction, the thickness of the LC layer, and the wavelength of the incident light, respectively.

Equation 4–27 describes the transmissivity when LCs are twisted which are signified as a voltage is not applied yet. When the value from the sine function on the right–hand side becomes nπ (n = natural number), the minimum value condition of transmissivity is determined. From the function, the smallest positive number u, allowing the minimum transmissivity, should be given as $u = \sqrt{3}$ and it means dΔn = 476 nm for light with the wavelength of 550 nm. However, when an electric field is applied to the LC cell, LCs rise up vertically to the substrates, as shown in Figure 4.11(b), finally it makes the actual value of u gets closer to 0 and the transmissivity of Equation 4–27 becomes one. In this way, if polarizers are attached on the upside and downside of the TN cell parallel to each other, the TN cell would have the minimum transmissivity when no voltage is applied, i.e., dark state, and with a voltage applied, it would get brighter, which is called normally black mode. It is hard for a TN cell in the normally black mode to have a high contrast ratio (= the highest transmissivity / the lowest transmissivity) and thus, this mode is not applied to commercialized products. Meanwhile, if polarizers are attached on the upside and downside of the TN cell perpendicular to each other, as shown in Figure 4.11, the TN cell gets bright when a voltage is not applied, and it gets dark when a voltage is applied; this mode is called normally white mode, the transmissivity of this mode is expressed as:

$$T=1-\frac{\sin^2\left(\frac{\pi}{2}\sqrt{1+u^2}\right)}{1+u^2} \qquad \text{(Equation 4-28)}$$

As shown in Figure 4.11(a), when the applied voltage is 0V in the normally white mode, the incident light from the backlight is polarized linearly (linear polarization) while passing through the lower polarizer, but fully penetrates the upper polarizer after being rotated 90 degrees through the LC layer, which makes the cell bright. However, when the voltage applied to the LC layer is high enough, the LC molecules are arranged vertically, as shown in Figure 4.11(b). In this case, the incident ray passes through the LC layer without any change in its polarization, failing to penetrate the upper polarizer, which makes the cell dark.

The distribution of the director of LC depending on the applied voltage can be described by a polar angle (θ) and an azimuth (ϕ), as in Figure 4.12. In the inactive state (where V=0), all LC molecules are arranged parallel to the substrates and thus, the polar angle is zero and the azimuth is continuously twisted 90 degrees from the bottom to the top, i.e., −45 degrees to +45 degrees. As the applied voltage increases, the LCs in the middle rotate in a vertical direction, increasing the polar angle, but the LCs on the surface of the substrates get fixed due to the high anchoring energy.

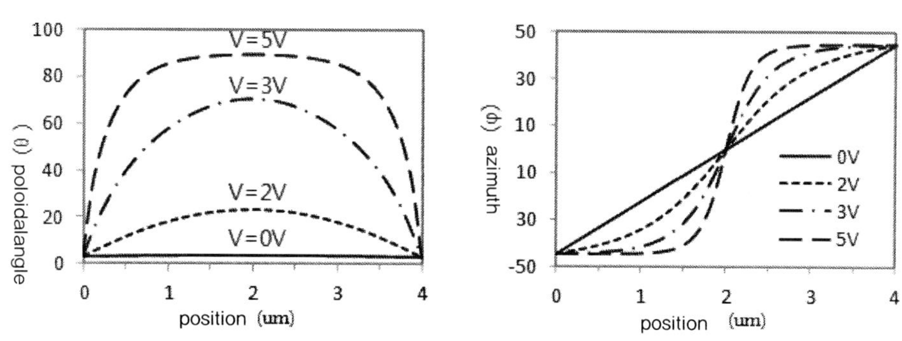

Figure 4.11

Changes of Polar Angle and Azimuth of Director depending on the Applied Voltage in the TN LC Layer

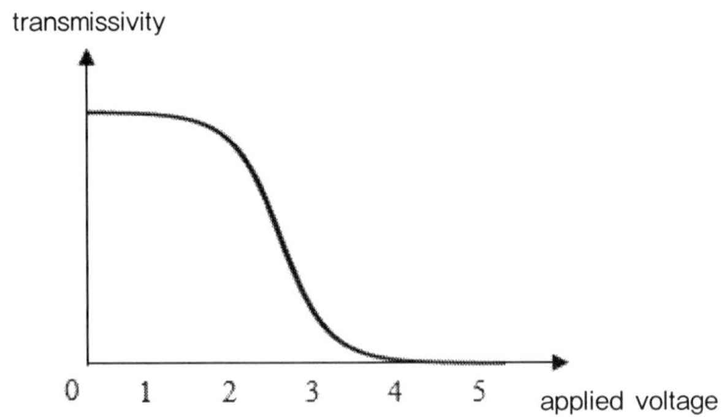

Figure 4.13
Transmissivity depending on Applied Voltage in TN Cell

The light transmittance curve in accordance with the applied voltage in the normally white state is shown in Figure 4.13. We see the highest

transmissivity in the state without an electric field, and as the applied voltage gets gradually higher, the transmissivity decreases. When the applied voltage is higher than 5V, the transmissivity becomes zero, bringing the dark state.

When the transmitting axis of the polarizer is parallel to the long axis of LCs (e–axis) in the normally white TN mode, as shown in Figure 4.11, is called e–mode. Conversely, when transmitting axis of the polarizer is parallel to the short of LCs (o–axis) is called o–mode. The optical characteristics of the front view in the two optical modes are the same to each other, but their viewing angle characteristics could be different from each other; normally, e–mode is more commonly used.

Figure 4.14
Image Distortion depending on the Viewpoint of TN LCD

The viewing angle characteristics of TN mode are described in Figure 4.14. The left–right viewing angles are wider and symmetric, whereas the up–down viewing angles are asymmetric; especially in the up viewing angle, gray–scale inversion appears as if we see the film, as the brightness of high gray–scales and that of low gray–scales are reversed. This is basically because of the twisted structure of the LC molecules – due to this structure, effective refractive index changes depending on the viewing direction. Accordingly, the retardation of the incident ray changes. To realize a wide viewing angle such optical anisotropy needs to be offset, for which a compensation film including the discotic LC layer is used. The compensation films used for TN modes are called wide–view films. The viewing angle is greatly improved with the use of compensation films, but the asymmetry of the up–down angles cannot be overcome completely; thus, TN LCDs are generally not used for high–end display products.

TN modes normally show faster responsive speed than other LC modes. This is because the LC with positive dielectric anisotropy tends to have low viscosity, the strength of the electric field applied to the LC is quite strong, and such LC has a pretilt angle of approximately five degrees.

2.2 Vertical Alignment (VA) Mode

VA mode, which was suggested by Schiekel and Fahrenschon in Germany in 1971, adopts vertical alignment layers and negative LCs with negative anisotropy that tends to arrange vertically to the direction of the electric field. This is discernable from the planar alignment layer used in TN mode or IPS mode (which will be introduced in the next chapter) or the LCs with positive anisotropy (positive LCs). The operating principle of VA mode is described in Figure 4.15. When V is zero, all LC molecules are arranged vertically, and the linearly polarized incident ray proceeds along with the optic axis of LCs, resulting in no polarization change. Thus, the upper polarizer blocks all incident rays, regardless of the wavelength or cell–gap, making the VA cell dark. As there is no residual retardation by a pretilt angle, the cell looks very dark especially from the front view and thus, the VA mode can get much higher contrast ratio than any other LC modes. When a voltage is applied to the middle of the upper and lower substrates of a VA cell, the long axis of the LC molecules with negative dielectric anisotropy is tilted toward the direction vertical to the electric field direction. Accordingly, the light penetrating the LC layer is retarded by the refractive anisotropy of LCs, and the strength of the retardation is adjusted by the applied voltage. As we discussed in the previous chapter, the transmissivity of the light penetrating the LC layer can be expressed as:

$$T=\frac{1}{2}\sin^2(2\psi)\sin^2\left(\frac{\pi\Delta n_{eff}d}{\lambda}\right) \quad \text{(Equation 4-29)}$$

The transmissivity varies depending on the effective retardation, which is calculated by the LC azimuth component (ψ) on the flat surface parallel to the substrate and $\Delta n_{eff}d$. In the case of VA mode, LCs rotates with the applied voltage in the direction vertical to the substrates and thus, ψ is fixed as 45 degrees. However, the transmissivity is modulated by the last term of Equation 4-29, i.e., $\sin^2(\pi\Delta nd/\lambda)$. In order for a VA cell to have the highest brightness, the value for $\Delta n_{eff}d$ needs to be multiplied by and odd number of integers of the length of half wavelength ($\lambda/2$); in practice, Δnd of a VA LCD, which is the multiplication of the refractive anisotropy of the LC material by the thickness of the LC layer, ranges from 300 to 310 nm. The reason why the Δnd value is higher than the half of the mean-wavelength of visible ray, i.e., 550/2 = 275 nm, is that even if the highest voltage is applied, the actual value of effective retardation would be lower since the LCs near the substrates remain standing up or being tilted.

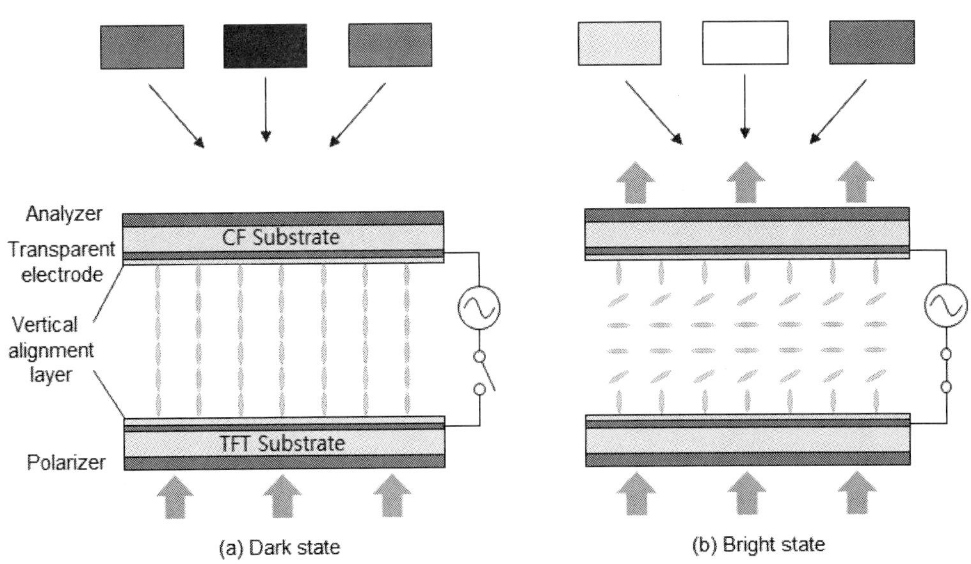

Figure 4.15
Structure and Operating Principle of VA LC Cell & Change of Brightness by Viewing Direction

The effective refractive index of LC layers in VA mode, similar to TN mode, changes by much depending on the viewing direction and thus, VA mode has a viewing angle issue that the brightness becomes different from the side view; especially in moderate gray-scales, the gray-scales look asymmetric from the left-right viewing angle. Thus, to apply VA mode for the products requiring a wide viewing angle such as TV, the viewing angle issue has to be overcome. Normally, so as to realize VA mode with a wide viewing angle, it is required to employ compensation film technologies that help to offer the consistent dark state depending on the viewing angle and multi-domain technologies that change the direction of which LCs lie in each domain.

For the compensation film used in VA mode, biaxial films are used, which can be produced by orienting polymer films. Biaxial films are used for VA mode, as they are cheaper and easier to manufacture than the compensation films for TN mode using discotic LCs. Meanwhile, a usual polarizer consists of a PVA (polyvinyl-alcohol) film with the polarizable iodine compounds arranged thereon and two TAC (tri-acetyl-cellulose) films attached to both the front and back of the PVA film; however, the polarizer comprising the compensation film for VA mode consists of a PVA film with a TAC film on one side and a compensational biaxial film on the other side (closer to the LC layer) and thus, the thickness of the polarizer or its production costs rarely changes compared to other polarizers.

The multi-domain technologies allow an equal viewing angle from every direction by offering the average gray-scale of all of the divided domains when viewing from the side. The major domain division methods are Multiple-Domain Vertical Alignment (MVA) and Patterned Vertical Alignment (PVA). MVA is to properly arrange protrusions on the ITO surface of the upper and lower substrates, as shown in Figure 4.16, by which the LC molecules are tilted vertical to the protrusions when a voltage is applied. By placing protrusions properly in a pixel, the LC molecules can lie down evenly in four directions at an angle of 45 degrees from the axis of the polarizer. However, in the state without an electric field, the LCs on the surface of the protrusions are tilted, which could cause a slight light leakage and lower the contrast ratio. The technology called Advanced MVA creates more domains MVA LCD's and reduces the variation of transmittance in oblique angles. It helps improving colour washout and provides better image quality in oblique angles than conventional VA LCD's.

Unlike MVA, PVA uses a patterned ITO for multiple domains. In the state without an electric field, all LC molecules are arranged vertically, like the normal VA mode. Thus, there is no retardation by the tilting of LCs, and the highest contrast ratio is available. When a voltage is applied, an electric field is generated between the upper and lower substrate, the line of electric force of which is slightly bent by the patterned ITO, as in Figure 4.17; it is called Fringe field. The LC molecules in each domain tend to lie in the direction vertical to the boundary of the domain, based on which the multiple–domain structure is formed.

As explained above, the VA technologies using compensation films and multi–domain technologies provide much–improved viewing angle characteristics. However, in comparison with IPS technology (which will be covered in the next chapter), which is the most competitive technology among VA product–groups, the former is inferior in terms of viewing angle characteristics. This inferiority is caused

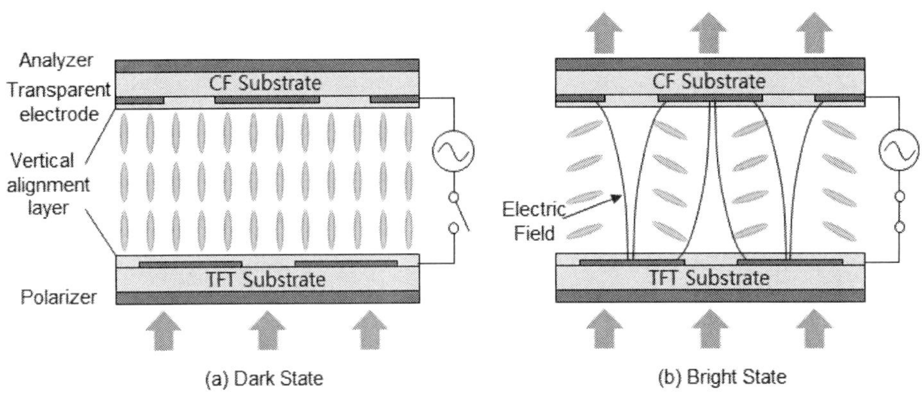

Figure 4.16
The Structure of MVA LC Cell: LC arrangement of (a) Non–electric Field State (b) Electric Field State

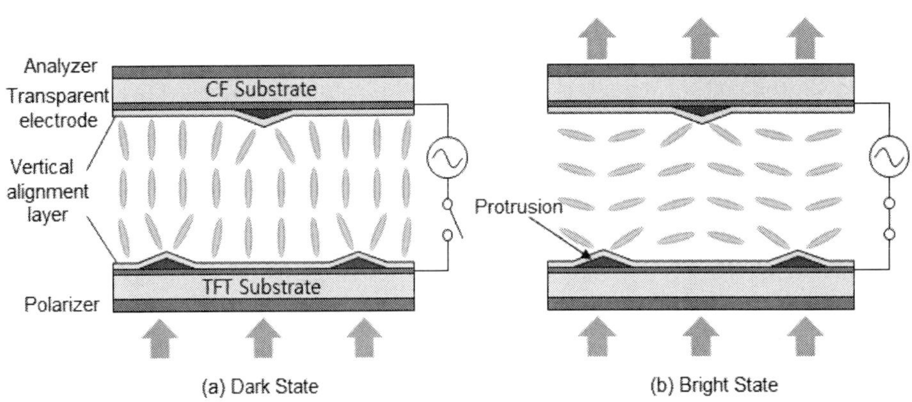

Figure 4.17
The Structure of PVA LC Cell: LC arrangement of (a) Non–electric Field State (b) Electric Field State

by the dissimilarity between the gray-scale curve of the front view and that of the side view. That is, the distortion in the gray-scale curve by the viewing angle lowers the image quality of the side view in the multiple-domain VA mode, fading the images. To improve the side visibility of VA mode, new technology has been developed, which is called SPVA mode technology. According to this technology, a pixel is completely divided into two different pixels, for which two different TFTs are used or two separate embedded circuits are introduced, by which different strength of voltages is applied to each of the divided pixels. In SPVA mode, there are two divided pixels with different applied voltages, and each pixel has four domains. Thus, pixels have eight domains in total, unlike the pixels in PVA or MVA with four domains. Therefore, SPVA shows greatly improved side viewing angle characteristics compared to PVA or MVA technologies.

Recently, to improve the productivity and transmissivity of VA mode, PSVA (polymer stabilized VA) mode technologies are widely used. The LCs used in PSVA mode are a mixture of VA LCs with negative dielectric anisotropy and a small amount of reactive mesogen (RM). When ultraviolet ray is incident to such light-reactive chemical, the RM molecules fix some LC molecules to the surface of the substrates through polymerization. Also, the pixel electrode used in PSVA mode is in the form of fish-bone; the width and the interval of the fish-bone shaped pixels are about 3~4 μm.

The LC alignment process of PSVA is as follows: first, a mixture of LCs and RM is filled in between the two substrates, and an electric field is applied to both ends. At first, the LC molecules lie down randomly, but as time passes by, they are slowly arranged towards four different directions, along with the fish-bone shaped pixel electrodes (as shown in Figure 4.18). If the ultraviolet ray is applied while maintaining the electric field, the RM molecules mixed in LCs are fixed to the surface of the substrates, through polymerization. In the meantime, some LC molecules, being tilted towards the electrode direction by the applied electric field, get fixed to the substrates together with the RM molecules, with a pretilt angle of about 0.5 ~ 2.0 degrees, this remains even after the electric field is removed. The LC molecules in the panel where a pretilt angle is formed lie down towards each domain direction at fast speed forming a multiple-domain LC cell, when the electric signals of each pixel are applied. The alignment in PSVA mode is similar to the photo-alignment method in that a rubbing process is not needed and ultraviolet ray is used for the alignment;

however, the photo-reactive chemical is mixed with LCs in PSVA mode, not included in the alignment layer. Also, the radiation method of ultraviolet in PSVA mode is simpler than other photo-alignment methods, which allows high productivity.

The greatest merit of PSVA mode is that each domain has a pretilt angle in the same direction without separate structure and thus, the length of the boundary areas dividing the domains is shorter than that of other VA modes. Thanks to the transmissivity of PSVA mode that is a 10 ~ 20% higher than that of other VA modes. Another merit is fast responsive speed. A pretilt angle of LCs reduces the ON/OFF responsive speed, which will be useful for the products requiring high-speed driving. The high-speed driving technology is applied to the technologies to improve video displays as well as the time-sharing stereoscopic 3D technologies.

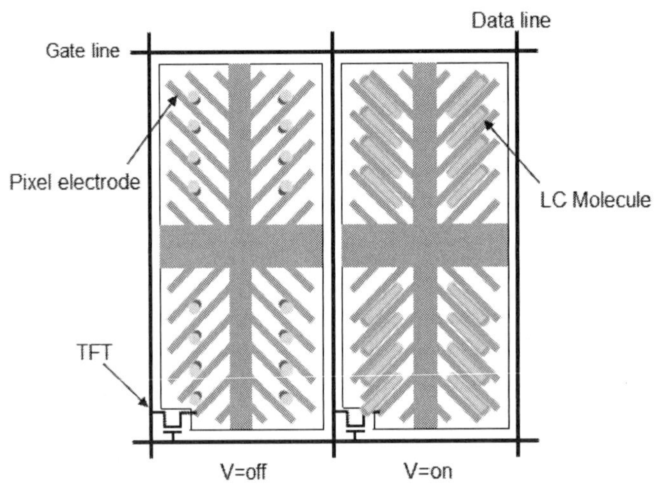

Figure 4.18
Structure of PSVA Pixel and LC Arrangement Direction by Applied Voltage

The response of VA mode is very fast between a low gray-scale and a high gray-scale, but is slower between low gray-scale. To address this issue, DCC (dynamic capacitive compensation) technology, which improves the responsive speed through image processing, is being applied to most of VA mode products.

To sum up the characteristics of VA mode, it has a wide viewing angle and high contrast ratio. As it does not require a rubbing process, it is good for the manufacturing of large-sized panels. Also, the manufacturing process of VA mode is very similar to that of TN mode and thus, it offers good productivity. However, side view distortion of images is the biggest weakness of VA mode, for which many technologies have been developed. In addition, VA mode with fast ON/OFF response is suitable for time-sharing stereoscopic 3D technologies and thus, many time-sharing 3D VA products have been produced.

2.3 In-Plane Switching (IPS) Mode

TN mode or VA mode controls the rotation of LCs by forming signals and the common electrodes on the upper and lower substrates and applying a vertical electric field. Unlike this, the LC mode controlling the polarization of light by forming the electrode only on the lower substrate and rotating LCs parallel to the substrates (in-plane) with a lateral electric field is called In-Plane Switching (IPS) mode.

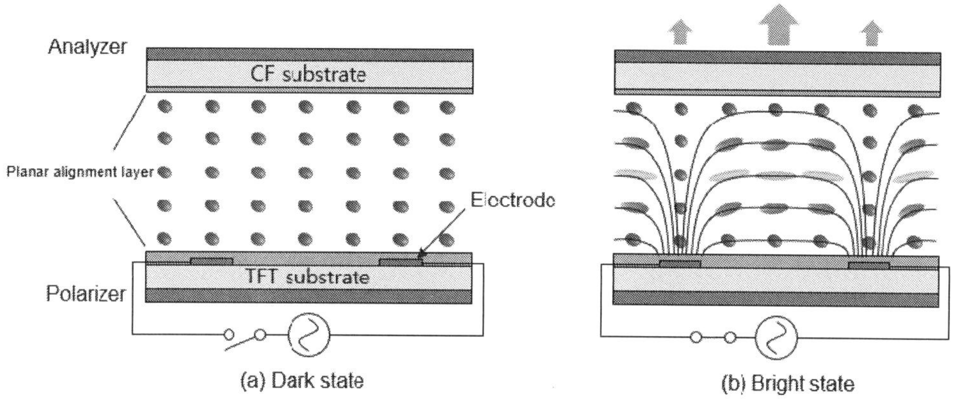

Figure 4.19
LC Arragement under Non-Electric Field State (a) and Electric Field State (b) in IPS Mode

The LC cell structure of IPS mode and the behaviors of LCs by an electric field are described in Figure 4.19. The internal surfaces of the upper and lower substrates are coated with planar alignment layers, being rubbed in the opposite way to each other, to arrange LC molecules on the surface. The pixel electrode and the common electrode for applying planar electric fields are formed in a interdigitated shape, and the long axis of the electrodes are tilted 7 to 15 degrees from the rubbing direction. The rubbing direction of the substrates and the penetrating axis of the lower polarizer are parallel to each other or cross at right angles. When there is no voltage applied, the optic axis of LCs are arranged in parallel to the rubbing direction. The incident light entering through the lower polarizer proceeds without changing its polarizing direction and can not pass through the upper polarizer, making the LC cell dark state. When a voltage is applied to the IPS cell, the LC molecules with positive dielectric anisotropy rotate in the direction of the electric field applied, and accordingly, the polarizing direction of the light penetrating the LC layer is modified. The rotation of LCs by the electric field is done on a flat surface parallel to the substrates, which can be expressed as changes of the azimuth (ψ) in Equation 4–29. Thus, unlike VA mode, the transmissivity of IPS LCs is adjusted by the term, $\sin^2(2\psi)$, on the right in Equation 4–29. When a high voltage is applied, LCs rotate following the direction of the planar electric field, with the effective azimuth angle of 45 degrees, resulting in the maximum brightness.

The angle between the electrode direction and the rubbing direction plays a key role in determining the sufficient strength of the electric field for having LCs rotate 45 degrees against the polarizing axis. Normally, the products aiming to operate with a low voltage are designed to have a smaller angle between the electrode direction and the rubbing direction, so that the maximum brightness is realized at a lower voltage. The products requiring a fast response are designed to have a larger angle between the electrode direction and the rubbing direction, so that they have higher brightness and faster response, although the necessary voltage is higher. If the rubbing angle becomes higher, the V_T curve becomes gentle, weakening the degree of stains that could be generated due to the inconsistency of the relevant electric field.

Unlike TN or VA modes, liquid crystals in IPS mode show a twist–elastic distortion in a direction parallel to the surface of the substrate, not rising up in the vertical direction, when a voltage is applied. Thus, there is no difference in image quality regardless of the viewing direction, unlike TN or VA modes, where the image quality is low depending on the viewing direction, as either the head or body part of the liquid crystals is seen. VA mode is an LC mode adopting wide viewing angle technology, whereas IPS mode was invented basically as a wide viewing angle LC mode. The merit of IPS is the excellent side image quality.

Figure 4.20(a) shows the electrode structure and liquid crystal movement of a basic IPS mode using a single domain, which is viewed from the above. The LC molecules are arranged parallel to the rubbing direction in the inactive state, and then get rotated on the parallel surface when a voltage is applied between the signal and the common electrode. However, if LCs rotate in one direction, the retardation effect appears asymmetrically between the left and the right view and thus, each color (wavelength) moves to its opposite side, causing color changes. To compensate for this disadvantage, S-IPS mode with a V-shaped chevron electrode structure, as shown in Figure 4.20(b), was developed. The direction of the electric field on upside and downside of the point where the electrode is bent is different from each other, and thus domains are formed in a way that the LC molecules in each area can rotate in the opposite direction to each other. Thus, the color change from a wide view can be compensated by each domain, which helps to improve the side visibility compared to a normal IPS mode.

Also, IPS shows an excellent wide-viewing angle characteristic without any compensation film. Thus, IPS mode without compensation film has a merit in terms of the production expenses. However, high-end IPS mode requires a compensation film to minimize a light leakage from the side at low grayscale. If compensation film is not applied, there will be a noticeable light leakage when viewing dark images from the side. The best compensation film suitable for the high-end IPS mode is an anisotropic film whose refractive index in the vertical direction to the film is the average value of the refractive indices of two axes on the film surface. However, such compensation film, which is called Z-plate, is expensive to produce and thus, is used for very limited occasions. Meanwhile, various polarizing films which are cheaper for IPS have been researched and developed, despite their limited light compensation ability.

Another advantage of IPS mode, compared with VA mode, is that the arrangement of LCs is relatively less affected by an external physical impact. In VA mode where LCs are arranged vertically when the screen is on impact, because of the LC molecules tilt in a moment and get back to the original state, the screen flashes. However, in IPS mode, image quality hardly changes despite such impact. This advantage could be a great merit in developing an LC panel embedded with a touch panel. However, in most cases, the touch panel is separately equipped with tempered glass, and thus any LC panels, no matter what the LC mode is, are endurable against external impact.

One of the weaknesses of IPS mode is low transmissivity due to low aperture ratio. As shown in Figure 4.19(b), the LCs between two adjacent electrodes can be easily twisted by a lateral electric field, but the LCs right above the electrodes cannot be twisted due to the weak lateral electric field. Such LC molecules do not twist in the direction parallel to the substrate, but tilt in the direction vertical to the substrate, and the LC layer in this part does not contribute to the transmissivity. Due to this, the pixel electrodes in IPS mode are mostly non-transparent metal electrodes, which induce low aperture ratio and eventually lead to the transmissivity 50% less than that of TN mode.

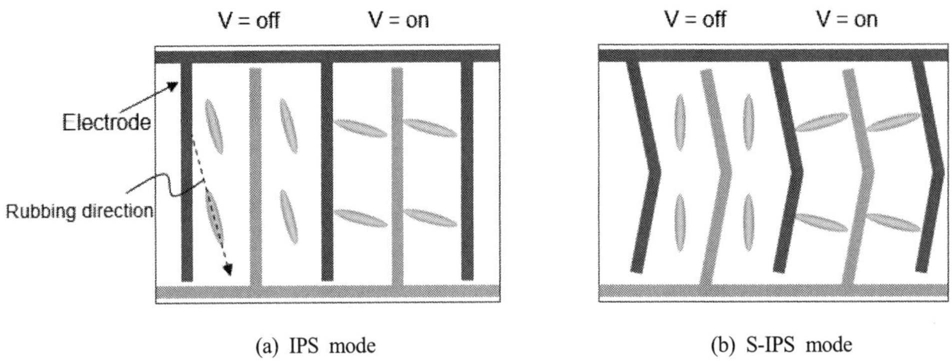

Figure 4.20
(a) Electrode Structure of Normal IPS Mode (b) Electrode Structure of S-IPS Mode

The contrast ratio of IPS is another weakness that has been studied intensively. VA mode in which the rubbing process is not required and the initial arrangement is vertical to the substrate has very high contrast ratio, but IPS mode in which the rubbing process is required and LCs arranged parallel to the substrate has relatively low contrast ratio. The rubbing process leaves rubbing scratches on the substrate, around which light can be leaked. Also, uneven electrodes formed on the substrate can cause the rubbing effect to be inconsistent. There are many cases where light leakage occurs due to the insufficient rubbing in the part adjacent to the electrode. Also, IPS LCs arranged horizontally scatter light by the dynamic shaking of the LC molecules, by which they are depolarized, causing light leakage by the light scattering. Such depolarized light scattering also occurs in VA mode. It is not significant enough to affect the contrast ratio. However, the depolarized light scattering is a major reason for reducing the contrast ratio in IPS mode. Overall, the highest contrast ratio of IPS mode is known as 1500:1, which is only one-tenth of the highest contrast ratio of VA mode. Although a contrast ratio of 1500:1 does not pose any problem with displaying high-definition images, it could be uncompetitive in displaying dark images.

The response speed of IPS mode has both advantages and disadvantages, compared to VA mode. The response speed of IPS mode is relatively consistent between any grayscales. However, in VA mode, the response speed between a low grayscale and a high grayscale is very fast, whereas the response speed between low grayscales is slow. Thus, the consistent response speed can be an advantage of IPS mode. However, VA mode with fast response speed between a low grayscale and a high grayscale is useful in driving high-speed frames or time-sharing 3D technology, which are hard to apply to IPS mode with relatively slow response speed between a low grayscale and a high grayscale. Various technologies came out from initial IPS, such as S-IPS, H-IPS, e-IPS, p-IPS, AH-IPS. These technolgies tried to improve response speed and contrast ratio. (http://www.tftcentral.co.uk/articles/panel_technologies.htm#intro)

2.4 Fringe Field Switching (FFS) Mode

In FFS LC mode, the electrode is generated only on the lower substrate like IPS mode. But the contrast is that the electrode and the counter electrode are generated on the same surface in IPS mode, whereas the pixel electrode and the common electrode are generated on different layers in FFS mode. In Figure 4.21, the pixel electrode is created on the pixel across the board, above which an insulating layer is deposited, and then a comb-shaped common electrode, similar to that of IPS mode, is generated on the insulating layer. For the pixel and common electrodes in FFS mode, transparent electrodes (ITO) through which light can penetrate are used, and the pitch of the common electrodes is approximately 4.5 μm, which is narrower than the gap between the pixel electrode-the common electrode in IPS mode, i.e., 8 μm to 10 μm. For the alignment layer, a planar alignment layer is used like IPS mode.

When a voltage is applied to the pixel electrode of an FFS cell, a fringe field is generated between densely and repeatedly formed pixel electrode and the common electrode, and the strength of the fringe field is much stronger than that of IPS where the distance between the electrodes is wider. Similar to IPS mode, LCs in FFS mode rotate in the direction parallel to the substrate surface, due to a strong lateral electric field generated around the boundary between the common electrode and the pixel electrode. In addition, LCs above the electrode rotate in a similar way, due to the elasticity of the adjacent LCs, and thus all the LC molecules in the pixel rotate being parallel to the substrate surface. In IPS mode, the LCs above the electrode do not rotate due to the insufficient lateral electric field, causing low transmissivity, but in FFS mode, light penetrates the entire pixel, and thus the transmissivity is higher than that of IPS mode.

Also, the common electrode is generated on the very top of the TFT substrate in FFS mode, and the electrode is usually covered with data bus line. Normally around the data bus line of an LCD, LCs show abnormal behaviors, being affected by the signals of the data bus line, and light leakage occurs accordingly. In order to prevent this, a light blocking film called BM (black matrix) is generated. The width of BM is generally 20 μm to 30 μm, which causes a lower aperture ratio of the pixel. However, there is no light leakage around the data bus line in FFS mode, and thus the width of BM can be very narrow. This can be a critical factor in raising the aperture ratio of high-resolution products. The pixel and common electrodes generated on the lower substrate form a storage capacitance (Cst), without any additional structure. Such structure also helps FFS LC mode have a high aperture ratio. The relatively shorter distance between the signal and the common electrode, compared to that of IPS mode, also helps LCs in FFS mode drive with a relatively low voltage. Such attributes as a wide viewing angle, a high aperture ratio, and the operation with low voltage play a key role in FFS mode used for small/mid-sized display devices requiring low power consumption and high resolution, e.g., high-end smart phones, tablet PCs, laptops, etc. Meanwhile, FFS mode is not suitable for large display devices, since the high storage capacitance of FFS mode increases the RC delay, decreasing the driving margin.

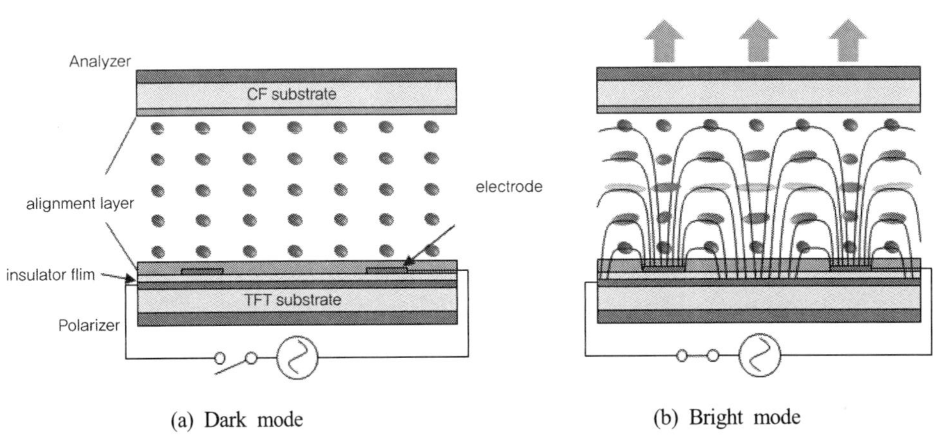

Figure 4.21

The structure and the operating Principle of FFS LC Cell

In recent years, FFS LC mode using LCs with negative dielectric anisotropy was developed. Such FFS mode with negative LCs has a transmissivity 15% higher than that of normal FFS mode. This is because all the LCs move in the direction parallel to the substrate in a negative LC cell, whereas LCs could also rise in the direction vertical to the substrate by the fringe field in a positive LC cell.

2.5 Comparisons about each LC mode

The characteristics of each LC mode that we have covered so far are summarized in Table 4-1 below. VA mode has the best contrast ratio from the front view by expressing dark state, which is higher than 1000:1. Other LC modes have a contrast ratio of thousands to one since a residual retardation occurs on the surface of the alignment layer or around the electrode.

The viewing angle is defined as the angle from the side to the point where the contrast ratio is 10:1. The viewing angle of TN and VA modes is very low, as they look different depending on the viewing direction. In particular, TN mode also shows the gray-scale inversion where the brightness of high gray-scales and low gray-scales is inverted to each other, from the bottom viewpoint. In order to improve such characteristics related to viewing angles, TN mode uses a compensation film based on LCs with panel structure, which is called Wide Viewing TN mode (WV-TN mode), and VA mode generates domains in pixels so that the LCs rotate in different directions.

The response speed of LCs is related to the coefficient of elasticity of the LCs. Normally the coefficient of elasticity is related to inequality:

$$K_{33}(bend) > K_{11}(splay) \geq K_{22}(twist)$$

The major characteristics of TN mode, VA mode, and IPS mode can be described as splay, bend, and twist, respectively. Therefore, the response speed of VA mode is the fastest, whereas that of IPS mode is the slowest in this group. The response speed of TN mode is the average response speeds of VA and IPS modes. However, there are also other factors affecting the response speed, such as the thickness of the LC cell (d), the viscosity of the LCs, the applied voltage, etc., so the above relationship could not be applied to all cases.

Nowadays, viewing angle and display bruising by the touch are considered to be important, along with the development of portable display devices with a touch function, such as smart phones and tablet products. As mentioned earlier, TN and/or VA LC modes have a narrower viewing angle than that of IPS and/or FFS LC modes. Also, LCs that are supposed to be arranged vertically in TN and/or VA modes could be disarranged on external impact, which causes flow and modification, so it has distortions on the screen, and thus TN and/or VA modes are not suitable for the above–mentioned portable devices. However, the LC modes where a parallel electric field is used and LCs have mobilities in–plane, like IPS and/or FFS, have a good viewing angle and are less sensitive to external impact, compared to other LC modes.

As explained above, each LC mode has different display characteristics, depending on which it is applied to other electronic devices. Cheap and productive TN mode is applied to personal laptops or monitors that do not require an outstanding viewing angle. VA and IPS modes with a wide viewing angle and good productivity are used for large TVs. FFS mode with a high transmissivity at high resolution is widely used for smart phones and tablet PCs that require a wide viewing angle and low power consumption, as well as for some high–end monitors.

Table 4.1 Display Characteristics of Various LC Modes and applications

Display mode	TN	VA	IPS	FFS
Contrast	○	◎	△	△
Aperture ratio	◎	○	○	◎
Viewing angle	△	○	◎	◎
Response speed	◎	○	○	△
Main applications	Note-book, monitor	TV, (cell phone)	TV, (monitor)	Tablet PC, cell phone, (monitor)

Table 4.2 Electrooptic Characteristics of Various LC Modes

Parameter	τ_{on}	τ_{off}	V_{th}	K_{eff}
TN	$\left(\frac{d}{\pi}\right)^2 \frac{\gamma_1}{K_{eff}} \left(\frac{1}{(V/V_{th})^2-1}\right)$	$\left(\frac{d}{\pi}\right)^2 \frac{\gamma_1}{K_{eff}}$	$\pi\sqrt{\frac{K_{eff}}{\epsilon_0}\Delta\epsilon}$	$K_{11}+\left(\frac{K_{33}-2K_{22}}{4}\right)$
VA	K_{33}			
IPS	$\left(\frac{\pi \cdot l}{d}\right)\sqrt{\frac{K_{eff}}{\epsilon_0 \Delta\epsilon}}$	K_{22}		

PRACTICE

1. Explain what liquid crystals are in terms of the combination of liquid and crystals, and how different/similar liquid crystals are from/to usual liquid and solid substances with a lattice structure.

2. List four physical attributes of LCs that are the most highly related to the operation and characteristics of LCDs. Then explain what characteristics of LCDs are the most highly related to each of the physical attributes.

3. Assuming a square–wave of alternating currents, where positive (+) and negative (–) voltages are applied repeatedly, in turn, is applied, explain behaviors of LC molecules and the reasons.

4. Explain why the concentration of an ion in LCs should be very low, and what will happen if ions are collected on the surface of an alignment layer.

5. Explain what problem could be caused if an alignment layer is too thick.

6. Unlike other display technologies, there are various modes (methods) in LCDs. Explain your thoughts on the reasons for this.

7. List four major LCD modes and describe the advantages and disadvantages of each mode, taking examples of the major products based on each mode.

8. Explain what characteristics will be the most important for displays to be used for smart phones.

9. Infer how to improve the response speed of an LCD from the response speed equation, and explain it in terms of the physical properties of LC materials and the variables of LC cells.

10. Argue what should be taken into consideration when designing an LCD for vehicles.

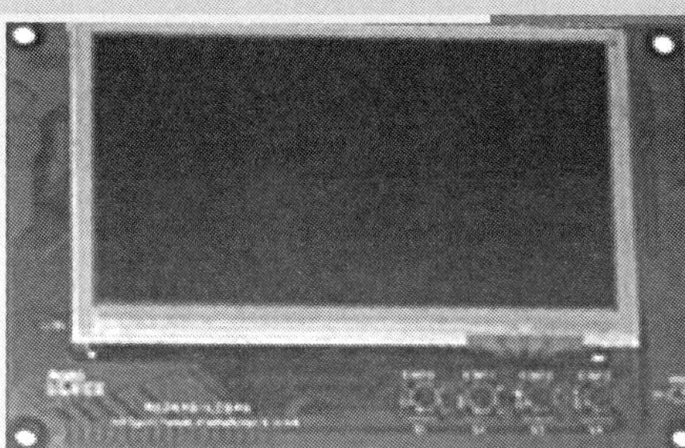

Display system drive

Chapter 04→05

1 The operation process of TFT-LCD

2 TFT LCD operating system

3 The operation principle of OLED

1 The operation principle of TFT-LCD

1.1 The operation process of TFT-LCD

All pixels of the TFT LCD have a TFT as shown in Figure 5.1. The gate is connected to a scanning line in the horizontal direction, one electrode of the source or drain is connected to a data line in the vertical direction, and the other electrode is connected to the pixel electrode. If the sufficient positive voltage is applied to the scanning line in the horizontal direction as in the right of Figure 5.1, all the TFTs of the scanning line are held in the ON state. At this time, because the pixel electrode on the scanning line is connected to the data line in the vertical direction, the image data is transmitted through the data line and the pixel electrode is charged with a proper voltage. Then when a sufficient negative voltage is applied to the scanning line, the TFT is closed to the OFF state and stores the electronic charge to the liquid crystal capacitor until the next signal is applied. Once image data of the entire screen is applied in this order, starting from the first scanning line to the last, a signal from the first line will start to be applied again.

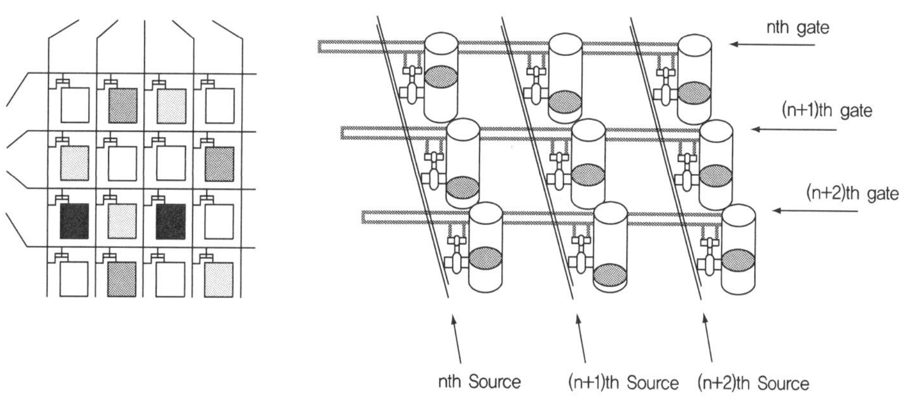

Figure 5.1
Operation schematic of TFT-LCD

1.2 Polarity Inversion

The momentum of the liquid crystal molecules in the electric field is proportional to the square of the electric field. This momentum is used to control an array by overcoming the elasticity of liquid crystal. The electric field applied to the liquid crystal molecules has directivity.

When an electric field is applied to the liquid crystal in a certain direction at a point of time, and another electric field is applied in the opposite direction at a different point of time, it is referred to as Polarity Inversion. In the LCD, the distance between the top and the bottom electrode is constant and the direction of the electric field is determined by the polarity of the potential difference. Therefore, the polarity inversion means applying an opposite potential difference to the liquid crystal. There are two reasons to drive the liquid crystal with the polarity inversion.

1.2.1 DC Blocking Effect of Orientation layer

To control the arrangement of the liquid crystal, the surface of the liquid crystal should be covered with an organic thin film such as polyimide (PI) and then it should be rubbed by a textured cloth or a groove should be cut on the material by the ultraviolet light so that the liquid crystal molecules are fixed in the desired alignment. This thin film with grooves is referred to as an orientation layer (OL) or an alignment layer. As shown in Figure 5.2, the OL has a sandwich-like structure where the liquid crystal is placed between two OLs. The voltage is applied to the liquid crystal through the electrode of the OL. Thus, the equivalent circuit consists of a capacitor and a parasitic resistor connected in parallel.

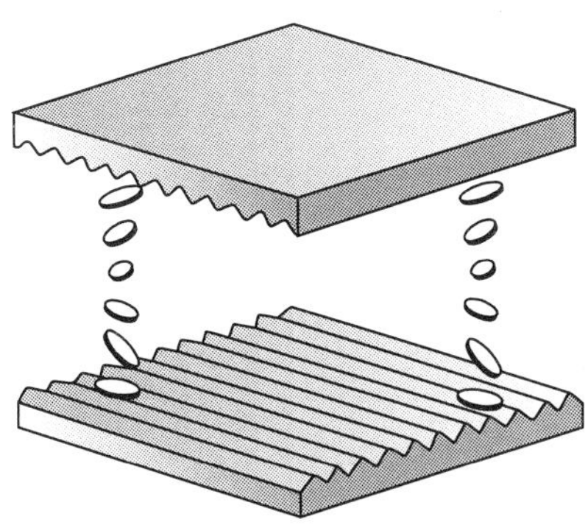

Figure 5.2
Direct-current blocking effect of the alignment film

In the OL, charges gather during the operation. These charges offset the external voltage applied to the liquid crystal by blocking a part of the external voltage, changing the orientation of the liquid crystal.

1.2.2 Mobile Ions and DC Residue

It is impossible to create a completely pure liquid crystal in the manufacturing process. Inevitably, certain mobile ions remain in the liquid. The mobile ions move to the positive or negative electrode when a voltage is applied to the OL. If the average value of the applied voltage is zero, the motion of mobile ions toward the two electrodes will be balance out, and thus the essential distance traveled by the mobile ions becomes zero. However, if the average value of the applied voltage is not zero, the mobile ions move to the surface of the liquid crystal to its polarity opposite to that of the mobile ions. The internal electric field influences the transmittance of the liquid crystal. Eventually, even when a voltage is not applied at all, the arrangement of the liquid crystal can affect the transmittance because it could be changed from the original arrangement depending on the internal electric field. This situation is called a DC Residue, which should be reduced in display devices.

To reduce the DC Residue, the average value of the applied voltage should be made to zero by applying the driving voltage as positive and negative alternately. Alternately reversed polarity of driving voltage is called the polarity inversion. Through the polarity inversion, the average value of the voltage of the positive and negative voltages should cancel each other. That is, the voltage to be applied should be free of the DC component of the applied bias.

1.2.3 Flicker

Flicker is a phenomenon of LCD malfunctioning. Because of the applied voltage to the liquid crystal it changes at a rate at which our eyes responds, due to which our visual perception changes, leading to discomfort in eye. Although there are individual differences, human eyes feel uncomfortable when it detect changes of the brightness within 90 Hz. Flicker occurs when the asymmetric symptom of the voltage is serious at the time of polarity reversal, and it is also a serious defect of a display device. The human eye has different acceptance for the flicker phenomenon depending on the brightness of the environment. The human sensitivity of the flicker depends upon background light, the individual perception, or the brightness of the display device.

1.2.4 Polarity Inversion method of Pixel array

As mentioned earlier, all of the pixels of liquid crystals are driven by polarity inversion. Usually, the polarity inversion method of a pixel array include frame inversion, column inversion, line inversion, and dot inversion, as shown in Figure 5.3.

When signal of one frame is applied and the next frame waits to be applied, if the polarity of the electrode stored in the pixels of the entire frame is equal, it is referred to as frame inversion. If the polarities of the electrode stored in the pixels of the same column are equal and the polarities of the electrode stored in the pixels on the left and right of the column are different, it is referred to as column inversion. Also, if the voltage polarities of the pixels in the same line are equal, and the pixel electrode of the line is opposite to that of top and bottom lines, it is referred to as line inversion. Finally, if the electrode polarities of the pixel are different from those of the electrodes of the adjacent pixels in every direction, it is referred to as dot inversion.

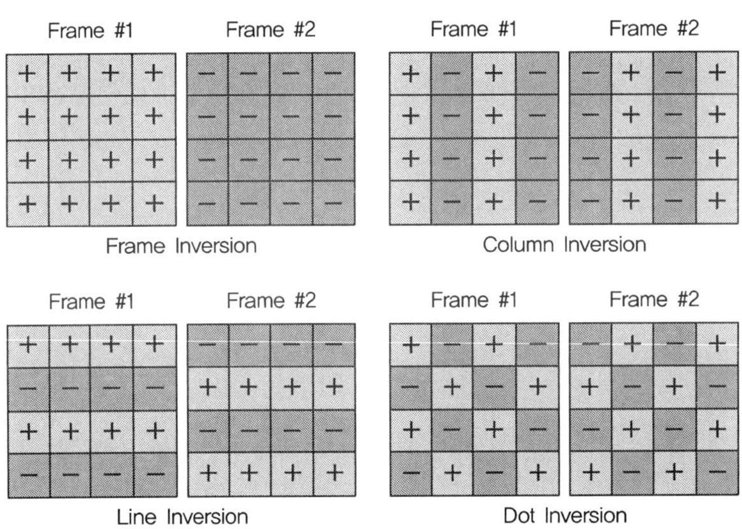

Figure 5.3
Polarity inversion of Pixel array

1.3 Holding

1.3.1 Charge and discharge current

The relationship among the charge or discharge amount, the current, and the voltage in the liquid crystal capacitor is as follows.

$$dQ(t) = Idt = CdV$$

For a polarity reversal, the setting of the pixel electric potential starts at the electric potential set during the previous operation of the display device. If the previous polarity of the potential was negative, the next potential should be positive. A liquid crystal capacitor must be "charged." Conversely, if the previous polarity of the potential was positive, the next-one should be negative. At that time, a liquid crystal capacitor must be "discharged."

For the charging and/or discharging in the TFT LCD, the gate of all the TFTs is set to a constant voltage to supply electric current, and the pixel electric potential is set to the same potential as the data line. In the charging and discharging process, the potential of the data line is set to the constant voltage corresponding to the gray level of the display. The output of the data from the driving IC, controls the charging or discharging of the pixel electrode. The voltage of the pixel electrode is gradually getting closer to the voltage value set in the data line during the charging and discharging process. Therefore, as the charging and discharging are in the process, the drain-source voltage difference and the current of the TFT are gradually lowered.

1.3.2 Voltage Range

(1) Voltage range of the data driving IC

The data driving IC sets the gray level by accurately controlling the voltage. The data driving IC supplies and controls the voltage. Therefore, a data driving IC is usually considered as the analog IC. To get the polarity inversion, as mentioned earlier, setting the voltage of the common electrode to approximately 5V, the positive voltage of 5 to 10V to the anode and the negative voltage of 0 to 5V to the cathode in an actual TFT LCD results in a voltage of 5V applied to liquid crystals. Therefore, the range of the maximum charging voltage of the data line is 0 to 10V.

1) Reduction in the output voltage range of the data driving IC

For the capacitor C, the power consumption for charging or discharging with the square wave is as follows.

$$P = (1/2)fCV^2$$

Here, f is the frequency of charging or discharging and V is the range of the voltage to be charged or discharged. If the voltage fluctuation range in the liquid crystal is low, the power consumption is also low.

A data driving IC is considered to belong to an analog IC, which should accurately control the voltage. The IC produced by the general semiconductor manufacturing process can only provide voltage lower than 5V. If the output voltage range of the data driving IC could be lowered to less than 5V, it will be advantageous in broadening a range of selection of the data driving IC and lowering the power consumption.

2) Common Modulation (Common Toggle)

Using "Common Modulation," the output voltage range of the data driving IC can be lowered. The operation principle is as follows. When the liquid crystal pixels need to be applied by (+) polarity, the common electrode should be set to the level of 0V. In this case, the output voltage of the data driving IC is set in the range of 0 to 5V in accordance with the gradation, and the pixel voltage is set in the range from +0V to +5V. When (−) voltage needs to be applied to a liquid crystal pixel, the common electrode voltage should be set to approximately 5V. In this case, the output voltage of the data driving IC is in the range of 0 to 5V, depending on the gradation. For example, to have a 0V applied to the pixel voltage, the output voltage of the data driving IC should be set to +5V. Conversely, to have −5V applied to the pixel voltage, it should be set to 0V. In this way, the pixel voltage can be set in the range from 0V to −5V.

(2) A common electrode voltage range

In theory, in order to prevent DC residual potential from shifting the symmetrizing of the positive and negative polarity of the liquid crystal voltage, the voltage of the common electrode should be set to the symmetric central value of the data line voltage. However, due to the parasitic capacitances of the TFTs, the capacitive coupling effect actually results in the pixel voltage being affected by the gate voltage change, deviating from the voltage input to the data line when the TFT is closed, as shown in Figure 5.4. To compensate the change of the voltage, the common electrode voltage should be adjusted to a lower value than that of the symmetric center of the data line voltage. This is referred to as the common electrode voltage compensation.

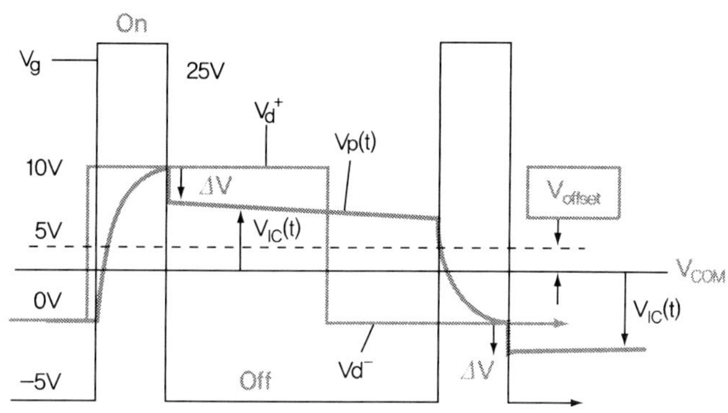

Figure 5.4

A common electrode voltage range

1.3.3 Pre-Charging

When a TFT LCD is operating, the process of charging the data line and the pixel electrode from negative to positive and discharging from positive to negative is repeated. This process should be completed within the charging time. Also, a time gap occurs when a frame and a row are switched and replaced. If the charging and discharging are performed as a Pre-Charging using the time gap, it is possible to quickly set the voltage and save time. For the Pre-Charging, two-pulse scanning and the charge sharing methods are used.

(1) Two-pulse Scanning

Two-pulse scanning is a Pre-charging method in which two scan lines are opened at the same time, as shown in Figure 5.5.

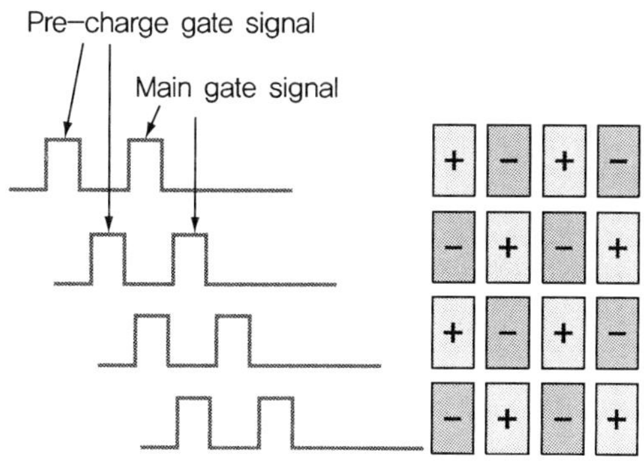

Figure 5.5

Double-pulse scanning

One scan line is opened twice; It first 'pre-charges' the other pixel voltage, having the same polarity as that of the line, by applying the voltage to its own pixel electrode, and then the pixel voltage is actually set to the pixel electrode. The TFTs on the two scan lines are opened sequentially; when the second scan line is opened, it enters the pixel voltage that is actually to be set up. On the other hand, when the first scan line is opened, it 'pre-charges' the other pixel voltage, having the same polarity as that of the line, by applying the voltage to its own pixel electrode.

(2) Charge Sharing

In addition to the above Pre-charging method using a two-pulse scanning, the effect of the Pre-charging can be achieved through the charge sharing. The charge sharing can reduce the energy consumption for the purpose of improving the energy efficiency.

When the TFT LCD is driven by line inversion method, a pixel voltage of every frame is applied, the polarities of adjacent data lines should be opposite. In this case, if the charge sharing switch is used, the adjacent data lines are short-circuited together, so that the positive and negative charges of both the data lines are mutually neutralized, causing their voltages to be close to each other. As this process does not require charging or discharging through the data driving IC, it does not consume additional energy, and the data line voltage, which is originally positive or negative, can get closer to the target voltage, even if it cannot directly reach the target voltage it tries to reverse the polarity. The gap there between is achieved through charging and discharging of the data driving IC. In this way, as the data driving IC lowers the range of the voltage to be charged or discharged, the effects of partial pre-charging and energy consumption can be reduced. If a TFT LCD is driven by a dot inversion, once pixel electrodes of one line are applied, the polarity of the electrode becomes opposite to that of the electrodes for the data line adjacent to the next line. Based on this, the data line is short-circuited with the adjacent data lines, providing the same effect.

1.4 Holding

After charging and discharging, TFT switch is opened by the next scan line. Usually, the duplicating frequency of a regular scan line is 60 Hz and the time for the holding is 16.67 ms. In theory, a pixel potential should be able to keep the voltage ON, once the charging and discharging periods are set up. However, in practice, the voltage can slightly change, that is influenced by the leakage current, which causes the RMS voltage applied to the liquid crystal capacitor to be changed and it affects the transmittance. The holding is a way of trying to lessen the influence of the leakage current on the potential and the effect on the transmittance to be less than the acceptable range.

1.4.1 Path of the leakage current

(1) Leakage current of liquid crystal capacitor

Pure liquid crystal material can be viewed as an insulator because of its very large resistance. However, the leakage current can occur because of the reduction of the resistance caused by the debris that could come in while processing liquid crystals. The voltage charged in the capacitor decreases exponentially with time, and the decrease is proportional to the product of the resistance and the capacitance values. Therefore, the resistance component of the liquid crystal should be reduced.

In the early processing of the liquid crystals, the voltage drop of 0.7 V occurred during the potential holding time. It now meets the requirements of the holding potential through the voltage drop of 8 mV during the potential holding time because the current liquid crystal processes are more stable.

(2) The leakage current of TFT

Another path of causing the leakage current is a TFT which serves as the pixel switch. Even in the case of applying a negative voltage to the gate (OFF state), the TFT has the leakage current of 0.1~1 pA level. The reason to be careful of the leakage current of the TFT is that the leakage current is not a fixed value. It changes with the gate–source voltage. Also, it should be noted that when TFT panel is actually operating, electron–hole pairs are generated by the light energy of the backlighting, causing the leakage current to increase. To prevent this, the light from the external light source should be blocked with a black matrix.

1.4.2 Storage Capacitor

The storage capacitor has two roles. It reduces the leakage currents of the liquid crystal capacitor and TFT, which is a direct and effective method of maintaining the potential–holding. When the condition of holding potential is not satisfied even though the leakage current is reduced as much as possible, the potential needs to be maintained through other means. If a liquid capacitor and a capacitor that does not leak current, are connected in parallel, this capacitor can be used to help charge accumulation. So, the capacitor element is referred to as the storage capacitor.

The larger the area of the storage capacitor is, the more advantageous the potential holding becomes. However, since the storage capacitor can not transmit light, the larger the storage capacitor, the wider the area for blocking light and the aperture ratio is lowered and transmittance of the light is reduced. Also, in the process of charging by opening the TFT, the liquid crystal capacitor and the storage capacitor should be charged at the same time. However, if the capacity of the storage capacitor is too large, it is not charged completely.

1.4.3 The leakage current of point defect type

Another current leakage process occurs due to the defects. This leakage current causes a defect in the display device; such a defect is displayed on the pixel point of the panel at the position where the defect occurs. It is called a point defect. Such leakage current by the point defect can occur in a scan line, a common electrode, a data line, an adjacent pixel electrode, and so on.

1.5 Coupling

1.5.1 The principle of coupling

(1) Charge Conservation

Each capacitor of Figure 5.6 is connected to one electrode. First, the voltage of each capacitor is fixed. Therefore, the potential of all connected electrodes is the same as in Figure 5.6 (a). If the potential of the electrode of a capacitor goes down, the potential of the other electrode of the capacitor gets lower than those of the electrodes of other capacitors at that moment, since the voltage of the capacitors is continuous across the capacitors as shown in Figure 5.6 (b). However, as the potential of the common electrode should be the same, charges flow from the other capacitors into the capacitor with a lower potential and the potential goes up again. Eventually, the potentials of the connected electrodes become equal to each other again as shown in Figure 5.6 (c). However, the value of the potential gets lowered than that of Figure 5.6 (a). This phenomenon is referred to as a coupling effect. Again in Figure 5.6, there is no increase or decrease of the charges in the common electrode connecting the capacitors, This is a prerequisite called "charge conservation." If there is any path for the charges to flow in or out, the charges in the electrodes connected in common could increase or decrease, which could affect the potential of the connected electrode; in this case, it is not proper to take only the "coupling effect" into consideration.

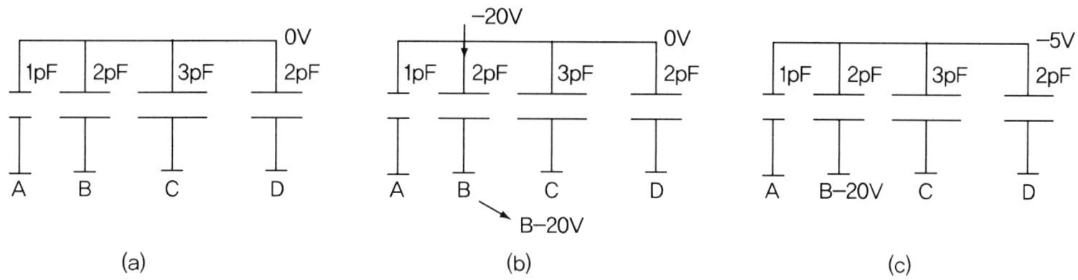

Figure 5.6

Pixel capacitive coupling diagram (a) First voltage of each capacitor (b) Variation of the capacitor voltage (c) Final voltage of each capacitor

(2) Partial capacity

If the sum of the capacitance value is ΣC in the circuit consisting of a number of capacitors, when all of the capacitors are connected to one electrode and the other electrodes are connected to different voltage sources respectively (i.e., Charge Conservation), the electrodes connected to the capacitor with the value of ΣC does not have inflow or outflow of charges through a direct current route. Under this premise, when the voltage source connected to a capacitor C changes by ΔV, the voltage variation of the connected electrode is as follows.

$$\Delta V \times \frac{C}{\Sigma C} \qquad \text{(Equation 5-1)}$$

1.5.2 Capacitors in pixels

When TFT is OFF and the leakage current of the TFT is ignored, the electrode of the pixel enters the state of charge conservation. Since the pixel electrode controls the liquid crystal voltage, it occupies most of the pixel area. Capacitors associated with the electrode referring to a pixel layout are described in [Figure 5.7], which will be explained later.

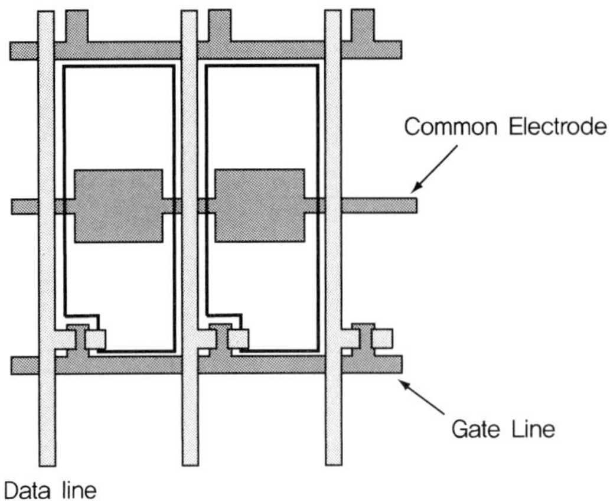

Figure 5.7
An example of pixel layout

(1) Self-parasitic capacitance of TFT

For the proper operation of the TFT, the overlap of the gate and the source/drain is required, by which the parasitic capacitance occurs. This parasitic capacitance is generated on both the source and drain sides. The TFT source electrode and the drain electrode of the pixel vary depending on the polarity of the voltage difference between the data line voltage and the pixel voltage. The parasitic capacitance at the other end of TFT is connected to the data line and the pixel voltages, thus it cannot enter the conservation state of charges. Therefore, in this case, the capacitive coupling effect does not need to be considered; instead, the effect of load should be considered.

(2) Parasitic capacitance between the pixel electrode and the scan line

The pixel electrode part controls the rotation of the liquid crystal and determines the brightness of the pixel. However, the part without the pixel electrode should be blocked because it is a portion where the liquid crystal transmittance can not be controlled by applying a voltage. In addition to covering the portion by a black matrix, it is required to set the overlap of the pixel electrode and the scan line or the partial overlap of the pixel electrode and the data line, so that the light leaking between each line and the pixel electrode can be blocked. As the black matrix is on the other substrate with a color filter, its misalignment with the pixel electrode is relatively large; thus, when using the black matrix, the blocking range should be wide. On the other hand, the scan line and the pixel electrode are on the same TFT substrate, by which light can be accurately shielded, and thus the aperture ratio is large. This method is referred to as integrated BM. The pixel electrodes and the scanning lines are disconnected from each other in the manufacturing process by an insulating layer. Thus, there is no possibility of short–circuiting, but the parasitic capacitance could exist.

(3) Parasitic capacitance between the pixel electrode and the data line

In the conventional TFT process, the part between the data line and the pixel electrode may be short–circuited because it is not blocked by the insulating layer. The short circuit should be prevented by avoiding the overlap of the pixel electrode and the data line. However, light in the region, that is not controlled in accordance with the direction of the data line, should be shielded. When shielded by the black matrix, the aperture ratio–loss is large because of the long pixel in the data line direction. Therefore, the TFT LCD with small pixels and large PPI directly uses integrated BM with the data line itself, which causes the parasitic capacitance to occur as shown in Figure 5.7. However, in this case, an insulating layer should be inserted separately, and the pixel electrode (typically a transparent ITO electrode) should be placed on top to prevent a short circuit of the data line and the pixel electrode. This special process is called 'Top ITO'.

(4) Pixel equivalent circuit with added capacity

Applying the parasitic capacitance to the pixel equivalent circuit is shown in Figure 5.8. Among the capacitors connected to the pixel electrode are the storage capacitor (C_s) led to the common electrode line located at the bottom, the liquid crystal capacitor (C_{lc}) connected to the upper common electrode, the parasitic capacitance of the TFT (C_{gd}), the capacitor (C_{pd}) connected to its own data line, the capacitor ($C_{pd'}$) connected to the adjacent data lines, the capacitor (C_{pg}) connected to its own scan lines, the capacitor ($C_{pg'}$) connected to the adjacent scanning lines, and the capacitive coupling effect needs to be taken into account in this circuit. If the integrated BM method is selected for the TFT, the values of C_{pd}, C_{pd}', C_{pg} and $C_{pg'}$ are negligible. Also, the load capacitors are C_{x1} between scan lines and data lines, C_{x2} between common electrode lines at the bottom and data lines, the parasitic capacitance of TFT C_{gs}, C_{d0} between the data line and the upper common electrode, C_{g0} between the scanning line and the upper common electrode.

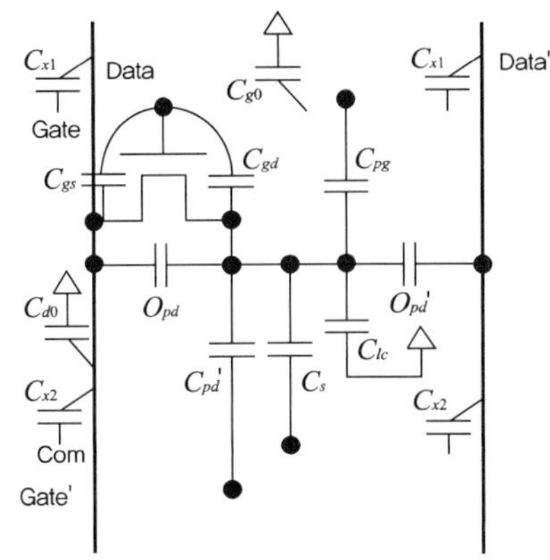

Figure 5.8
Pixel equivalent circuit with added capacity

1.5.3 Capacitive coupling effect of the scan line

(1) Capacitive coupling of the scan line and the pixel electrode

From Formula 5.1 if the voltage variation of the scanning line is $V_{OFF}-V_{ON}$ with charge conservation, then the change in pixel voltage (ΔV) will be,

$$-\Delta V=(V_{OFF}-V_{ON})(C_{gd}+C_{pg})/(C_s+C_{lc}+C_{gd}+C_{pd}+C_{pd'}+C_{pg}+C_{pg'}) \quad \text{(Equation 5-2)}$$

when ignoring C_{pd}, $C_{pd'}$, C_{pg} and $C_{pg'}$

$$-\Delta V=(V_{OFF}-V_{ON})(C_{gd})/(C_s+C_{lc}+C_{gd}) \quad \text{(Equation 5-3)}$$

As the ON voltage of the TFT, V_{ON}, is larger than the OFF voltage of the TFT, V_{OFF}, the magnitude of the voltage change is negative. This change in the voltage is directly proportional to the change of $V_{OFF}-V_{ON}$ as well as the parasitic capacitance of the TFT. However, it is inversely proportional to the sum of the storage capacitance and the liquid crystal capacitor.

Under the assumption of charge conservation, when the TFT is changed from OFF to ON, it starts to charge and discharge. At this time, the electrons move between the pixel electrodes and the data line, the conservation of charge is not maintained, and capacitive coupling effect does not occur. However, when the TFT is changed from ON to OFF, the charge of the pixel electrode does not move to the data line.

Thus, the charge conservation and capacitive coupling occurs.

(2) The storage capacitor of the scan line

Figure 5.9 and Figure 5.10 show the layout and equivalent circuit of this design. In this pixel, there is no common electrode line at the lower substrate, so there is no capacitance C_{x2} between the common electrode at the lower substrate and the data line. Also, the storage capacitor C_s and the capacitance between the pixel electrode and an adjacent scan line changes to a single capacitance C_s. Therefore, Formula 5.3 may be modified as follows.

$$-\Delta V=(V_{OFF}-V_{ON})(C_{gd})/[C_s+C_{lc}+C_{gd}]+(V'_{scan1}-V'_{scan2})(C_s)/[C_s+C_{lc}+C_{gd}]$$

(Equation 5-4)

V'_{scan1} and V'_{scan2} are the voltage before and after a change of the adjacent scan lines, which affect the pixel voltage through the storage capacitance C_s. As shown in Formula 5.4, the effect of changes in the storage voltage on the adjacent scan lines ($V'_{scan1}-V'_{scan2}$) on the pixel voltage is (C_s/C_{gd}) times the change in the self-scan line.

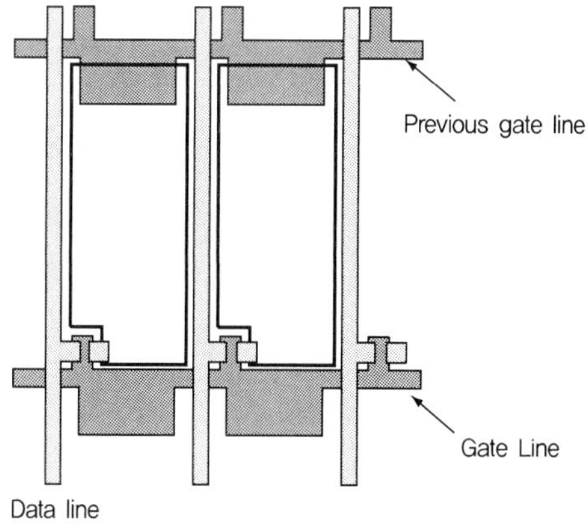

Figure 5.9

Storage on the gate pixel layout

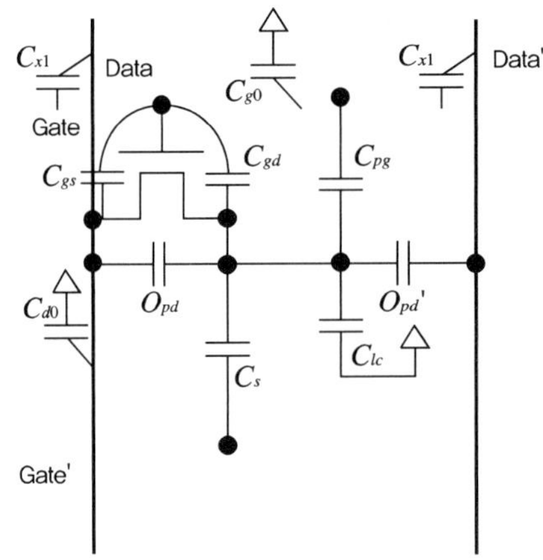

Figure 5.10
Storage on the gate pixel equivalent circuit

(3) The storage capacitor of the previous scan line and the next scan line

In the design of storage capacitor on the gate, electrodes of the storage capacitor are made of the pixel electrode and the scan line of the adjacent pixel. There are two adjacent pixels in the vertical direction; one is on the previous scan line, and the other is on the next. As the storage capacitor, placed at a distance, on the gate pixel layout shown in [Figure 5.9], it is difficult to connect the storage capacitor to pixel scan lines. [Figure 5.11] shows an equivalent circuit of the storage on the gate structure.

The calculation of the state where a storage capacitor is associated with the previous scan line is as follows. First, when the previous scan line is in ON state due to, the voltage change from V_{OFF} to V_{ON}. At this time, the pixel voltage changes as follows, through the storage capacitor C_s.

$$+\Delta V_1 = (V_{ON} - V_{OFF})(C_s)/[C_s + C_{lc} + C_{gs}] \quad \text{(Equation 5-5)}$$

When the previous scan line is in OFF state due to, the voltage change from V_{ON} to V_{OFF} the pixel voltage changes as follows, through the storage capacitor C_s.

$$-\Delta V_1 = (V_{OFF} - V_{ON})(C_s)/[C_s + C_{lc} + C_{gs}] \quad \text{(Equation 5-6)}$$

When a scan line of the pixel itself is ON, the voltage also changes from V_{OFF} to V_{ON}. However, the pixel voltage is set to the voltage of the data line by the TFT regardless of the pixel voltage. When the scan line of the pixel itself is OFF again, it changes the pixel voltage due to the parasitic capacitance C_{gs} of the TFT.

$$-\Delta V_2 = (V_{OFF} - V_{ON})(C_{gs})/[C_s + C_{lc} + C_{gs}] \quad \text{(Equation 5-7)}$$

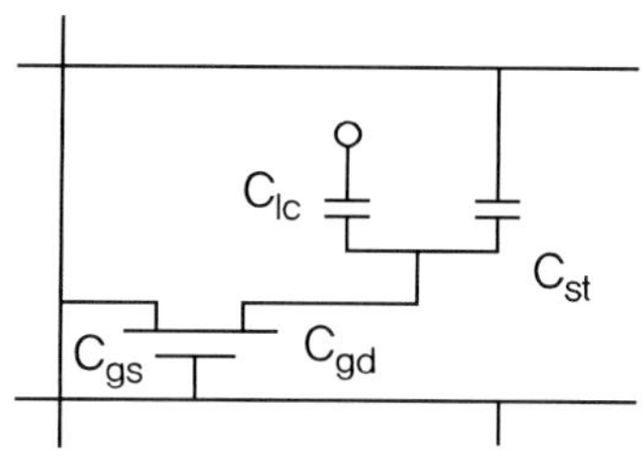

Figure 5.11
An equivalent circuit of storage on the gate pixel

(4) 3-Step driving method of the scan line

If the effect of the previous scan line on the capacitive coupling of the pixel voltage is used for the storage in gate pixel, the range of variation $\Delta(\Delta V)$ of the capacitive coupling due to the changes in capacitance of liquid crystal can be compensated. This is not possible with a single common configuration. The method of scanning the electrode voltage is referred to as a 3-step driving method of the scan line because of the three levels as shown in [Figure 5.12]. The largest voltage makes the TFT ON; otherwise, the TFT is OFF.

Referring to Figure 5.12, the change in the coupling of the voltage before the TFT is turned ON, is set again by the voltage of the data line during the time ①. At this time, the pixel voltage V_{pixel} is equal to the data line voltage V_{data}. During the time ②, the change in the scan line voltage is $-(V_{ghl} + V_{gc})$, so the change in pixel voltage is $-\Delta V_1$,

$$\Delta V_1 = (V_{gh1} + V_{gc}) C_{gs} / (C_{gs} + C_s + C_{lc} + ...) \quad \text{(Equation 5-8)}$$

During the time ③, the voltage change of the previous scan line is $+V_{gc}$ and the pixel voltage is changed by ΔV_2,

$$\Delta V_2 = V_{gc} C_s / (C_{gs} + C_s + C_{lc} + ...) \quad \text{(Equation 5-9)}$$

During the time ④, the scan line voltage change of the pixel voltage is $+V_{gc}$ so the change in pixel voltage is ΔV_3,

$$\Delta V_3 = V_{gc} C_{gs} / (C_{gs} + C_s + C_{lc} + ...) \quad \text{(Equation 5-10)}$$

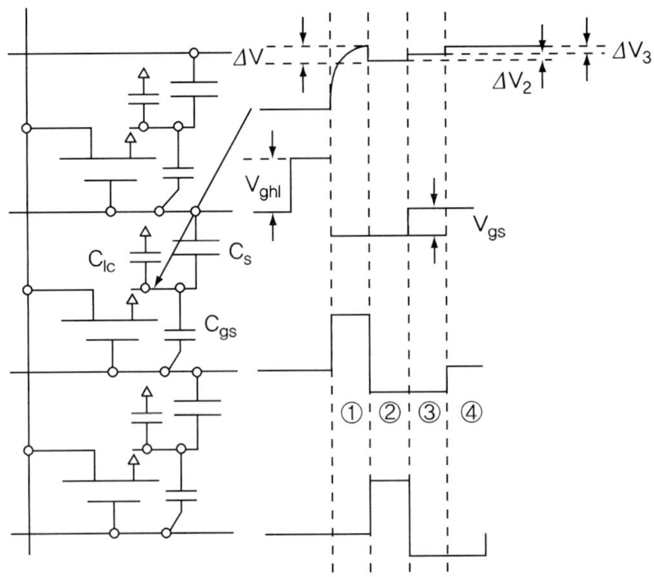

Figure 5.12 The equivalent circuit and the corresponding waveform of the three adjacent pixels in the vertical direction, according to the 3-step driving method of the scan line

After the time ④, the relationship of the pixel voltage V_{pixel} and the data line voltage V_{data} is as follows.

$$V_{pixel} = V_{data} - \Delta V_1 + \Delta V_2 + \Delta V_3 \quad \text{(Equation 5-11)}$$

In order to make $(-\Delta V_1 + \Delta V_2 + \Delta V_3)$ to be zero, the V_{data} needs to go through a series of the capacitive coupling effects for the time steps ②③④ and then the pixel voltage V_{pixel} must return to the same level as the data line voltage V_{data} when the TFT is OFF. When $-\Delta V_1 + \Delta V_2 + \Delta V_3$ is substituted with (Formulas 5.8), (formulas 5.9), and (formulas 5.10), the denominators of the three expressions are the same and the liquid crystal capacitance changes the value of the denominators. Therefore, the design should consider the total changes as zero if the numerator is zero. In other words,

$$V_{gs} C_s = V_{ghl} C_{gs} \quad \text{(Equation 5-12)}$$

In this way, it is possible to exclude the effect of the liquid crystal capacitor and fully compensate the effect of the capacitive coupling with the scan line. The main concept of the 3-step driving method can be summarized as follows.

A. If there are a large number of scan lines, the pixel voltage does not have much impact on the RMS voltage, and the direct current applied to the liquid crystal in the frame implementation.

B. Regardless of the pixel voltage before the TFT is ON, the pixel is charged or discharged to the data line voltage after the TFT is ON.

C. When the current of the TFT is as small as that of the holding potential the OFF voltage of the TFT is not unique, but remains within a range.

According to this concept, by using the rising voltage of the scan line, the pixel voltage can be raised through the storage capacitor coupling, compensating the decrease of the pixel voltage.

* 2 TFT-LCD operating system

2.1 Operating system overview

TFT-LCD operating system mainly includes the operation of scan lines or data lines. Its basic circuit consists of function blocks such as a shift register, a level shifter, an operational amplifier, etc. This TFT LCD operating system is shown in Figure 5.13. IT contains the entire operating system including the TFT LCD panel and the backlight system.

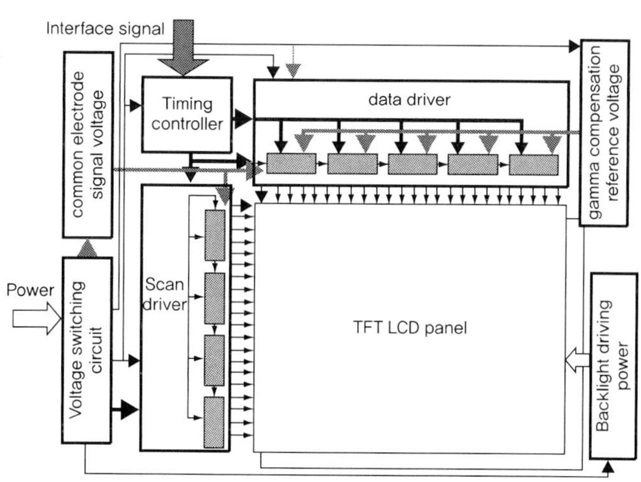

Figure 5.13
TFT LCD drive system overview

2.1.1 Timing Controller

The central part for controlling the operation procedure of the entire display device is when a timing controller starts the horizontal scan in accordance with the timing implemented in each frame; it converts the image signal from the interface to the data signal format, that is used by the data driver to transfer the signal to the memory of the data driver; and it controls the data operation at a suitable time sequence according to the horizontal scan. The basic functions of the timing controller are to provide the image data according to the required timing control signals for the scan driver subsystem and the data driver subsystem.

As shown in Figure 5.13, the external data signal is input to the timing controller and the timing controller generates a signal that is used in the data driver or the scan driver according to the resolution of the TFT LCD panel or the operation frame frequency. The images can be displayed properly only if timing among signals matches to one another.

2.1.2 Scan Driver

A scan driver sends appropriate ON voltages or OFF voltages to a particular scan line under the control of the timing controller in order to drive the scan lines of the TFT LCD panel.

2.1.3 Data Driver

Digital image data applied with high frequency, is stored in the memory under the control of the high frequency timing controller. The digital image data is converted to analog image signal voltage that is to be sent to the pixel electrode and then to be sent to the data line. The image signal voltage supplied to the data line is supplied to the pixel electrode through the transistors of the scan lines. The transistors are driven by scan voltage.

2.2 Scan Driver

2.2.1 The basic function blocks of the scan driver

The role of the scan driver is to determine the ON/OFF state of the TFT connected to the scan line and its structure is shown in Figure 5.14.

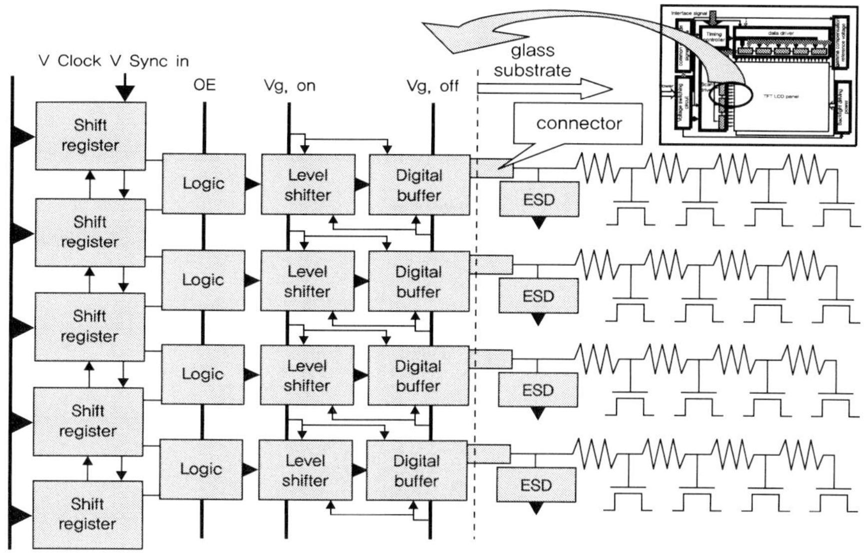

Figure 5.14

Basic functional block diagram of a scan driver circuit

The function blocks are explained as follows.

(1) Shift register

In general, the shift register transfers the logic state of the input stage to the next stage when it completes a clock cycle, by using the D–type flip–flop. The TFT LCD operates the scan lines sequentially and supplies the video signals to the pixels of each line sequentially. If a frame is started, the pulse is supplied from the first scan line to the next line sequentially. Thus, a signal is needed to indicate the start of a frame. If the start signal comes in, a pulse is supplied to the first scan line. The start signal supplied to the scan line is for the vertical synchronization. As shown in Figure 5.15, when a scan synchronizing signal is sent to the first shift register and the clock signal in the vertical direction is applied to the circuit, output of each shift register is controlled by the clock and the logic state to determine the ON/OFF of a scan line is generated sequentially. Since the screen can be turned upside down, the scan direction should be able to be reversed depending on the situation.

The above process does not need a high voltage because it only decides the logic state of ON or OFF. The voltage ranging from 3 V to 5 V is enough. The low voltage is advantageous for low power consumption.

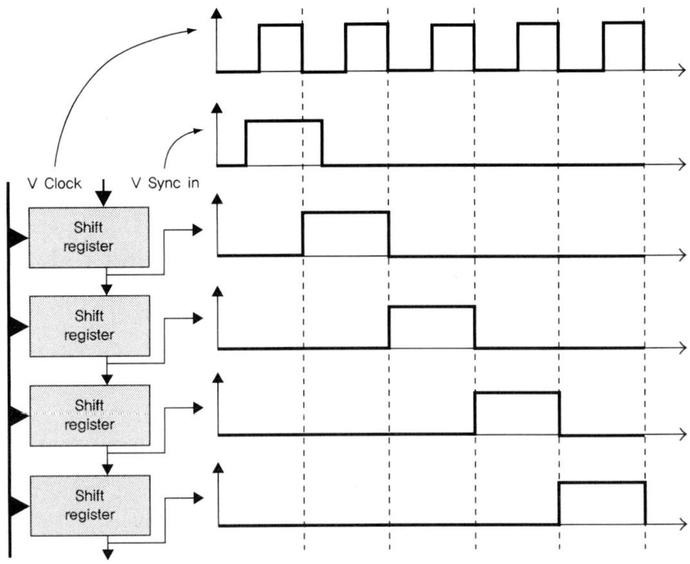

Figure 5.15

The shift register input/output waveform

(2) Logic

This block can implement the necessary logical operations. The voltage from 3V to 5V is usually enough because it operates the logic state only. It is used when there is a need to change the waveform, and it is possible to change the shape of the pulse by a control signal supplied to the line output enabled (OE) as shown in Figure 5.16. This change in the shape of pulse varies the width of the gate pulse and thus, is useful to prevent the distortion of the image quality caused by the signal delay.

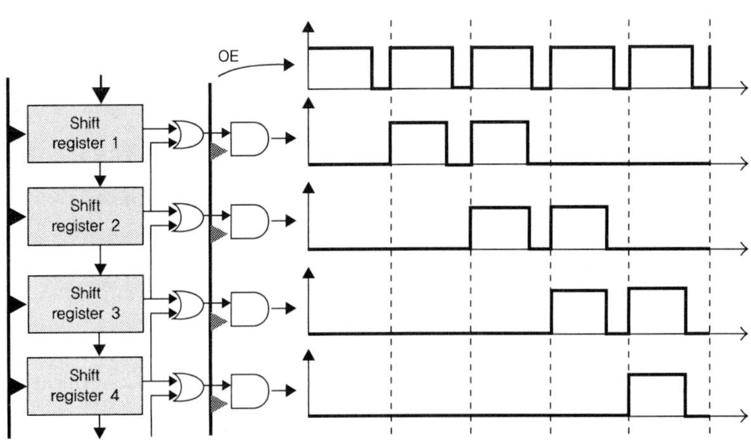

Figure 5.16

A logic operation circuit waveform

(3) Level shifter

The level shifter converts the low voltage logic level of 3V/0V or 5V/0V to ON−voltage of more than 20V and the OFF−voltage of less than −5V necessary to open or close the pixel TFT. The example of input and output waveforms of the level shifter is as shown in Figure 5.17. The voltage required to ON or OFF the TFT is different according to the composition and type of material and is converted to the ON−voltage and OFF−voltage, suitable for the characteristics of the TFT.

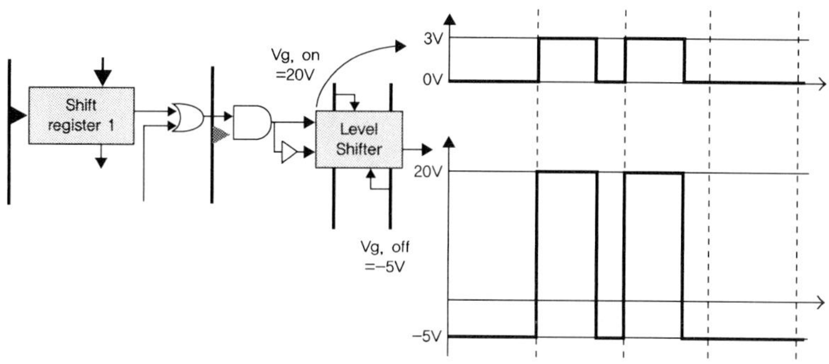

Figure 5.17

Input and output waveforms of the level shifter

(4) Digital buffer

If the scan lines are driven directly from the output of the level shifter, the driving voltage becomes insufficient due to the high resistance and capacitance of the scan lines, therefore the signal has to be increased by using a buffer. As shown in Figure 5.18, the signal from the level shifter passes through the buffer that is composed of inverters, The current capacity of each inverter increases as the process is repeated. and the buffer is designed to have the current capacity that can change the voltage of the gate line fast.

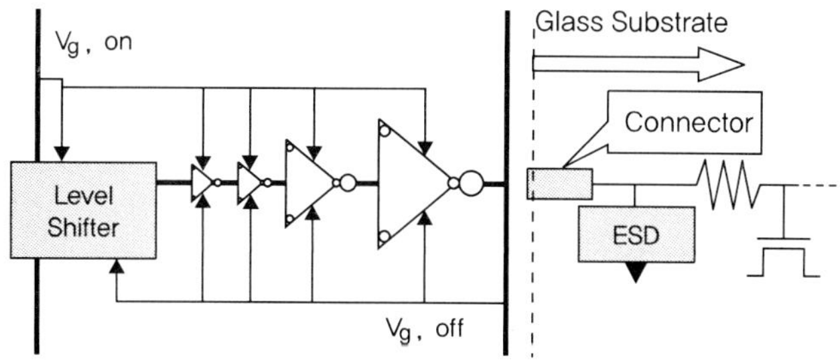

Figure 5.18

An example of digital buffers

2.3 Data Driver

2.3.1 The basic function block of the Data Driver

The role of the data driver is more complex than the scan driver, and the structure is shown in Figure 5.19.

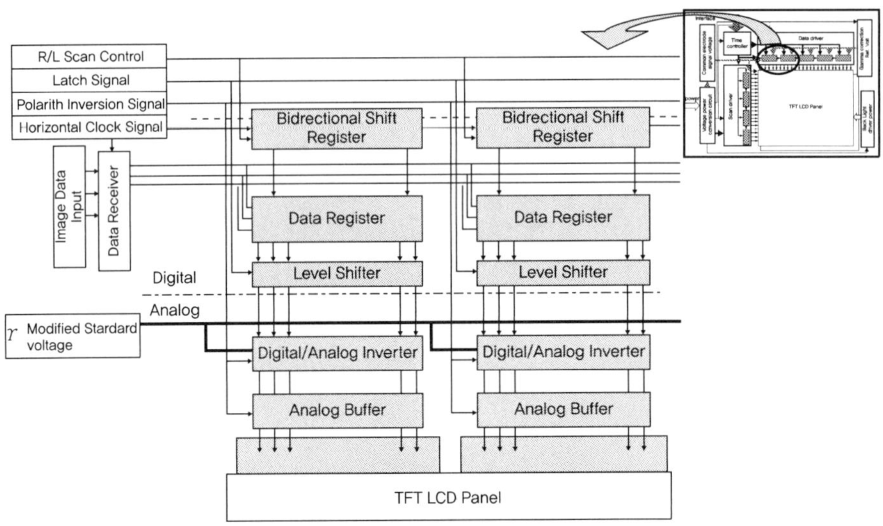

Figure 5.19
The basic functional block diagram of a data driver

As shown in Figure 5.20, it is possible to explain the driving method of the data driver by unfolding the function blocks in detail. The operation of the shift register(S/R) is controlled by the clock signals in the horizontal direction (H clock) and the scan synchronizing signals in the horizontal direction (H sync). Once the clock signals in the horizontal direction are applied, it generates the pulse sequentially according to the clock cycle. It sequentially opens the first data register driven by the latch method and the digital video data which is implemented by the pixels of $(n-1)$th scan line is stored in order. It stores the data to be implemented by the pixels of the entire scan line in the row in order in the first data register, and then stores this data in the second register (Latch 2) at a time, according to the next synchronizing signal in the horizontal direction (H sync). At the same time, these digital signals are converted to the pixel voltage by using the digital/analog converter (DAC), and the driving voltage is amplified by using the analog buffer again. Thus, the voltage is applied to the pixel electrode through the data line of the panel of the pixel array.

The time of applying this pixel voltage is charging time. When the voltage is applied, the shift register receives the next Hsync and stores the video data of the n–th scan line pixel simultaneously. The n–th data update is stored in Latch 1 and the $(n-1)$–th voltage applying can be done simultaneously because it goes for Latch 2. By repeating this operation, it is possible to complete the TFT LCD operation.

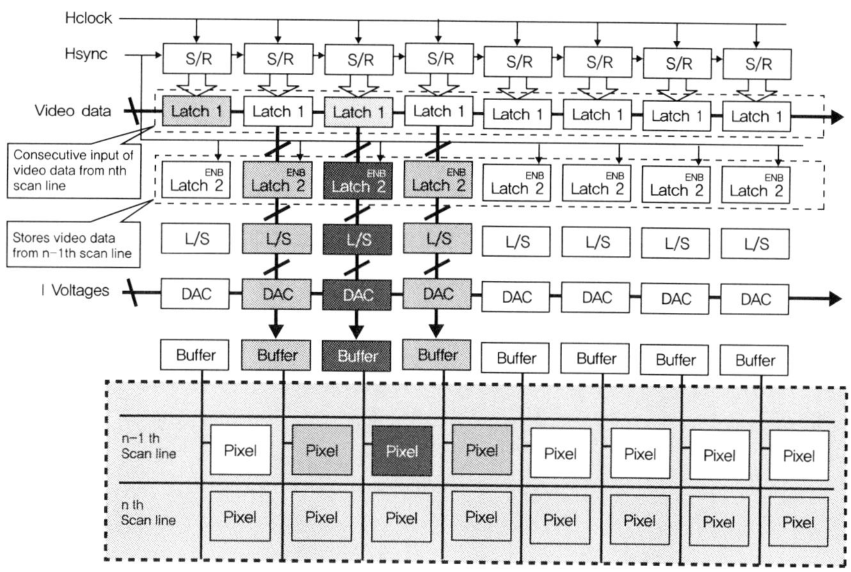

Figure 5.20

Operation of the data driver

The detailed explanation on the function blocks of the data driver is as follows.

(1) Shift Register

This shift register is equivalent to the one used in the scan driver, but the operating frequency is about few hundreds times faster. The practical operating frequency is related to the data to be transferred and the number of channels of the data signal interface.

(2) Data Register

The stored data is digital data that can be memorized through the Latch which is composed of two inverters, whose inputs and outputs are connected to each other.

(3) Level Shifter

The L/S block in Figure 5.20 is a level shifter. The logic state is decided at the function block. If there is a need for the digital/analog converter (DAC) to raise the voltage of this digital signal and use the same as the switch of the reference voltage, the reference voltage range will be about 10 V including the positive and negative polarity driving the liquid crystal. Therefore, the level shifter raises the reference voltage, used by the shift register and the data register, by 5 V.

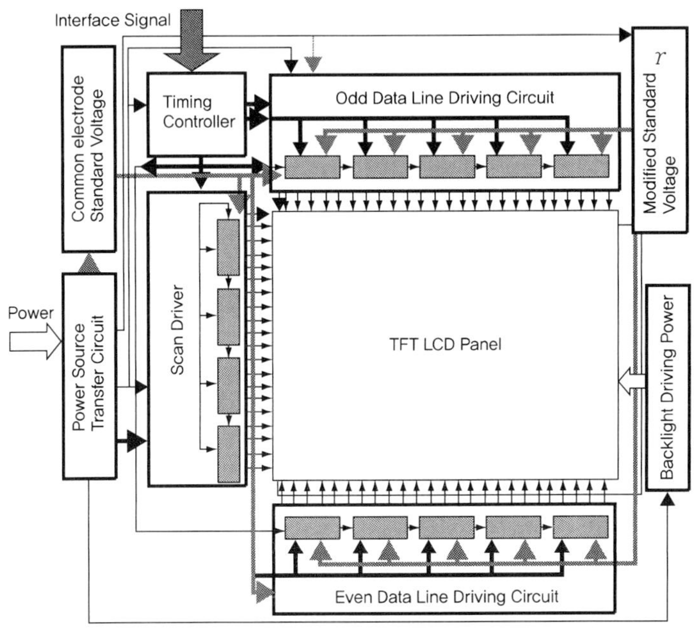

Figure 5.21
System with a data driving IC arranged in the upper and lower part of the TFT LCD

(4) Digital to Analog Converter(DAC)

The structure of the DAC is as shown in Figure 5.22, i.e., DAC generates the reference voltage and selects the voltage according to the input data. For example, when the digital signal data is 100110, V_{out} is connected to V38, and when the digital signal data is 010011, V_{out} is connected to V19. It converts the digital signal to the analog voltage for driving of the pixels by choosing the voltage divider according to the digital data.

When driving the liquid crystal, polarity inversion is required. Thus, the positive and negative polarity of the applied voltages have to be provided. The driving voltage is changed periodically for the polarity inversion. The $+V_{out}$ and $-V_{out}$ are made by changing the polarity of the voltage of V0~V63. One-to-two multiplexer is used for choosing and outputting the proper voltage.

D0~D6 in Figure 5.22 are digital signals, but because the voltage is raised by the shift register, the reference voltage in the positive or negative polarity range can be opened or closed.

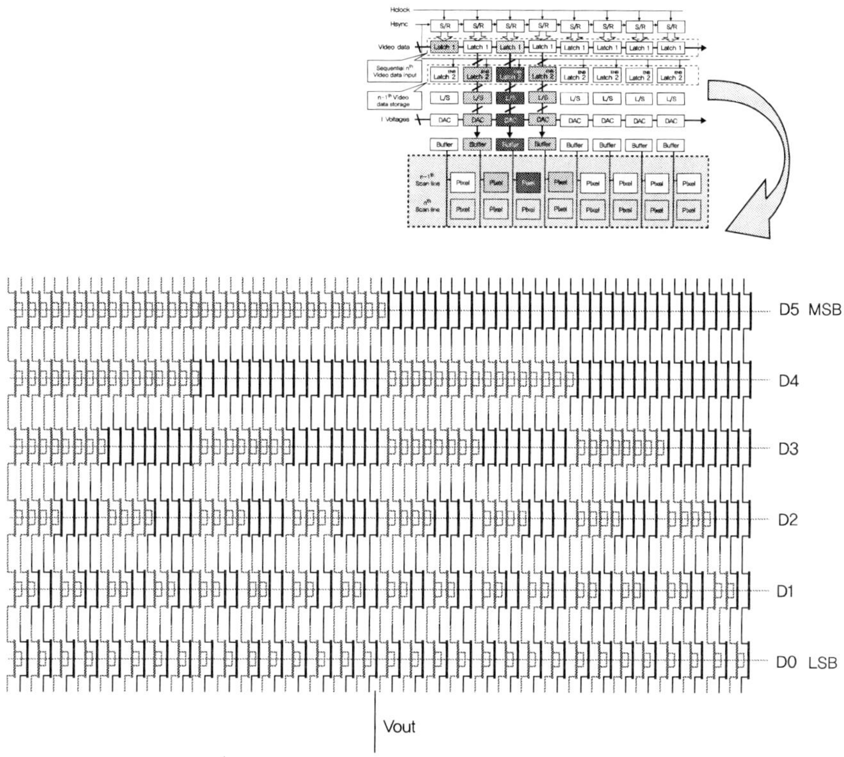

Figure 5.22

Example of Voltage selector-type DAC structure

(5) Analog Buffer

An analog buffer is similar to a digital buffer. It drives the large load of the pixel array with the enhanced output after receiving the signal source from the low-load input. However, the biggest difference from the digital buffer is that the driving capability needs to be extended as the analog buffer deals with analog signals, not digital signals. In other words, this analog buffer is needed because 25 V or 25.2 V can be acceptable as ON-voltage for digital signal but not for analog signal. It is usually implemented using the operational amplifier as shown in Figure 5.23. The output voltage is equal to the input voltage but has greater driving power.

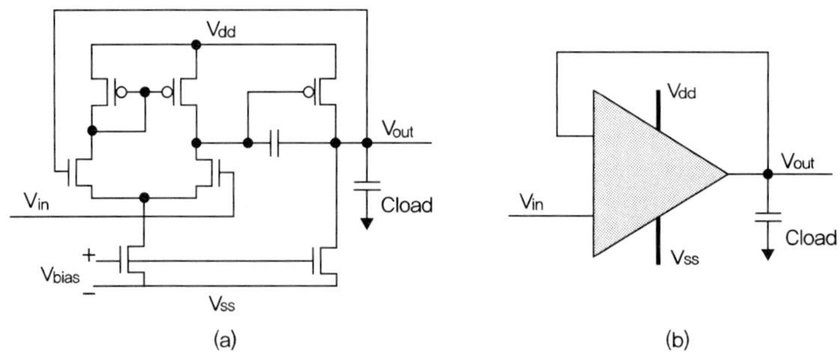

Figure 5.23
Examples of Analog buffer (a) Circuit Diagram (b) Sign–map Diagram

However, according to the voltage division theory, the absolute value of the resistance is not critical, but the ratio of the resistance determines the output voltage. The change in the temperature or the IC process does not affect the voltage. However, the absolute values of resistance affect the power consumption and driving power. That is, the bigger the absolute value of the resistance is, the smaller power the resistive voltage consumes. In this case, the reference voltage will be changed easily as it is affected by the load. To reduce the power consumption, the absolute value of high resistance needs to be used, and the load of the reference voltage needs to be reduced. Therefore, the analog buffer that has the high input resistance should be used, as shown in Figure 2.12.

The structure of Figure 5.24 shows that since each data driving IC has a voltage–dividing resistor, the divided voltage across these resistance in these ICs is connected to each other in parallel, up to the analog buffer output used in the reference voltage source, if the reference voltage source drives several data driving ICs simultaneously.

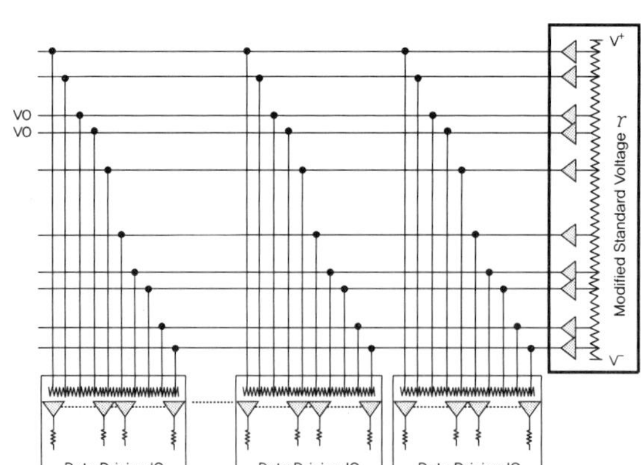

Figure 5.24
Appliance of analog buffer to the reference voltage circuit

3 The operation principle of OLED

3.1 The definition of OLED

OLED stands for Organic Light Emitting Diodes and is also called organic light emitting diode or organic EL. It is a self-emitting organic material that emits light by using an electroluminescence phenomenon that emits light when a current flows through a fluorescent organic compound. It can be made like a thin film and driven at a low voltage. The image quality does not change or there is no afterimage even if viewed from the side, unlike the conventional LCD, due to a wide viewing angle and fast response speed. It also has a more favorable price competitiveness due to the image quality better than that of LCDs and the simple manufacturing process, when it is used for a small screen.

There are three kinds of primary color implementation methods, the three-color (red, green, blue) independent pixel method, the color conversion method (CCM), and the color filter method. Depending on the molecular weight of the organic material contained in the light-emitting material to be used, it is called a low-molecular OLED or a polymer OLED, and divided into a passive driving method or an active driving method.

3.2 Operating principle of OLED

3.2.1 Basic Structure

OLED, whether it is a low molecular weight compound or a polymer, is based on the same light emission principle. Holes injected from an anode and electrons injected from the cathode in the organic EL device and form a hydrogen like atomic state. Typically, this state is called as 'Exciton.' The exciton emits light energy after a certain lifetime. Since this light emission has a wavelength corresponding to the energy of the exciton, it is possible to change the color of the emitted light by changing the molecular design. This is a great advantage of using the OLED display device.

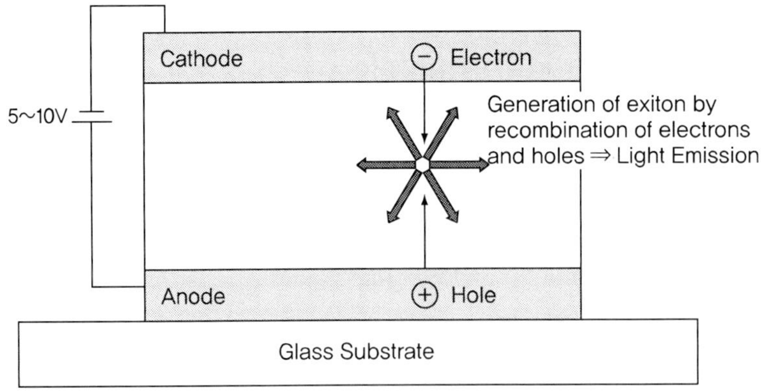

Figure 5.25
The basic principle of the OLED light emitting

The materials for forming the OLED light emitting device are divided into low molecular materials and polymer materials. A low–molecular material has a relatively low molecular weight it assumes after the film formation by deposition. A polymeric material is designed with a relatively higher molecular weight because it has one part emits light while the other part taking the fusibility, for the purpose of the solution process such as printing.

3.2.2 Emitting mechanism

OLED light emission is based on the mechanism of injection and transport of holes, exciton generated by the recombination of electrons and holes, and exciton energy transfer.

(1) HOMO and LUMO

Among the orbits filled with the electrons in the ground state, an electron orbit that has the highest energy is called the highest occupied orbital or HOMO (Highest Occupied Molecular Orbital). Among the orbits that are not filled with electrons, an electron orbit with the lowest energy is called the lowest occupied molecular orbital or LUMO (Lowest Unoccupied Molecular Orbital). E_V is HOMO and E_C is LUMO.

(2) Two electrons arrangement

Before discussing the mechanism of the light–emission of OLED, it is necessary to understand what kind of electron arrangement should be made in the ground state and exciton state. If the electronic wave equation is solved with the assumption that two electrons go into two different states, two wave functions can be derived for each electron, corresponding to the spin quantum number $\pm 1/2$. A total of four wave equations can be generated, considering practical molecular state.

Here,

$$\frac{1}{\sqrt{2}}\chi_A(1)\chi_B(2)+\chi_B(1)\chi_A(2) \cdot \frac{1}{\sqrt{2}}\alpha(1)\beta(2)-\beta(1)\alpha(2)$$ is called a singlet state;

and

$$\frac{1}{\sqrt{2}}\chi_A(1)\chi_B(2)-\chi_B(1)\chi_A(2)\alpha(1)\beta(2)$$ is called a triplet state.

The emission resulting from the energy transfer, from the singlet state to the ground state, is the fluorescence and the emission resulting from energy transfer, from the triplet state to the ground state, is the phosphorescence.

(3) Singlet exciton and triplet exciton

In the ground state, an electron exists as a spin-up and a spin-down like S_0 in Figure 5.26. In the excited case, the electron entering the top level is excited to the same spin state (S_1) or the reverse spin state for some reason (T_1). The former is called a singlet exciton, the latter is a triplet exciton.

When two electrons enter the same orbit, the spin state of the two electrons has to be different from each other. Because of this reason, the spin direction needs to be changed in the process of energy transfer from T_1 to S_0. Since it takes some time for this process, the lifetime of exciton of phosphorescence is relatively longer than that of fluorescence. There are many cases where the triplet excited state T_1 cannot be recognized by visible light because the energy of the triplet excited state is lower than that of the singlet excited state, and it becomes long wavelength.

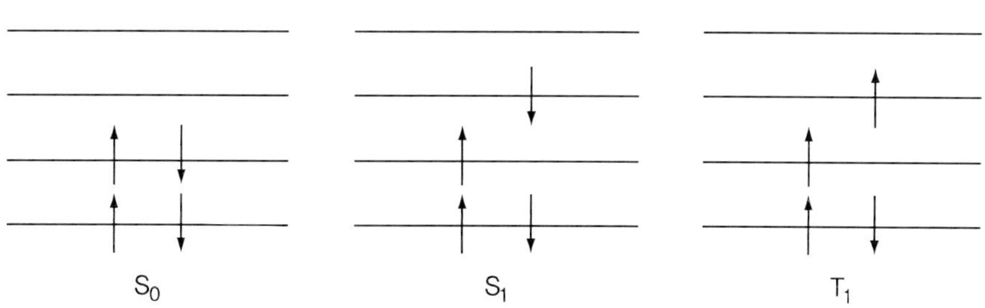

Figure 5.26
Electron arrangement of ground state and excited state

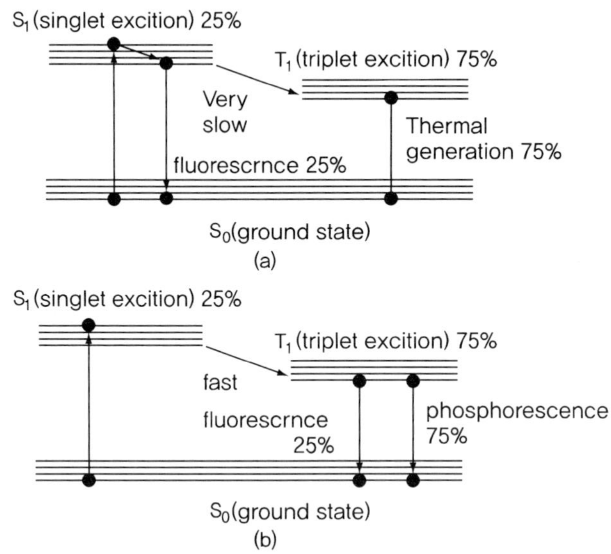

Figure 5.27

(a) Fluorescence type (b) Jablonski Diagram

(4) Injection from the electrode of the charge

The OLED device emits light by injecting electrons and holes into the device and recombining them. By a strong electric field, the electrons are injected into the LUMO energy level of the ETL (Electron Transportation Layer) layer from the cathode, and the holes are injected into the HOMO energy level of the HTL (Hole Transportation Layer) layer from the anode. In order to understand the charge injection mechanism in the OLED device, it is important to understand the Schottky thermionic emission, tunneling injection and the vacuum level shift model.

(5) Transportation and recombination of the charge

1) Transport of charge

Considering a case that charges are injected into the electrode in OLED when all HOMO are already full, the injected electrons enter the LUMO. The electrons entering the LUMO are transported by an electric field, forming the electric current. If the holes are injected into the electrode, the holes enter the HOMO. These holes are transported by an electric field, forming an electric current.

2) Generation and recombination of exciton

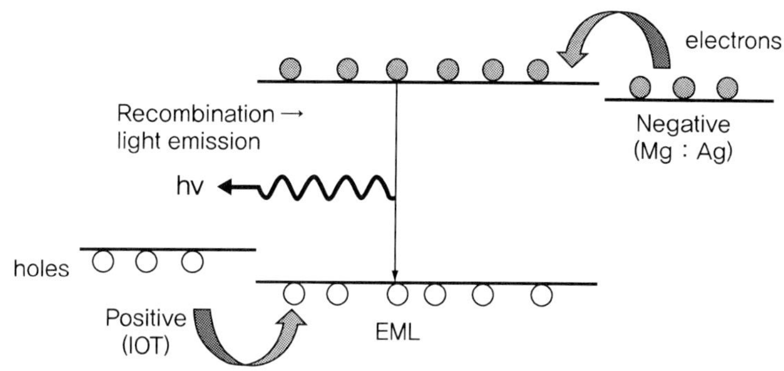

Figure 5.28

OLED band diagram by a single organic layer

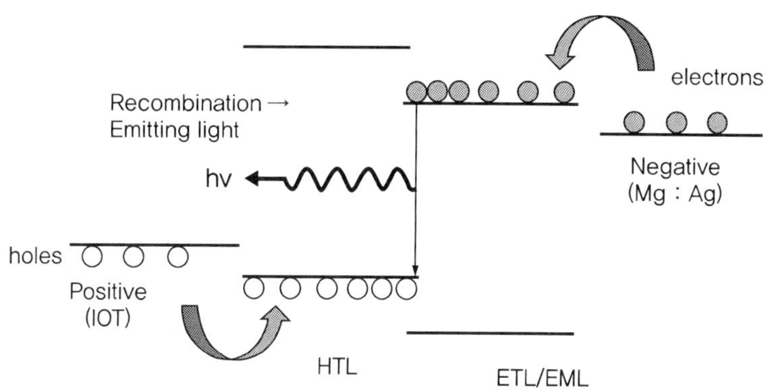

Figure 5.29

Efficient hole–electrons recombination by the multi–layer structure

The injected electrons and holes to the OLED device is shown in Figure 5.28 and forms the exciton. When the exciton lifetime ends, the light of a wavelength corresponding to the energy difference is emitted. If the probability of recombination of electrons and holes is increased, more light is emitted. If the energy difference between the HOMO/LUMO bands using the multi–layer thin film made of various materials interrupts the flow of the electrons or holes. as shown in Figure 5.29, the position where the electrons and the holes recombine can be controlled. If the charge is trapped, the recombination probability of electrons and holes increases and thus, a high efficiency can be implemented.

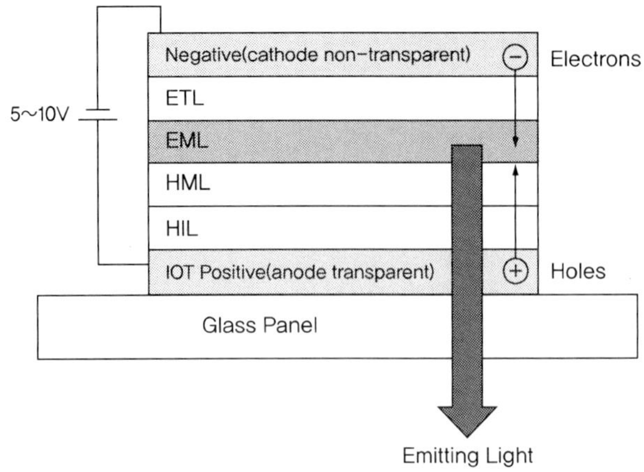

Figure 5.30

The OLED schematic diagram of the multi-layer structure sharing functions

〈OLED under a multi-layer structure sharing functions 〉

Material layers having each function used in the OLED are displayed as shown above.

- An electron injection layer (EIL) injecting electrons from the cathode
- An electron transport layer (ETL) transporting electrons injected from the EIL to the light-emitting layer
- A hole injection layer (HIL) injecting holes from the anode
- The hole transport layer (HTL) transporting holes injected from the HTL to emission layer
- A light emitting layer (EML) taking the emitting function by using the Exciton generated by the electron-hole.
- A hole blocking layer (HBL) blocking the movement of holes

The electron injection layer and the hole blocking layer are sometimes used to control the location of generation of the excitons, by blocking the movement of holes. The choice of the electrodes also become an important factor.

(6) Doping

Finally, light emission efficiency of the OLED can be increased or the color can be changed by adding a small amount of luminescent material to the organic layer, which is called doping. The material used for the doping is called a dopant. This doping method is effective to realize low power consumption and high luminance.

3.3 Drive system of PMOLED and AMOLED

3.3.1 PMOLED

In Figures 5.31 and 5.32, there are OLED devices between the vertical data line and the horizontal scan line in the PMOLED. R, G, B pixels are listed, and the data line and the scan lines are driven by the data driver and the scan driver, respectively. If a voltage is applied to a specific data line and the scan line, the current flows to the pixel where the two lines meet, emitting light.

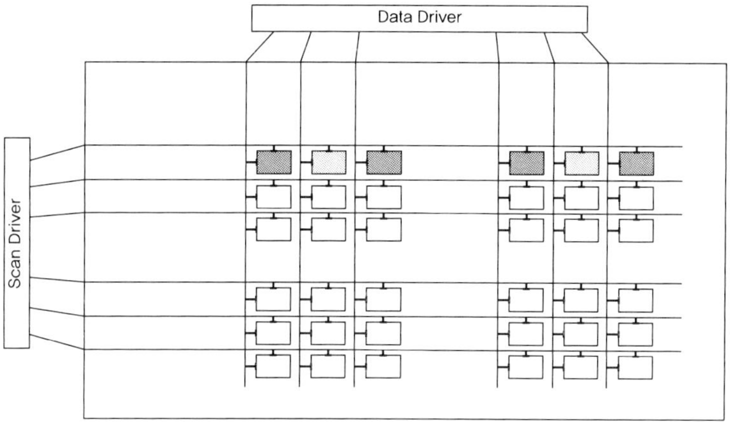

Figure 5.31
Matrix driven form of passive OLED display

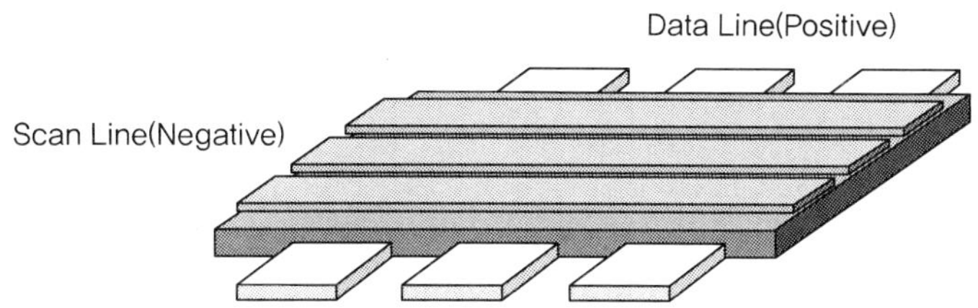

Figure 5.32
The structure of the passive OLED displays

Figure 5.33 shows the equivalent circuit of the PMOLED. The PMOLED is shown in the form of being attached to a parasitic capacitance in parallel with the diode. The current used to charge and discharge the capacitance does not contribute to light emission and thus, leads to the power increase and the reduced lifetime. Therefore, it is important to prevent unnecessary charge or discharge in the design of the PMOLED. The PMOLED can secure the charging time by driving a voltage lower than the extinction voltage with respect to lines that do not emit light when applying a voltage to scan lines and data lines. In addition, the crosstalk (or the difference in pixel brightness) affecting a display image is caused by the difference in the charged amount between the pixels that are not charged in advance and those pre-charged in the previous stage. To avoid this crosstalk phenomenon, the precharge needs to be reset by connecting the data lines and the scan lines.

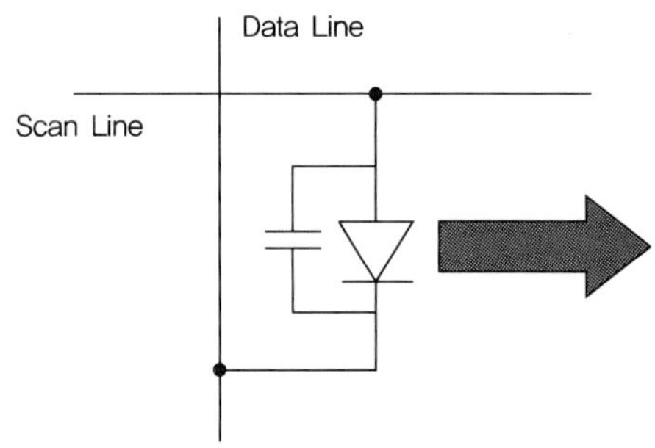

Figure 5.33
Pixel circuit of a passive OLED

3.3.2 AMOLED

When realizing a large, high-resolution display in the OLED, the AMOLED is needed, unlike the PMOLED.

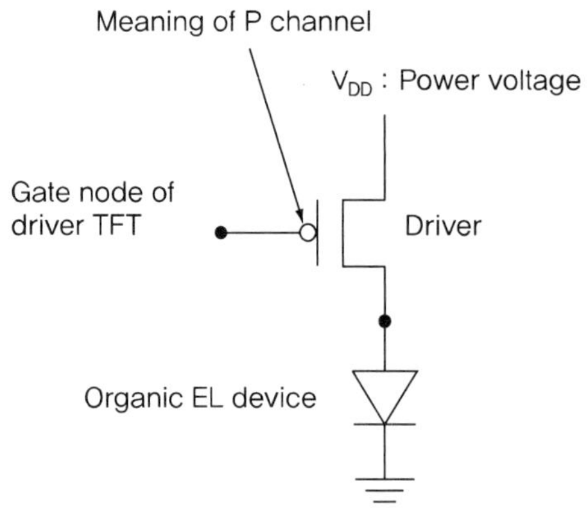

Figure 5.34
Current supply to the OLED element by the driver TFT

Figure 5.34 shows the state that the driver TFT supplies a current to the OLED device. The current of the driver TFT controls the luminous intensity of the OLED by changing the voltage of the gate node.

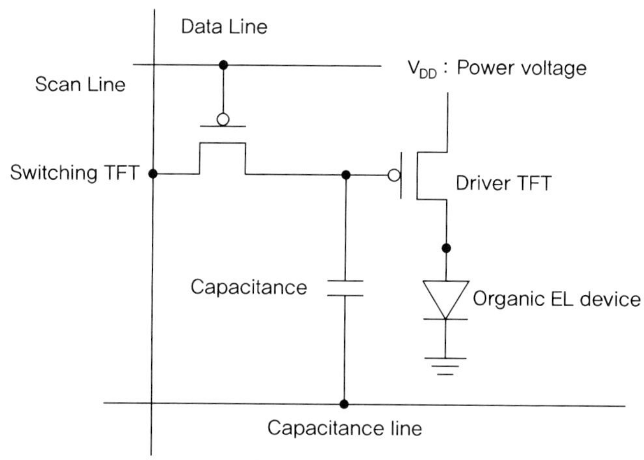

Figure 5.35
2T1C Drive circuit

Figure 5.35 is called 2T1C, which is the most basic pixel circuit of AMOLED display. The reason it is called 2T1C is that two transistors and one capacitance exist. This circuit adjusts the gate node units of the driver TFT that modify the luminance of the OLED device by using a switching TFT. The switching TFT and the method of writing the signal using the capacitance are the same as that of the LCD pixel circuit.

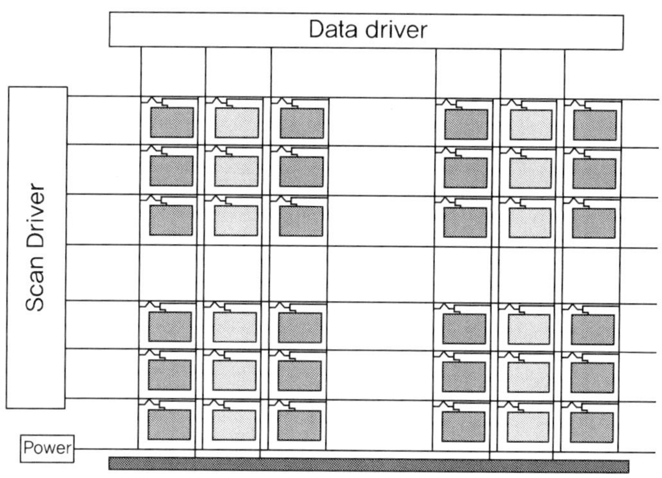

Figure 5.36
Active OLED display

Figure 5.36 shows pixels along the vertical axis on the matrix. Since it is possible to use three primary colors of the RGB, it is used to express the desired color as the combination using each sub-pixel of the RGB. It is the strength of AMOLED that various colors can be displayed without the extra backlight.

The 2T1C circuit consists of a two-step process that the voltage is stored in the capacitor, and the stored voltage is converted to the current. Compared to LCDs, the pixel circuit is a little more complicated because it consists of two parts: a switching transistor operated by a scan voltage and a driver transistor converting the stored voltage to the current as shown in Figure 5.35. The current of the driver transistor is determined by the power supply voltage as well as the voltage stored in the storage capacitor Cst. Compared to the LCD pixel, The 2T1C circuit has one more transistor and requires more V_{DD} lines as the power supply of V_{DD} is connected to the drive transistor. The other side of the drive transistor is connected to the OLED.

The current determined by the voltage applied to the gate of the transistor and the characteristics of the current and voltage of the OLED continues to flow for one frame. Depending on the magnitude of the voltage applied to the gate of the driving transistor, the current is varied, changing the brightness of the OLED. Thus, the desired brightness can be obtained by adjusting the voltage.

3.4 The pixel circuit of AMOLED

3.4.1 TFT threshold voltage variation and compensation

The brightness of LCD is determined by the voltage applied to the liquid crystal when driving the voltage. On the other hand, the brightness of OLED is determined by the current flow in the device when driving the current. Although the TFT characteristics somewhat lack uniformity, the image quality is not affected for LCD as shown in Figure 5.37.

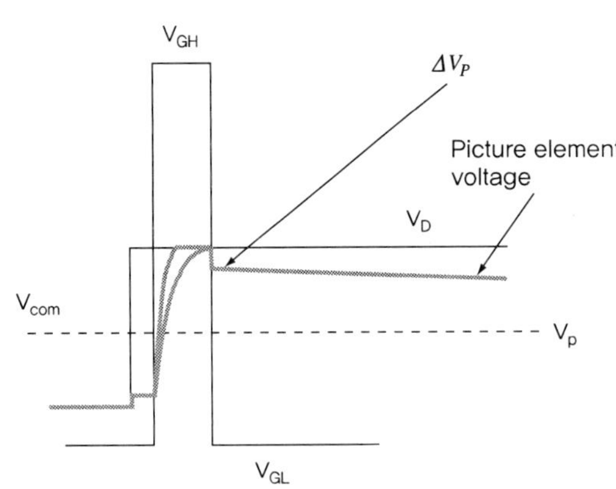

Figure 5.37
The characteristics of LCD according to the TFT characteristic change

Even if the charging time is changed, the brightness does not change unless the voltage is changed as shown in Figure 5.37. Therefore, there is no significant change in the image quality when the characteristic changes of the transistor are insignificant. The current flowing through the driving transistor is changed by the voltage applied to the gate of the driving transistor, so if the characteristic of the driving transistor is changed, the current and the brightness of the pixel will also be changed, resulting in a problem in the image quality.

There could be many characteristic changes of the transistor; the change of threshold voltage is a major one among them. If the change of mobility is not large, the characteristic change is mostly caused by the variation of the threshold voltage. The problem is that the current of the OLED varies with the threshold voltage. To solve this problem, several methods compensating the changes of transistor threshold voltage have been suggested. This method keeps the current, passing through the OLED, as consistent, by changing the applied voltage to the gate of the driving transistor as much as the variation of the threshold voltage.

As the threshold voltage of the driving transistor has to be extracted to reflect the change of the threshold voltage, the diode structure using a transistor is used. The diode configuration for extracting the threshold voltage used in the P-channel TFT is as shown in Figure 5.38, where the gate is connected to the ground and a power source is connected, to which the positive voltage is applied. If the power supply voltage is positive, the voltage of the gate electrode is lower than the power supply voltage, i.e., the negative voltage region. If this voltage becomes lower than the threshold voltage, the transistor will be turned ON; otherwise, the transistor remains OFF. Therefore, if the power supply voltage increases gradually, the initial state is OFF, and if it becomes higher than the absolute value of the threshold voltage, the transistor is turned ON.

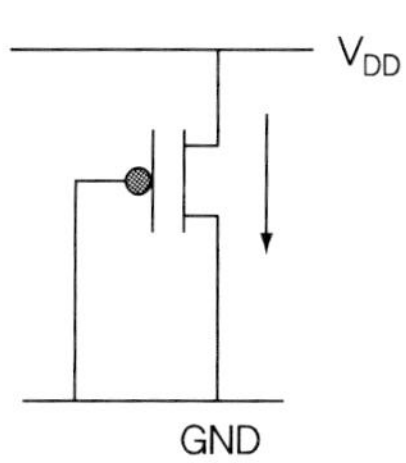

Figure 5.38
Diode configuration of P-channel transistor

Figure 5.39 shows the principle of a circuit extracting the threshold voltage.

If a high V_{DD} is applied, the total current flows in the direction of the arrow in the figure. In this state, the switch connected to GND is turned OFF, and then the current flows to the point P, increasing its voltage and gradually decreasing the difference between the point P and V_{DD}. When the voltage of the point P reaches a certain value or the voltage difference between the point P and the V_{DD} reaches a threshold voltage, the transistor is OFF. If the switch is turned OFF as shown in Figure 5.39, the supply power is cut off, by which the switch connected to the point P where the charge in the capacitance is stored will also be OFF, and the extracted voltage will be stored at the point P.

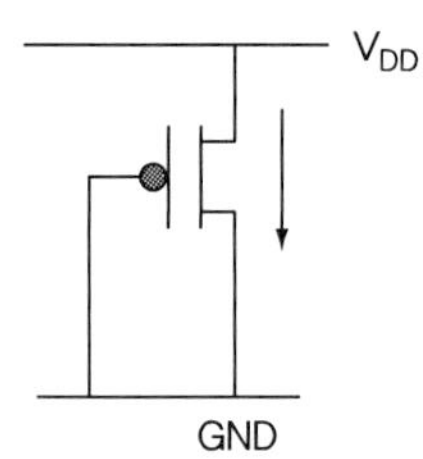

Figure 5.39
Threshold voltage extraction of a P-channel transistor

If each switch is replaced with a transistor and the OLED is connected, the circuit is configured as shown in Figure 5.40.

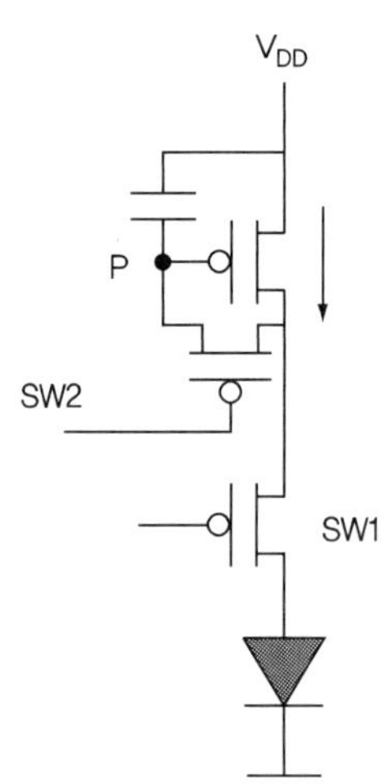

Figure 5.40
The threshold voltage compensation and OLED

After the threshold voltage is extracted, the data have to be loaded. If the data signal is directly connected to the point P, the extracted data will disappear. Thus, the voltage should be added or subtracted through the capacitive coupling, as shown in Figure 5.41. The C1 plays that role, and the extracted threshold voltage is changed in proportion to the voltage variation of the left node of C1. Therefore, as the extracted threshold voltage increases or decreases as much as the variation of the data voltage, and the voltage is adjusted automatically according to the threshold voltage, allowing the current, that is flowing through the OLED, to be constant.

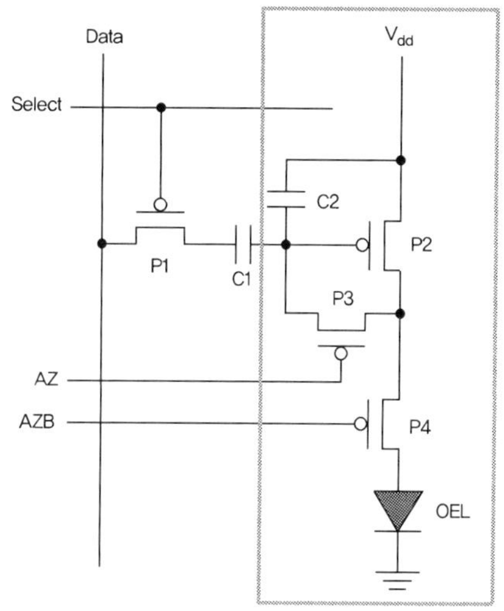

Figure 5.41
The threshold voltage compensation AMOLED pixel circuit

The voltage supplied to the data line when driving the OLED is converted to current by the driver transistor, then it is applied to the OLED. Since the signal is the voltage, it is called the voltage writing method. When current is used instead of the voltage, it is called the current writing method; in this case, the current is used as the signal instead of the voltage.

3.4.2 The current writing method pixel circuit

Figure 5.42 shows an example of the current writing method. If the data current is supplied to the pixel, the current flow through the transistor in the OLED is equal to the current applied from outside. Then a constant voltage, which can allow the data current, is set up in the gate of the driver transistor. This voltage is set up by reflecting the characteristics of the transistor such as the threshold voltage and the carrier mobility in the transistor, and thus all such changes in the voltage or charge mobility can be compensated. In the meantime, the transistors T_3 and T_5 remain ON, and T_6 is OFF.

While the data current flows, once the voltage is set up at the gate of T_2 and then T_3 becomes OFF, the setup voltage remains the same by the capacitance C. Because the setup voltage can produce the data current, the data current is kept constant. Then once T_5 is turned OFF and T_6 is turned ON, the V_{DD} voltage is applied to the T_2, with the data current remaining the same. Although the drain-to-source voltage of T_2 is changed, the data current is kept constant in the OLED, as applied from the outside, because the transistor is driven in the saturation region.

If the threshold voltage or the carrier mobility in the transistor is changed, the voltage setup is changed accordingly and amount of the current that flows through the transistor is the same as that of the data current. Thus the characteristic changes of the transistor are compensated. However, if the parasitic capacitance of the data lines is large and the data current is small, the time to charge the parasitic capacitance becomes longer causing the driving time to be longer.

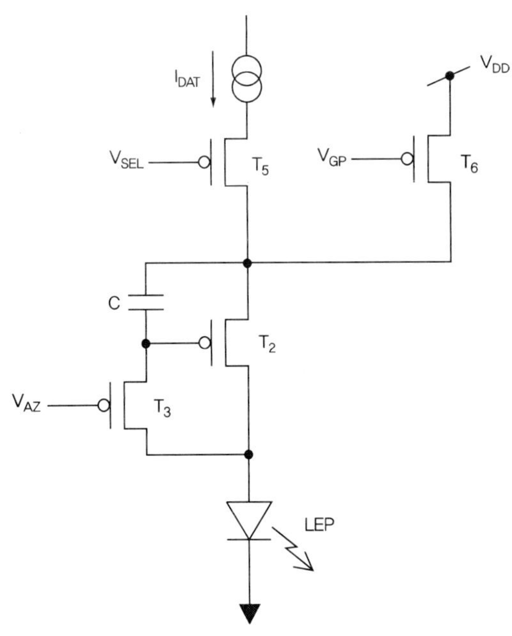

Figure 5.42

Configuration of the current writing pixel circuit

The basic compensation circuits have been explained so far, but other various compensation methods have also been suggested. Furthermore, there are circuits to make the driving easier and reduce the wiring and methods that compensate for the characteristic changes from outside.

3.4.3 IR compensation

The current flowing in the OLED decides the brightness of pixels. The current flows in the OLED pixels by the power supply voltage applied to the pixel and the voltage drop occurs due to the resistance of the V_{DD} line by the current flowing along the V_{DD} line. As the voltage drop will be the resistance multiplied by the current, the farther the V_{DD} line is away from the entrance, the more the voltage drop is. The longer the distance is, the higher the resistance is. Therefore, with the increase in distance, it is necessary to compensate for the change in brightness due to the reduced voltage than the applied voltage. The circuits that compensate for the voltage drop by the product of the current and the voltage (IR drop) are also being used.

3.4.4 The digital driving pixel circuit

In addition, the digital driving that uses only two tones of the brightness of OLED, black and white, was proposed in order to improve the uniformity of the image quality. That is, the current flowing through the OLED is only of two modes, the black mode and white mode, and the brightness is adjusted by controlling the current–flowing time. It takes an advantage of the fact that the deviation of brightness according to the signal voltage is large in the medium brightness, but small at both ends of a bright signal or a dark signal, as shown in Figure 5.43. In such a digital driving, one frame is divided into several sub–frames, and the mid–range brightness is implemented by controlling the black and white brightness in each frame. In addition, a number of different digital driving methods have been proposed.

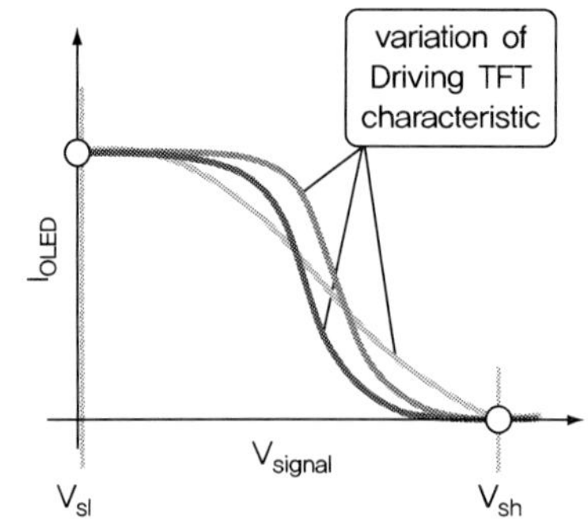

Figure 5.43

The relationship between the signal and the brightness applied to the OLED

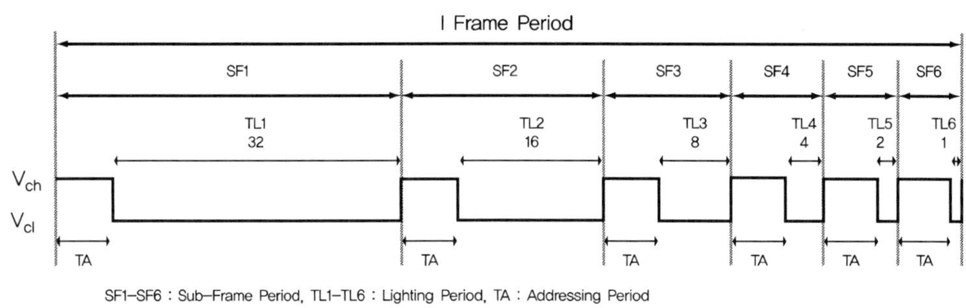

Figure 5.44

The example of the composition of several frames

References

[1] Diasang. Design and Operation of TFT-LCD Panels. Trans. Sang-Jik Kwon, et al. Seoul: Duyangsa, 2009.

[2] Wei-Chung Cheng, et all. "Power minimization in a backlit TFT-LCD display by concurrent brightness and contrast scaling." Proceedings of the conference on Design, automation and test in Europe-Volume 1. IEEE Computer Society, 2004.

[3] Inseok Choi, et all. "Low-power color TFT LCD display for hand-held embedded systems." Low Power Electronics and Design, 2002. ISLPED'02. Proceedings of the 2002 International Symposium on. IEEE, 2002.

[4] Sung Min Cho. "OLED displays and lighting Technology Status." Lighting·electrical equipment 23.1 (2009): 25-32.

[5] Dae gyu Moon. "OLED Displays and Lightings." Electronics journal 35.8 (2008): 49-59.

[6] Chang Hee Lee. "[Special: Display Ⅱ] OLED display technology trends." Optics and Technology 8.1 (2004): 13-21.

[7] Korean electricity journal. Color TFT liquid crystal display.Seoul: Intervision, 2007.

[8] Takatoshi Tsujimura. OLED display fundamentals and application. New Jersey: Willey, 2012.

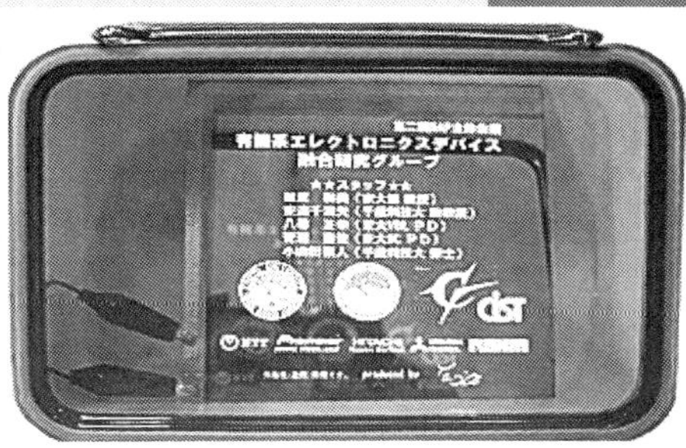

Organic Light-Emitting-Diode OLED

Chapter 05 → 06

1. Introduction and definition of Organic Light-Emitting Display
2. Luminescent Materials and Properties of Organic Light Emitting Diode
3. Structure and Operation Principle of OLED
4. Operating Principle of OLED
5. Trend of research and task of Organic Light-Emitting Display

1 Introduction and definition of Organic Light-Emitting Display

CRT, which had been the mainstream in the market until 2000, has been replaced with LCD due to the increasing interest in Flat panel displays, along with the needs and demands for the transportability, lower power consumption, reduced weight, and flat surface of displays. Meanwhile, [6]organic EL has a competitive edge in the small display market and is emerging as a competitor to conventional LCDs.

EL in Organic EL is a self-light emitting phenomenon of luminescent materials when an electric field or a current is applied to a fluorescent substance. The first EL was found in inorganic EL in the 1920s, and the light emission of organic EL was first observed in early 1953. After the light emitting phenomena from direct current was discovered in 1963, Organic EL became popular. In 1987, American Eastman Kodak Corporation obtained many basic patents on the organic EL device featuring high brightness and low voltage driving method. Such devices were commercialized by Pioneer Corporation. They produced and sold FM text broadcasting receiver for vehicles, using east Pioneer Corporation's organic EL, in 1997, for the first time.

Figure 6.1 shows the basic structure of a light emitting display which uses organic matters. It consists of transporting layer of electrons and holes having a thickness of about 200 nm between the cathode and the anode metal electrode, and a luminescent layer emitting light by the recombination of the electrons and holes. Organic EL uses the luminescence of light at a certain wavelength; the holes and electrons injected into the organic thin film through both the anode and the cathode freely move around and recombine with each other in the organic material. By the recombination precede formation of exciton, from which a specific wavelength of light will be emitted due to the energy of the exciton.

6) Designation of organic EL
The various terms to refer to organic EL in Korea are as follows:
- Organic Light Emitting Diode
- Organic EL (OEL ; Organic Electroluminescence)
- Organic EL (OLED ; Organic Light-Emitting-Diode)

As shown above, there are various names for OLED in Korea, but "Organic EL" is prevalent in Japan. However, it is often called "Organic LED" in Europe and America, especially among those who treat polymer materials.

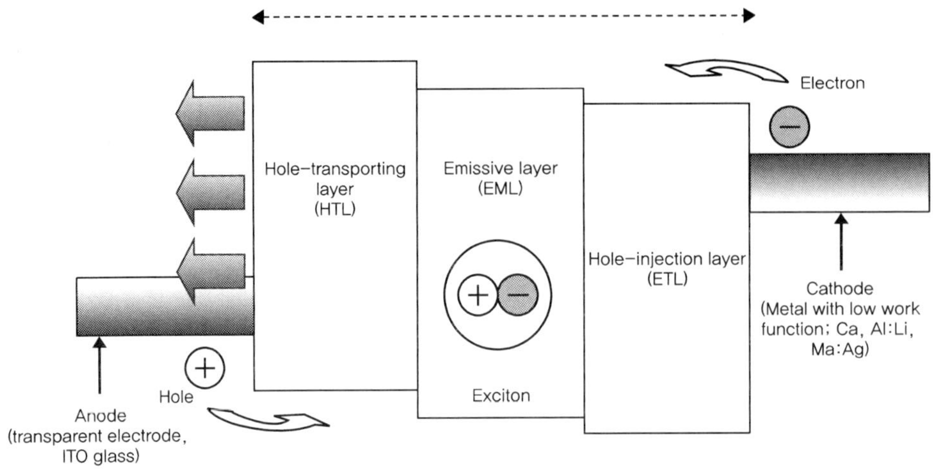

Figure 6.1
basic structure of organic EL

Organic layer can be classified into Low molecular weight type and High molecular weight type depending on the material. Low molecular weight type is capable of early mass production and is easy to develop as its properties are well-known. However, it had a short lifetime, and it was hard to use in a large screen due to low luminous efficiency. On the other hand, the high molecular type has higher thermal stability than the low molecular type, has excellent mechanical strength, has the same color tone as natural color, and has a low driving voltage, which is advantageous for display applications. However, the reliability of the materials is insufficient and active research on polymer materials is underway. In addition, the organic EL can be divided into a passive mode (PM) and an active mode (AM) depending on the driving method of the device. Passive driving instantaneously emits light from the organic EL element at a high brightness, but the active driving can continuously emit light from the organic EL at the desired brightness. Thus, in the case of PM, the instant brightness has to get higher as the resolution increases. The PM is not suitable for a large screen because electric power consumption for the shorter time of operation. On the other hand, AM is capable of operating at a low current, as it emits light continuously. Its pixel-shaping process is easier than that of PM, and it is possible to manufacture high-resolution panels. However, AM needs to use more than one transistor per organic EL, which requires the more complex manufacturing process and with a greater expense.

Table 6.1 Classification of organic EL

Classification		Pros	Cons	Original technology Owner
Organic EL luminescent material	Low Molecular Weight	Capable of early mass production	Complex manufacturing process	Eastman Kodak
	Hight Molecular Weight	- Capable of using Flexible substrate - High color quality	Low reliability of material	CDT
Driving Method	PM	- Low price - Simple manufacturing process	High power consumption	Pioneer
	AM	- Suitable for Large screen - Low power consumption	High price	Sanyo

Organic EL began to be used for small/medium-sized displays under 10 inches. Especially, it is developing more rapidly than FED, due to its unique properties and advantages. That is, organic LED has great durability as it is a complete solid-state device, whereas LCD, PDP and FED use liquid, gas, and vacuum medium, respectively. As it is a self-light-emitting element, it has high luminance and efficiency, excellent contrast, wide viewing angle and it does not require a backlight unit. Also, it has merits of high operating speed and low operating voltage. In comparison, inorganic EL demands large operating voltage and is hard to be used in fabricating a large screen. Furthermore, since inorganic EL is unable to emit blue light, it is difficult to make a full-color display which is a requirement of modern displays. However, for organic EL, various luminescent materials and colors are available, which makes organic EL as excellent in full-color display than in-organic EL. Besides, organic EL can be made into a large screen, with great luminance and excellent emitting efficiency. The injection layer and luminous layer can be composed of a small molecule organic matter or a polymer molecule organic matter. Table 6.2 summarizes the pros and cons of organic EL and inorganic EL.

Table 6.2 Pros and cons of organic EL and inorganic EL

Classification	Low molecular weight organic EL	High molecular weight organic EL	Inorganic EL
Materials	Athracene, Alq3	PPV, PPP, PT	ZnS, GaN, SiC
Strength	- Excellent luminous efficiency - Full-color display - Formation of multi-layer - Easy to pixelation - Suitable for Large screen	- Low driving voltage - Excellent mechanical strength - Polarized light emitting - Suitable for Large screen - Simple thin film process	- Impact resistance - Durability - Environmental resistance
Weakness	- Crystallization - Low magnetic stress - Diffusion through multiple-layer film	- Hard to pixelate - Hard to form multi-layer	- High AC driving voltage - Hard to emit blue light - High price
Prospect	Expansion of commercialized market	Expansion of commercialized market	Using special field

2 Luminescent Materials and Properties of Organic Light Emitting Diode

Organic EL materials can be divided into the small molecule organic materials and the polymer organic materials. The classification and strong/weak points of organic EL displays are summarized in Figure 6.2

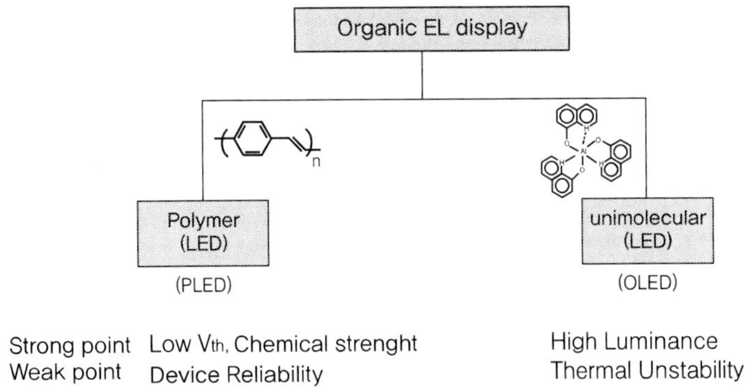

Figure 6.2
Classification of Organic EL

(1) Small molecule organic materials

Small molecular organic materials include blue coloring andracene, phenyl substituted cyclopentadiene derivatives, and DVPBi. 1,2–phthalo–perinone derivatives emitting yellow (580 nm) are among the most representative materials, and there is Alq3 emitting green (550 nm) light (see Table 6.3). By doping this material with organic pigments, it can emit light in a wide range of colors, from green to red. For example, it is possible to manufacture a material that emits red light by doping DCM dye in Alq3.

Table 6.3 Small molecule organic materials

Type	Alq3	Anthracene	Phenyl-substituted Cyclopentadiene	Perylene
Light Color	Green (510nm~530nm)	Blue	Blue	Red

As mentioned above, these small molecule organic EL have merits such as low driving voltages and relatively high brightness. However, there are also problems like continuous light emission, stability and quantum efficiency.

The low–molecular system has the merit of being able to develop the organic luminescent material without difficulty because it is easy to understand the characteristics of the organic luminescent material. However, it is difficult to make it into a large screen and the method has low productivity because it has to go through the dry process using a vacuum equipment, such as thermal evaporation and vapor phase deposition etc. Eastman Kodak Co. holds the patent for this method, and Asian countries including Japan, Korea, and Taiwan have been actively developing the method.

Small molecular materials can be divided into hole injecting/conducting materials, electron conducting materials, hole–restricting materials and light emitting materials depending on their function. The light emitting materials can be divided into fluorescent materials caused by the singlet excited state and phosphor materials caused by the triplet exciton, depending on the luminescent mechanism. Also, they can be divided by the luminous color into blue·green·red emitting materials and yellow·orange emitting materials, which are additionally needed for the realization of better natural colors.

The desirable properties of all materials to be used in organic EL are high purity and capability of vacuum deposition. It should also have high heat stability at a glass transition temperature and thermal decomposition temperature. Further, it must be amorphous to prevent the destruction of a device by crystallization caused by joule heat during the device operation.

1) Hole injecting/conducting materials

The hole injecting/conducting materials are those that make it easy to inject holes from the anode. To reduce the hole injection barrier, the ionization potential should be similar to that of the anode, i.e., ITO (Indium Tin Oxide), and its interfacial reliability with the ITO should be high.

To increase external quantum efficiency, the absorption into the visible light region should be prevented as much as possible. The materials known to be currently used are CuPe (Copper phthalocyanine) and PEDOT. In the case of CuPe, it has high heat stability, but it has a problem with absorption in the visible light region. In the case of PEDOT, because it uses spin casting, it is hard to be applied to the process of single molecular organic electric light emitting diode that has vacuum deposition as the main process.

Hole conducting materials easily conduct holes and restrict electrons within the light emitting region, by which they increase the probability of forming an exciton. Thus, it is desirable for hole conducting materials to have a high conductivity, in addition to the basic properties mentioned above. Among these materials, an aromatic amine can stabilize cation radical generated when holes are injected is mainly used. R&D has been aimed to increase the glass transition temperature, which is the most important property.

For electron conducting materials, the compounds containing electron-drawing elements, which can stabilize anion radical generated when electrons are injected from the cathode, or the metallic compounds that greatly accommodate electrons are mainly used. For example, the compounds containing a functional group to draw electrons by resonance, such as cyanide, oxadiazole, triazole, etc., and the compounds like Alq3 are widely used. However, Alq3 is most commonly used and very effective electron transport layer. An electron injection layer or an insulation layer is often placed between the cathode and Alq3, and a hole confinement layer between Alq3 and the light emitting layer.

2) Light emitting materials

Generally, light emitting materials can produce almost all colors that we want if it has red, blue, and green, the three primary colors. However, in the case of compound light, the combination could turn almost white, decreasing chromaticity. Thus, in order to realize perfect natural colors, it is good to have yellow and orange as well. Also, when only one light emitting material is applied, there are demerits of the low purity of colors and the low efficiency of light emitting.

Among the electroluminescent materials used, Alq3, which is a green light emitting material, is a material that has been developed. It exhibits sufficient luminescent properties for practical use. Alq3 is a material released from Kodak in 1987 and its luminous efficiency is more than 15 Cd/m^2. It is known that modified Alq3 among these derivatives has the highest luminous efficiency. In addition, several green emitting organic materials have been reported, but they are reported to be less than the luminescent properties of Alq3.

The material known to have good light emitting efficiency is a distryl compound of Idemitsu, DPVBi, but it is said that the modified DPVBi compounds have better light emitting property.

Among organic blue-light emitting materials developed so far, the material used for the 13-inch prototype of SONY is known to perform better than any other materials in terms of light emitting efficiency and color purity, although the structure has not been known. But these organic blue light emitting materials don't have enough lifetime for practical use due to some problems with long-term thermal stability of color purity and efficiency.

Blue light emitting materials of organic metal compounds have been developed by Japanese and Korean companies. However, they have disadvantages such as lack of chemical stability, difficulty of purification, low whiteness, especially whitening due to wide emission wavelength band at high luminance.

Light emitting materials have some limitations such as low efficiency in light emission. For example, even though quantum efficiency is 100%, the converted light intensity at 620 nm wavelength, and at 5 V bias is 5 1m/W. Some of the limiting factors for the quenching effect are, the molecular interactions through the highly concentrated pi electrons, low color purity due to a broad band light emission etc. Thus, in order to improve efficiency, the host material is doped. This doped host material can be used as a highly efficient red light emitting material.

But this method also poses some problems such as light emitting of the host material and the increase of trapping effect in the red light emitting material at a high voltage. Due to this, the red light emitting materials have been recognized as the most problematic light emitting materials for commercialization.

In the case of using one light emitting material only, the maximum light emitting wavelength shifts to the longer wavelength region, and there will be hillock peaks in the long wavelength region, causing color purity and efficiency to be reduced. To increase color purity and light emitting efficiency by energy transfer, the host/dopant are used.

Among the excited states of electrons in a compound, there are singlet state and triplet state. The probability of the singlet state is one–fourth, and the probability of the triplet sate is three–fourth. Falling from the singlet state to the ground state is called fluorescence. Falling from the triplet state to the ground state is called phosphorescence.

Therefore, the internal quantum efficiency limit of the fluorescent light emitting material is 25%, which is 75% for a phosphorescent light emitting device. Also in the system where energy transfer occurs from the excited triplet state to the excited singlet state, the theoretical internal quantum efficiency limit reaches to 100%. (There are different opinions as the internal light–emitting efficiency of some fluorescent devices is sometimes over 25%). These are phosphorescent light emitting materials. The desirable electric phosphorescent material is an organic metal compound, the central atom of which is transition metal with a high atomic number, where the energy transfer or the intersystem crossing easily occurs from the singlet/triplet excited state to the triplet excited state. Those materials known to date are Iridium (Ir), Platinum (Pt), Europium (Eu), and Terbium (Tb) compounds. The Europium and Terbium compounds are excellent in that the light emitting range is very narrow, but their stability is low which makes it difficult for vacuum deposition, and their charge mobility is also low, causing low efficiency.

In the case of a red light emitting device with Iridium compound as the host and Kodak's red light emitting material as the dopant, the external quantum efficiency increases from 1% to 4%. In the case of a phosphorescent device, the biggest demerit is that although the efficiency at low current density is high enough to be commercialized, the efficiency at high current density drops sharply because of extinction phenomenon by saturation of the triplet excited state. To improve this demerit, the lifetime of the triplet/singlet excited state of the luminous object for the dopant has to be short, the lifetime of the triplet/singlet excited state of the host should be long, and the material suitable for the inner system transfer is desirable.

Considering the properties of the light emitting materials developed so far and the light emitting property needed for commercialization, the light emitting material that can be commercialized is green, unlike red with light emitting efficiency problems or blue with lifetime problems. In the case of red, it is possible to gain enough efficiency by using electrical light emitting with phosphorescence, but at high brightness where extinction of phosphorescence occurs, the light emitting efficiency decreases sharply. Thus, there is a problem in applying red to passive devices requiring high brightness, but red can be used for active devices that do not need high brightness.

(2) Polymer organic materials

Traditionally polymer materials have been mainly used for their insulating property as in packaging materials or photoresist materials of semiconductor devices. From a traditional point of view, the invention of conductive polymers was very much surprising and ground breaking.

Conductive polymers started to be known by a team led by a professor Hideki Sirakawa when they synthesized poly acetylene by accident in 1977 and discovered that the polymer could have conductivity almost like metal by oxidation doping. After that, in 1987 Eastmann Kodak discovered EL in a coloring matter with a π-conjugation structure called aluminaquinone, and in 1990 Cambridge university observed EL in a π-conjugate polymer called poly (p-phenylenevinylene). Accordingly, the research on EL using polymer has rapidly become active. Polymer organic matters are mechanically stronger and more thermally stable than small molecule organic matters. There are various synthesization methods for the polymers to change the luminescent color form blue to red or white. Since the synthesis route is many, as in the case of organic materials, the properties of the synthesized molecule are represented thereby. Thus, the blue light emission is easy and full range of color is achievable. Displays using electroluminescence devices are said to be the best example of commercialization of polymer materials. To distinguish between the electroluminescent displays based on polymers and small molecules, a term PLED (Polymer Light Emitting Diode) is used instead of OLED.

Polymer materials that are used for organic ELs have different properties than those of small molecular materials. The molecular weight of polymer materials is usually ten thousand times higher than that of small molecular materials. The polymer materials are more thermally stable and mechanically stronger than small molecule materials. Two materials show a big difference in thin film processing as well. Small molecular materials go through a dry process that uses a vacuum equipment such as vacuum thermal evaporation or vapor phase deposition. However, polymer materials go through a wet process that forms a thin film by spin coating or ink-jet printing after melting in an appropriate solvent. Therefore polymer materials can be used as a technology that can solve the problems such as increase of the investment costs due vacuum equipment and limitation of enlarging display sizes, etc.

The driving voltage of polymer electroluminescent material is lower than that of the device made of small molecules and its mechanical property is better than that of small molecule. Thus, the polymer electrical light emitting device is advantageous for display applications. There are various ways to change colors of emitted light when using polymer materials. It is possible to display various colors from blue to red or white, by adopting various substitutions to the polymer chain, adjusting the conjugate length in the chain, adopting various shapes of rings in the chain, blending light emitting materials for different colors, and so forth.

However, there also is some weakness of polymer materials. It is not easy to enhance the purity of polymer materials, unlike small molecule materials. This is because either the catalyst added during the polymerization process of synthesizing polymers is not completely removed, or there are remaining impurities that are generated by the formation of variants with slightly modified chemical structure. Despite these problems, polymer materials continue to draw attention as organic EL materials because the polymers have a high potential in terms of large screen application.

1) Light emitting materials

π conjugate polymers have a chemical structure that single bond (or σ bond) and multiple bonds (or π bond) exist alternately. Thus, it has a π-electron that can move relatively freely along the polymer chain without being localized by these chemical bonds. Figure 6.3 shows the conjugate polymers and their chemical structure.

The major polymer materials include PPV(poly (p-phenylenevinylene), PPP(poly(p-phenylene), PT(polythiophene), and their respective derivatives.

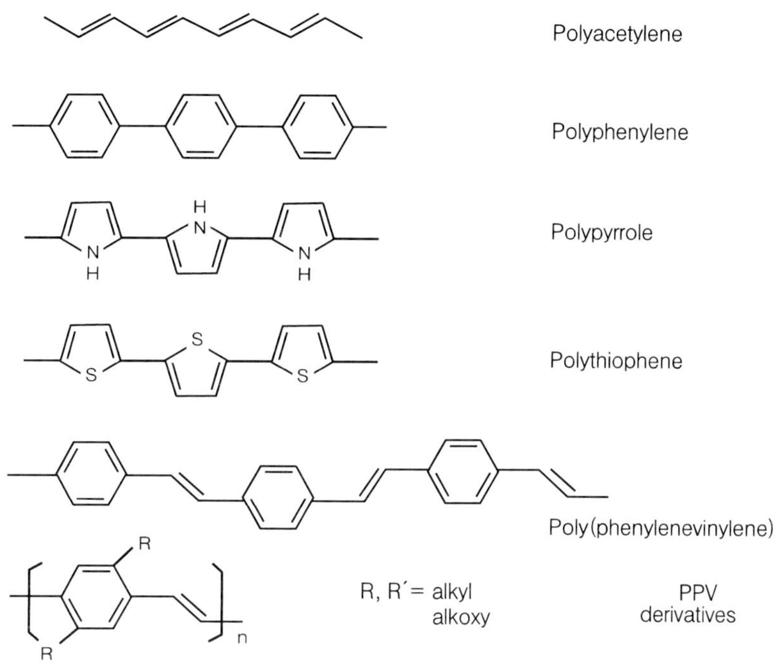

Figure 6.3

Conjugate polymers and their chemical structure

- poly(p-phenetlenevinylene) : PPV

PPV is a method to gain polymers that have large molecular weight and are soluble in organic solvents. With the maximum PL peak of 540 nm, PPV emits green light, and its light emitting efficiency is 0.1%; once about 14 V of voltage is supplied, electricity starts to flow.

Recently, MEH–PPV (poly(2–methhoxy–5–(2–ethylhexylxy)–1, made by replacing phenyl group of PPV with an alkyl group or alkoxy group to increase solubility, and draws high interest.

MEH–PPV can be made into a thin film as it is soluble in common organic solvents. Its maximum absorption wavelength is 510 nm; its PL spectrum moves toward long wavelength as compared to the PPV, and the maximum peak appears at about 590 nm. The electric current starts to flow at a driving voltage that is about 4 V, lower than the current–voltage characteristics of the MEH–PPV. This derivative's light emitting efficiency is about 1%.

To increase the relatively low light emitting efficiency, CN–PPV was developed by adopting CN that has large electron affinity to PPV. It has become possible to emit blue light, which had been a major problem of PPV and its derivatives, by developing CNMBC (conjugated–nonconjugated multibook copolymer) with the conjugate length evenly adjusted.

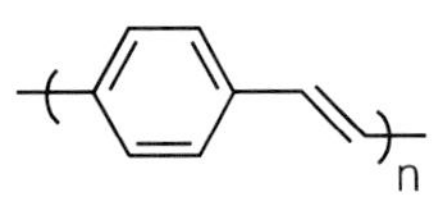

Figure 6.4
PPV

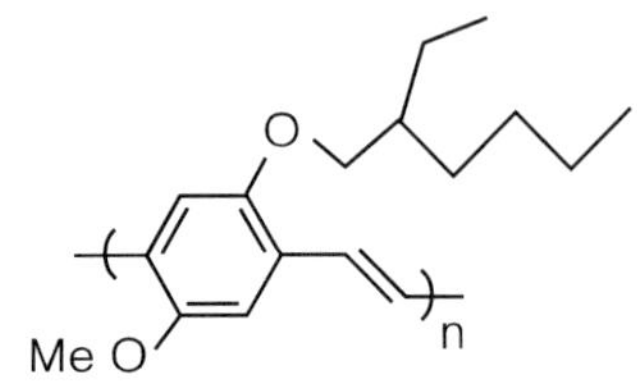

Figure 6.5
MEH–PPV

– poly(p–phenylene) : PPP

PPP is a polymer that has high durability and heat resistance, bearing one single form. The property of PPP can be adjusted to some extent depending on the synthesization method. It has very high crystallizability. It is hard to handle and not easily dissolvable. It remains stable up to 450 degree celcius in the air. Regarding the current–voltage properties of a device made of the PPP precursor, it requires about 10 V of the driving voltage, and its light emitting efficiency is about 0.01%~0.05%. The light emitting wavelength of the PPP is about 460 nm.

In the case of the PPP generated by the precursor, its processability of making various structures is difficult, machinability is not good as the final product is insoluble. In order to improve this, PPP synthesis by using soluble materials has been developed in a way that replaces alkyl group or alkoxy group with PPP material, and thus light emitting efficiency and organic solvent solubility of PPP have been enhanced. Also, by producing a device using polymers of PPP group, it is possible to generate green light; the device can operate at a low driving voltage, and it is possible to control the color of emitting light as blue and yellow by controlling the pi–conjugate.

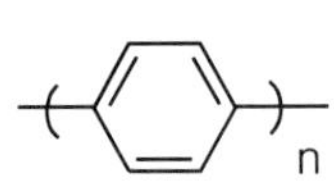

Figure 6.6
PPP structure

– polythiophene : PT

PT and its derivatives can be synthesized by chemical and electrochemical methods. They are very stable in air and water. PT has different light emitting wavelength and emission intensity depending on the length of an alkyl group, temperature, and solvent. For example in the case of poly(3–thiophene), as the temperature increases, the absorption peak moves to a short wavelength from ultraviolet absorption spectrum, and light emitting intensity increases as well.

Pi–pi metastasis of PT and its derivatives ranges from about 480 nm to 520 nm, and the absorption peak is in between 600 nm and 650 nm. It is possible to change the absorption peak wavelength with the various types of substitution products attached to the thiophene ring.

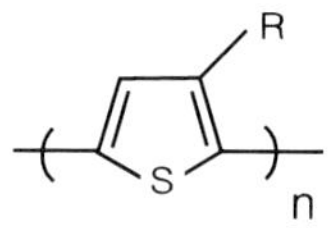

$R=C_6H_{13}$

Figure 6.7 PT structure

Below is a classification table of polymer organic EL materials.

Table 6.4 Polymer organic EL materials

Type	PPY	MEH–PPV	CN–PPV	CNMBC	PDSiPV	SiPPV
Maximum Peak	PL(540nm, Green)	PL(590nm, Red)	PL(710nm, Red)	EL (Blue)	PL(470nm, Blue-green)	EL(440nm, Blue)
Luminescence efficiency(%)	0.1%	1%	4%			
Driving voltage (V)	14V	4V			23V	10V

Type	SiPVKPPV	PPP	PPP–Alkoxy group substituted	PDHFMPV PDHFPPV	polyalkythi ophene
Maximum Peak	EL(450~480 nm, Blue)	PL(460nm, Blue)	PL(420nm, Blue)	PL(440~550nm) EL(426~482nm)	PL(590nm, Red)
Luminescene efficency(%)		0.01~0.05%			
Driving voltage (V)	10V	10V	10V	10V	

Development of polymer light emitting materials is an important part of the commercialization of polymer electrical light emitting devices. Many chemists have developed polymers with various structures, and based on these researches materials that have high potential for utilization start to appear. More specifically, the most major problem of polymer light emitting devices, i.e., lifetime issue, has been dealt with and resolved to some extent, with the high purification of materials. This outcome is playing a key role in the development of devices with better performance. When polymer EL elements are used in a display device, it is easier to enlarge the size of the display up to few meters, flatten the surface achieving a wider viewing angle, and to curve the screen. The screen can be much brighter. Also, if light emitting efficiency is enhanced, there is a possibility that polymer EL elements will be used for laser diodes, not limited to display devices.

3 Structure and Operation Principle of OLED

3.1 Structure and Operation Principle

(1) Layer structure

OLED basically has a structure, as shown in Fig. 6.8, consists of substrates (made of glass, plastic, etc.), the upper and lower electrodes (anode and cathode), where an organic emission layer between the two electrodes. The layer structure of organic EL is divided into a single-layer and a multi-layer; Fig 6.8 (a) is the single-layer where one organic layer exists, and Fig 6.8 (b) is the multi-layer where an electron transport layer and hole transport layer are laminated on the upper and lower parts of the emission layer for more active carrier injection. The multi-layer structure can lower the driving voltage by a two-level ejection process passing through the transport layer, not directly injecting the carriers. The electrons and holes are injected to the emissive layer, some of them may travel to the opposite electrodes, where they are blocked by the opposite transport layer and kept confined to the emissive layer. This makes it possible to adjust recombination of the electrons and holes, and to prevent reduced light emission caused by the formation of singlet exciton at the interface of the electrode and emission layers.

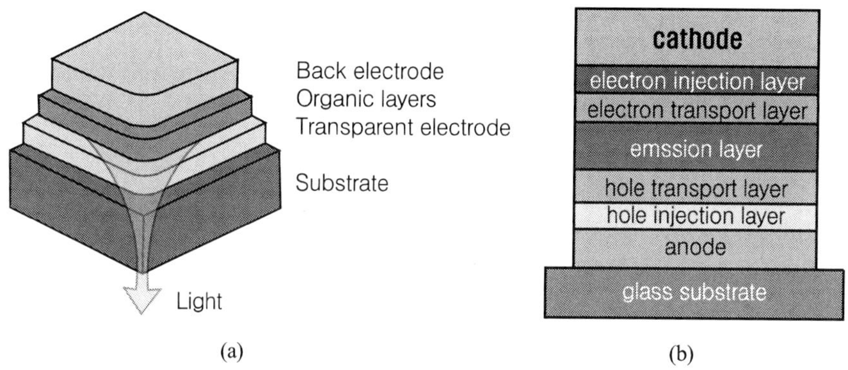

Figure 6.8

Organic ELD with Single–layer and Multi–layer Structures

(2) Light emitting mechanism

EL in the polymer can be explained as the following. First, the simplest structure of an Organic EL device is the one with light emitting conjugated polymer thin film that is placed between a high work function metal electrode and a low work function metal electrode metal electrode, as shown in Fig 6.9.

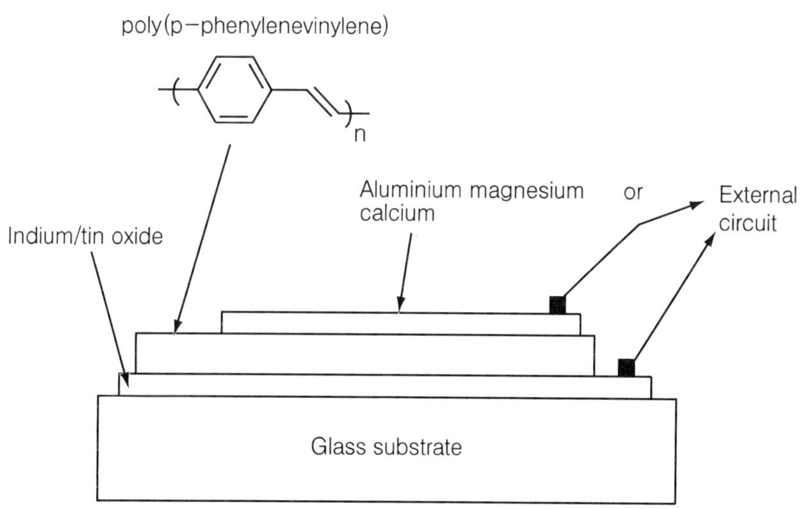

Figure 6.9

basic structure of polymer EL device

While high work function metal is used as a hole injection electrode, low work function metal is used as an electron injection electrode. To let the emitted light pass through the device, the substrates and one side of the electrode need to be transparent so that the light is not absorbed. For this purpose, ITO is commonly used as transparent electrodes, whose work function is high (5 eV) so that it can be used as a hole injection electrode.

By applying a forward bias to the electrode of high work function anode electrode and low work function cathode electrode, holes and electrons are injected into the emission layer. At this time, electrons move from the cathode to LUMO (lowest unoccupied molecular orbital), and holes move from the anode to HUMO (highest occupied molecular orbital); this is called carrier injection stage. Next, electrons and holes create negative and positive polarons respectively by electron–lattice interaction within the emission layer, which is called carrier alleviation stage. The created polarons move along polymer chains to the opposite electrode (carrier transport stage), and when they meet each other in the chains, singlet polaron excitons are created (excitons generation stage). When singlet polaron excitons vanish after emitting light, light with the same as the energy gap of the polarons is generated. Therefore, for the purpose of efficient light emission, there should be a balance between the injection and transport of holes in the anode and those of electrons in the cathode. This process can be expressed schematically in comparison with PL as shown in Fig. 6.10.

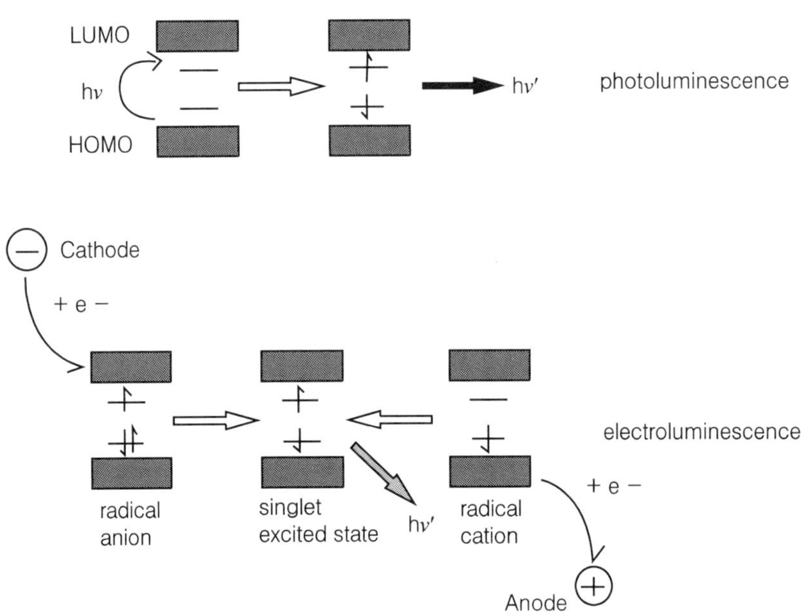

Figure 6.10
Comparison with photoluminescence and electroluminescence

(3) Luminescence efficiency

Luminescence efficiency of a device is divided into internal quantum efficiency and external quantum efficiency, and it can be adjusted by the current density of minority carriers of the device. The external efficiency is determined from the light intensity—current characteristic curve.

Internal quantum efficiency is expressed by the ratio of the number of photons created inside the device to the number of carriers injected from the outside electrode.

$$\eta_{int} = \eta_{injection} \times \eta_{recombination} \times \eta_{fluorescence}$$

$\eta_{injection}$ is the ratio of the number of electrons created inside the device to the number of excitons flown in the circuit, $\eta_{recombination}$ is the ratio of the number of singlet excitons to the number of excitons created in the device. Finally, $\eta_{fluorescence}$ is the ratio of the number of excitons consumed by light emitting to the number of singlet excitons.

External quantum efficiency is an experimentally measured luminescence efficiency, which is influenced by the refractive index of the luminescence material.

$$\eta_{ext} = \eta_{int} / 2n^2$$

The quantum efficiency can be maximized in two ways. One is matching LUMO level of polymers and Fermi level of the cathode each other, and the other is by adjusting the movement of electrons and holes in a way so that their recombination takes place at the center of the emission layer, not around the electrodes.

Table 6.5 Change in quantum efficiency depending on the cathode metal's work function in MEH-PPV emission single-layer devices

Metal	Work function(eV)	Quantum efficiency(%)
Ca	2.87~3.00	4×10^{-3}
In	4.12~4.20	1.6×10^{-4}
Ag	4.26~4.74	1.8×10^{-4}
Al	4.06~4.41	8×10^{-5}
Cu	4.65~4.70	8×10^{-6}
Au	5.1~5.47	5×10^{-7}

The former can be achieved by using a cathode with low work function, or using a polymer with high electron affinity so as to lower the barrier height. A minor change of barrier height can highly influence the operating voltage and luminescence efficiency. Therefore, if we choose a luminescent polymer and an electrode metal properly to match the polymer's HOMO and LUMO with the Fermi level of the hole injection electrode or the electron injection electrode, it is not difficult to obtain lower voltage than the operation voltage of 5 V. According to Table 6.5, by using a cathode with lower work function, it is possible to obtain higher quantum efficiency.

However, a metal with low work function has a defect that it is unstable to the air and humidity, which limits its use. Therefore, when a light emitting polymer having a high electron affinity is used while using a conventional metal such as Al as a cathode, the quantum efficiency can be increased.

The latter method is using two other polymers which have different band gaps that can compose a heterostructure. Because the electron mobility is lower than the hole mobility under this structure, in order to efficiently recombine electrons and holes, the movement of holes has to be suppressed to enlarge the charge density locally so that electrons can easily be injected by the space-charge field; and using high mobility electrons raises the probability of its recombination with holes, which will in turn enhance luminescence efficiency. A heterostructure can be made by putting a charge transport layer between the emission layer and the electrodes(Fig 6.11).

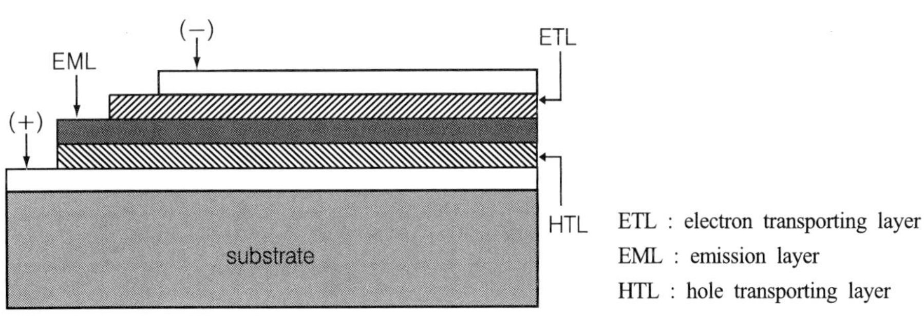

Figure 6.11
Structure of device including a transporting layer

A transporting layer includes an ETL (electron transport layer) and a HTL (hole transport layer); its key effects are as follows:

- An ETL efficiently transports carriers to luminescence polymers, by which it raises the probability of luminescence recombination within the polymer.

- It suppresses the movements of carriers as the HOMO or LUMO level of the material transporting electrons mismatch those of luminescence polymers to some extent. For example, an ETL blocks the movement of holes in the interface with the luminescence layer, by which the electric field within the ETL is increased, and electron injection from the cathode is improved.
- Since the hole mobility in the light emitting polymer is better than that of the electrons, the light emitting recombination for forming singlet excitons occurs mainly in the vicinity of the cathode, and the probability of recombination is greatly reduced by acting as a source of extinction near the cathode. When an ETL is introduced, it assists luminescence recombination between the emission layer and the transport layer, which is distant from the metal electrode, and thus it leads to a reduced number density of excitons.
- An HTL inserted between the hole injection electrode and the emission layer can lower the operating voltage of a device. This is attributed to a decrease separation in the Fermi levels between HOMO and ITO.

There is another way of increasing quantum efficiency. That is prolonging the lifetime of luminescence excitons; suppressing excitons reduces the probability of recombination with those causing non-luminescent recombination. This can be achieved by surrounding the conjugate part of luminescence polymer with non-conjugate parts within the same chain.

Quantum efficiency can also be enhanced by using polymer blends. Luminescence polymer blends are a good way of controlling colors and increasing the quantum efficiency. Polymer blends have the strength of component polymers, but also act as a new material different from component polymers. The blend of luminescence polymers and polymers with high electron transportability increases quantum efficiency in this way: the transportability of the injected carriers is improved, and the excitons generated by dilution effect are trapped; accordingly, the extinction rate due to the collision with defects will decrease. The quantum efficiency also highly increases when luminescent polymers with two different band gap energies are blended.

3.2 Manufacturing Process and Major Technology

The basic manufacturing process of an organic LED is shown in Fig 6.12. After depositing the anode on the substrates, proceed with the patterning against a glass substrate, that is mostly covered with ITO transparent electrode. Next step is the formation of organic emissive layers. From this step, the process is going on under the inert gas condition such as vacuum, nitrogen, argon, and so on. After forming emissive layers, the cathode material is deposited and patterned. In this step, in order to prevent the organic layer from damage, a dry process such as shadow mask or lift-off is applied. When formation of thin film and patterning process is done, the element is encapsulated with a metal can or shield, transported to outside, and connected to a driving circuit.

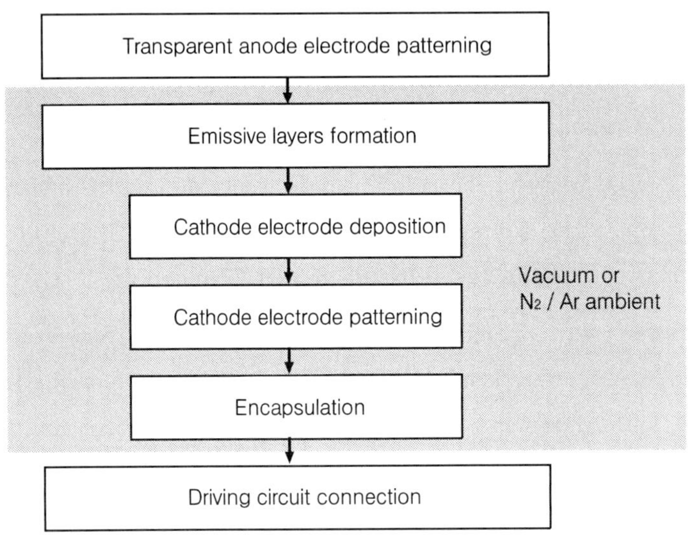

Figure 6.12
basic manufacturing process of organic ELD

Since the manufacturing process is relatively simple and the processing temperature is low, there are various advantages in productivity and applications. However, when looking into details of the process, there are also some limitations and aspects to improve, including the following: the uniformity of the surface of ITO thin film (the transparent electrode); the characteristics of the emission layer, electron-hole transporting layer, etc.; the preciseness of the lift-off process in patterning and elimination of the contaminants such as the remaining materials of the lift-off process; the reactivity and sealing performance of the encapsulation to improve lifetime; and invention of ways to increase productivity. In order to complete the technology for organic LED devices and systems, it is important to efficiently connect the same with such technologies as material technology focusing on organic materials and substrates, micro processing technology led by thin film formation and patterning, device technology relating to color formation and integration of active circuits, packaging technology considering sealing performance and productivity, and system technology applying driving ICs and PCB circuits.

(1) Color

Among technologies related to organic LED devices, there are five ways of forming R(red)–G(green)–B(blue) colors, including the followings: using emission layers for each of R–G–B (side–by–side method), forming a white luminescence layer with a R–G–B color filter, forming a blue luminescence layer and with color converting materials, turning it green or red, using optical mechanism with a large color spectrum layer and changing it to R–G–B (micro–cavity method), and forming R–G–B by single sub–pixel selectively. Typically, Fig 6.13 is the structure of forming a luminescence layer between two glass substrates. Among those emitting blue color are anthracene, a cyclopentadiene derivative substituted with phenyl, and DVPBi; 1, 2–phthalo–perinon derivatives emit yellow color (580 nm); and Alq3 emit green color (550 nm). In order to transmit light from the bottom substrate, ITO is used as both the top and bottom electrodes, and on the bottom ITO electrode, a supply electrode is formed.

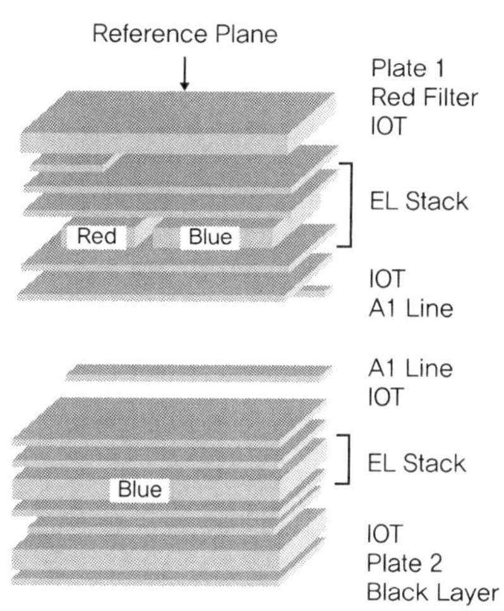

Figure 6.13
Double substrate stacking structure of full color ELD

1) Pixel processing technology

The ultimate goal of studying organic materials or polymer electroluminescent devices is to realize color displays. There are several research going on to realize color display using organic matters or polymer luminescent elements, some of which are close to commercialization. Followings are the major four methods.

The first method is arranging R–G–B pixel in parallel (side–by–side: Fig. 6.14 (a)). At this time, each pixel becomes an individual pixel of R, G, and B and the three pixels compose one color pixel. The problem with this method is that the three RGB pixels need to be formed on the same substrate, which is difficult since three times of formation need to be done for each of R, G, and B pixels. Also, it is difficult to pattern the organic layer precisely as the organic material used for the emission or transporting layers is vulnerable to organic solvents.

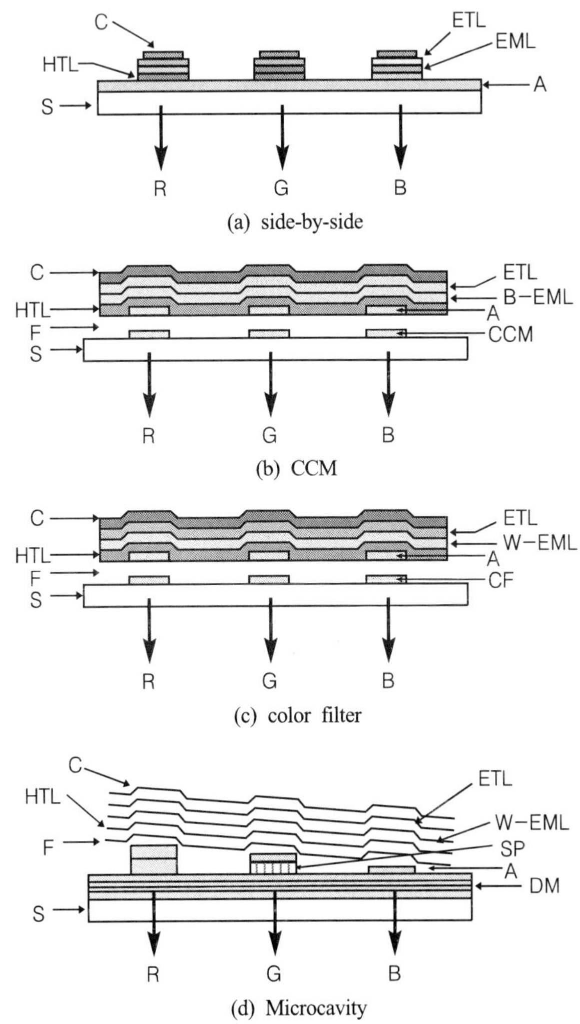

Figure 6.14

Typical methods of colorization using electric luminescence elements

Second method is forming R–G–B pixels from blue light with a color changing matrix (Fig 6.14 (b)); more specifically, forming color pixels by applying a color changing matrix of R, G, and B with excellent photoluminescence (PL) efficiency, that emits bright colored light. With this method, the process of handling an organic film can be omitted, enabling micro patterning.

The third method (color filter) is forming the R–G–B color pixels in the same way as in color liquid crystal panels, by applying a color filter to white–light electroluminescence element containing RGB colors (Fig 6.14 (c)). It is possible to realize a high–resolution panel with TFT and active driving method, just like liquid crystal panels. However, the light output is less intense as a color filter is used to get R–G–B from white light.

Fourth, R–G–B pixels are formed using light emitted by white light emitting element with a micro–resonance structure (Fig 6.14 (d)). This method is similar to the third method, but the difference is in using the micro–resonance structure instead of a color filter. RGB pixels are divided from white light from the light source by controlling micro–resonance length with a spacer and a dielectric mirror. It has an advantage of getting R–G–B pixels with narrow luminescent wavelength. But its luminescence efficiency is still low and emitted R–G–B pixels show the directional nature, narrowing the viewing angle. Comparisons in performance of these four methods are listed in Table 6.6

Table 6.6 Characteristic of several pixel manufacturing processes

	side–by–side	Color filter	CCM	Microcavity
color purity	Depend on Active Layer	Good	Good	Excellent
Output Efficiency	Normal	Low	Fair	Excellent
Fabrication	Very Difficult	Easy	Easy	Easy
Cost	High	Low	Low	Moderate
Drawback	Different Aging for R, G, B	Low Efficiency		Color shifr dep. on Angle
Application	Large area FPD	Low-cost display	Md-size	Personal Panel Viewfinder

2) Color Upgrading Methods

EL radiation of organic matters and polymers has a very wide spectral range, the Full–Width Half Maximum of which is up to 100~150 mm. However, a wide spectrum has a disadvantage that it is not possible to display clear colors when three original colors are mixed for full colorization. Thus, to show good quality colors, the color purity of each of the three original colors should be good, which is called saturation of color. This means that the monochromaticity of emitting light should be good. For this, the radiation spectrum should be narrow, and materials with a very narrow spectrum need to be used. Although there is no such material for now, the radiation spectrum of organic matters containing rare earth metals tends to be narrow [Figure 6.15]. Using a dopant with those organic matters can enhance the purity of red color.

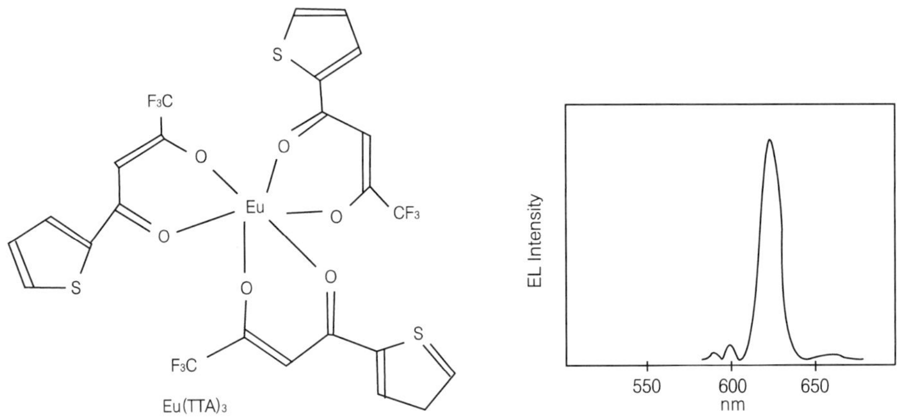

Figure 6.15

(a) Chemical structure of Organic matters including Rare earth metal Eu (b) EL light emitting spectrum

The color made in the way mentioned above should be close to the NTSC standard value (the RHr spot of the triangle) when a full-color display is driven by video signals. The expected values in the standard color coordinate of the color gained from organics are x=0.624 and y=0.374 for red, x=0.269 and y=0.628 for green, and x=0.137 and y=0.084 for blue. In terms of color purity, the colors located on the curve outside the triangle of CIE color coordinate are considered to have the best color purity.

(2) Technology about packaging

Packaging technology is very important in organic LEDs; more specifically, organic materials such as a light emitting layer are highly vulnerable to moisture and oxygen and thus, the packaging in the vacuum or inactive gas is needed to improve its reliability and the lifespan. At the same time, the manufacturing process should be simple and short so as to improve productivity. In other words, packaging technology covering inactivity–sealing–collective process needs to be introduced. The current method, as shown in Figure 6.16(a), is that a metal can with moisture absorbent such as BaO is attached to the back of organic LED in inactive gas. However, there are some problems with this method; not only the weight and volume are problematic, but also stress can easily be generated as the metal can is attached to a substrate that is made of different materials, and productivity is poor because of low throughput. Thus as shown in Figure 6.16, a method of forming a protective layer, not using structure types like metal or plastic cans, is being developed. In order for an organic film to be used for a protective layer, the development of organic materials which can completely block moisture and oxygen is required. Also, in the case of inorganic materials, packaging performance technologies such as using low–temperature process SiO_2 and/or SiN_x based on PE–CVD (plasma Enhanced–Chemical Vapor Deposition) need to be developed.

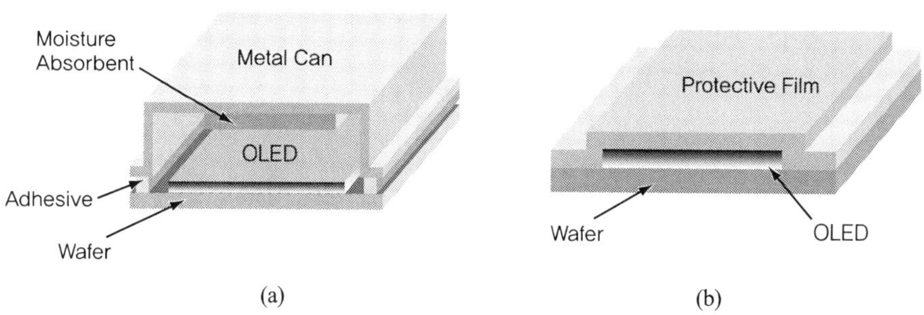

Figure 6.16
Organic LED packaging using Metal can(a) and protective film(b)

(3) Technology related to Micro process

Forming and patterning a negative electrode are performed after the formation of an organic layer, and thus in order to prevent the organic layer from being damaged by moisture and chemical components, the process undergoes dry etching, not wet etching, such as lift–off process or shadow mask. Figure 6.17 shows the lift–off process that first patterns PR (Photo Resist) inversely inclined and then attach metal electrodes above that, insulating each line.

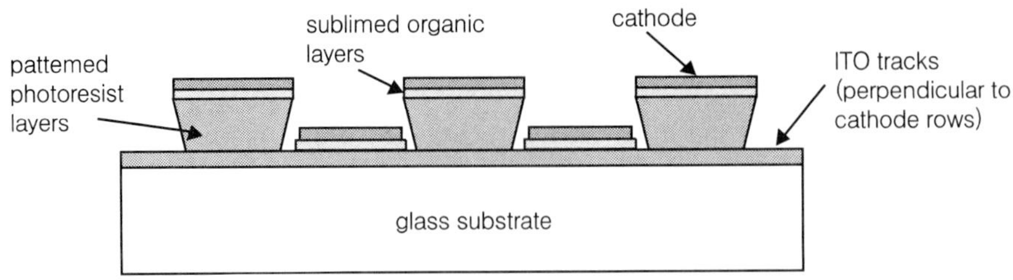

Figure 6.17
Cathode line Formation by Lift–off process

Recently, the method of forming a light emitting layer based on ink–jet printing method is being developed to enlarge the size of organic LEDs, as shown in Figure 6.18. That is, insulating polymide structures are regularly arranged on a transparent electrode, and then the polymers dissolved in an aqueous solution only in the selected areas are ink–jet printed. According to this method, regardless of the size, it is possible to form the light emitting layer in the selected area. Since there is no damage to the material, such a method is advantageous in terms of enlargement and productivity; and for this reason, the realization of this method is very promising.

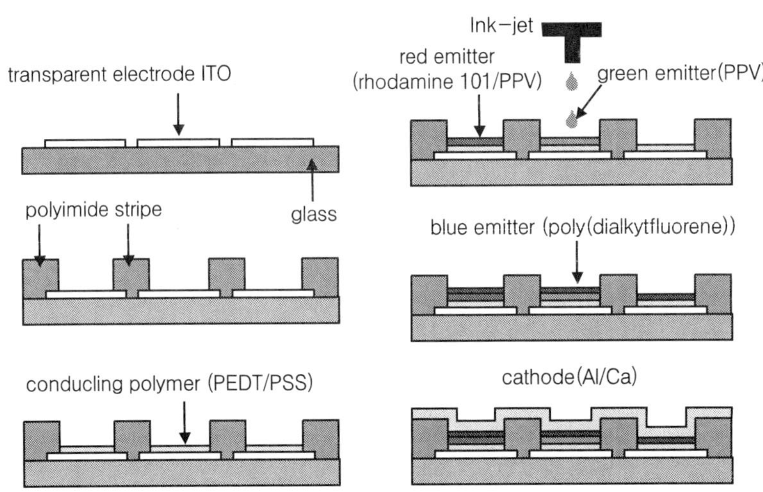

Figure 6.18
Manufacturing process of organic LED with ink–jet printing applied

(4) Material–related Technology

Among the materials for organic LED substrates, currently, glass and plastic are mostly used. Characteristics of each material are summarized in Table 6.7. In terms of TEC (Thermal Expansion Coefficient) which has a massive impact on adhesion of organic materials, plastic is relatively better than glass; further, plastic is also advantageous in terms of price, flexibility, weight, and so forth. On the other hand, glass substrates are advantageous in terms of prevention of moisture or oxygen invasion, physical hardness, the process temperature, durability against heat, and transparency. The material to use should be selected depending on the usage, but the technology to supplement defects of each material also needs to be developed. Further, to apply organic LEDs to various areas such as those using electronic papers or wall papers, more flexible and light substrates will be needed.

Table 6.7 Comparison of Glass vs. Plastic as substrate materials for organic LED

	Glass	Plastic
Adhesion(TEC)	W(Weak)	S(Strong)
Cost	W	S
Flexibility	W	S
Permeation (moisture, oxygen)	S	W
Physical strength	S	P(Possible)
Process temperature	S	W
Thermal durability	S	W
Transparency	S	M
Weight (for large size)	W	S

In addition, in terms of organic films, the researches related to light emitting layers as well as transporting layers of electrons and holes to improve the efficiency of light emitting layers are getting much more important. More specifically, for monomolecular LEDs, thermal deposition is mostly used; for polymer LEDs, such methods as spin coating, roll-to-roll, doctor blade, dip coating, ink-jet printing, and so forth are employed.

4 Operating Principle of OLED

In order for organic LEDs to operate as displays, matrix driving is required, which can be divided into passive and active driving methods. As to the passive driving, although the manufacturing process is simple when the screen gets larger and the number of pixels increases, there will be some problems such as RC delay, cross–talk, power consumption, and so forth. We will first review the differences between the two methods and move on to the problems with the active driving method as well.

As shown in Figure 6.19 (a), (b), one of the differences between them is the basic structure of the element. A passive element is composed of horizontal and vertical electrodes crossing each other in a matrix form. On the other hand, in an active element, a TFT (Thin Film Transistor) to open/close each pixel is placed with each pixel; the TFT is operated as a switch, turning ON/OFF each pixel in the bottom electrode, and the opposite electrode is used as the common electrode.

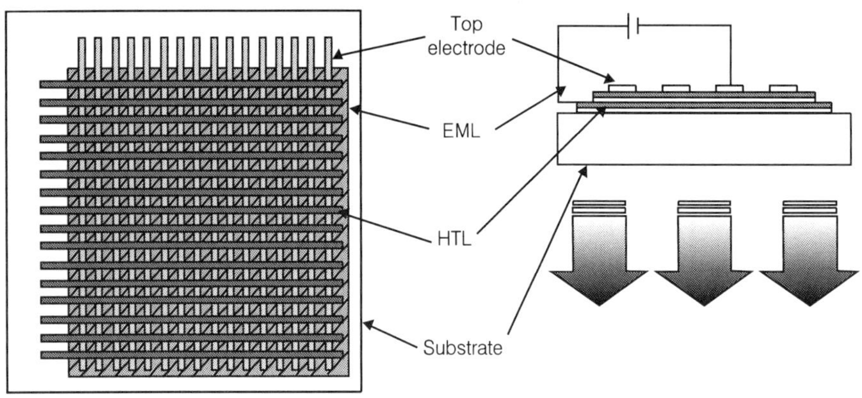

(a) Passive driving method

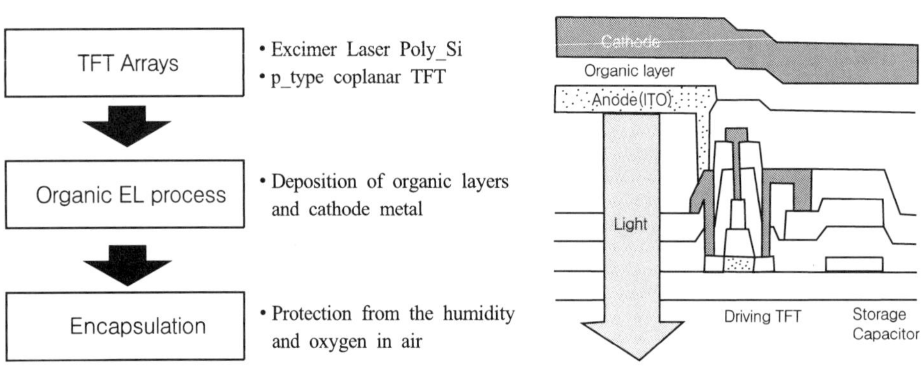

(b) Active driving method

Figure 6.19
Difference of element structure by the driving method

Forming a negative electrode with a regular linear pattern is essential for the manufacturing of an organic EL element using passive drive system.

Considering that it is generally difficult to use wet etching to organic materials, the technology of forming a cathode–separator using photolithography, before stacking the organic matter on the ITO anode electrode, has been developed, and the limitation has been overcome. Through continuous deposit process of the organic layer and the cathode electrode after the cathode separator is formed, about 256×64 of resolution can be realized even without forming another cathode pattern.

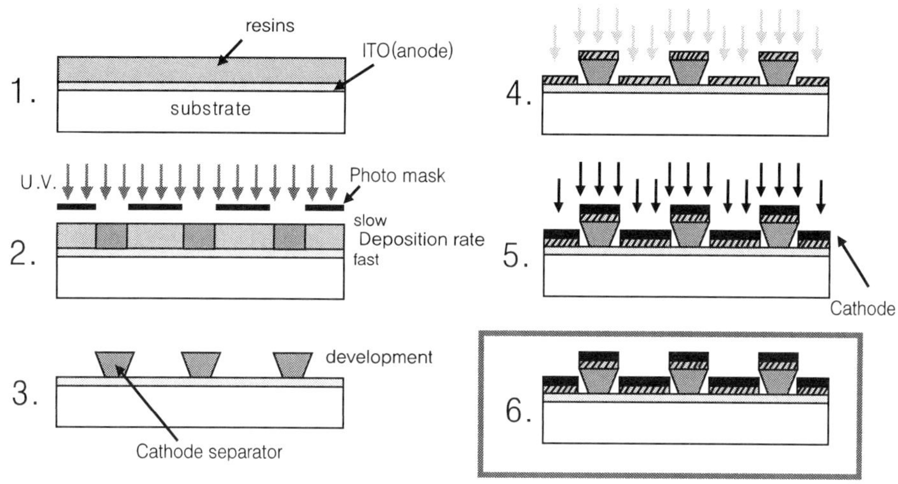

Figure 6.20

Cathode–separator formation by Photolithography and cathode patterning

Also, passive drive system structure can be applied to both monochromatic and full–color organic EL displays, for which basically line–sequential scanning and constant current driving methods are adopted. When driving each pixel, passive drive method drives scanning lines by time sequence. Thus, in order to achieve the required average luminance, the instantaneous luminance should be as much as the average luminance multiplied by the number of the lines. Therefore, the higher the number of lines, the higher voltage and more current should be applied instantaneously, which accelerates the deterioration of the device and increases the power consumption, which is not suitable for high–resolution, large–area display.

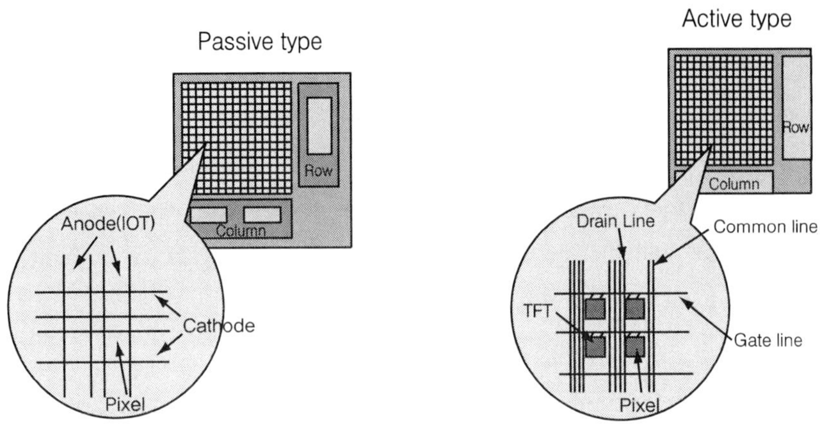

Figure 6.21
Circuit structure of passive drive and active drive pixels

In order to solve this problem, active drive method is used as shown in Figure 6.21; under the active drive method, each pixel is integrated with TFT (Thin−Film Transistor) so that it can be fully ON−OFF, and by maintaining the pixel current at a constant value, it is possible to improve contrast ratio and power consumption. Figure 6.22 shows the simple structure of the passive drive method and the active drive method. Among the active matrix driving methods, instead of amorphous silicon low−temperature poly−Si (LTPS) is preferred, which has low resistance and can be formed directly on a glass substrate at a lower temperature. The LTPS does not affect the organic thin film material whereas it is not stable at a higher temperature.

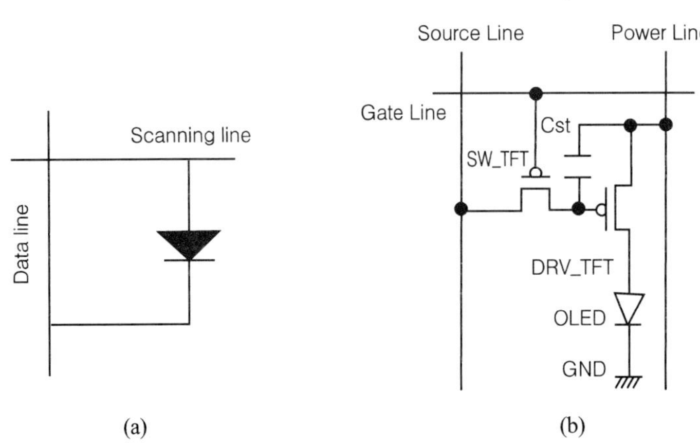

Figure 6.22
Circuit of passive drive method (a) and active drive method (b)

Figure 6.22 (b) shows the basic pixel structure of an active drive organic EL. Let's take a look at the driving mechanism. When applying signals to the selected electrode depending on the select signal, SW−TFT is open, and then the data signal which is given by the data electrode passes through SW_TFT, reaching DRV_TFT and the reservoir capacitor. When DRV_TFT is open, the current flows from the power line (V_{dd}) through DRV_TFT to the organic EL element, finally emitting the light. How long DRV_TFT remains open depends on the size of the data signal, and thus it is possible to display gradation by adjusting the current flow through the DRV_TFT.

The data stored in the capacitor can be applied to DRV_TFT, and thus it is possible to emit light continuously from the organic EL device until the signal for the next display is applied. Due to this mechanism, it is acceptable to apply lower voltage and current than that is used in the passive drive; and regardless of the number of the selected lines, it is possible to continue to drive the organic EL element while displaying one image. Thus it is advantageous in terms of low power consumption, high resolution, and enlarged image. Meanwhile, the active drive method bears the structure that the current should flow through TFT. As the existing amorphous silicon based TFT is difficult to use because of low carrier mobility, poly−silicon TFT can be an alternative with better carrier mobility.

Under the active drive method, the voltage applied to a pixel is charged in the capacitor, supplying power until the next frame signal is applied, thus, regardless of the number of the scanning lines, the element keeps driving while one image is displayed. Therefore, with the active drive, it is possible to achieve the same luminance even with a lower current; that is, low power consumption, high definition, and enlargement can be realized with this method [see Table 6.8]. An Organic EL is a current driven element. In order to obtain consistent display characteristics, it is important to have a consistent TFT characteristics. That is, if the threshold voltage and electron mobility of the TFT are inconsistent, a poor quality gray level will be displayed because of the inconsistent luminescence. For this reason, an organic EL needs more consistent TFT characteristics compared to TFT−LCD using low temperature poly−silicon.

Table 6.8 Comparison between active drive element and passive drive element

Division	Active Matrix	Passive Matrix
Driving	Active, TFT	Simple Matrix
Auxiliary electrode	unnecessary	necessary
Cathode separator	unnecessary	necessary
Luminance	Luminescent maximum Luminance	(Luminescent maximum Luminance)/number of lines
Power consumption	Low	High
Large area	Easy	Difficult

Currently, the efforts of crystallization at a low–temperature compensation for the unevenness of TFT characteristics have been made in various ways. The compensation circuits published so far can be divided into voltage mode programming method and current mode programming method, as shown in Figure 6.23, depending on the method of writing data to the reservoir capacitor of the basic pixels.

While the voltage mode programming method maintains a consistent voltage at the reservoir capacitor of a pixel while writing data, the current mode programming method flows a constant current through the organic EL element regardless of the characteristics of the TFT, by applying the voltage corresponding to the relevant current to the storage capacitor. Typical examples of the voltage mode is a circuit co–developed by SAMSUNG SDI and Hanyang Univ research team and Sarnoff circuit. The characteristic shared by these two circuits is compensation of changes in the threshold voltage by using two adjacent TFTs for driving. Meanwhile, typical examples of the current mode are NFC circuit and SONY circuit; these methods make the current flowing in an organic EL equal to the data current, based on the current mirror mechanism.

The current mode method has an advantage of compensating the current difference caused by the inconsistency of the carrier mobility as well as the threshold voltage of the driving TFT. But, it also has disadvantages that there is a long time delay in writing data to the reservoir capacitor and that the current drive IC cannot be built in a panel, requiring a separate installation.

No matter which method is to be used, when using these compensation circuits, the number of transistors placed in each pixel increases, narrowing the actual area of light emission and lowering luminescence. Thus, a transparent electrode at the front structure is required.

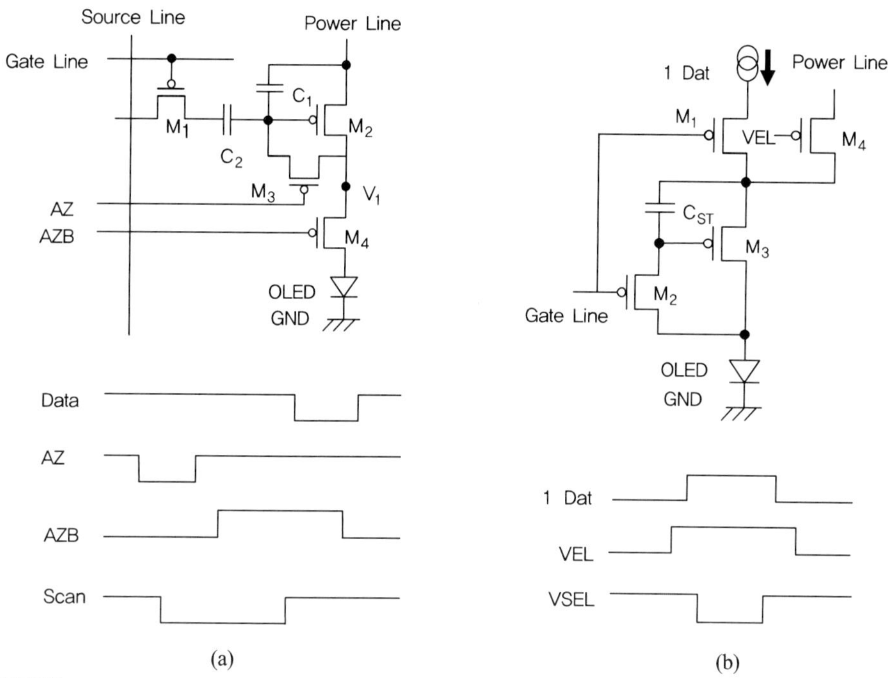

Figure 6.23
Compensation circuits of voltage mode programming method and current mode programming method

Sophisticated gray scale is essential for displays to show delicate color changes. Gray scale of the organic ELs are scaled basically by controlling the amount of the current flowing in the organic EL. In the active drive method, the amount of the current flowing in the transistor is adjusted by applying data in a way that the gate voltage of the transistor supplying power to the organic EL is maintained in accordance with the gradation value. In this way, the emission intensity changes, by which the gray scale is displayed. This method is called analog gradation marking method. When using the compensation circuit mentioned above, the inconsistency of the threshold voltage of TFT can be overcome, and a certain level of high gradation can be achieved.

Another method is a digital method that uses only the complete OFF and ON states of the driving TFT supplying the current to the organic EL. In this case, even if the characteristics of the TFT are not consistent, the difference in current is not very large from each other, and thus it is possible to resolve the inconsistency of the TFT. Two methods can be used for this purpose; one is the area division method adjusting brightness according to the number of light emitting unit pixels by splitting them from one single pixel. This method has a problem marking many gradations in small-sized high-resolution displays because it is hard to divide each pixel into several unit pixels. The other method is time division method that changes brightness by adjusting the ON/OFF time. In this case, as the number of gradations increases, the ON/OFF times of data increases and the operating speed of the TFT is fast enough to precisely display gradations. In addition to this, a hybrid of a combination of the above two methods can also be used.

The active drive is essential for organic EL to be applied to large displays with high resolution. To commercialize active drive organic ELs, there are more areas to improve, i.e., enhancing consistency of TFT and developing more efficient compensation circuits. Also, active drive organic ELs have advantages of further downsizing small displays (such as mobile phone displays) and reducing power consumption by embedding the drive IC in a panel with low-temperature poly-Si circuit. Embedding compensation circuits in each pixel, however, could lower the aperture ratio. In order to prevent this, the conventional back surface light emitting structure should be changed to the front surface light emitting structure. In this case, technologies on transparent electrodes and transparent encapsulation need to be developed.

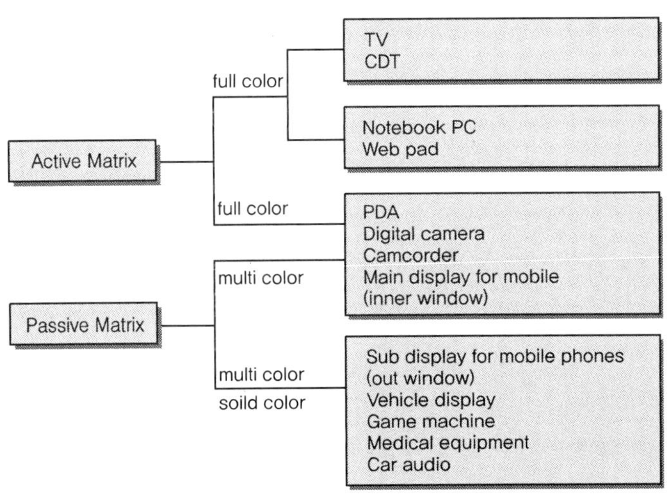

Figure 6.24

Application of organic EL

5 Trend of research and task of Organic Light-Emitting Display

5.1 Prospect and trend of research

(1) Domestic trend of research

In Korea LG Electronics corporation, Samsung SDI corporation, Orion Electronic Company, and so forth have been working on research and development for commercialization of OLEDs, the trend of which is summarized in [Table 6.9].

Table 6.9 Trend of research of materials in Korea (source: Electronic Parts Researcher, "Industrial trend about organic EL components", 2002.12.)

① Samsung SDL	② LG Chem.	③ SKC
④ Elia Tec	⑤ NESS Display	⑥ Smart Display
⑦ Hyundai Display Tech.	⑧ Neo Display	⑨ LG Electronics
⑩ Acedigitech	⑪ Samsung Corning Precision Materials	

Key Materials	Industrial Company										
	①	②	③	④	⑤	⑥	⑦	⑧	⑨	⑩	⑪
Assembly	■			●	■		●	■	■		
Luminescence materials	●	●			■				●		
Luminescence doping materials	●	●			■				●		
Electron injection/cathode electrode materials	●			●	●		●		●		
Electron-transporting materials		●									
Hole-blocking materials									●		
Hole-injection materials		●			●						
Encapsulation substrate											
Encapsulation					●						
Moisture absorbent											
Polarizing plate										●	
Glass substrate											□□

Note : ■ Production , ● R&D , □□ Sales

Regarding low molecular weight EL materials, LG Chemistry, LG Electronic Technology Institute, Samsung SDI, NES Display, Gyeongsang National University, Hongik University, Seoul National University, etc. have been working on the research and development to some extent, but most of the patents of original technologies are retained by Japan and America. KIST and KAIST hold domestic and international material patents for polymer materials; however, most of the original technology–related patents are dominated by the foreign enterprises like Dow, Covion, Unix, CDT. Although the efforts for commercializing organic ELs in Korea are being made with vigorous and multiple patent applications, however, it is too early to compete with other foreign companies. In Korea especially research and development on new materials are limited until now.

Among the companies who are vigorously trying to commercialize the organic ELs are, major companies including Samsung Display, LG Display, and other electronic companies.

(2) Foreign trend of research

There are several foreign companies actively taking part in R&D and production of organic ELs, such as CDT (Cambridge Display Technology), Covion, eMagin, Philips, Pioneer, Sanyo, Sony, Uniax, Universal Display, and so forth. CDT developed an active matrix organic LED in collaboration with Seiko–Epson, and is carrying out research on ink–jet printing method for producing a large screen. Covion focuses on developing light–emitting polymer materials and flexible displays, and eMagin implements micro displays by integrating organic LEDs on Si substrates. Philips carries forward developing light–emitting polymer materials and displays using the materials. Figure 6.25 (a) to (d) show the active matrix element of CDT and Seiko–Epson, the flexible element of Covion, the micro display of eMagin, and the light–emitting polymer of Philips, respectively.

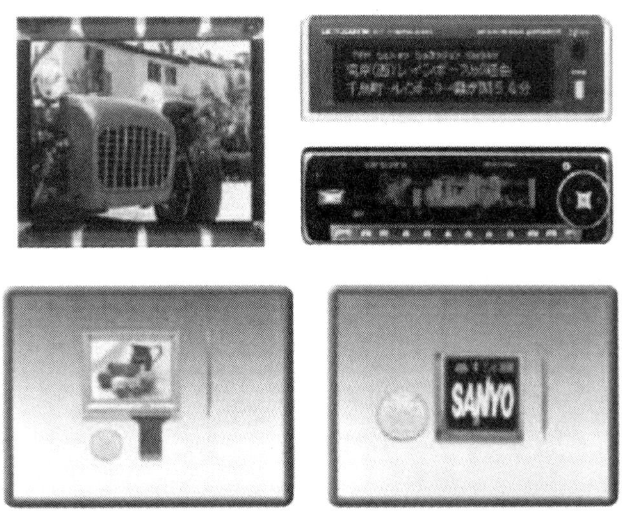

Figure 6.25

Development trend of organic materials

Also, Pioneer who put organic ELs to practical use for the first time in the world produces a million of the element, capable of displaying characters or simple images on a car audio. Sanyo in partnership with Eastman Kodak also has been producing simple and active matrix organic ELs. Figure 6.26 shows the prototypes of organic ELs made by Pioneer and Sanyo.

(a) CDT

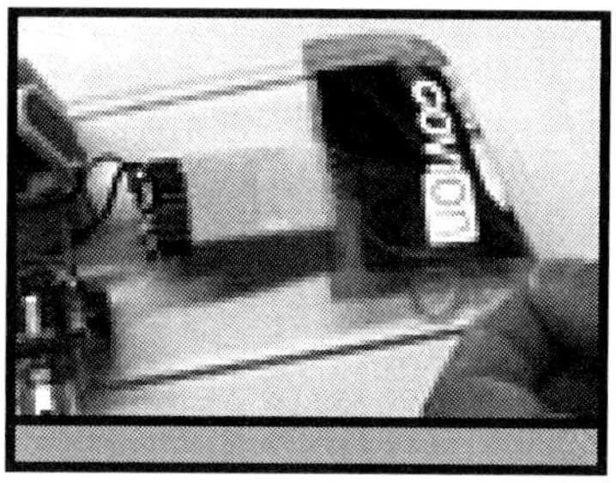

(b) Covion

(c) Organic LED element of eMagin

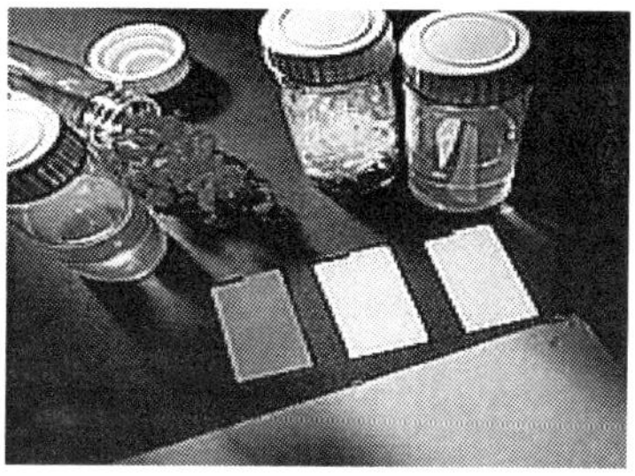

(d) Light-emitting polymer of Philips

Figure 6.26
Organic EL Prototypes of Pioneer (a) and Sanyo (b)

Universal Display is in the middle of developing three types of organic ELs, i.e., TOLED (Transparent Organic Light Emitting Device), SOLED (Stacked pixel OLED), and FOLED (Flexible OLED), in collaboration with Princeton University and University of Southern California. TOLED can be simultaneously observed from the top and bottom, SOLED enhances brightness and resolution using the laminated structure, and FOLED can be bent as it is fabricated on a flexible substrate made of plastic or the like. Figure 6.27 shows the conceptual diagrams and shapes of these organic ELs.

Recently, six Chinese companies (BOE, CSOT, Visionox, EDO, Tianma, Trolley) which grow quickly in LCD market, started to develop Organic ELs. They organized R&D teams for both large-scale and cellular phone OLED panel. Also, they are considering to organize R&D teams for FOLED (Flexible OLED).

Japan and Taiwan companies also have stated to develop Organic ELs. Japanese display companies that have lost their competitiveness in LCD market integrated the mobile panel business and newly established JDI with government-funded private fund. OLED professional display company JOLED also launched.

Apple has also begun to participate in the flexible OLED smartphone and smart watch market, which are dominated by Samsung Electronics and LG Electronics.

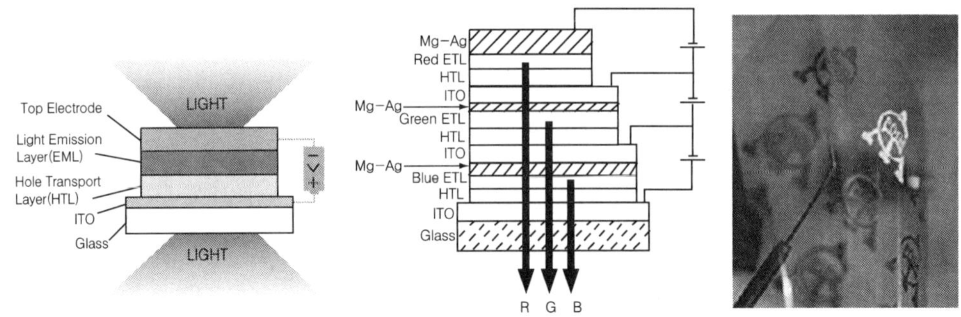

Figure 6.27
Conceptual diagrams of TOLED and SOLED / Shape of FOLED

(3) Technology level and road map

According to the survey of Korea Development Bank, the technology level of organic ELs in Korea maintains the world top level.

Table 6.10 Comparison of Technical standard in organic ELs

Classification	Advanced country	Korea	Technology gap (year)	Total		
				Advanced country	Korea	Technology gap
Holding Original technology	90	50	5	80	50	2
Reliable Inspection Technique	85	70	1.5			
Self-sufficiency of materials and parts	85	50	3			
Production Technique	85	80	0.5			
Human Resources	85	75	1			

Data : Korea Development Bank, "Technoreport", 2003.3.

According to National Technical Road Map published by the National Science and Technology Committee in 2002, a technology road map is created focusing on the technical challenges for medium/large-sized organic ELs using the AM system, for which a large demand is expected. Tables 6.10 and 6.11 show the technical road map of color organic EL products published by Korea Development Bank and the National Science and Technology Committee.

Table 6.11 Technical road map of organic ELs

	~2004	2005	2006	2007	2008	2009 ~ 2010
Organ ic EL	Development of small-sized AM ELs - Pixel 100PPI - Yield: 50% - Size of substrate : 400×400		Development of medium/large-sized AM ELs - Pixel 200PPI - Yiel: 50% - Size of substrate : 730×920			Flexible organic ELs - Pixel 300PPI - Yield: 90% - Size of substrate: 1,500×1,800

Data : National Science and Technology Committee, "National Technical Road Map", 2002.11.

(4) Application and prospect

Among the various features of organic LEDs, such properties as low driving voltage, flexible, thin, light weight, and low price, in particular, will likely be able to derive attractive applications. Roughly speaking, there are many applications such as automotive displays including CNS (Car Navigation System), dashboard, and front glass panels as well as information communication devices, portable displays such as PC and flexible large displays including electronic papers and billboards. Detailed applicable functions within these categories are indeed infinite.

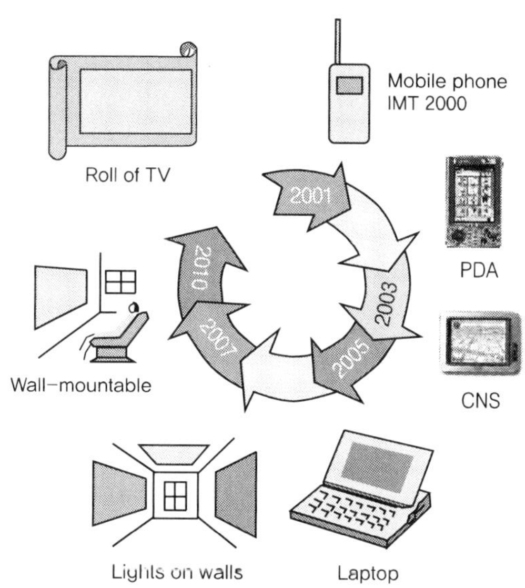

Figure 6.28
Development Trends of organic LED Applications

Figure 6.28 predicts the developmental concept in the application of organic LEDs. It starts off with small displays ranging from one to three inches such as mobile phones, IMT–2000, and view finders. And then it is expected to be extended to 5 to 10–inch products such as CNS, and PDA (Personal Data Assistant). As the relevant technologies are accumulated, it is expected to reach the core of the display area, i.e., 10–inch or higher class, such as laptops and Palm–PCs. In addition to this, specialized areas using unique organic LED features such as electronic papers, displays for wallpaper, super–large signboard will be open.

For such visions to come true, several elemental technologies regarded as a bottleneck in development of OLEDs should be improved. For instances, for the purpose of improving operation characteristics, high purity, high brightness, high efficiency, and long lifetime of red and blue materials should be achieved; for the purpose of ensuring the reliability, packaging technology should be established, and the drive unit should be optimized; for the purpose of building high definition and enlargement technologies, precision in large screens and printing technology should be improved; for the purpose of securing productivity, throughput should be improved, and the ways of handling large substrates and developing mass production machines should also be considered.

5.2 Market Situation and Prospects

OLED technology has recently gained significant market share in the display market. OLED displays are now mass produced for mobile phones, tablets, TVs, and wearables. OLED displays will reach nearly $16 Billion USD this year of 2017 and will grow to $57 Billion USD in 2026. The latest evolution is plastic and flexible displays. Compared to conventional glass–based displays, plastic AMOLED panels are much thinner and lighter, enabling either slimmer devices or bigger batteries. Future flexible displays will also make foldable mobile devices a reality. The two main segments are currently smart phones and wearable devices such as smart watches. However, as the technology matures it will be possible to use those displays in other applications, such as automotive displays.

Both Samsung Display and LG Display have recently announced significant investment to expand their production capacity. Now we expect plastic and flexible displays to grow rapidly from a $2 Billion USD market this year of 2017 to $18 Billion USD by 2020. The rise of plastic and flexible displays will be accompanied by a shift from glass substrates to plastic substrates such as polyimide. However, thin glass–based displays will remain an important technology, especially in the TV segment where scale–up and cost reduction are still the main challenges. New 4K OLED TVs were recently launched by LG and Panasonic to critical acclaim. However, some TV manufacturers are hedging their bets by investing in LCD panels enhanced with quantum dots. These so–called "quantum dot LCD" TVs will be positioned as a cheaper upgrade from existing sets. New production technologies will make OLED more competitive, allowing the market for OLED TV panels to grow at 26% over the next decade. New applications in wearable devices such as augmented reality (AR) and virtual reality (VR) are also coming to market and provide new opportunities for suppliers of OLED displays. Sony, Oculus, and HTC have already announced new VR headsets based on AMOLED technologies. For AR glasses, OLED microdisplays are a major contender against existing LCoS (liquid crystal on silicon) technology. OLED display market in eight segments: Mobile phone displays, Tablet and notebook displays, TV panels,

Automotive and aerospace, Wearable electronics, Industrial and professional displays, Microdisplays, and Other applications.

Probelms.

1. Considering the potential of the OLED market expansion as well as the technologies relating to the manufacturing process, what role can OLEDs take other than displays?

2. Briefly, explain the basic structure and principle of OLED displays.

3. Briefly, describe the difference between active driving method and passive driving method. Also, describe some problems to solve in relation to the active method.

4. Think about the potential status of OLED in the display market in the future and the difference of OLED from other displays, based on the features of OLED.

References

[1] Jae Woo KIM, et al.: "Organic Electroluminescent Display," *Korea Institute of Science and Technology Information: Technological and Industrial Information Analysis* (2003).

[2] Hyo Gyun JUNG: "Active Driving Organic EL Technology using Low Temperature Poly-Si TFT", *Electronic Times Themed Lecture* (2001).

[3] Tae Hyung JUNG, et al.: *Display Engineering 1*, Cheong Beom Publishing Co. (2000)

[4] Tae Hyung JUNG: "Organic/ Polymeric Electroluminescence Display", *Electronics and Telecommunications Research Institute: Semiconductors.*, Original Technology Research Institute Device Applications Division (2002).

[5] Soon Ki KWON: "Development Trend of Organic EL Materials," Display Technology Education Center Short-term Course material (2004).

[6] Hyeong Rak LEE: "Understanding of organic EL elements," Gyeongbok National Univ. Physics, nano-technology nano-physic lab: Display Technology Education: Organic EL Elements (2004).

[7] Woo Young KIM, Chan Hoon SHIN: "Development Trends & Prospects of Organic EL (Part 1)," *Weekly Electronics Information* (2000).

[8] Jee In KWON: "Information Communication Industry Trends: Information Communication Devices: Organic EL," *Korea Information Society Development Institute* (2003).

[9] Soon Ki KWON: "Organic EL Unimolecular Material," *Electronic Times Themed Lecture* (2001).

[10] Electronics Display: "Latest Trends of Organic EL." *Electronics Display Manuscripts* (April).

[11] Jun Sin LEE: "Thin Film Growth Methods," *SAMSUNG SDI Lecture Note* (2000).

[12] Byeong Kwon JOO, et al.: "Appearance, Current State, and Prospects of Organic LED Technology," *KIST Information Material Device Research Center* (2001).

[13] Byeong Kwon JOO, et al.: "Next-Generation Display dreaming of Infinite Applications: Organic EL," p.3-8 (2001).

[14] *Polymer Science and Technology* Vol. 7. No. 6, December 1996

[15] *Polymer Science and Technology* Vol. 11. No. 11, April 2000

[16] K. G. Chittibabu, et al.: "Pure and Applied Chemistry," A33, 1283 (1996)

[17] The Korea Display Research Association (http://www.edirak.or.kr/)

[18] Korea Institute of Science and Technology, Display & Nano-Device Lab (http://diana.kist.re.kr/)

[19] Korea Information Society Development Institute (http://www.kisdi.re.kr/)

[20] OLED Madang (http://www.oledmadang.com/)

[21] Next-Generation Information Display Technology Development Project (http://www.display.re.kr/)

[22] Integrated Systems Design & Test Lab. (http://isdnt.chungbuk.ac.kr/research/el.htm)

[23] Seong Kyu LIM: "Electoluminiscent Dispalys el. 1, el. 2", p. 11, p.10

[24] The Korea Information Display Society (http://www.k-ids.or.kr/)

[25] Korea Institute of Science and Technology Information (http://www.kisti.re.kr/)

Flexible(E-paper) Display

Chapter 06→07

1 Overview of the flexible display

CRT played a key role over the past 60 years in the area of the display. However, with the advent of flat panel displays (FPD), it's position is threatened. Even portable devices, as well as the largest market of the display(TV and monitor), have been replaced with the flat panel displays. Among flat panel displays, LCDs are growing rapidly with light weight, thinness, and low power consumption, which are convenient to carry while maintaining almost the same image quality as conventional CRTs. PDP and OLED have entered the market afterward. In 1988, Mark Weiser of Palo Alto Research Center, Inc. used the term, ubiquitous, for the first time. He argued that ubiquitous computing will lead to a third wave of information revolution following mainframes and personal computers (PC). It is anticipated that not only networking between PCs on the desk, but also all devices such as mobile phones, TV portable terminals, car navigation systems and sensors will be networked, and an era where anytime, anywhere, anyone can use a large capacity communication network will come soon. Just as this prospect proves that the era will not be long, U-city, U-healthcare, USN (ubiquitous sensor network) and the likes are easily accessible. What is required in the upcoming ubiquitous era is an information display device (display) that connects human and information.

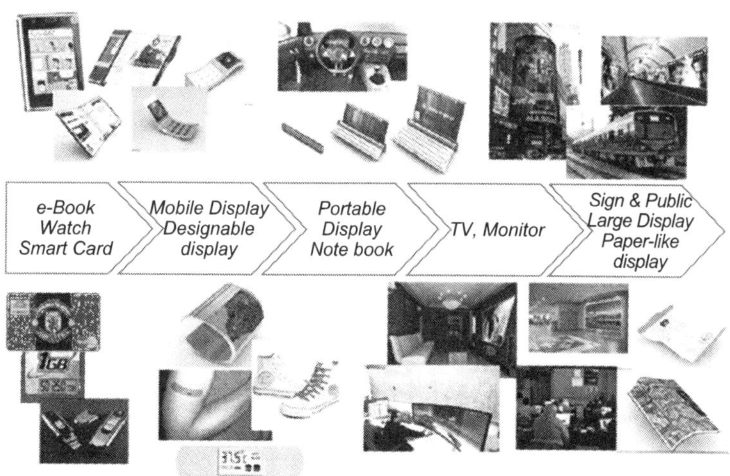

Figure 7.1
Flexible display.

It requires a new type of information display and communication system free from time and space constraints. The flexible display is expected to play a role in this. The display is evolving with the changing needs of consumers, the creation of new applications, and the development of display-related technologies.

Flexible displays are rollable, foldable or foldable like paper, free of exterior design, portable, flexible and non-breakable one. It requires a display made on a flexible substrate that can be bent or folded without loss of characteristics. The present state of the art is at the level of the rugged display which can withstand large mechanical shock. The future technology will be the bendable display that will eventually lead to the implementation of a rollable display. Flexible displays such as LCDs, PDPs, and OLEDs etc are a not new concept. The existing glass substrate-based display is implemented on a flexible substrate such as a metal foil or a plastic film. In other words, this means a change in hardware technology, and the quality of the display should be the same as that of the conventional glass substrate-based display or its improved version. Flexible OLEDs, flexible LCDs, and electronic paper are currently the most promising areas of implementation. Among these, the flexible electronic paper is the most likely to be implemented shortly in future. When the primary definition of a flexible display is the substrate changing to a flexible material other than glass, it can be classified into four types according to the development and use of this display technology.

1) Durable: Thin and light unbreakable display. As the earliest form of flexible displays, the only substrate changed to a flexible material from the glass. It uses a robust outer part of an existing portable one. It is a type of display in which the thickness, weight and reliability of the product are innovatively improved.
2) Bendable & Conformal: Display that wraps and design-free. The key is that the design can be freely implemented. It can continue to maintain a curved shape. By freely designing the exterior of the display, it is possible to create a new product that maximizes the use of non-flat space including a new type of mobile product with fashion concept
3) Flexible: The ultimate form of flexible display is a display that can be freely rolled up like a scroll or a freely foldable display. It is possible to create a space-saving new product that can realize a large screen while utilizing a minimum space.
4) Disposable: The cheapest display that can replace paper. Cheap implementations can become available based on printing process by a roll-to-roll process

technology. Large scale in-line production results in low cost and creates a new product that can replace the normal paper. It may be possible to implement a display that can be easily bought and discarded or recycled.

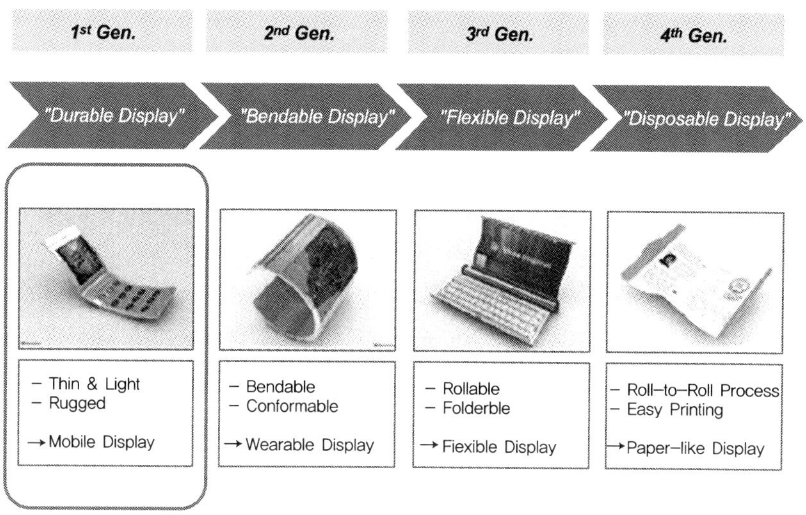

Figure 7.2
Types of flexible displays.

For implementing the flexible display technology, various flexible substrates, driving module technologies and display technologies have been studied. These technologies can be divided as follows.

1) Substrate Technology (Substrate) : plastic film, metal foil, thin glass
2) Driven elements Technology (TFT Array) : a-Si TFT, poly-Si TFT, Organic TFT
3) Display Mode Technology (Display Mode): LCD, OLED, Electrophoretic

Part I	Part II	Part III
Substrate	TFT Array	Display Mode
1. Thin Glass	1. a-Si TFT	1. E-paper
2. Metal Foil	2. Poly-Si TFT	2. LCD
3. Plastic Film	3. Organic TFT	3. OLED

The flexible substrate is the fundamental starting point for the flexible display. By replacing the current display based on the glass substrate with a flexible plastic film or metal foil type substrate, the primary flexible display is implemented. Although this may appear simple, but accurately forming active matrix driving devices of the semiconductor-based technology into micro-units on flexible substrates needs protection against moisture, oxygen and thermal expansion or contraction. Particularly, flexible substrates satisfying these requirements and new low-temperature drive element forming technologies that easily form drive elements within the limits of physical properties of currently available flexible substrates, are closely related to each other. This is a major obstacle to the commercialization of flexible displays and requires much research. For flexible substrates, a very thin glass plate of 0.1 mm or less, a metal foil and plastic film have been mainly examined. Substrate satisfying the required properties for implementing a flexible display is not standardized yet. The research has been mainly focused on the plastic film and the metal foil. For active driving elements, a-Si TFT and LTPS technology mostly used in the drive element for the TFT-LCD or AMOLED. These are is being developed in line with the existing technology. Organic thin film transistor (OTFT) technology uses low processing temperature, and/or solution processing, has been actively studied as active element for the flexible display technologies. However, the real performance and reliability of the device are insufficient to achieve the required display quality. In principle, the flexible display mode can generally use all the display modes of the active-passive devices. In the case of LCD using existing liquid crystal, the most stable technology is display quality and process, or the technique of keeping constant intervals of liquid crystal cells formed between flexible substrates between the upper and lower plates and flexible backlight technology. For OLED using an organic luminescent material, the innovative improvement of a protective film (passivation) and encapsulation techniques are required because OLED uses a flexible substrate vulnerable to moisture and oxygen contamination. However, it is more suitable for flexible display than LCD. The electronic paper technology, which combines the characteristics of the paper and electronic display, is a suitable technology for e-book due to the reflective and memory storage function. Since, here the display quality is worse than that of the LCD or OLED, high-resolution technology such as color implementation needs to be developed.

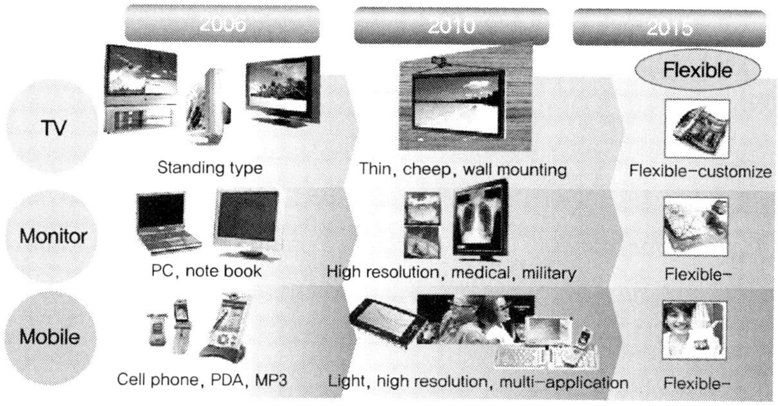

Figure 7.3

Flexible display prospects for application development.

The most appropriate technological development of flexible displays will depend on the performance of the display, productivity and prices. The universal plastic substrate is thought to be beneficial in order to have major characteristics of the flexible display such as lightness, thinness and flexibility. An organic thin film transistor (OTFT) device is most suitable driving element because it can be mass produced in the printing process at room temperature and normal pressure. OLED which can be thin and can be bent without degrading display properties, therefore, is expected to be used for advanced products that require dynamic image and natural colors. In commodity products that need a low price, the color electronic paper (E-paper color) with an improved performance of the color and video expression is expected to be employed. In this case, all organic display (the board, the driving device and display mode) can be implemented. However, development efforts should be made for each technology because it has lots of problems that should be solved.

D i s p l a y E n g i n e e r i n g

* 2 Material characteristics and requirements of flexible displays

2.1 Plastic Substrate

As a display for visualizing various information a display which is thin, low power, light and flexible as a paper is attracting much attention. The ultimate goal of flexible displays is paper–like displays that are easy to carry, lightweight, and space saving. In order to realize the low price, it is aimed to secure mass productivity through the printing–based roll–to–roll process. To realize a flexible display, flexible substrate, organic/inorganic materials for the low–temperature process, flexible electronic device, protective film, barrier film technology, and packaging technology are required. Among them, flexible substrates are recognized as the most important parts for determining performance, reliability and price of flexible displays. Among the flexible substrates, plastic films have been extensively studied for its ease in processing, light weight, and suitability.

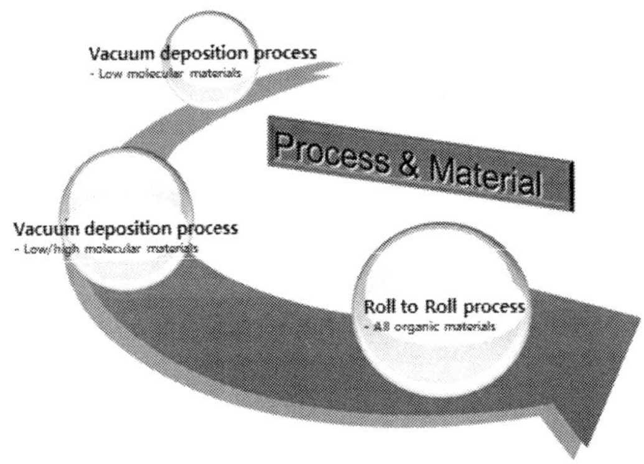

Figure 7.4

Development Direction of Flexible Display Manufacturing Process.

This chapter explores the materials and technologies that are essential for the implementation of flexible displays and looks at what areas should be followed by technological innovations for the flexible displays to be implemented in the future.

2.1.1 Need for a plastic substrate

The three types of substrates are metal foil, thin glass, a plastic substrate that has been considered for the implementation of the flexible display. There is a lot of differences in applicability due to their advantages and disadvantages. A thinner glass (50~200 μm) is required, than that was previously used (700 μm). Among three kinds of substrates, thin glass has the advantage of being transparent, good surface flatness and resistant to moisture and oxygen contamination. However, it has the disadvantage of being fragile and difficult to apply in a continuous process. Where shock resistance is required, the metal foil may be applied. The stainless steel is usually used to avoid corrosion problems.

There is a disadvantage that a process of coating an insulator and polishing a rough surface to coat the organic flat film is necessary. When using the metal foil, it has the advantage that there is no constraint on the process temperature. However, due to the thermal property of organic planar film, there will have restrictions on the process temperature. It is also applied for only reflection type display or a light emitting direction (top emission) mode because of non-transparency. A plastic substrate has been receiving attention for a long time as a material best suited for implementing a flexible display. The plastic substrate has the advantage of almost no restrictions on the form and lightness. Industrially, it is the most suitable material that features the ink-jet printing. However, it has the problems like lack of thermal stability and high permeability of moisture and oxygen.

(1) Classification of the plastic substrate

Figure 7.5 shows the classification of plastic substrates according to thermal behavior and T_g (glass transition temperature). Due to the difference in thermal behavior, the manufacturing method of the plastic substrate is slightly different. The T_g of the plastic is a starting temperature of the molecular chain's local segmental motion. The physical and mechanical properties of the plastic changes occur rapidly in the vicinity of T_g. Therefore, the T_g determines the operating temperature of the plastic substrate.

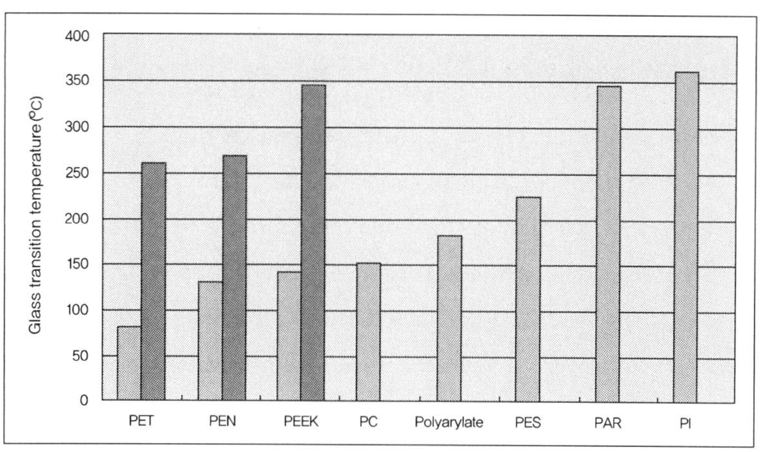

Figure 7.5

Types of plastic substrates.

A thermoplastic semicrystalline polymer such as PET(polyethylene terephtalate), PEN(polyethylene naphtalate) and PEEK(polyethylene ether ketone) has T_g (~140°C) and T_m (melting temperature, ~340°C). The plastic melts at a temperature above T_m. After melting, PET and PEN substrate form a substrate by stretching in two directions. They have higher T_g than thermoplastic semicrystalline polymer's T_g. The thermoplastic amorphous polymer can be formed as optical substrate by solvent cast method in addition to the stretching method. Finally, PI (polyimide), polyarylate and PAR (poly(ADP–ribose)), etc. having a relatively high heat resistance is difficult to manufacture the substrate by stretching after melting. These substrates are produced using a solvent caster method. The forming process is possible as a low cost when manufacturing the plastic substrate by melt extrusion method. When the substrate is stretched MD (machine direction) and TD (transverse direction) directions, the thermoplastic semicrystalline polymer has different optical (refractive index, transmittance, etc.) and mechanical properties of the substrate. Therefore, such a substrate is difficult to apply to use as a display using a polarizing film.

The thermoplastic amorphous polymer has processing and equipment difficulties in forming substrate through stretching. Therefore, the solvent caster method is used for such a polymer. Although this method has the advantage of obtaining a substrate having no directionality, it has a disadvantage in that it is less mass–producible than the melt extrusion method and requires separate equipment for recovery, purification, and reuse of the used solvent.

(2) Kinds of plastic substrates

Table 7.1 shows kinds of plastic substrates and manufacturer of each of the substrates. Advanced chemical companies led the commercialized research in plastic substrate materials. In particular, the study to improve the thermal, optical and mechanical characteristics of the material are primarily important. It has focused on differentiated improved barrier film coating technology development and heat resistance property in order to increase resistance of water, the oxygen–barrier, that is required in the flexible display. Physical and chemical properties can have a different point of view in accordance with the intended use and application range. From the standpoint of a substrate for implementing a flexible display, the following is summarized.

Table 7.1 Kinds of plastic substrates and manufacturer

Types of plastics	color	manufacturer
Polyethylene terephtalate	Transparent	General Atomics(USA) Mitsubishi Chem.(Japan)
Polyethylene naphtalate	Transparent	DuPont Teijin Film(Japan)
Polycarbonate	Transparent	Teijin Film(Japan) GE(USA) Mitsubishi Eng.(Japan)
Polyethylene sulfonate	Transparent	Sumitomo Bakelite(Japan) i-Component(Korea)
Arylite	Transparent	Ferrania Image System(Italy)
Polyimide	Yellow/Transparent	DuPont(USA) Mitsubishi Gas(Japan)
Polynorbonene	Transparent	JSR(Japan)

1) PC(polycarbonate) Substrate: PC (polycarbonate) is a plastic with excellent optical and mechanical properties, but it is known that its chemical resistance (especially ketones) is weak (particularly ketone type). There are many restrictions on the solvent used (photoresist developing solution, the photoresist remover, metal etching, cleaning reagents, etc.). In order to solve a problem, it is necessary to separately form a chemical-resistant layer or a hard-coating layer in order to ensure chemical resistance on both sides of the substrate. It has the disadvantage that the coefficient of thermal expansion (CTE) is 10 times greater compared to the inorganic glass and low-temperature process (up to operating temperature range, 150 ~ 180°C) is needed.
2) PET Substrate: PET (polyethylene terephtalate) has a low melting temperature (T_m) and it is easily molded as a plastic substrate at a low manufacturing cost and process. Semi-crystalline plastics like a PET, have excellent chemical resistance than amorphous polymers and low moisture absorption. So, it is advantageous in shortening the processing time. However, practical application is difficult because of poor thermal stability(low glass transition temperature, and melting temperature). In addition, it is not applicable as the substrate of the display that uses polarizer for the LCD application because the PET substrate is optically anisotropic. Furthermore, due to whitening phenomenon, it has a problem that the optical transmittance is easily reduced by the low-temperature heat treatment.

3) PES Substrate: PES (polyethylene sulfonate) has first been commercialized by Sumitomo Corporation of Japan as a plastic substrate. Unlike the other plastic substrate, it does not have a noticeable disadvantage. However, it has a low CTE and low process temperature (180°C) compared to glass, similar to PC and PET substrates, and has a relatively high hygroscopicity. Thus, in the case of contacting with water or leaving it in the air for a long time, there is a disadvantage that a dehydration process is additionally required. There are not many companies that manufacture and supply PES pellets. The PES are raw materials for substrate production. Therefore, the price of PES substrate is considerably higher than that of glass. However, it is suitable for the LCD device using the polarized light because there is no optical anisotropy.

4) PI Substrate: Since the commercialization by the DuPont in the late '60s, PI (polyimide) is widely applied to electric and electronic components due to an excellent chemical resistance and heat resistance. Recently it was applied as a flexible substrate noting the physical properties of PI. PI is required to introduce a suitable dehydration process since it has larger moisture absorption (about three times) than the PES. Amide groups in the molecular chains of the PI can reduce optical properties of the material as well as make it a heat-resistant. General PI substrate is markedly dark yellow due to this amide groups. Therefore, the light transmission of the PI is ~ 30% for visible light of 550 nm. Such optical properties have become a major constraint on the use of a PI as a substrate material for a display. Recently PI substrate which is optically clear and has high T_g(〉 300°C) are being developed by changing the molecular chain.

5) PEN Substrate: PEN (polyethylene naphthalate) is a plastic having an optical anisotropy similar to the PET. Although there are many problems in terms of thermal stability, recent developments have been made and various research results using PEN have been published. Its price is low and although the CTE is high compared to the inorganic material, it has the lowest value compared to other plastic materials. Therefore, PEN is considered as the best material for flexible display. When the processing temperature of the substrate is constrained to less than 180°C because the thermal shrinkage of the material is 1 to 2%, however, when temperature increases over 200°C the thermal shrinkage increases dramatically.

6) Arylite Substrate: Arylite optical substrate is a substrate developed by Ferrania Image System of Italy. It has received a positive assessment from the domestic institutions and companies. Compared to the other plastic substrate material, it shows excellent thermal, optical, and chemical properties. In particular, CTE increases with an increase in temperature but shows excellent thermal property which increases linearly up to 200°C. However, the thermal stress of the barrier layer, that is deposited on the surface, tends to warp out of the substrate, which causes cracks on the surface of the substrate.

Table 7.2 Comparison of physical properties by plastic substrate.

	Tensile Strength (Mpa)	Young's Modulus (Gpa)	T_g(°C)	CTE (ppm/K)	Dielectric Const.	Resistivity (Ωcm)	color	Chemical Resistance	
								Acetone	Acid
PET	55	2.5~3.0	76.5	79.2	3	2×10^{15}	Transparent	bad	good
PC	52~72	2.1~2.4	150	70.2	2	7.3×10^{16}	Transparent	bad	good
PES	83~100	2.4~8.6	228	49.1	3.8	2.9×10^{16}	Transparent	bad	good
PEN	275	6.1	200	23	–	2×10^{17}	Transparent	bad	good
PI	221	2.8	385	20	3.5	1.5×10^{17}	Colored	bad	good
Poly arylate	100	2.9	330	50	2.6	1.0×10^{16}	Transparent	bad	good
Glass(1737)	–	73	620	5	5.7	–	Transparent	good	good

(3) The plastic substrate structure

The plastic substrate structure is differentiated and patented by several manufacturing companies. Generally 50–200 μm thick PET, PES, PEN and Arylite base films are used. In order to ensure the chemical stability of the base film which is vulnerable to ketones or the like, a hard coat or an undercoat layer of a polymer layer with a thickness of several micrometers is laminated on both sides of the base film with a crosslinked acrylate. Unlike a glass substrate, a barrier layer is formed by inorganic or organic/inorganic hybrid multi–layer to prevent deterioration of the image characteristic of the display due to the contamination by moisture and oxygen. To protect the barrier layer, the plastic substrate is formed like of a laminated structure in which a transparent polymer layer and an overcoat layer are organically combined with each other on the barrier layer. The type and thickness of each layer laminated on a plastic substrate are selected through an optimization of the optical and thermal properties of the substrate. There have been two attempts to develop plastic substrates for flexible displays. The two efforts are barrier coatings and the

the other is barrier layer coating technology. GE is developing a substrate with graded ultra high barrier (UHB) coating technology on a PC film with a T_g of 240°C. The core of Vitex system technology is to maximize the barrier properties of gas permeation. This is achieved through repetitive coating of organic and inorganic layers, as shown in Figure 7.6. This is a specialized technique of flash evaporation, which coats and cures the organic layer in vacuum.

Symmorphix is attempting the gas permeation using a high-quality film formation of the inorganic layer. Dow Corning has developed a film by Silicon-Carbon alloy to be deposited by Plasma Enhanced Chemical Vapor Deposition technique, on a plastic film, to reduce its gas permeability. On the other hand, there is a wide range of development going on in Japan. Sumitomo Bakelite developed a PES optical film using the PES resin, that was originally developed by Sumitomo Chemical. They have been in the process of development using a composite film in recent years. Although there has not been a big improvement in T_g, a film using the composite has been reported where CTE is greatly improved.

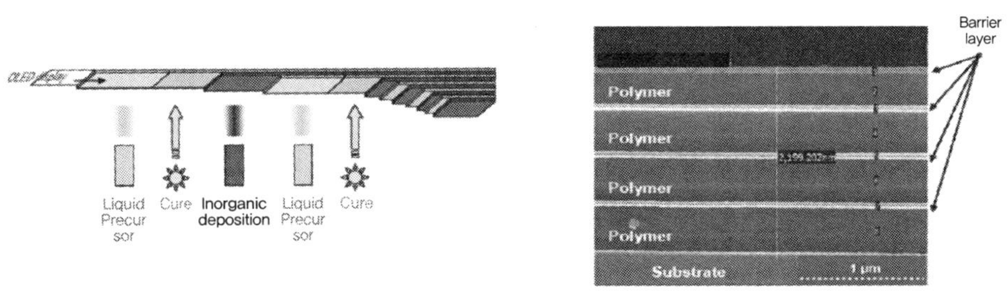

Figure 7.6
Flash evaporation technology developed by Vitex.

Dupont-Teijin has announced the development of a PEN film usable at a temperature of 160°C. Through appropriate heat treatment, it was shown that it is possible to improve the film properties. Separately, the company Teijin developed a highly heat-resistant PC board. JSR Corp. is trying to process films using a COC (cyclic olefin copolymer). In addition, various studies for the highly heat-resistant polymers are going on in Europe. Promerus has announced the development of a film series Polynorbornene (T_g~330°C). Fraunhofer Institute, has developed Ormocer, an organic-inorganic hybrid polymer, that has excellent gas barrier properties. Ferrania has also developed a substrate having a high T_g (about 320°C) of the series arylite and supply a plastic substrate for a high-temperature process for the research and development in small portions.

(4) Required characteristics of the plastic substrate

In order to use the plastic substrate instead of glass, certain thermal, optical and mechanical properties are required. Changes in the dimensions and shape of the substrate due to expansion and contraction of a plastic substrate according to the heat treatment have a major effect on the displayed image characteristics. Therefore, it requires a low CTE at a high process temperature. The flatness and cleanliness of the substrate surface are factors that affect the reliability and stability of the thin film (organic or inorganic) formed in the subsequent process. When the substrate is formed, its related characteristics should be improved by optimizing the material composition and the forming process. In order to ensure a process–ability, chemical resistance to a solvent over a certain range is required. The transmittance of the substrate has a direct impact on the performance of the display. It is particularly important in the case of LCD, OLED (bottom emission) and E–paper, which utilize the substrate as a visual window, and requires a transmittance of 85% or more at a wavelength of 400 to 800 nm.

1) Dimensional stability

Dimensional stability of the substrate is generally defined as the degree to which a material maintains its original dimensions when subjected to changes in temperature and humidity. The unit length of the substrate to be evaluated differs by manufacturer and is mainly evaluated using a TMA–7 thermo–mechanical analyzer (Perkin Elmer). If a plastic substrate is exposed to a high temperature, it shows a phenomenon of expansion and contraction. This phenomenon is the biggest obstacle to the progressive multi–layer process in the implementation of the flexible display. In order to achieve the thermal stability of the substrate, there are two approaches. First, the material company uses a material having excellent heat resistance as a base substrate material for the thermal stability of the plastic substrate itself or it minimizes the thermal history of the substrate and difference in thermal behavior (mismatch) between the adjacent layers during the production of the substrate, by minimizing the thermal history. Secondly, set makers are making efforts to reduce the process temperature as much as possible in order to minimize the thermal history of the plastic substrate while securing the reliability of the process by carrying out the process under the constant temperature and humidity.

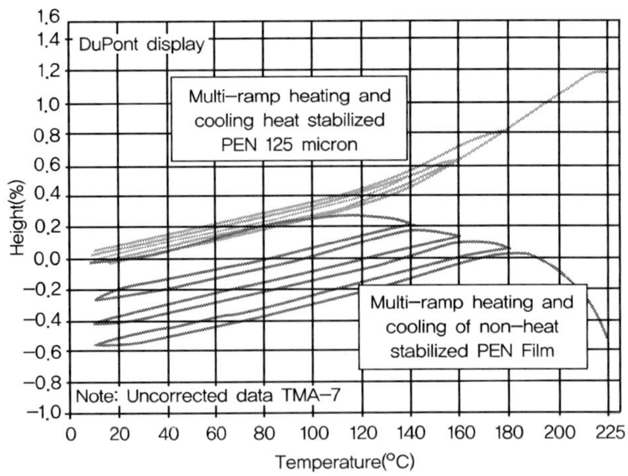

Figure 7.7

Change in dimensional stability of PEN substrate.

In addition, the plastic substrate which has been thermally deformed due to the high process temperature, is cooled down to room temperature, and then the process is carried out so that each process proceeds in the same substrate state. The rate of change of the substrate due to the heat treatment is also dependent on the heat treatment but also depends on the time. Due to these steps, when a proper thermal stabilization process is achieved, a plastic substrate exhibiting much improved dimensional stability can be obtained. Figure 7.7 shows the thermo–mechanical analysis of PEN with and without thermal stabilization starting at 8°C and heating to 140°C, 160°C, 180°C and 220°C with repeated heating and cooling. In the case of a substrate without thermal stabilization, the dimensional change is reduced by heating and cooling, but permanent deformation occurs in the dimensions of the substrate. On the other hand, in the case of the substrate subjected to the thermal stabilization process, the substrate was not permanently deformed. In addition to the thermal factors, the dimensional stability of the substrate may also be a problem due to the swelling phenomenon caused by the solvent used during the manufacturing process of the display device, but efforts are being made to reduce this effect by introducing an appropriate desolvation process in between two successive process step.

2) Low coefficient of thermal expansion (CTE)

There are two ways of evaluating the thermal characteristics of the substrate, as previously mentioned: one is the method for heating the substrate and then cooling to the initial temperature and assessing the rate of dimensional change (shrinkage usually occurs) to the substrate, and the other is evaluating the degree of the elongated substrate in constant temperature heating. The latter defines the evaluation results of a coefficient of thermal expansion. In order to realize a flexible display, several layers of inorganic materials are laminated on a plastic substrate. In this case, when the difference in thermal expansion coefficient between the substrate and the laminated layer is high, cracks may occur in the deposited layer or peeling may occur in a high-temperature process.

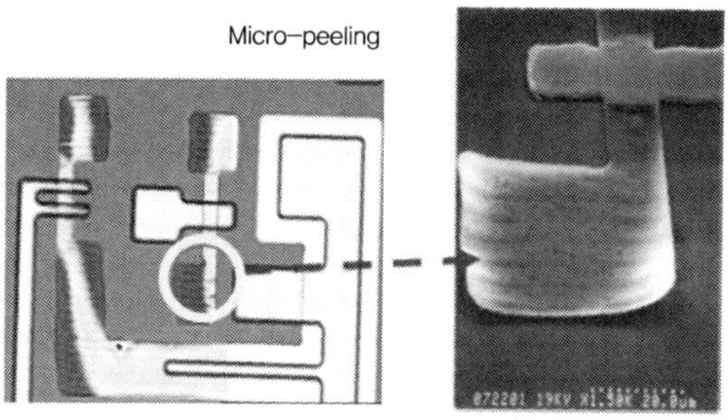

Figure 7.8

Crack and peeling phenomenon of thin film due to difference of thermal expansion coefficient.

Therefore, the coefficient of thermal expansion of the plastic substrate is required to be 20 ppm/°C or less, but no substrate satisfies this requirement at present. Also, since the coefficient of thermal expansion tends to increase sharply in the vicinity of the T_g of the plastic, a material having a high T_g and a low coefficient of thermal expansion is a suitable combination for a flexible display.

3) high barrier properties

In order for a plastic substrate to be used instead of conventional glass, barrier properties of glass are required. Moisture permeability of thermoplastic semi-crystalline plastics (PET, PEN, etc.) tends to be slightly lower than that of amorphous polymers (PES, PC, etc.). On the other hand, the plastic based substrate alone shows a high moisture permeability, that is about 1 to 100 g/m^2 [day]. However, in order to prevent deterioration of the image characteristics of the display such as the contamination of the liquid crystal, the deterioration of the driving element, and the oxidation of the OLED cathode metal layer and so on, a thin barrier film is required. The required quality of the barrier film may vary slightly depending on the sensitivity to the pollutant, but it is required to be ~ 10^{-2} g/m^2 [day] for LCD and less than 10^{-6} g/m^2 [day] for OLED. Research on thin barrier films of plastics begun to develop food packaging materials from the 1950s. However, in the case of a barrier for plastic substrates, a thin barrier film having a water vapor permeability of 1/1000 or less than that of a conventional food packaging material, is required in order to reduce deterioration of light transmittance of the substrate. The quality of the thin barrier film is a combination of the cleanliness of the deposition surface, the denseness of the thin barrier film, and the interfacial characteristics with adjacent layers. Recently, a barrier film having a structure in which organic/inorganic composite films are laminated has been studied. As a method of evaluating the characteristics of the thin barrier film, the moisture permeability and oxygen permeability are evaluated using an equipment of MOCON. However, because of the measurement limit of the equipment (in the case of the water vapor permeability, the measurement is not possible in the case of $5x10^{-3}$ $g/m^2.day$), the thin barrier film with very low permeability to oxygen and moisture requires other forms of evaluation means.

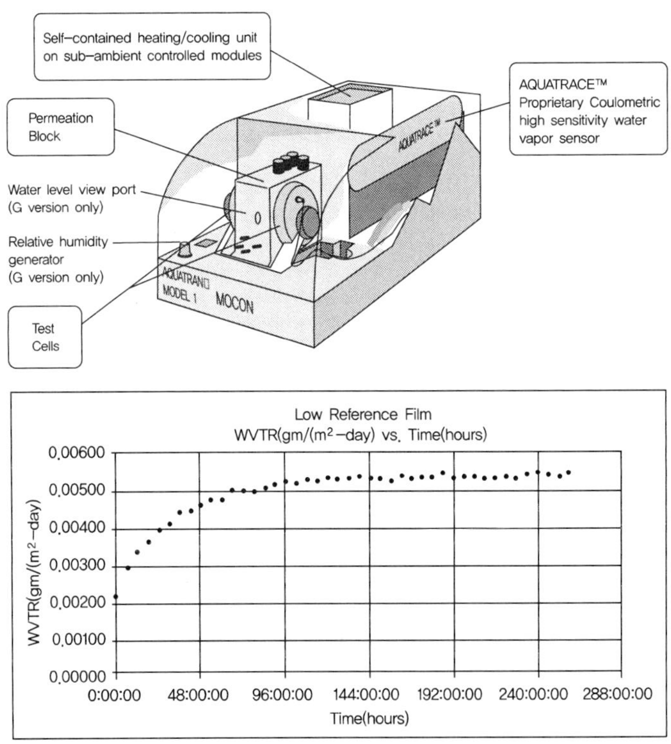

Actual data from test performed on AQUATRAN low reference film

Figure 7.9

Conceptual diagram of MOCON's moisture permeability evaluation equipment.

There are various evaluation methods, but the barrier characteristics that can not be evaluated by MOCON equipment are usually evaluated by the calcium (Ca) tester method.

4) Rigidity

The rigidity of the plastic substrate has a significant importance in evaluating the process suitability of the substrate. It is defined as a function of Young's modulus (E), thickness (t) and Poisson's ratio (υ).

$$\text{Rigidity} = \frac{E \times t^3}{12(1-v)} \quad \text{(Equation 7-1)}$$

The Young's modulus of thermoplastic semicrystalline plastics are three times as large as amorphous polymers. Therefore, with the same thickness of the substrate, thermoplastic semicrystalline plastics are three times harder than amorphous plastics.

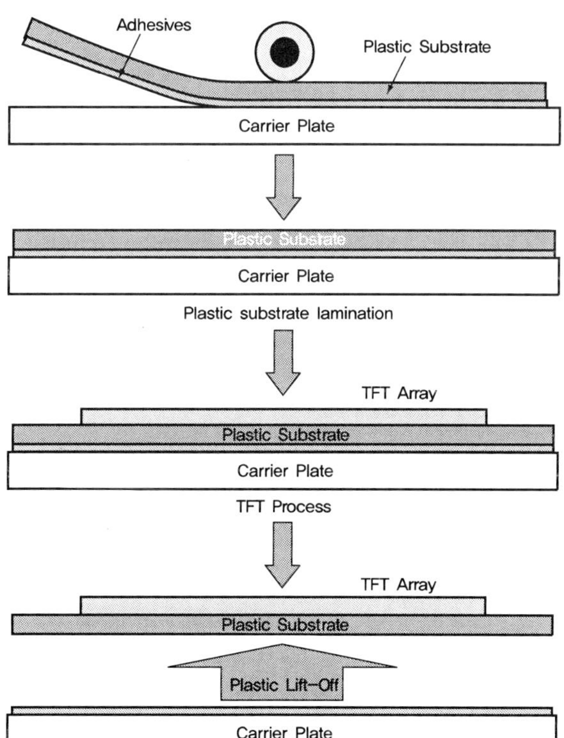

Figure 7.10

Plastic lift-off process.

Poisson's ratio is the ratio of the horizontal and vertical strain, which is about 0.3~0.4 for ordinary plastic. The rigidity of the plastic substrate plays an important role in the process. When the substrate is warped during the process, if the vacuum equipment is not used smoothly and the force is applied to the plastic substrate, cracks may occur due to stress to the thin film. Therefore, in order to solve this problem, a method used for attaching a plastic substrate to a planar carrier substrate such as glass during the process as shown in Fig. 7.10 and then removing it when the process is completed.

5) Transmittance

The optical transmittance of the plastic substrate is an important material property which has great influence on the characteristics of the displayed image. High transmittance is required in display devices and there should be no changes in the degree of crystallization by process change (heat and solvent treatment process), the scattering and reflectivity changes with varying crystallization and surface roughness. In the method of evaluating the transmittance, the optical characteristics (transmission and reflection) of the mixed light in the visible wavelength region (400 to 700 nm) or the light characteristic at a specific wavelength (550 nm) is measured. Most of the optical substrates show a high transmittance (at least 85%) at 550 nm and generally do not show a significant difference when compared to the optical characteristics of the laminated films, including the base film and the thin barrier film.

6) Durability

In order to realize a display that can bend like paper, it is necessary to understand the mechanical properties of the component layers constituting the substrate. The plastic substrates are of a multilayer structure formed by stacking organic and inorganic materials such as a base substrate, barrier, hard-coat, overcoat, and the transparent electrode. Cracking any of the constituent thin films and delamination of layers may occur during bending or during it lifetime. The structural problem of such a plastic substrate is due to difference in Young's modulus between the organic material and the inorganic material (by a factor of approximately 100), but it is inevitable, unless a suitable organic material developed to replace the inorganic material. Therefore, in order to realize a reliable display device, capable of bending while securing the mechanical characteristics of the substrate, it is necessary to design an appropriate substrate materials based on the mechanical characteristics of each layer as well. When a plastic substrate that is laminated with various layers (organic/inorganic materials) is bent until mechanical breakage occurs, internal stress is generated on the inner substrate side and in the thin film deposited on the substrate, resulting in a failure of adhesion or cohesion within the deposited thin film layer. The internal stress (σ_i) applied to the plastic substrate by bending is calculated by substituting the radius of curvature (*R1, R2*), Young's modulus, Poisson's ratio and thickness before and after coating with the thin films, following equation.

$$\sigma_i = \frac{E_s \times h_s^2}{6(1-v_s)h_c}\left(1+\frac{h_c}{h_s}\left(4\frac{E_c}{E_s}-1\right)\right)\times\left(\frac{1}{R_2}-\frac{1}{R_1}\right) \quad \text{(Eq. 7-2)}$$

E_s and E_c are Young's modulus of the substrate coated with a thin film. υ_s is the Poisson's ratio of the substrate. h_s is the thickness of the substrate. h_c is the thickness of the coating film. The failure of cohesion of a deposited thin film can be evaluated by a bending method and a fragmentation method. As shown in Fig. 7.11, the bending method is observing the cohesive failure of the deposited thin film on the plastic substrate while stressing the plastic substrate on both sides. It is widely used method because of its simplicity. The fracturing method uses a CCD camera to observe the cohesive failure of the deposited thin film while stretching the substrate.

Considering the failure of cohesion of the deposited film (the transparent electrode, ITO, IZO) and the barrier (SiN_x, SiO, SiC) as phenomenological, a foreign matter (particle) or micro defects such as pinholes could be found below the surface of the thin plastic substrate even if it is not subject to strain. As the strain is applied to the substrate, it can be seen that the crack starts to appear around the micro-defects already existing on the substrate or on the deposited surface. As the strain further increases, cracks having a length of about 100 to 500 times of the deposited film thickness start to appear. At this point, the growth of the cracks becomes more acute, and as shown in Figure 7.11, the deformation appears. If a larger strain is applied to the deposited film, tensile failure is observed on the entire deposition surface. Figure 7.12 shows the results of the cohesive failure of 50 and 100 nm deposited ITO films on plastic substrates using bending method. As the thickness of ITO increases, the sheet resistance decreases. In particular, a low-resistance transparent electrode is required for OLEDs.

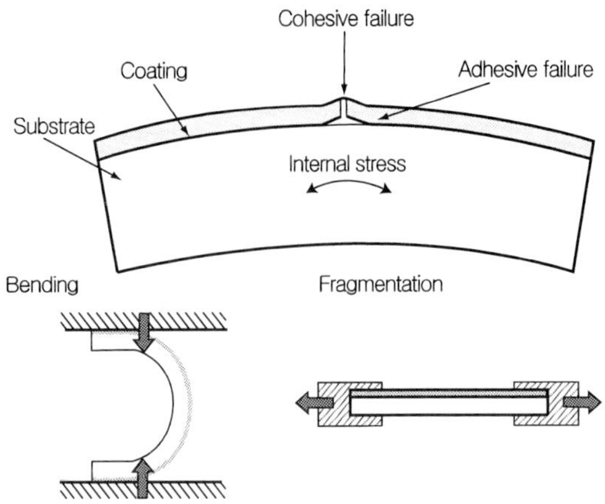

Figure 7.11

Mechanical testing of plastic substrates.

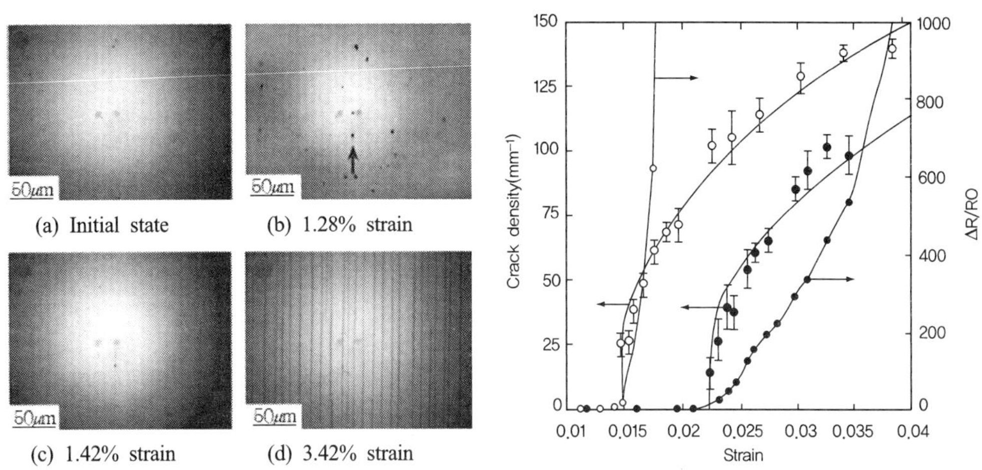

Figure 7.12

Crack progression on 100 nm ITO coated plastic substrates (○: 100 nm, ●: 50 nm).

Figure 7.12 shows that the resistance value of 100 nm ITO thin film increases rapidly from relatively low strain compared to 50 nm ITO thin film. These results show that increasing the thickness of the ITO thin film can lower the sheet resistance, but it can lead to problems in the application related to bending. This result can be similarly observed in inorganic thin films deposited with a barrier. Therefore, when a currently used inorganic transparent electrode or a barrier film is directly applied to a plastic substrate, a bendable display may be realized, but it is almost impossible to realize a rollable display in this way.

7) Required Characteristics by purpose

Required characteristics of plastic substrates may vary slightly according to the purpose of use. The organic material of OLED requires a very low water vapor permeability and oxygen permeability, as previously mentioned, and they are the most important scheme of management for stability of the OLED panel. When using the OLED in the form of bottom emission, high optical transmission is required because it recognizes the light through the plastic substrate as a visual image. In addition, the surface flatness, thickness uniformity, and environmental resistance of the surface on which the light emitting layer is deposited have a significant influence on the image characteristics. To apply plastic substrate in a semiconductor process, many separate control parameters are required such as thickness uniformity, the adhesive strength of plastic substrate with an inorganic thin film formed on the substrate and its good mechanical properties. The most important thing is the heat-resistance of the substrate that affects transparency and reliability. Conventional semiconductor process largely deals with the inorganic thin film deposition and this is a high temperature process. When too low process temperature is used to the plastic substrate, the quality of the deposited inorganic thin film degrades and the performance of the device deteriorates. The heat resistance of the plastic substrate is recognized as a factor directly affecting the quality and device characteristics. For smooth semiconductor processing, CTE below 20 ppm/°C and thermal stability (dimensional stability) at 180°C should be secured. Also, the shrinkage rate of the substrate should be within 0.003% in the thermal cycle between the maximum process temperature and room temperature. Depending on the type of display used, even though there may be small differences, but the light transmittance, moisture permeability, and the optical transparency required by the OLED of the prior art are required. In addition to heat resistance, chemical resistance to all solvents used in the process is also important. Figure 7.13 summarizes the required characteristics of the barrier membrane by the field of application.

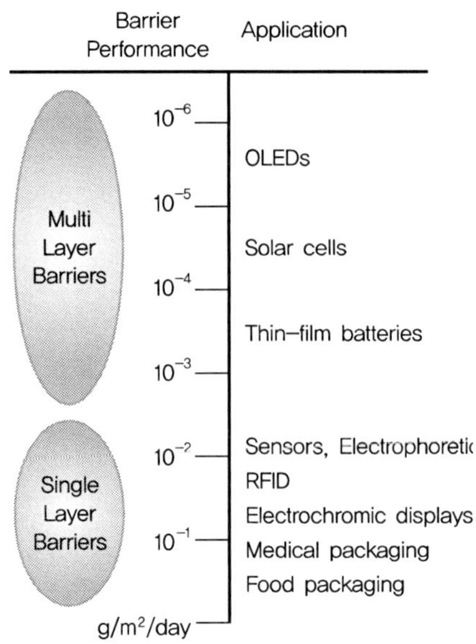

Figure 7.13
Characteristics of barrier film required by application field.

Until now, the success factors for the glass substrate-type display companies have facility, investment, time and size based on the size of the original glass, but the next generation displays are expected to dominate the market by diversifying applications and responding to various types of products. In this market, the importance of parts and materials will increase even more than ever. Many plastic substrates have been studied mainly by different companies. However, there are no currently available substrates to replace the conventional glass. For this reason, several prototypes have been announced, but flexible displays with image characteristics and reliability have not yet been reported. In view of the development trend of the display, the flexible display is expected to form a single market trend in the future. Considering the current technological level, it is difficult to secure the desired characteristics of the flexible display compared to the conventional glass-based displays, without progressing material development and process development. The quality of plastic substrate, which is the key material of flexible display, is the material field that has the greatest influence on display image characteristics and process yield, but the domestic research base is weak. In order to secure market dominance of the display market in the future, there is a need to continuously develop the related materials and solve the related problems with a long-term vision for marketing.

2.2 Printing Process Technology

Printing technology can be divided into direct printing and indirect printing methods. An ink–jet printer using contactless ink dispersion, touch screen, flexography, and gravure type roll printer represent direct printing. Indirect printing prints the substrate via other media. For example, contact imprinting technology, larger transfer patterning, etc are indirect printing. A roll printer using roll/blanket is a continuous process. This technology provides higher–speed to the printing process. Recently, various technical development are in progress focusing on the LCD manufacturing process. Roll method printing technology has the advantage of high–speed printing. However, the viscosity of used ink and resolution differ depending on the method, and the equipment research for the production process of the organic electronic element is not yet sufficiently progressed.

Table 7.3 History of Major Technological Advance of Flat Panel Displays Comparison of the viscosity of ink, the resolution and the production speed according to printing method.

Printing method	Viscosity (Pas)	Layer thickness (μm)	Feature size (μm)	Registration (μm)	Throughput (m^2/s)
Gravure	0.01~0.2	<0.1~8	75	>20	3~60
Flexography	0.05~0.5	0.04~2.5	80	>200	3~30
Offset	5~100	0.5~2	10~50	>10	3~30
Screen	0.5~50	0.015~100	20~100	>25	2~3
Ink–jet	0.001~0.04	0.05~20	20~50	5~20	0.01~0.5

Therefore, the importance of high–speed printing technology in the manufacturing process of flexible display devices is very high, and various materials for LCD, OLED, and electrophoretic paper that can be implemented on a flexible substrate have been developed. In addition to commercialization, it is expected that the market will be expanded. Especially, if the establishment of the industrial base is achieved by the expansion of the roll–to–roll system/process and infrastructure, it can contribute to the improvement of economic development of the next generation display industry. Figure 7.14, discloses an organic polymer thin film transistor by a high–resolution ink–jet printing method. In addition, a source/drain electrode pattern is formed on a polyimide insulating film by using a water–soluble PEDOT: PSS, (9,9–dioctyltrifluorene–co–bithiophene, F8T2) thin film of active organic semiconductor layer. Its passivation (PMMA / PVP) was formed by using an organic solvent free from solvent interference. In general, a fine pattern using an ink–jet can be obtained with a range of 20 to 25 μm, and a separate surface treatment method is required to obtain a finer pattern. From the viewpoint of equipment development, the United States, Japan, and the United Kingdom are actively pursuing research. In particular, efforts to obtain patents and cross licensing are being carried out. Seiko Epson manufactures prototype product by their own-developed technology. In Korea, MEMS-based piezo ink-jet print head (MEMS) devices are under development by Samsung.

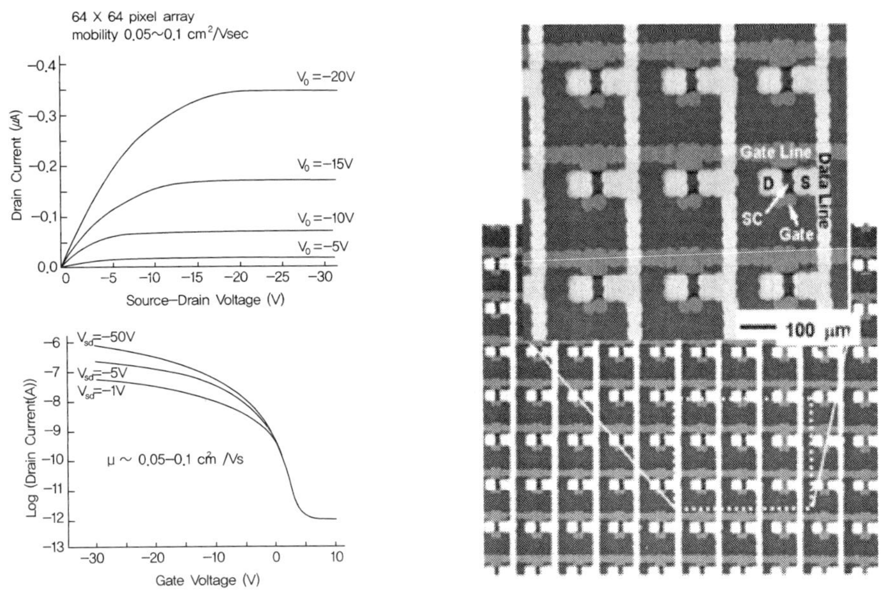

Figure 7.14
OTFT printed array of 300 μm polythiophene (Xerox/PARC) by ink-jet printing.

The laser transfer patterning method was developed by 3M and Samsung SDI using LCD color filter transfer method. Research and development for applying a color pattern method for the purpose of developing a large area (fourth generation or more) low molecular weight organic EL device of an OLED are actively performed. In the patterning process of an organic semiconductor device, studies are being conducted to transfer/print the active layer material of an organic semiconductor device, a gate electrode, and a conductive polymer for a source/drain electrode by a laser pattern method. Transfer of the donor film onto the pixel of the substrate is achieved and a pattern transfer with a pixel resolution of 5 μm and a width of 250 pixels was reported. The advantage of this dry-patterning is that it can have a high resolution and patterning of a multilayer thin film is possible. For high-resolution transfer quality, the control of adhesion and cohesion between substrates, films, and materials of the transfer material is required. In this section, we will briefly describe the major printing methods that can be applied to the roll–to–roll process. These are the core of the low–cost large–area manufacturing process for the futuristic flexible display.

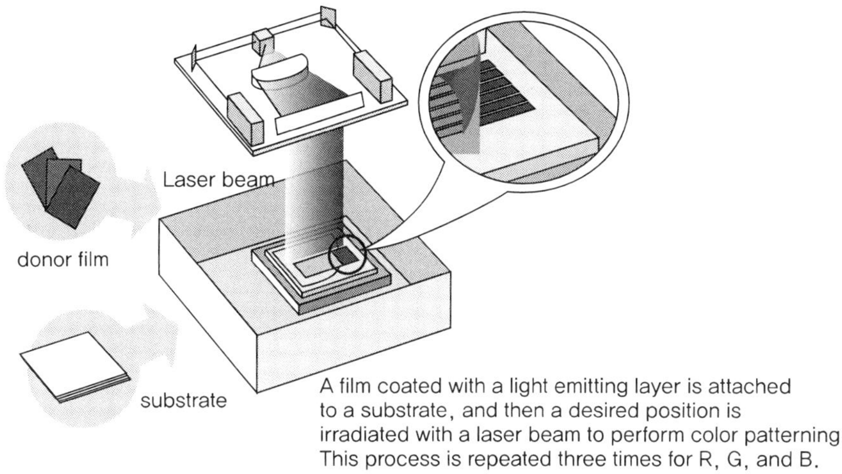

Figure 7.15

Schematic of the laser transfer patterning process.

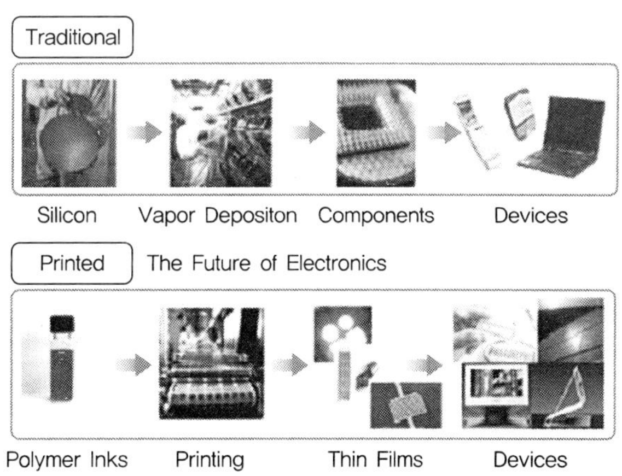

Figure 7.16

The traditional manufacturing process and future printing manufacturing process.

1) Screen Printing

Screen printing is a non-pressurized printing process using a stencil, which shows relatively low process speed. However, it is advantageous in that it is not affected by ink and substrate. It is suitable for thick ink deposition on a large surface area of the substrate. It is a printing method that can be carried out at one time from thin film formation to patterning in the normal atmospheric condition. In supplementary equipment, it can be carried out in a simple and inexpensive process compared to other equipment, which has a great advantage in time and cost. The main process elements are a screen, stencil, squeeze, ink, press and substrate. It is a relatively simple process, but there are many variables that can affect the printing process, including printing speed, squeegee angle enhancement, screen to substrate distance, mesh material, mesh number, and viscosity of the ink. As the viscosity of the ink increases, the thickness of the thin film increases. As the number of meshes increases, thin film deposition is advantageous. The thin film deposited by screen printing at the beginning of the technology development is not much different from the thin film obtained by spin coating. On the other hand, the pattern accuracy tends to reduce due to mesh effects of hundreds of nanometers high, 100 μm long, and 10 μm wide, which can negatively affect device characteristics.

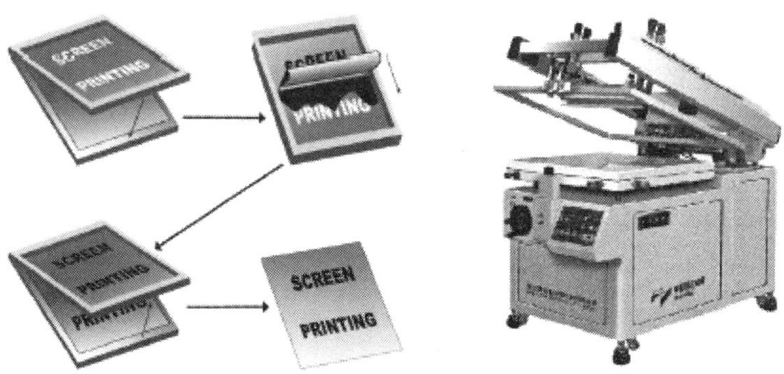

Figure 7.17
The screen printing schematic process and the screen printing equipment.

2) Gravure Printing

Gravure printing is a roller-type printing process using a copper cylinder with a pattern of 25 to 30 μm in depth. The pattern on the copper cylinder constitutes an image in the form of a square cell, usually 250 cells per inch. Although the mechanical pattern making method is dominant, a method of using the laser has been introduced, and the trend is towards achieving a more flexible and easy patterning technique. In the gravure printing process is first filled with ink in the cell, where the excess ink is continuously removed by a doctor blade. However, high pressure and direct contact with metal cylinders make them unsuitable for printing on rigid substrates such as glass. The application of gravure printing to the fabrication of organic electronic devices has not yet been extensively studied. It is possible to apply various ink systems, and it is possible to manufacture a thin film having a considerable thickness at a low process speed by using a high viscosity ink.

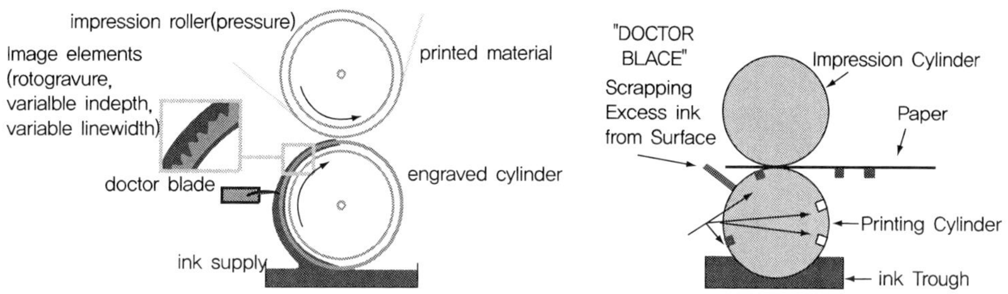

Figure 7.18

Gravure printing process schematic.

3) Flexography Printing

Flexographic printing is a technique of making intaglio or embossed patterns on a plate made of rubber or photopolymer and using it as a printing medium, and the ink is delivered via anilox roller. Control of process variables such as speed, ink viscosity, the pressure between rollers is very important to obtain optimum printing accuracy. This process technology has a precise line patterning just like lithography and has characteristics suitable for printing of electronic devices.

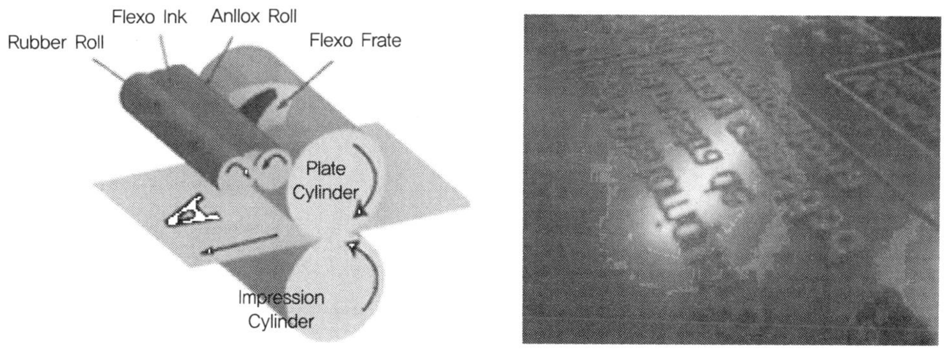

Figure 7.19

The flexography printing process schematic and flexography printing substrate.

However, it has technical problems such as forming a relatively thick thin film at the end of the pattern. Therefore, it is necessary to investigate process control and optimization such as pressure, viscosity and Anilox roller selection.

4) Offset Lithographic Printing

Offset lithography is a relatively well-established process for printing various types of materials and has the advantage of being able to produce circuits on both sides of the substrate. For continuous processing, the plates are made in cylinders and at the protruding parts, ink is transferred from plate cylinders to a rubber blanket and successively transferred onto a flexible substrate. When the printed image is dark, it is hydrophobic, it rejects water but absorbs the ink, and when the printed image is bright, it is hydrophilic, and water causes the ink to be rejected. Upon contact between the ink and the roller, the printing plate absorbs water and ink where necessary to form the image to be printed. The formed image is not directly printed on a substrate, but is transferred to a rubber cylinder and then the process is completed by pressing the ink film on a substrate supported by an impression cylinder. Since offset lithography printing technology is not a process of forming line width and line boundary, it shows very high pattern accuracy, and transfer rate compared to the traditional printing process. The key process factors include ink properties, printing plates, solutions, and equipment. Its current resolution is 10 ~ 100 μm. Offset lithographic printing has the advantage of being able to manufacture circuits on both sides of the substrate. The technological difficulties for this can be overcome by using existing double-sided printing technology.

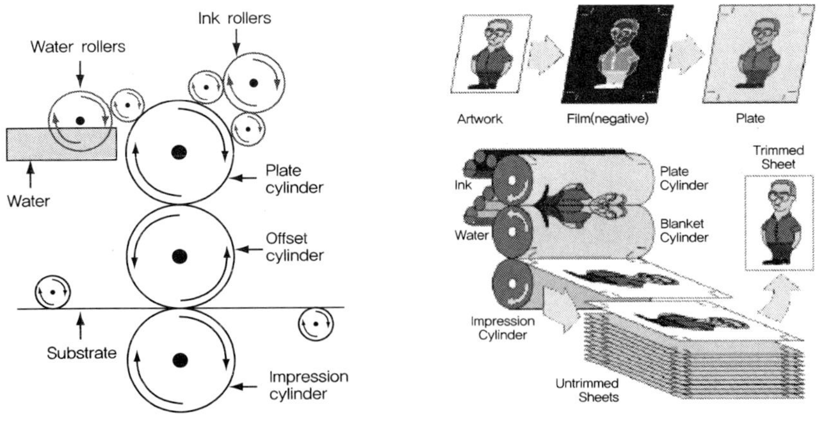

Figure 7.20
The offset lithography printing process schematic.

5) Ink-jet Printing

Ink-jet printing is a process for making fine ink droplets and patterning them at the desired positions on a substrate and is a technology suitable for realizing complicated shapes in a small volume by a non-contact manner. It is divided into the continuous process and drop-on-demand (DoD) process. The continuous process is excited by a vibrating transducer when ink is supplied to the ink chamber at a constant pressure as shown in Figure 7.21. Ink flows through the nozzle with an appropriate pressure, depending upon its appropriate viscosity, and ink droplets pass through the electrodes to get a charge. As a result, the charged ink droplets are deflected in a specific direction when an electric field is applied. By this principle, the movement of each ink droplet is controlled and synchronized. Adding a substrate motion at the same time, the ink droplet causes deposition of a continuous ink pattern, but the biggest limitation of the continuous process is the low resolution due to the droplet injection of 200–300 droplets per inch. Drop-on-demand ink jet injection quantity (less than 10 pico liter) and excellent positional accuracy are suitable for electronic device printing process technology, but there are limitations in the mass production. Despite the continuous development of printer-injector-head unit, the resolution of print remains limited to a few tens of micrometers, because of limitation in the channel resolution, droplet speed variations, spreading and splashing of ink etc.

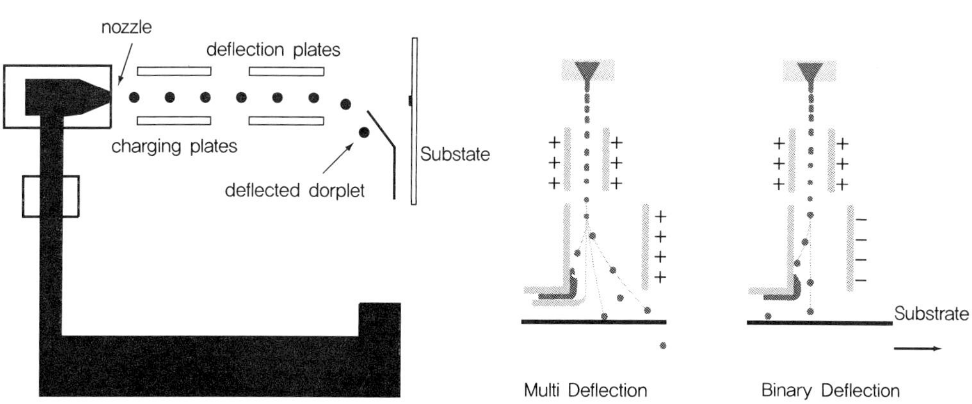

Figure 7.21

Continuous ink-jet printing schematic.

Recently, to solve this problem, a surface energy patterning process of a substrate is applied to selectively induce a flow of a conductive polymer droplet in a liquid state to improve the accuracy of the channel length to 5 μm.

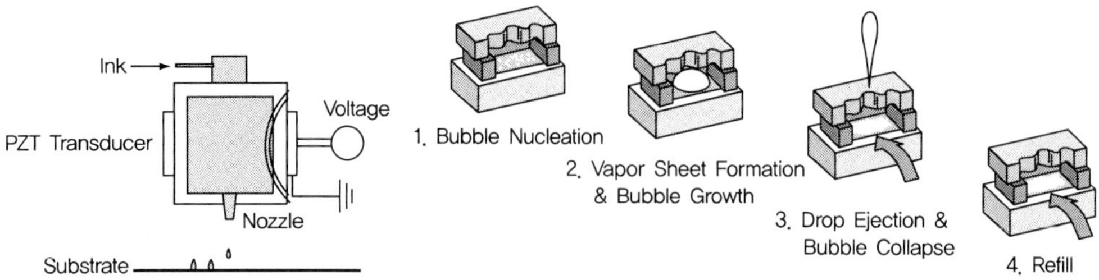

Figure 7.22 The principle of the Drop-on-Demand piezo ink-jet and thermal ink-jet

The roll-to-roll process of an organic electronic device means an innovative technological progress in process capability, in which a method of mass-production of the existing print media is applied to a display manufacturing process. Indeed, the cost reduction of existing displays and the development of fabrication processes for future flexible displays have enormous future-oriented value in this area. Paper-based technology such as paper, label, electronic ticket, processes and equipment for embedding RFID antennas as few of the methods of digital computerization have been actively researched and developed. These electronic tickets, flexible electronic devices/dynamic signage using plastic substrates are still in the early stage of development. This conveyor belt-based roll-to-roll production method requires solving some key technological problems in processes and systems, that is different from the conventional batch production methods. First, for an in-line process system, the handling of the belt is important. The tension condition, kind and speed of the belt should be determined for accurate positioning control and transport. Second, when the belt is continuously moving, the coating process must be able to cope with this shift. Gravure, offset, ink-jet printing, etc. must be designed for these conditions, since plastic substrates must continuously apply organic materials as they travel in the longitudinal direction. Third, since the substrate moves continuously, the vacuum of the process equipment can not be secured. Therefore, it is very important to secure the vacuum in the process. Various sealing needs to be studied, and in process equipment using chemicals, process chemicals should not leak into the hypothetical setting. Recently, the improvement of resolution and thickness of precision electronic circuit pattern, printing process and equipment research is aimed at solving the above technical issues. Figure 7.23 shows a simplified schematic diagram of the roll–to–roll process. As shown in the figure, the roll–to–roll process can be considered as a separate process in which individual processes are separated and the whole process in which all the processes ultimately pursued are progressed in real time as a continuous process.

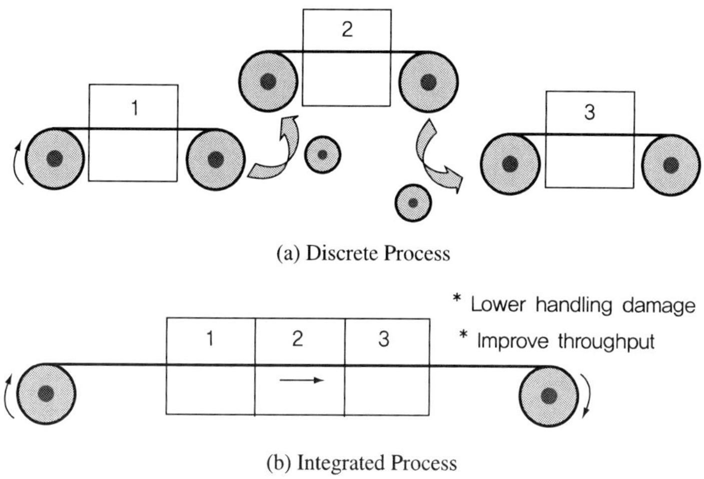

Figure 7.23

A simplified schematic diagram of the roll–to–roll process [1. Treatment & coating process 2. Pattern print (organic/metallic, etc.) 3. Inspection & monitoring] (a) Discrete process: each of the processes is connected as a whole by the assistant roll, (b) Integrated process: the individual process exists in the overall process continuously.

In the case of research carried out in other countries, the Institute for Print and Media Technology at the University of Chemnitz in Germany is conducting device fabrication and process research on organic electronic logic devices for line widths of several tens of micrometers. This is useful in organic semiconductor devices and ring oscillators, by continuous mass printing processes. As shown in Figure 7.25, the process is completed by printing the source/drain electrodes, the organic semiconductor, the protective layer, and the gate electrode in a printing unit such as flexography and offset in order to complete the logic device.

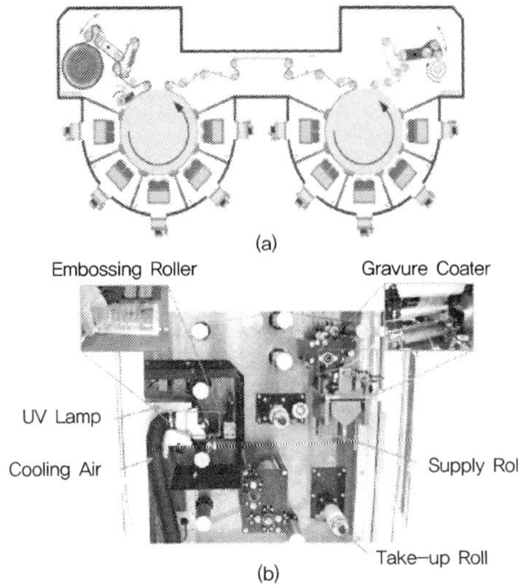

Figure 7.24

(a) Roll-to-roll vacuum deposition system of Applied Film, (b) Roll-to-roll imprint lithography facilities of Hewlett-Packard.

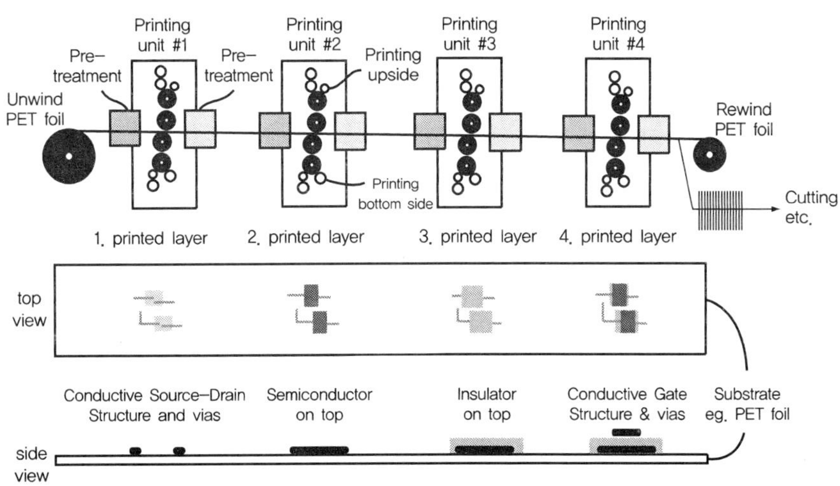

Figure 7.25

The making schematic of organic transistor element by mass printing continuous process.

Finnish national technology development center (VTT) announces device fabrication based on roll printing, and is developing related manufacturing process technology targeting a variety of materials such as a printable organic transistor, solar cell, OLED, organic memory, RFID, photonic micro system, sensor, indicator, etc. They are participating in joint research on material/device technology with Arizona State University/Army Flexible Research Center of US. At present, a conductive nano silver ink pattern having a line spacing of 20 μm and a line width of 20 μm or less (electronic circuit functional minimum line width of 75 μm) can be realized by gravure printing in Figure 7.26. The OLED device is manufactured by the gravure method. Here, the cell depth of the gravure roll is about 3.0 ~ 4.5 μm, and the thickness of the organic film can be adjusted. In the case of the electrodes, ink printing using aluminum particles is attempted to obtain a characteristic of about.

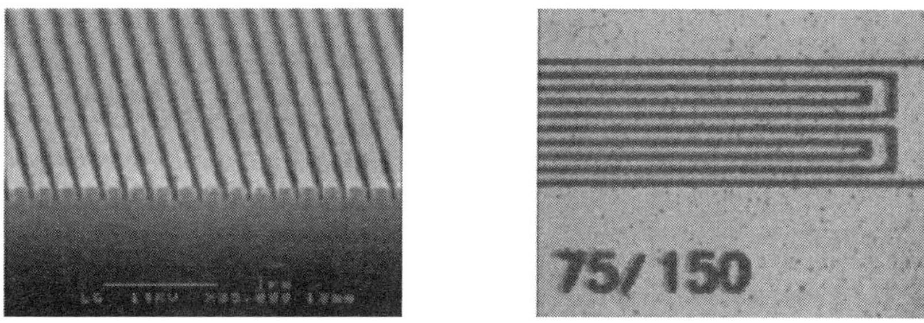

Figure 7.26

The gravure printing pattern implementing 20 μm line spacing and 75 μm of the line width of the functional electronic circuit (VTT).

1.0 Ω−cm. Although initial results are published, indicating the possibility to fabricate organic electronic devices by using continuous printing technology and to realize precise patterns, it is reported that the interface suitability between functional organic/inorganic is somewhat lacking. For example, it is possible to realize a luminescent pixel pattern of an OLED, an active semiconductor of an organic semiconductor device, and a source/drain pattern of several tens of micro−lines. However, a multi−layer printing technique has not been fully developed yet. In Korea, research institutes such as Konkuk University, KIST, and Korea Institute of Machinery and Materials have started basic researches for the production of organic electronic devices. While recognizing the necessity in the industry, they are working with academia and research institutes. Roll−to−roll process technology, which is closer to commercialization, is mainly performed by venture companies and universities with the support of the US Display Research Association (USDC). Kodak initiated manufacturing process with flexible cholesteric LCD. Sipix developed a new type of paper display of electrophoretic media films (used for electronic paper display). (Figure 7.27).

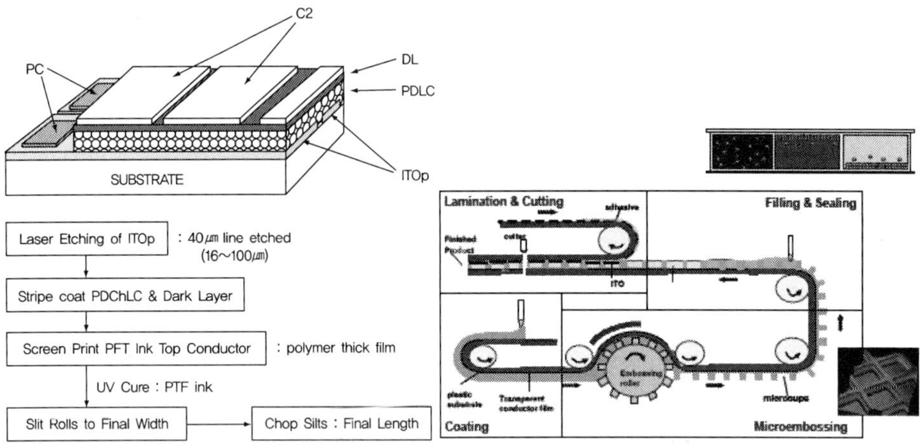

Figure 7.27

(a) Roll–to–roll manufacturing process of cholesteric liquid crystal display announced by Eastman Kodak company. (b) Roll–to–roll manufacturing process of electrophoresis film of SiPix company.

Conventional approaches to electronics manufacturing use complex processes to form patterns in circuits. First, a semiconductor or a metal film is formed on a substrate as a base. When the circuit area is determined, the remaining part is etched. At this time, the coating and etching process must be run in a vacuum, where bulky and expensive vacuum pumps are a major cause of increasing the cost of fabrication. Direct printing technology can solve these problems of conventional methods. Because the circuit pattern is drawn directly on the paper or by fetching the image from the printing plate, the three–step process, that is required by the existing method can be completed in one step. Vacuum equipment is also unnecessary because the process can be run at standard atmospheric pressure. In addition, the conventional method has a problem that a high cost is to be paid for the processing of the substrate (wafer, glass, etc.). Because the wafers are processed individually, they must be transported separately to other processes. Due to the required loader and unloader, the size of the processing system varies. However, the roll–to–roll method allows the substrate to pass through the processing system continuously, and the process equipment can be configured as a continuous line, which can greatly reduce the time, space, and equipment required for large production.

The combination of direct printing and roll–to–roll processing can significantly reduce equipment investment and increase productivity as compared to the traditional processes. This can lead to a reduction in manufacturing cost per unit area. This effect will vary depending on the type of equipment and the size of the production, but can generally reduce the cost by a factor of 10. It is anticipated that existing flat panel displays will be able to reduce manufacturing costs by more than one tenth when they are entirely produced by the printing process. These printing electronics will require materials completely different from those used for semiconductors, interconnects, and substrates. Plastics will be used as substrates and new organic semiconductors. Pure metal molecules will be used for circuit formation. The thickness, weight, and flexibility of new devices are primarily due to the nature of these materials. In particular, by using a plastic substrate, it is possible to manufacture light and thin electronic devices. When organic semiconductors are used, flexible displays can be manufactured. In the area of flexible display and electronic device research, development trends of plastic substrates and technological development of continuous process and printing technology have been examined. As described above, in order to realize a flexible display, the supply to the flexible substrate should pre-exist, but there is a limit to the current level of the plastic substrate. In order to develop a plastic substrate with excellent barrier properties and process suitability, it is necessary for the substrate development company and the device manufacturer to make technological progress by establishing a cooperation. Additionally, in order to secure excellent properties such as ON/OFF current ratio and high charge mobility of TFT device using polymer and organic semiconductor, not only alignment and orientation technologies at the molecular level of organic substance are necessary but also the material technology related to securing thin film uniformity should be supported. The development of a process that can be formed on the flexible array by a low-temperature process using appropriate micro-patterning should be effectively promoted. Although many reports on organic electronic elements that have been published to date, that are using vacuum deposition and semiconductor etching process, the efficient development of process based on printing is necessary in order to highlight the fundamental advantage of these new elements. Recently, solution process based research and development on advanced materials have been actively promoted and available, along with high-resolution printing process, technique of precise control of position, and roll-to-roll continuous process for multi-layer thin film coating. So, the future of organic electronic element based on various printing and outlook of the market are expected to be very bright.

3 Principle and research trends of Organic Thin Film Transistor (OTFT)

3.1 Technology of driving elements

Substrate technology for flexible display and technology of active driving elements are directly related to each other. These are areas in which many research institutes and companies intensively invest their research resources, since these are now the biggest obstacle to actualize the flexible display. In the case of the active driving elements, technologies of low temperature poly-Si thin film transistor and amorphous silicon thin film transistor (a-Si TFT) used mainly in backplane for TFT-LCD or AMOLED. Recently, organic thin film transistor (OTFT) technologies with the relatively low-temperature process and solution process of organic semiconductor have been actively studied as technologies of driving elements for the flexible display. However, the performance and reliability of elements necessary to improve the display quality are still insufficient. Therefore, this chapter explores the organic semiconductor technology focusing on the latest technologies.

3.1.1 Prototype flexible display

Studies of the display element formed on the flexible substrate have been focused on amorphous silicon, polysilicon, and organic thin film that make up the backplane of the main display panel such as EPD, LCD, OLED, etc. In recent years, reports about EPD produced on the plastic substrate has been announced. In 2005, Philips, Sony, and E-ink jointly commercialized the first e-paper display module. However, because EPD is driven in the reflection mode, it is difficult to realize full-color video display with the EPD. Since the shape of the product is limited to e-paper or e-book, LCD and OLED have become more interesting for flexible display. Although the plastic TFT-LCD has many advantages compared to the conventional LCD, bending or rolling display is a factor that greatly threatens the stability of cell gap and sealing material around LCD. In addition, it is not suitable to make the thin display that can bend the polarizer used on both sides of LCD. In this respect, display industry is focusing on the OLED, the self-luminous type display, for the flexible display. Although the OLED is advantageous for the flexible display production in that it consists of very thin and strong material, it remains a challenge that it should form a high-performance barrier layer to protect the OLED itself from diffusion of oxygen or moisture.

Sony announced prototype of 2.5 inches flexible OTFT-OLED color display that is produced by OLED on the PEN plastic substrate for the first time in the world. However, it is staying in prototype level, research and development should be continued in order to commercialize the device. The characteristics of the required driving element in accordance with display mode must be developed.

Figure 7.28

2.5-inch prototype display [16.8 million colors at a 120 x 160 pixel resolution(80 ppi, 318-mm pixel pitch). is 0.3 mm thick and weighs 1.5 grams without the driver].

3.1.2 The advantages of the organic semiconductor (OTFT)

The typical characteristic of the organic semiconductor is that it is essentially organic material, although it has semiconducting property. The organic semiconductor has advantages that it is soluble in organic solvents like other organic materials (including polymers), evaporates if the temperature is increased, doesn't change the electrical properties if bent, it is lightweight and resistant to mechanical shock.

First, since the organic semiconductor is easily crystallized even at the room temperature or at a temperature below 100°C, there is no need to increase the temperature of the substrate during film formation. Therefore, the maximum operating temperature is low, but it can be bent and can be used as a substrate that is thin and resistant to shock like plastic. This feature has a great advantage to the electronic element in which mobility is emphasized like a mobile phone. It is also very useful when it is necessary to install electronic elements such as display or sensor in the curved surfaces like a windshield of an automobile or a pillar of the building. After the fabrication in a planar state, it can be easily installed in any necessary surface.

Second, polymers and small molecule organic materials that are designed to melt, have the advantage of being dissolved in a suitable solvent. In the case of the polymer, dissolving in an organic solvent, applying it to the necessary place, and making the solid by evaporating the solvent again does not change the electrical characteristics of the raw material. In this way, thin films can be formed by various methods. The organic semiconductors dissolved in a solvent are the same as inks in a printing process. Therefore, the organic semiconductor thin film can use the existing printing process as it is. The desired pattern can be formed by using a kind of printing technique such as spin coating, dipping, ink–jet printing, offset printing, gravure printing, flexo printing and screen printing. Such a method can be performed at room temperature in the atmospheric pressure, and ultimately, a roll–to–roll process is possible, so that an ultra low–cost electronic device can be manufactured.

Third, organic materials are widely available not only as semiconductors but also in the form of conductive polymers and insulators. Since it is possible to synthesize a material having a desired function as needed, the degree of freedom in manufacturing the device is high. If electronic circuits could be made as all these materials are printed in ink form, anyone can design and print electronic devices of their own type.

3.1.3 The conduction principle of conductive organic material

The conductive organic material is an organic compound consisting of a covalent bond of light elements such as carbon, hydrogen, and nitrogen. The conductivity is attributed to the repetitive structure of interatomic single bonds and double bonds. Let's look at the conduction mechanism of a conductive polymer for an example.

A polymer is a molecule with a large molecular weight formed by repetition of a simple chemical structural unit. Although it is not specified, it can be said that the molecular mass is a substantial group of about one to several million. These macromolecules are linked molecules called monomers. This process of obtaining macromolecules is called polymerization. A monomer is a molecule or compound that has a relatively small molecular weight and a relatively simple molecular structure. It is a basic substance capable of synthesizing a polymer. It has at least two functional groups or a combination of two carbon–carbon bond or triple bond. The characteristics of the polymer are light and excellent in processability, which makes it suitable for mass production. On the other hand, it has less mechanical strength than metals and inorganic materials, weak to heat, low in crystallinity, and is an insulator. The reason why the polymer has good insulation is that the bandgap energy between the valence band and the conduction band is large. Polymers have begun to attract full attention as electronic materials since the discovery of the phenomenon of non–conductor–metal phase transition, in which electrical conductivity increases rapidly by doping halogen elements in polyacetylene film in the late 70s. In polymers such as polyacetylene, carbon atoms are called π–conjugated polymers because they have a chain structure in which single bonds and double bonds are alternately repeated and the π–electrons move freely to some extent. Chemically or electrochemically doping these polymers can also be referred to as a conducting polymer because the electrical conductivity can be controlled from the insulator to the metal, as shown in Figure 7.29.

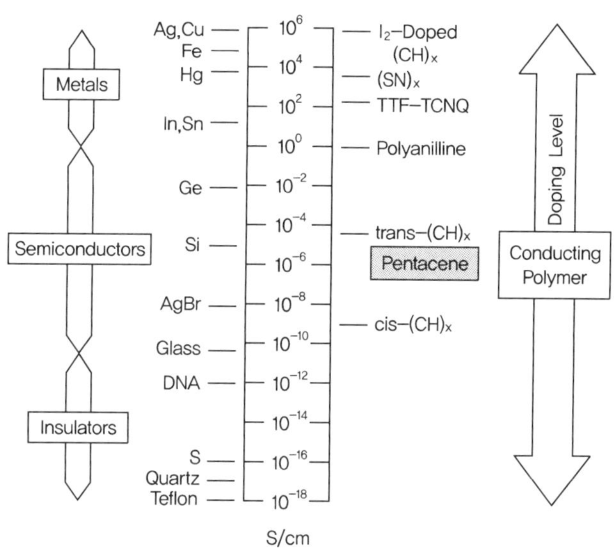

Figure 7.29

Conductivity by conductive polymer and material.

$(CH)_x$ film is the simplest polymer chain consisting of only carbon and hydrogen, and there are two types of polymer chains, trans and cis, as shown below. The electrical conductivity of cis–$(CH)_x$ is 10^{-9} S/cm at room temperature, close to that of the insulator, but trans–$(CH)_x$ is 10^{-5} S/cm. It is known that the energy bands are 1.9 eV and 1.4 eV, respectively. Since the π–conjugated polymer has one π–electron per carbon atom, if the distance of all the C–C bonds is the same without a twist of the conductive polymer chains, it can be said as the one–dimensional metal because the valence band is half filled.

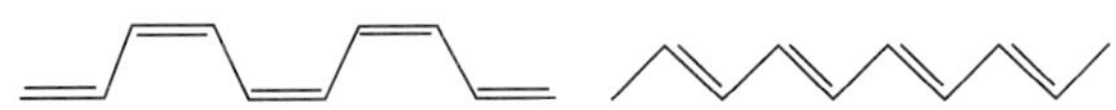

cis–$(CH)_x$ *trans*–$(CH)_x$

However, in a very long polymer chain, a bond alternation state in which alternating long and short alternating states is more stable than a state having the same bond length. The π–electron system causes distortion of the polymer chain by electron–phonon coupling, which is called the Peierls transformation. The Peierls strain has a chain structure in which short and long states of bond lengths between C and C are alternately formed, thereby forming a gap in the electron state density. As shown in Figure 7.30, in the case of a long polymer chain, when the bonding length is alternated, the lattice constant becomes twice that of the case of a uniform bonding length. This results in energy bandgap at the Fermi energy surface with uniform coupling length and phase transition from the conductor to the semiconductor.

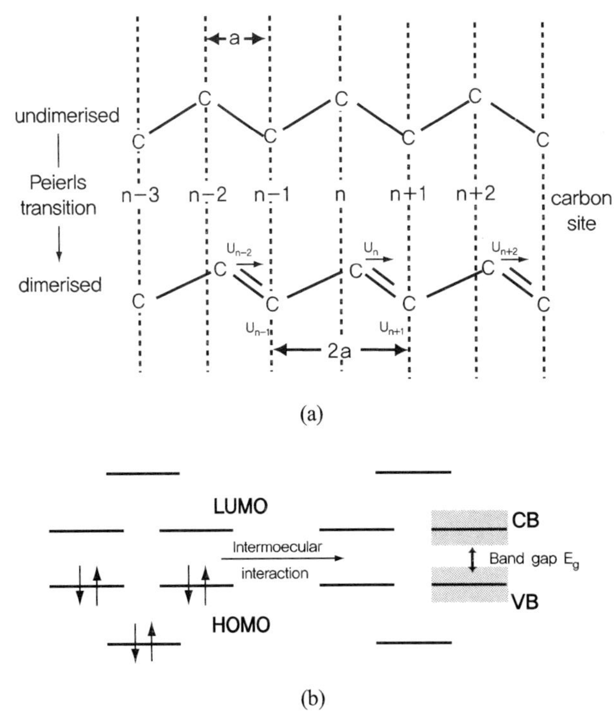

Figure 7.30
(a) Bond dimerization (b) energy bandgap by Peierls.

This is referred to as the Peierls interval or the bandgap between the valence band and the conduction band in the inorganic semiconductor. For organic matter, each band is called π–band and π*–band. The highest orbital of the valence band is called the HOMO and the lowest orbit of the conduction band is called the LUMO. The gap between HOMO and LUMO is called the π–π* gap. The π–π* gap of polyacetylene is about 1.4 to 1.7 eV (1 eV: ~ 1240 nm).

The fact that the concentration of electrons in a double bond is higher than that of a single bond and thus the bond length is short indicating that there is a relation between the π–electron and the polymer structure, which means that the electron–phonon interaction is large. In addition, since the difference in bond length and the energy gap of the polymer are closely related, the energy gap of the polymer semiconductor can be controlled by controlling the density of the π–bond. Unlike solids with three–dimensional structures, polymers have a one–dimensional structure. When a charge is injected, there is almost no lattice strain in the vertical direction of the chain, and the structure is easily deformed in a parallel direction to form a polaron. As described above, the non–linear excitons are formed by doping the π–conjugated polymer or by excitation by light or the like by strong electron–phonon interaction. In the case of polyacetylene, which may have two resonance structures with the same energy state at the bottom depending on the structure of the conductive polymer, a soliton is formed at the center of the energy bandgap. Most conductive polymers that have the same energy state of the resonance structure form bipolarons with two solitons combined. These solitons and bipolaron have been confirmed through various experiments and are presented as a theoretical basis for the semiconductor –conductor transition phenomenon in polymers.

To date, many experimental results on the electrical properties of conducting polymers have described the phenomenon of conduction using a model of charge hopping from molecule to molecule. Thus, if the overlap between the π–electron orbits of a molecule is large, the probability of hopping increases and the conductivity increases. The disordered arrangement of impurities or molecules limits the overlap of the electron wave function, thus decreasing the conductivity. Since the energy barrier between each molecule must be surpassed, the hopping process requires a phonon as a thermal excitation energy. However, according to the research results of the Bell Labs, it is suggested that electron conduction phenomenon in high purity organic semiconductors is not hopping between organic molecules but can be performed by conduction at energy interval like inorganic semiconductors such as GaAs. Therefore, it is difficult to understand precisely the charge transport phenomenon in the case of polymers with relatively high disorderliness compared to solid crystals. By studying the electrical conductivity characteristics of single polymer chains using nanotechnology, attempts to study quantum mechanical phenomena in devices with nanostructures have been actively conducted worldwide.

3.1.4 Operation Principle of Organic Thin Film Transistor (OTFT)

The operating region of the thin film transistor is divided into a linear region and a saturation region as in a field effect transistor (FET). Since pentacene, a typical organic semiconductor, exhibits p-type semiconductor characteristics, the operation principle of a p-type organic thin film transistor will be described. As shown in Figure 7.31 (a), if no voltage is applied to the source/drain and gate, all the charges in the organic semiconductor are uniformly distributed in the semiconductor. When a voltage (V_D) is applied between the source and the drain, an current between the source and the drain (I_{DS}) having a thin film conductivity as shown in Equation (7-3), which is proportional to the voltage, flows at a low voltage.

$$\sigma \cong \left(\frac{L}{Wt}\right)\frac{I_{DS}}{V_D}\Big|_{V_G=0,\ V_D\to 0} \qquad \text{(Equation 7-3)}$$

Where W is the channel width, L is the channel length, t is the semiconductor thickness, and V_D is the drain voltage.

When a negative voltage is applied to the gate as shown in Figure 7.31 (b), the barrier between the source-drain electrode and the organic semiconductor is lowered, holes are injected from the electrode, and an accumulation layer is formed between the insulator and the semiconductor. When the accumulated holes are not moved or trapped by the trap, and the voltage is applied to the source and the drain, current flows. When the gate bias is increased by δV_G, the hole charge of the channel is increased by $C_i\delta V_G$ and is increased by $WLC_i\delta V_G$ for all the channels. If the increased charge has a mobility of μ and a small amount of drain bias V_D is applied, the current increase δI_{DS} is given by Equation (7-4).

$$\delta I_{DS} = \frac{W}{L}\mu C_i V_D \delta l\, V_G \qquad \text{(Equation 7-4)}$$

Where C_i is the capacitance of the insulating layer and V_G is the gate voltage. If the drain current is measured as a function of the gate voltage at a low drain voltage, the transconductance is calculated as in Equation 7-5.

$$g_m = \frac{\delta I_{DS}}{\delta V_G}\bigg|_{V_D \to 0} = \frac{W}{L}\mu C_i V_D \qquad \text{(Equation 7-5)}$$

The field effect mobility is calculated as a function of V_G as in (Equation 7-6). This equation is the mobility calculation in a linear region.

$$\mu = \mu(V_G) = \left(\frac{L}{WC_i V_D}\right)\frac{\delta I_{DS}}{\delta V_G}\bigg|_{V_D \to 0} \qquad \text{(Equation 7-6)}$$

As shown in Figure 7.31 (c), when a positive voltage is applied to the gate, the energy barrier between the electrode and the organic thin film increases due to the electric field generated by the applied voltage. Thus the increased energy barrier blocks carrier injection from the electrode. Therefore, a layer having no conduction charge is formed near the insulator, and this layer is called a depletion layer. Applying a voltage to the source and drain in this situation reduces the number of charge carriers, so a current flow reduces as compared to the situation in Figure 7.31 (a). When the drain voltage is changed to a negative value, the energy of the drain electrode is increased, and holes in the peripheral region are rapidly introduced into the drain to form a depletion layer. Figure 7.31 (d), (e), and (f) show that the depletion region is extended from the drain electrode and the organic thin film is fully depleted. The current begins to saturate and the current–voltage graph appears as a transistor current–voltage characteristic.

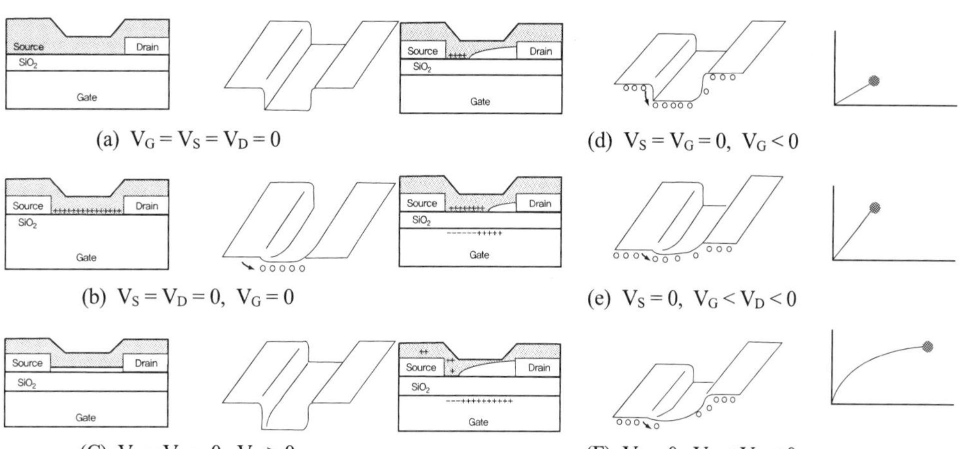

Figure 7.31
OTFT structure and operation mechanism

The I_D–V_D relationship of the OTFT can be derived from the conductivity relation of the organic thin film. As in (Equation 7–7), the conductivity of the organic thin film can be obtained by integrating the conductivity of the intrinsic region, the accumulation region, and the depletion region into the drain voltage. The approximate expression is given by (Equation 7–8).

$$I_{DS}=\frac{W}{L}\int_0^{V_s}[N_A qt\mu + C_O\mu(V_G-V_T-V_s)]dV_x + \frac{W}{L}\int_{V_s}^{V_d}[(t-w)N_A q\mu]dV_x \quad \text{(Equation 7–7)}$$

$N_A qt\mu$: intrinsic layer

$C_D\mu(V_G-V_T V_s)$: accumulation layer

$(t-w)N_A q\mu$: depletion layer

Where N_A is the charge density, q is the amount of charge, t is the semiconductor thickness, μ is the electric charge mobility, C_o is the capacitance of the oxide film, V_G is the gate voltage, V_T is the threshold voltage, V_g is the potential and W is the thickness of the depletion layer.

$$I_{DS}\approx\frac{W\mu C_o}{L}\left(V_G-V_T-\frac{V_D}{2}\right)V_D+\frac{\sigma Wt}{L}V_D \quad \text{(Equation 7–8)}$$

3.1.5 Organic Thin Film Transistor Performance Evaluation Factor

Figure 7.32 shows the transfer characteristic curves of the organic thin film transistor. From this curve, the performance parameters of the organic thin film transistor are extracted as follows. The electric charge mobility can be extracted from the current equation of the saturation region (Equation 7–9)

$$I_{DS}=\frac{WC_o}{2L}\mu(V_G-V_T)^2 \quad \text{(Equation 7–9)}$$

$$\sqrt{I_D}=\sqrt{\frac{\mu C_o W}{2L}}(V_G-V_T)$$

$$slope=\sqrt{\frac{\mu C_o W}{2L}} \quad \text{(Equation 7–10)}$$

$$\mu_{FET}=(slope)^2\times\frac{2L}{C_o W}$$

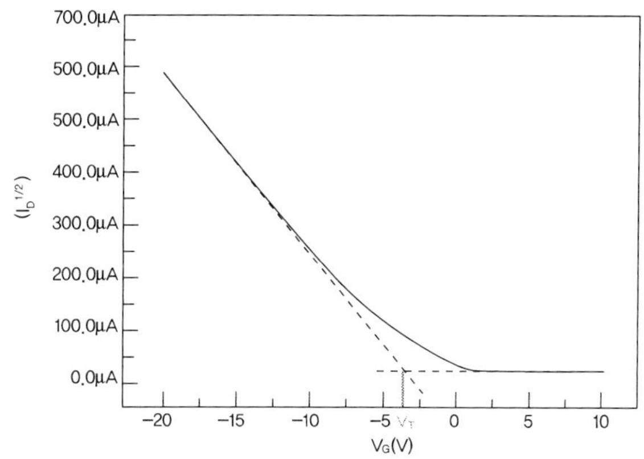

Figure 7.32

Graph of I_D–V_G output of OTFT device.

When the above equation is changed to $I_D^{1/2}$ V·s. V_G graph, the electric charge mobility is obtained from the slope as in (Equation 3–8).

The sub–threshold refers to the state before the organic thin film transistor reaches the threshold voltage, and the sub–threshold slope represents the change in the drain current with respect to the change in the gate voltage before reaching the threshold voltage. It is usually expressed as the amount of change in gate voltage required to increase the drain current by a factor of 10, and the unit is V/dec. The smaller the value, the better the capability to control channel conductivity by the gate voltage. The threshold voltage (V_{th}) can be obtained by the intersection of the slope of the linear part and the value in the OFF state in the $I_D^{1/2}$ Vs. V_G graph. The threshold voltage must be close to 0 V to consume less power. The current ratio is given by (Equation 7–11).

$$\frac{I_{on}}{I_{off}} = \left(\frac{\mu}{\sigma}\right)\frac{C_0^2}{qN_A t^2}V_D^2 \qquad \text{(Equation 7–11)}$$

In the above equation, the current ratio becomes larger as the dielectric constant of the dielectric film becomes larger and as the thickness becomes smaller, the type and thickness of the dielectric film become important factors for determining the current ratio and the OFF–state leakage current becomes smaller. The calculation method is obtained by the ratio of the maximum current value in the on the state to the minimum current value in the OFF state.

The blocking leakage current is a current that flows when it is OFF and it can be obtained from the OFF state to the minimum current at the ON/OFF current ratio. Recent *a*–Si TFT leakage current is very low, 10^{-12} Amp or less. Since the leakage current is directly related to power consumption and switching, it is necessary to minimize this leakage current.

3.1.6 Orientation of Organic Semiconductor

As a substitute for vacuum deposition type amorphous silicon, which is most widely used semiconductor material in a thin film transistor, in the case of a single crystal, a thin film deposited organic semiconductor, a solution type low molecular weight and a polymer semiconductor material, many performance enhancements have been announced, raising interest in various applications such as display devices, sensors, information tags, and smart media. For a low molecular organic semiconductor material such as pentacene, which can form a thin film by vacuum deposition at a low temperature, and a device using the low molecular organic semiconductor material, charge mobility has steadily advanced to 1 ~ 5 $cm^2/V{\cdot}s$. In the case of a device using a polymer type organic semiconductor, the practicality of a device using the active layer as a low-temperature film and a solution process, despite its merits such as fairness and economic efficiency, has not yet been sufficiently verified. In recent years, characteristics similar to or superior to the mobility of amorphous silicon have been reported. Poly-3-hexylthiophene (P3HT) is a representative example of a polymer material capable of solution processing. Micro-crystallinity and head-to-tail regio-regularity, and hole mobility of 0.05~0.1 $cm^2/V{\cdot}s$ are generally reported. With high regioregularity, when the direction of current flow is parallel to the π-π stacking plane, higher charge mobility can be obtained and thus, higher charge mobility can be obtained for P3HT thin films prepared by solution casting than by the spin coating.

The molecular structure and orientation of the polymer thin film have a great influence on the transistor characteristics. Various methods such as Langmuir -Blodgett method, the introduction of the mesogenic terminal group, the repetition of the main chain, and the provision of self-assembling of liquid crystal have been continuously studied. High positional regioregularity can be achieved when pendant alkyl groups are substituted for regular head-to-tail of poly-3-hexylthiophene (P3HT). It has been announced that such a regioregular structure imparts regularity through the two-dimensional sheet-like lamella structure to the polymer backbone, and orientation in the direction of the thiophene plane is regularly carried out by proper surface treatment of the substrate.

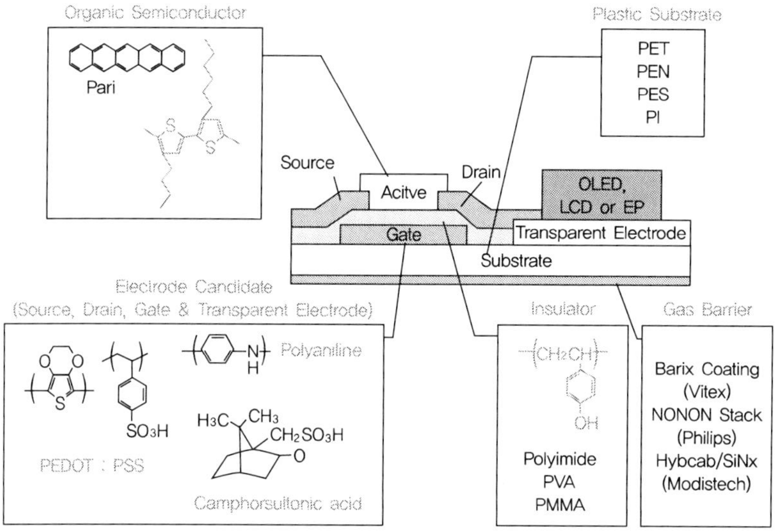

Figure 7.33

Organic Thin Film Transistor Structure and Related Organic Materials.

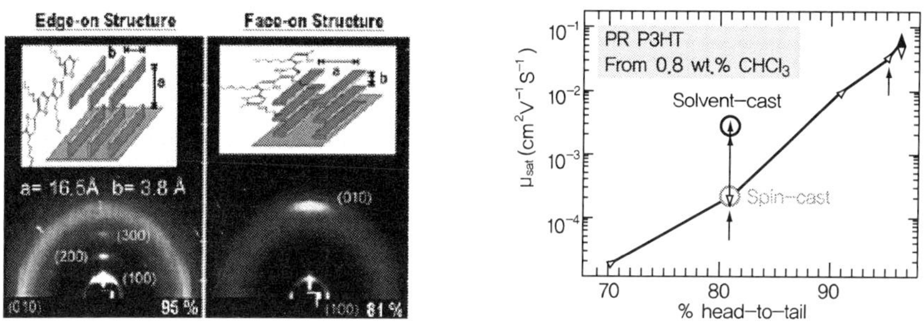

Figure 7.34

Relationship between P3HT regioregularity and charge mobility when WAXS (Wide Angle X-ray Scattering) image and silicon oxide gate insulating film is used for the orientation difference of P3HT

It has also been reported that the charge mobility and ionization potential (IP) increases by adding a thiophene repeating unit, which is not substituted in the main chain and by removing the regioisomer. In comparison with P3HT, it can help improve the stability to oxygen. The localized free volume increase between adjacent alkyl groups can raise the crystallinity, but the increase of the crystallization domain by this method is only about 10 to 15 nm even after the heat treatment at a temperature above the mesophase. The mobility is limited to 0.1 $cm^2/V \cdot s$.

Figure 7.35 summarizes the recent results of the improvement of the performance depending on the chemical structure and the type of the organic film material of the organic semiconductor active layer. In the case of pentacene, the mobility of amorphous silicon can be obtained, but there remain many problems such as the stability of the device, the ON/OFF ratio, and the degradation of device characteristics when fabricated with a pixel array. Understanding the interfacial phenomena with the dielectric material of the source/drain electrode and the insulating layer, as well as the properties obtained by controlling the microstructure and crystallinity of the organic material itself, and optimizing the device structure, is still a very important, yet it is a difficult problem to improve the level of device characteristics such as ON/OFF ratio and to secure uniformity.

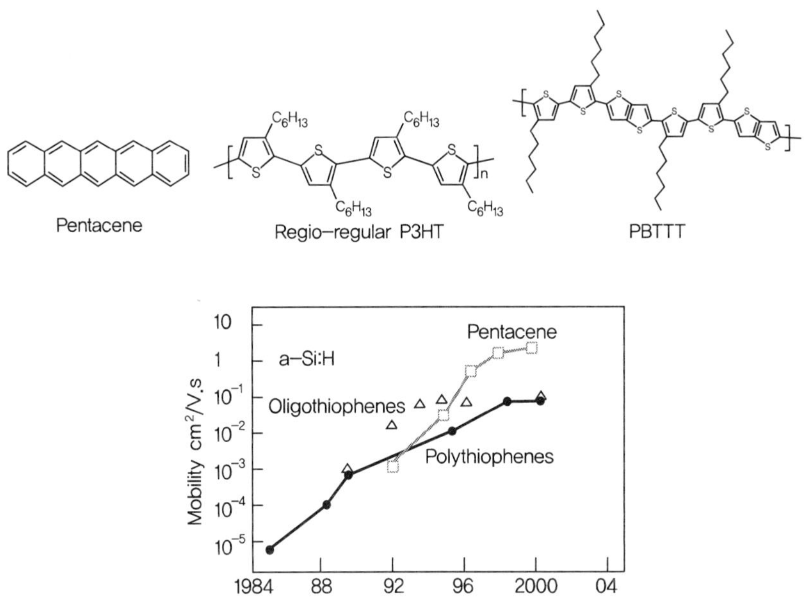

Figure 7.35

Organic semiconductor molecular structure and technology development trend.

3.1.7 Trend of organic semiconductor material technology development

In the field of organic semiconductor materials, research is actively conducted by various groups worldwide. Recent research trends are focused on the development of new semiconductor materials that can be used for solution processes rather than low–molecular vacuum deposition processes and have high electronic mobility as well as molecular stability. In the case of vacuum evaporated p–type materials, oligothiophene and pentacene have been studied mainly as low molecular weight and oligomer materials. According to the material and thin film forming method, maximum charge mobility of 0.01–0.1 cm^2/V·s was obtained in oligothiophenes and 2 to 7 cm^2/V·s for pentacene, surpassing that of amorphous silicon. This shows the possibility of practical application of the organic thin film transistor material.

Table 7.4 P-type vacuum evaporation material.

Material	Mobility ($cm^2/V{\cdot}s$)	ON/OFF ratio
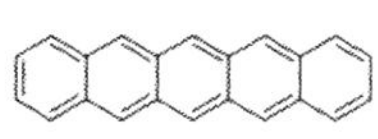	7	-
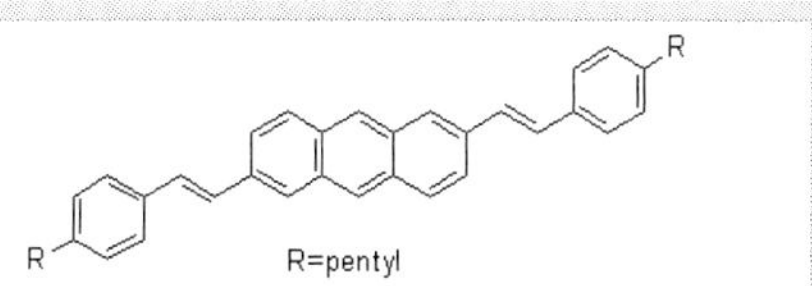	0.95	10^6
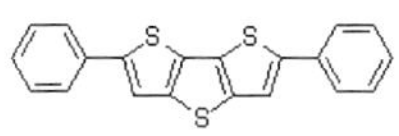	0.42	10^6
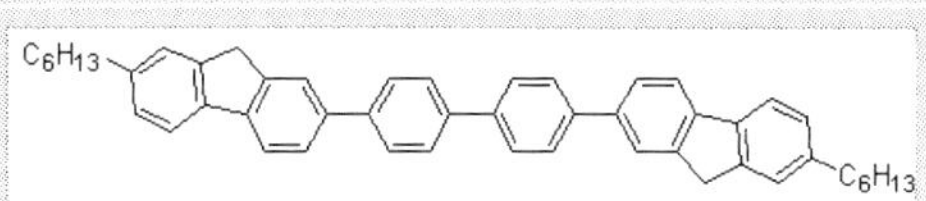	0.30	10^6

In the case of pentacene, the oxidation stability is not good and new materials need to be developed. In 2006, anthracene derivatives were studied to find that the charge mobility was 0.95 $cm^2/V{\cdot}s$ and a current ON/OFF ratio of 10^6. Princeton University studied fullerene derivatives and published in 2006 a material with a charge mobility of 0.30 $cm^2/V{\cdot}s$ and a current ON/OFF ratio of 10^5. In addition, dithienothiophene derivatives have been studied and materials with charge mobility of 0.42 $cm^2/V{\cdot}s$ and current ON/OFF ratio of 10^6 have been reported.

Table 7.5 Materials for p-type solution process.

Material	Mobility ($cm^2/V{\cdot}s$)	ON/OFF ratio
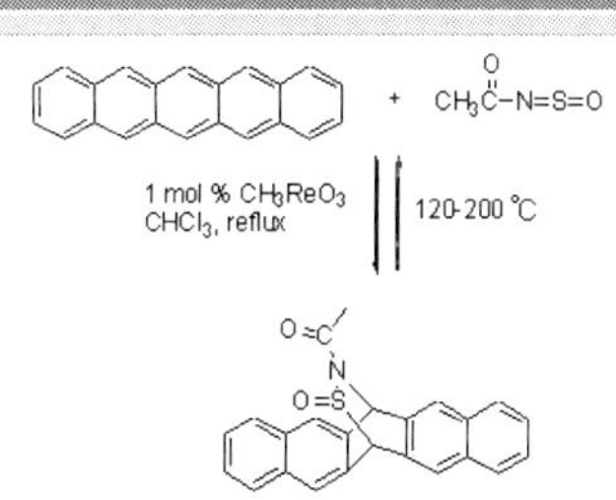	0.89	–
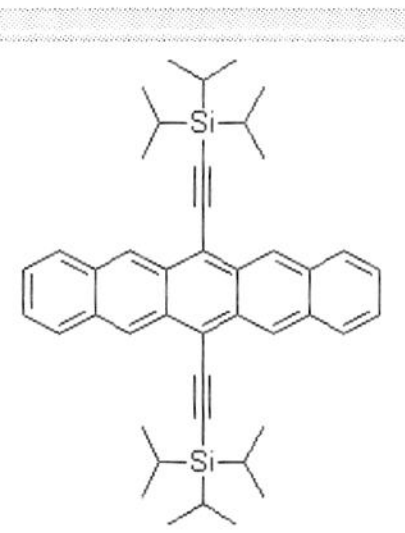	0.4	10^6

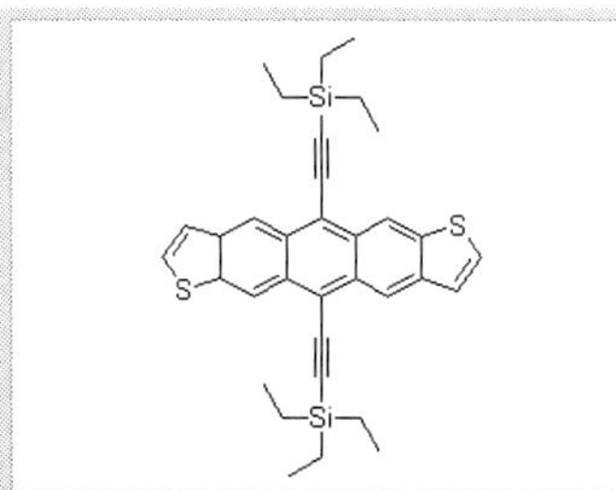	1	10^7
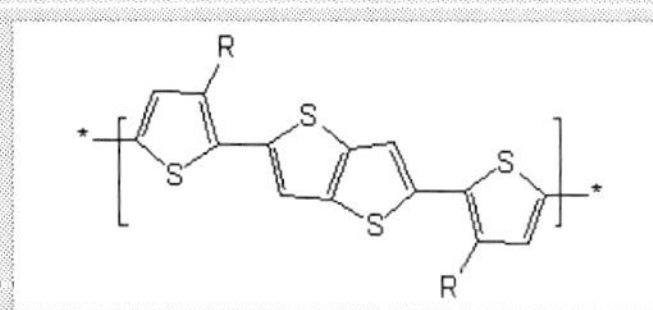	0.2 ~ 0.6	10^7
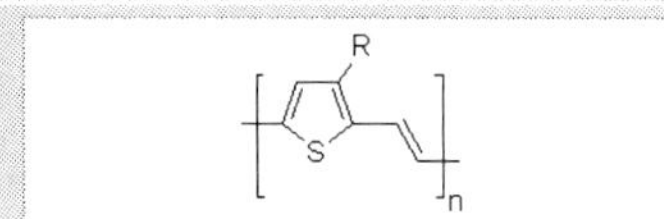	0.22	–

In the case of p–type solution processed materials, functional groups capable of dissolving in the pentacene base to enable solution processing, are introduced to pursue process convenience and device stability at the same time. First, the pentacene precursor is easily released at the 6 and 13 positions of the pentacene, and a functional group for the solution processing is introduced to form a semiconductor by a solution process. Then, by applying heat to form the pentacene on the substrate, the pentacene having charge mobility of 0.089 to 0.2 $cm^2/V{\cdot}s$ has been developed. In 2003, TIPS pentacene substituted with tri–isoproplysilylacetylene at positions 6 and 13 of the pentacene had a charge mobility of 0.4 $cm^2/V{\cdot}s$ and a current ON/OFF ratio of 10^6. In 2005, triethylsilylactylene was substituted at positions 5 and 11 of anthradithiophene TIPS pentacene (1.0 $cm^2/V{\cdot}s$) and a current ON/OFF ratio of 10^7 was reported. In addition, p–pentylphenylacetylene was replaced with p–pentylphenylacetylene at positions 6 and 13 of pentacene, resulting in the charge mobility of 0.52 $cm^2/V{\cdot}s$ and an current ON/OFF ratio of 10^5. As a p–type polymer material, polythiophene series is mainly used, achieving a maximum charge mobility of about 0.1 to 0.6 $cm^2/V{\cdot}s$. In 2005, Lain McCulloch developed a polymer with the thiothiophene –containing polythiophene derivatives with the mobility of up to 0.15 $cm^2/V{\cdot}s$ and current ON/OFF ratio of 10^5. Mitsubishi Group in Japan investigated, a polythiophene (μ = 10 $cm^2/V{\cdot}s$, ON/OFF = 10^2), polypyrrole (μ = 6.9 × 10^{-5} $cm^2/V{\cdot}s$) and polythienylenevinylene (PTV, ON/OFF =10^5). In the case of PTV, the reproducibility was a problem, and Lucent–Bell group of USA announced the mobility of poly (3–hexylthiophene) (P3HT) of μ = 0.045 cm^2/V·s. The Cavendish group used polyacetylene (μ = 10^{-4} cm^2/V·s) and P3HT (μ = 0.1 cm^2/V·s, ON/OFF = 10^6). At the Xerox Research Center in Canada as a terthiophene or quaterthiophene monomer with alkyl side chains has developed a polythiophene derivative that has a maximum mobility of 0.14 cm^2/V·s and a ON/OFF rate of 10^7. The regioregular polythiophene (μ = 0.07 to 0.12 cm^2/V·s, ON/OFF = 10^7) was studied at the Xerox Research Center since 2004 and found that thienothiophene substituted poly (2,5–bis –thienyl) –3,6–dipentadecylthieno [3,2–b] thiophene (μ = 0.25 cm^2/V·s, ON/OFF = 10^7). Merck has also published a polymer with thiothiophene–containing polythiophene derivatives with the mobility of up to 0.2–0.6 cm^2/V·s and ON/OFF rate of 10^7. In addition, 10, 12–pentacosadiynic acid (PDA), a diacetylene derivative, was deposited and its polymerisation (μ = 0.8 cm^2/V·s) was developed by UV irradiation.

In the case of blend systems, since the low molecular weight materials that can be used in vacuum vapor deposition, are not usually soluble in common organic solvents, the good solubility of polymers is maintained through blending with polymers, and high performance can be expected by introducing high performance low molecular weight materials. As a representative example, polystyrene and TIPS pentacene were blended to have a charge mobility of 0.5 to 1.5 cm^2/V·s and a current ON/OFF ratio of 10^7. The KAIST blended F8T2 and DH4T, it showed a charge mobility of 0.012 cm^2/V·s and the current ON/OFF ratio of about 10^5.

Table 7.6 N–type vacuum deposition material.

Material	Mobility (cm^2/V·s)	ON/OFF ratio
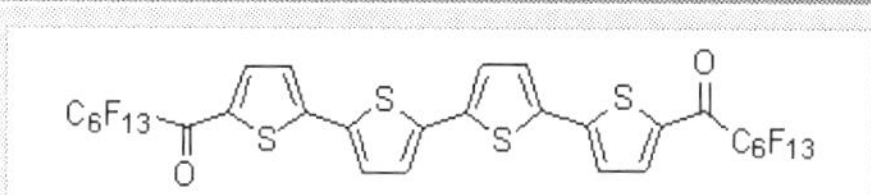	0.6	10^7
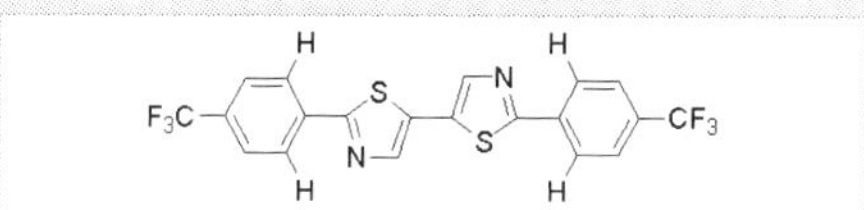	1.83	10^4
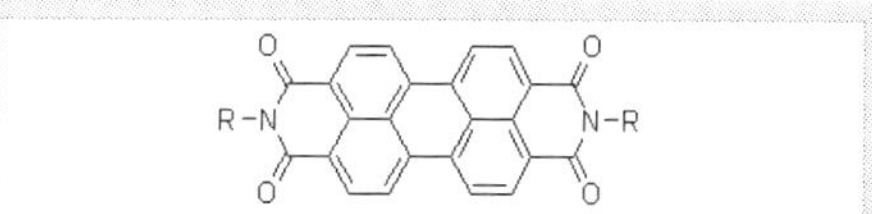	0.6	–
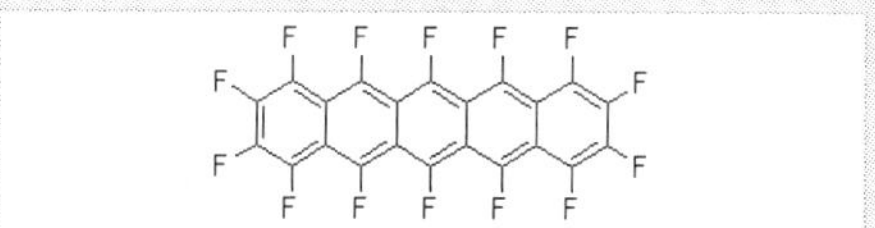	0.11	10^5

There are Phthalocyanine (PC) and fluorinated PC (FPC), tetracyanoquinodimethane (TCNQ), C60 (bucky ball), naphthalene tetracarboxylic dianhydride (NDTCA) and perylenetetracarboxylic dianhydride (PDTCA) derivatives as n-type organic thin film transistor materials. The charge mobility of FPC is 0.03 $cm^2/V{\cdot}s$ and 0.6 $cm^2/V{\cdot}s$ for PDTCA derivative. In 2005, thiazole derivative was synthesized and its surface was treated with OTS (octadecyltrichlorosilane). This lead to have a charge mobility of 1.83 $cm^2/V{\cdot}s$ and a current ON/OFF ratio of 10^4.

IBM developed a material with a charge mobility of 0.1 to 0.62 $cm^2/V{\cdot}s$ using an organic-inorganic hybrid material and developed organic thin film transistor materials by an ultra-thin Langmuir-Blodget (LB) or self-assembly method. In the case of n-type low-molecular materials, instead of moisture and air-sensitive n-type organic semiconductors, fluorine-substituted materials are also investigated into conventional p-type organic semiconductor materials to improve air and moisture stability. Materials where electrons can easily be injected from electrodes and have a maximum mobility of 0.6 $cm^2/V{\cdot}s$ have been reported, but there is still much room for development. In the case of the n-type polymer material, the developed examples are rare, and the ladder type polymer, poly (benzimidazobenzophenanthroline, BBL) and polymers through blend with BBL have been reported. BBL has a mobility of 0.1 $cm^2/V{\cdot}s$ and a current ON/OFF ratio of 10^5.

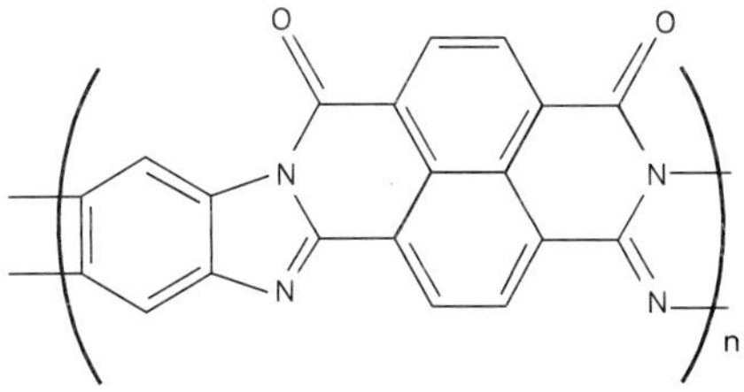

Figure 7.36
BBL molecular structure.

3.1.7 Polarization of organic insulators

The organic thin film transistor can be driven more efficiently by the polarization of the dielectric. As for the kinds of polarization phenomena and their principles, the polarization phenomenon is divided into electron polarization, ion polarization, orientation polarization and interface polarization according to the mechanism of forming the dipole. Figure 7.37 (a) is a schematic diagram illustrating electronic polarization. The net charge of a neutral atom is zero. Moreover, generally, the negative charge centers of the electrons are in agreement with the positively charged atomic nucleus, resulting in no net dipole moment. However, when the atom is placed in an electric field, it induces a dipole moment. The electrons are much lighter than the positively charged atomic nuclei and can easily be displaced by the electric field. As a result, the negatively charged center is separated from the positive charge center. Ultimately, the electric polarization takes place by the relative displacement of the electron cloud and the nucleus due to the external electric field. Figure 7.37 (b) is a schematic diagram illustrating ionic polarization. Ion polarization is a polarization in which an external electric field is applied to the ionic crystal and the cation and the anion are displaced in opposite directions to each other. In a solid, in the absence of an applied electric field, the dipoles of the same size are oriented as head to head, tail to tail, so that the net dipole moment of the solid is zero and the net polarization is also zero. When an electric field is applied, the dipole moment increases and the net dipole moment is no longer zero. Figure 7.37 (c) is a schematic diagram illustrating the orientation polarization. Unlike the two displacement polarizations described above, the orientation polarization has a permanent dipole moment even if there is no electric field. It is a polarization generated when the dipole is oriented in the direction of the applied electric field. Figure 7.37 (d) is a schematic diagram illustrating interfacial polarization. Interfacial polarization is a polarization that occurs in two different ways, one way is when charges of different polarities accumulate in one region or one phase, and the other way is when opposite charges accumulate at an interface between two phases.

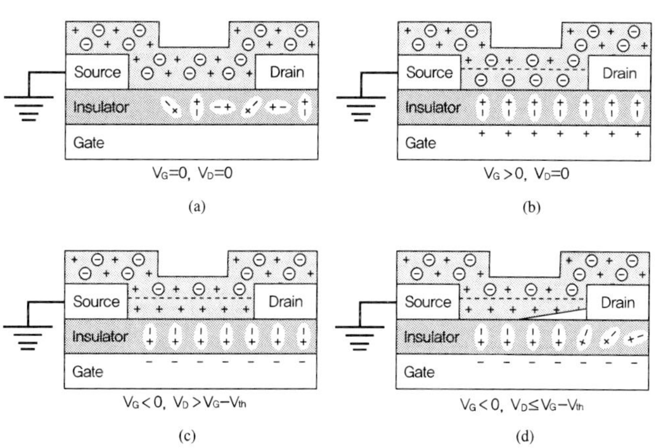

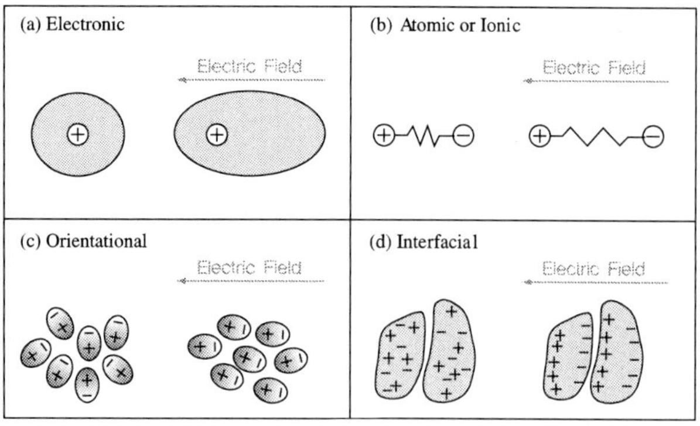

Figure 7.37

Interfacial polarization model (phase) of OTFT device and kind of polarization phenomenon (bottom)

3.1.8 Required properties of organic insulators

As an organic insulating film that can be used for an organic thin film transistor, it is required to develop a new dielectric material having the following characteristics. First, the dielectric constant should be high. The principle of operation of a field-effect transistor is that a voltage is applied to a gate to collect charges at both ends of the dielectric due to the applied voltage, and an electric field due to these charges forms an accumulation layer and a depletion layer. Therefore, a large dielectric constant means that a large amount of charge can be collected at both ends of the dielectric even at a low applied voltage so that the device can be driven at a low voltage. The dielectric constant is the ratio of the capacitance with the dielectric material between the two electrodes and that without it (strictly vacuum). Capacitance can be defined as the ability between two conductors to store the charge generated per unit applied voltage. Next, the insulation properties should be good. Leakage current acts to neutralize the charges collected at both ends of the dielectric, which causes driving problems that must be frequently recharged because the charge can not be maintained for a long time. It should also be possible to form thin films at low temperatures. This is because the glass transition temperature is low due to the characteristics of the plastic substrate so that low-temperature firing can be performed in order to realize a flexible display. The thermal expansion rate should be equal to or close to that of the substrate. Due to the characteristics of the device, heat is inevitably generated due to the current flow. Materials with similar thermal expansion rates would be ideal to withstand the heat. Finally, for use on bendable plastic substrates, there should be no cracking due to film stress by a certain degree of bending.

3.1.9 Trend of organic insulator material technology development

In the field of organic dielectrics, research and development are still at an early stage, although their importance is relatively high compared to organic semiconductor materials. First of all, if you look at domestic research trends, Korea Institute of Chemical Research (KRICT) is focusing on organic insulator. Organic thin film transistor (OTFT) devices were fabricated using low temperature processing of photosensitive polyimide as an organic insulator and pentacene as an organic semiconductor. The device fabricated after the formation of the organic insulating film by the exposure process has an electric charge mobility of about 0.1 $cm^2/V{\cdot}s$. However, unlike the conventional etching method used in photolithography, the organic insulator itself is patterned to form an OTFT array and the manufacturing process is simplified. In addition, a novel high dielectric constant organic insulating material that can meet the performance and required characteristics of organic insulators for organic displays has been developed. In the design of high dielectric constant photosensitive organic insulators, a technique for producing functional monomers and polymers for imparting functionality to a high-rate organic insulator has been established.

Currently, researches on low temperature curable polyimide insulators with minimized degradation of organic semiconductor in solution process type are being actively conducted. The research team of Kyunghee University used crosslinked polyvinylphenol (PVP) as an organic insulator and organic thin film transistor (OTFT) using pentacene as an organic semiconductor. Most studies using pentacene did not consider the fine patternability of pentacene. However, by attempting a fine patterning by using selective pentacene growth technique of controlling the surface properties of the part to be deposited with pentacene, ON/OFF current ratio of 10^7 and charge mobility of about 1.2 $cm^2/V{\cdot}s$ are obtained. However, in the case of polyvinylphenol organic insulator, due to hysteresis and high curing temperature of device driving, there is a problem that it is difficult to apply to a flexible device. In the case of SNU research team, a research group which is studying and applying the organic insulator to organic thin film transistor device recently, PVP and polyvinyl acetate (PVAc) were used as organic insulators and fine pattern with the organic semiconductor to make organic thin film transistor. Especially, the effect of dual organic insulator on device characteristics was investigated. The charge mobility of the device is about 0.1 cm^2/V·s and shows the same problem as in the case of PVP. Pohang University of Science and Technology research group is actively studying surface treatment rather than gate insulator itself. They are studying the effect of the molecular structure of self-assembled monolayer (SAM) used for surface treatment on device characteristics. In particular, polymer organic semiconductors are more interesting in solution process than the small molecule and have recently published a number of interesting results. In addition, they are developing ultra-thin insulator using self-assembled monolayer itself and ultimately trying to develop low-voltage driven organic thin film transistor device. In the case of Sungkyunkwan University, the effect of the hydroxy group content in the polymer insulator on the hysteresis of the organic thin film transistor was investigated. As the content of hydroxyl group in the dielectric increased, electron trapping is increased, thereby revealing hysteresis and an increase in gate leakage current. The mobility was about 5.5 cm^2/V·s due to high hole mobility. However, rather than developing a new insulator, this investigation is on the role of the insulator in the device and its relationship with the device characteristics.

The importance of organic insulators is more recognized in companies that manufacture actual display devices than academia, and organic insulators are being actively developed at Samsung Advanced Institute of Technology (SAIT). In particular, in order to increase the dielectric constant, a resin type organic insulator is developed by combining with an inorganic material, and it is known that it has a remarkable property although the details is not disclosed. However, it is known that there is a definite hysteresis problem in an insulator when the device is driven. As in Korea, studies on organic thin film transistors (OTFTs) have been actively conducted overseas, but unlike organic semiconductors, the development of organic insulators with new structures and functions is insignificant in the case of organic insulators. Table 7.7 shows the examples of organic insulators combined with organic semiconductors for vacuum deposition. As shown in the table, the research and development are done in a variety of ways but organic insulators are limited to polyimide, polyvinylphenol (PVP), polymethylmethacrylate (PMMA) and polystyrene (PS). Recent research have the tendency towards the formation of the ultra-thin organic insulator itself or the increase of the dielectric constant (hybridization with an inorganic insulator or double thin film) for low voltage driving. In this direction, the critical problems to be solved are leakage current and device hysteresis of the insulator itself occur.

Table 7.7 Characteristics of Organic Insulator (Applied to Organic Semiconductors for Vacuum Deposition).

Organic insulator	Semiconductor	Mobility (cm^2/V·s)	Substrate
Polyimide	Pentacene	0.27~0.71	PES
Polyimide	Pentacene	0.27~0.71	Polyimide
Polyimide	Pentacene	0.3	PEN
Cross-linked PVP	Pentacene	0.06	PEN
Cross-linked PVP	Pentacene	1.2	PES
Crosslinked PVP	Pentacene	1.3	PES
PS	Pentacene	0.16	Glass
PS	Tetracene	0.2	Si wafer
PVP	Pentacene	0.2	Paper
PVP	NTCDA	0.02	Si wafer
PMMA	Pentacene	0.01	Si wafer
Poly(chloro-p-xylene)	Pentacene	$10^{-4\sim5}$	Glass
Polyimide / SiO_2	Pentacene	0.75	Si wafer
polyacrylate / Al_2O_3	Pentacene	0.01	PEN

Table 7.8 summarizes the research trends of organic insulators used when applying organic semiconductor for solution process. An excellent chemical resistance of the organic insulator is required for forming the organic semiconductor thin film devices, because of which the selection of the organic insulator material is further narrowed. Therefore, a polyimide that is excellent in chemical resistance or as a material conventionally used in the semiconductor industry as a photoresist or a hardened polyvinyl phenol has been used. However, a process temperature more than 200°C needed to form an organic insulator thin film, that is a problem when using the flexible substrates such as a plastic. Recently, researches on organic thin film transistors have focused on new organic semiconductors, and studies on organic insulators have been less than half of those of organic semiconductors. In particular, insulators used

in organic thin-film transistor (OTFT) insulator analysis include aluminum oxide (Al_2O_3), polymethylmethacrylate (PMMA), polyvinylidene fluoride and trifluoroethylene fluoride and trifluoroethylene, P(VDF-TrFe), polyimide, polymethylmethacrylate / tantalum oxide (PMMA/Ta_2O_5), poly (vinyl acetate), cross-linkable organosiloxane, Self-assembled monolayer (SAM), Cross-linked PVP, Mylar, Polydimethylsiloxane (PDMS), cross-linkable perfluorocyclobutane. Researches on a combination of inorganic and organic semiconductors are ongoing. Researches on an organic insulator aren't deviating from the level of the existing organic insulator(cross-linked PVP, PMMA, general-purpose high-temperature processing polyimide, and etc).

Table 7.8 Characteristics of Organic Insulator (Application of Organic Semiconductors for Solution Process).

Organic insulator	Semiconductor	Mobility ($cm^2/V{\cdot}s$)	Substrate
Polyimide	Polythiophene	10^{-5}	Glass
PHEMA	P3HT	0.1	Glass
PVCN, Epoxy/ PMMA	PEDOT, PPy	NA	PET, PP
PMMA, cross-linked PVP, PHEMA, PS	P3HT	10^{-3}	Glass
CYPEL, PVP, PVOH	P3BT	0.04	Polyimide
PMMA	P3HT	0.001	PET
PMMA blend	P3HT	0.02	PET
BCB	Polyfluorene copolymer	10^{-4}	Glass
Poly(4-methylstyrene)	PTPTP, F15-NTCDI	0.02	Glass
Polymer photoresist	P3HT	0.003	Polyimide
Cross-linked PVP	F8T2	0.1	Glass

As shown above, some of the research groups have announced the unit device of the organic thin film transistor which is relatively good for the mobility and overall performance parameter for using the organic insulator and the organic semiconductor. However, materials satisfying all the required properties such as the performance which can be applied to the organic display, chemical stability, low temperature processing have not been developed. Also, for the case of the organic insulator, the types of development was limited to a few material. As mentioned above, some groups are using organic/inorganic hybrid materials as organic insulators to improve the performance of organic thin film transistors, but their device characteristics are insufficient. Another group is attempting to control the characteristics of organic thin film transistors by controlling the surface characteristics and the bulk characteristics using a multi-layered organic insulator rather than a single layer organic insulator. Also, the blending of two or more organic insulators has been studied. However, such attempts also have problems such as leakage current and process suitability.

In recent organic semiconductors, the development of materials with excellent oxidation stability is a key issue. In order to secure the stability of the organic insulator and organic thin film transistor, there should be no problem in long term reliability of the device due to moisture and temperature. Development of materials with excellent stability is indispensable. From this point of view, although polyimide is considered as an excellent candidate for the organic insulator, it is necessary to secure transparency and improve hygroscopicity. Most of the organic semiconductors used in organic insulator studies are dominated by pentacene, which is known to be the best in performance. In most cases, the electric charge mobility of organic thin film transistor, made of pentacene and organic insulator organic thin film transistors is reported to be about 0.1 ~ 1.0 $cm^2/V{\cdot}s$. In some cases, higher than 5.0 $cm^2/V{\cdot}s$ was reported but reliability is slightly low.

In the study of organic thin film transistors composed of organic semiconductors and organic insulators for solution processing, the cross-linked polyvinylphenol (PVP) and its derivatives, polyimide (PI), polymethylmethacrylate (PMMA) and derivatives, and polythiophenes and polyfluorene derivatives are mainly used as organic semiconductors. In most cases, the electric charge mobility of an organic thin film transistor is reported to be as low as about 0.001 to 0.1 $cm^2/V{\cdot}s$. It is suggested that the improvement of the properties of the organic semiconductor for solution process and the improvement of the organic insulator having the excellent bonding and interfacial characteristics with the organic semiconductor for solution process are required to improve the mobility. Based on a current technological level, it is considered necessary to develop and study the implementation of organic thin film transistor through joint research between organic semiconductor research group and organic insulator research group. In other words, it is considered necessary to develop a customized organic insulator suitable for the characteristics and development of organic semiconductors.

Ultimately, for the organic thin film device of the solution process without the vacuum process, the development of an organic insulator that is excellent in chemical resistance and stability is absolutely necessary. At present, organic insulators for organic thin film transistors are not developed in the countries such as Europe (Merck), USA (Dupont) and Japan, and the technology development level itself is not superior to Korean domestic ones. In the case of the organic insulator market, the potential for monopolizing the global market is expected to be high if securing the source material and technology can be achieved fast. Organic insulators for organic thin film transistors should be developed in such a way that they can be combined with organic semiconductor materials for optimum mutual effects. Ultimately, intensive research is needed from the viewpoint of the future flexible display market.

3.1.10 Organic thin film transistor using flexible substrate

Until recently, most organic thin film transistors have been fabricated using pentacene thin films. The carrier mobility of the pentacene-based OTFTs fabricated on a plastic substrate using a vacuum deposition process is in the range of 0.2 to 1.2 $cm^2/V \cdot s$, with this a ring oscillator and CMOS circuit composed of p-type devices have been disclosed. On the other hand, the carrier mobility of the pentacene organic thin film transistor fabricated by the solution process stays at about 0.05 $cm^2/V \cdot s$. When the polymer such as P3HT is used, the mobility value of 0.08 $cm^2/V \cdot s$ is obtained. In other words, although the operating performance of an organic thin film transistor fabricated by using a solution-based process is significantly lower than that of a vacuum process, it is still expected that the production cost will be greatly reduced by using a printing process. It has become one of the most important research fields. Generally, in the case of an amorphous silicon or polysilicon thin film transistor, a component portion of the device is patterned through a photolithography. However, organic thin film transistors using organic semiconductors often degrade greatly after the lithography because in this process step the organic thin films are damaged. Therefore, although a metal shadow mask is often used for patterning organic semiconductors in the fabrication of organic thin film transistors, the size of patterning is limited. The printing process which is currently in use may be a good alternative, but so far there has been little report on devices with good operating performance.

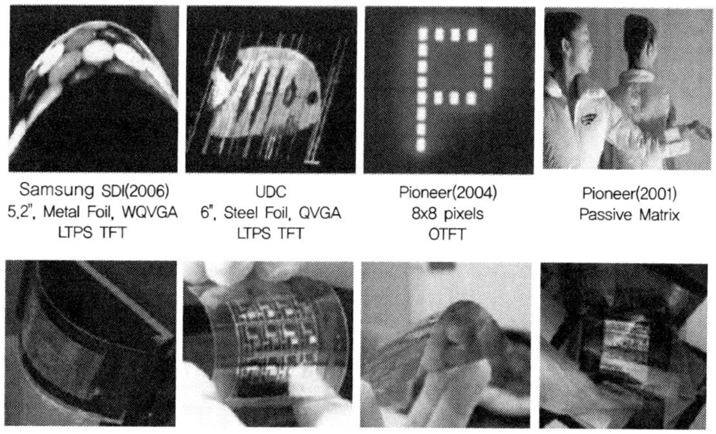

Figure 7.38

OLED demonstration using flexible substrate and thin film transistor.

From this background, some research groups are studying the selective formation of pentacene in the transistor region using the self-matching characteristics, and the characteristics of the device, fabricated by this method are reported. Research is also continuing to improve the operating life of the organic thin film transistor. It is known that the device operation characteristics before and after the passivation process are greatly changed. Therefore, in order to increase the longevity of the organic thin film transistor, the development of an adequate passivation process is required to prevent permeation of water or oxygen. It is a very important technological issue in the commercialization of the organic thin film transistor. In addition, the device degradation mechanism due to the repetitive operation should be clarified. Organic thin film transistor-based LCD or OLED panel technology is still in the early stage, and future development of organic thin film transistor backplane with an equivalent performance to TFT-LCD will be a challenge. The flexibility of the display panel itself is also a very important technological issue, which depends on the substrate thickness, device structure, construction materials, and substrate materials. It is also important to evaluate the driving reliability of the flexible display. In the future, a method to quantitatively evaluate the short- and long-term reliability of these devices should be developed to understand the operational aging mechanism of the device, and to improve the operational reliability of the device under the electrical and mechanical stress conditions in the application field.

4 Structures and Working Principles of E-paper Display

4.1 Characteristics of E-paper

In the case of flexible display, theoretically, it is possible to utilize any display modes that are running on the active devices. Liquid Crystal Display has assured display quality and mass-productivity; however, it requires a new technology to keep a certain distance between liquid crystal cells on two flexible display substrates. In the case of Organic Light Emitting Diode or OLED, it requires a quantum leap in passivation and encapsulation technology since flexible substrates are vulnerable to smearing of oxygen and humidity. Yet, OLED is more suitable than LCD for the flexible display. In the case of E-paper technology, which has characteristics of both papers and electric displays, it is able to store information and run as the reflective color display. Therefore, E-paper is the most appropriate technology for flexible displays on the E-books.

E-paper is an indicating element that is a hybrid of electric displays and papers. It can be bent or rolled and has a bistability so that the materialized writings or images will stay even after the device is turned OFF. To list some required details of E-paper, it must provide readability in different circumstances, have bistability so that it keeps showing the display without any power source, do not result in fatigue to user's eyes, and be rewritable so that it can update the information showing on the display. The advantages of E-paper are paper-like contrast ratio; simple material and simple process; extra low-cost implementation by mass production using roll-to-roll process; bistability, unique characteristic of E-paper that can't be implemented on LCD, PDP and OLED enables the device, that needs no additional power consumption for the implementation of fixed characters or patterns. In addition, it does not need a backlight as a reflective element and has little influence of ink on bending, making it suitable for flexibility. From the viewpoint of the display, electronic paper is a low-grade reflection type display, so it can be implemented in various ways. Nowadays many companies and laboratories do research on E-paper as shown in Table 7.9. A microcapsule type electronic paper uses electrophoresis of the E-ink, and it is the most popular of these.

Table 7.9 Research trend of technology for E-paper implementation.

Materials	Driving method	Display	Researcher
Particle technology	Electrophoresis	Microcapsule	E-ink
		In-Plane	Canon, IBM
		Microcup	SiPix
	Rotating motion	Gyricon	Xerox
		Magnetic twist ball	Keizai University, Meisei University
		Column twist ball	Oji Paper
	Transiating motion	Toner	Fuji Xerox
		QR-LPD	Bridgestone
	Magnetophoresis	Magnetophoresis	Pilot, Takara
		Magneto-heat sensing	
	Phase change	Thermal	Mitsubishi Paper
Liquid Crystal	Optical anisotropic	Cholesteric LC	Kent Univ.
		STN	ZBO
		FLC	NHK
	Pigment polarization	G-H type PD-LC	Dainippon Printing
Film	Actuated film	AFD	Toshiba
Thermal rewritable	Thermal sensing	Thermal paper	Richo
Fluid technology	Fluid motion	Electrowetting	Philips
Oxidation-reduction	Electrochromic	EC paper	ACREO, Uppsaie Univ
	Electrodeposition	EDD	Sony
OLED	Self-emissive	Organic EL	Pioneer

Table 7.10 Various types of E-paper.

	Plate	Book	Roll up	Paper
Style				
Rewriting unit	Included	Included / attached	Attached	Separate
Merits	Quick rewriting	Browsing ability	Compact	Thin like paper
Demerits	Rather high bulky	Rather high bulky	Difficult to browse	Bulky when content is considerable

Such electronic paper is expected to be applied to an electronic newspaper (E-newspaper), an electronic magazine (E-magazine), and an electronic book (E-book) replacing a newspaper, a magazine and a book. Applications in mobile communication devices such as mobile phones, personal digital assistants (PDAs), and portable multimedia players (PMPs) are also expected. In addition, potential application fields are expected to be various. As shown in Table 7.10, the form of the electronic paper can be plate, scroll, book and paper depending on the application field.

4.2 Technologies for E-paper implementation

The development of information society is based on a mixture of communication and computer technologies. It causes an increased demand and diversification of products related to IT along to diverse requirements of the information society. The development of information society recently based on computer and the internet leads a creation of new information society that demands development of new informative media for a human.

The next-generation display, which is currently being studied, is being developed as a display capable of realizing a lightweight, thin, highly efficient, and to display color video according to the advancement and portability of the information utilization environment such as Internet, IMT-2000 and PDA leading the information society. Furthermore, it is being developed into a thin flexible display. The electronic paper which is being studied with the aim of realizing high-quality flexible display in accordance with this demand is a next generation display which can be carried in paper texture and easy to display and erase. Recently, the rapid development of information science centered on computers and the internet has demanded a new level of transformation of information delivery media. Conventional displays have the disadvantage that visibility is lower than paper, bulky, and requires a power supply. The everyday life of the mobile communication environment that is rapidly proceeding, requires a display medium that enables the general public to use a large amount of advanced information at any time and anywhere with ease. The realization of paper and conventional displays maximizes the advantages of each, namely, the excellent contrast ratio and visibility, the fast response speed to display of moving images, the display of natural color, with low cost and portable like paper. It will revolutionize the infrastructure of society. Until now, electronic paper has not been fully commercialized yet, but it has been competing with various approaches (Table 7.9) in favor of each other. Three representative types of electronic paper are Xerox's twist ball type using electrostatic charge-filled hemispherical twisted balls (called 'Gyricon' balls with a Greek origin of rotating balls), electrophoresis and E -ink; electrophoretic display, and Kent Display's cholesterol liquid crystal display using cholesterol liquid crystal. The development of electronic paper technology based on the electrophoresis phenomenon which presents many prototypes as the most commercial approach, and an attempt to commercialize electronic paper will be described. Flexible displays that are as thin as paper, and can be seen when they are needed, can be realized not only in the imagination but also in the near future, and electronic paper will realize that dream. Let's look at e-paper, a display that can be used at any time and from anywhere with fast-changing information in the information society.

4.2.1 Principle of Operation of Electrophoretic Display

Electrophoresis is a phenomenon in which charged particles are moved by an applied electric field in a state of dispersed particles in a fluid. An electrophoretic display is a reflective display that can repeatedly write and erase a shape or letter using electrophoresis. In 1956, Metcalfe and Wright developed electrophoretic images for electrophoresis. Typical electrophoretic display panels consist of a transparent dispersion of charged particles in a dye fluid and are driven by applying a DC voltage of several tens to several hundreds of volts, in a form filled between a front electrode and a backplane made of pixel electrodes.

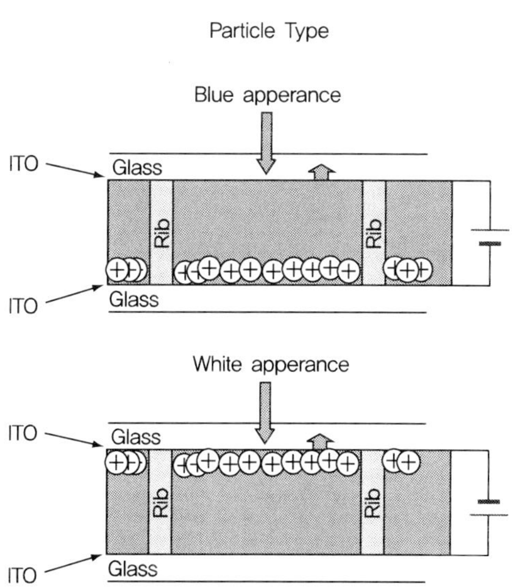

Figure 7.39

Electrophoresis principle.

The optical contrast ratio can be obtained by moving the charged pigment particles up and down between the electrode plates. When the particles are stacked or layered on the front electrode, the white state becomes a Lambertian scattering state, so that a bright state can be seen at all viewing angles. The switching speed of the electrophoretic display panel is calculated from the basic electrophoretic phenomena. The charged pigment particles move at a speed v, that is proportional in the fluid are proportional to the applied electric field (E). Assuming that the proportionality constant is μ and it is called electrophoretic mobility,

$$v = \mu E \quad \text{(Equation 7-12)}$$

Smoluchowski showed electrophoretic mobility as follows.

$$\mu = \zeta\varepsilon/\eta \quad \text{(Equation 7-13)}$$

ε is dielectric constant of electrophoresis fluid, η is viscosity of electrophoresis fluid and ζ is zeta potential of electrophoresis fluid. Zeta potential is electrostatic potential in the shear plane that surround charged particle and it is given by

$$\zeta = g\lambda_D/\varepsilon \quad \text{(Equation 7-14)}$$

g is an electric charge of a particle, λ_D is Debye length that occupied by a particle in an electrophoresis fluid. So switching time of electrophoresis particle, that is moving with velocity v at the cell gap h, can be quantified as

$$t_{total} \cong h/\mu E = h^2/\mu V \quad \text{(Equation 7-15)}$$

Where V is the applied voltage. These equations show how to adjust the physical properties of particles and fluids to reduce switching time. Based on this basic principle, various organizations are implementing electronic paper using various modes. Various studies are being conducted as the most likely technology for implementing flexible display. Among these technologies are EPD (Electrophoretic Display), microcapsule, microcup, in-plain type, Quick response liquid powder display (QR-LPD), Gyricon display, Electrowetting display and Cholesteric LC display.

(1) Encapsulated EPD

As electronic paper developed by E-ink Co. Ltd., the initial device form is a transparent microcapsule containing chargeable fine particles (mainly using rutile TiO_2)

and a colored fluid as described above in the driving principle. When the positive and negative voltages are applied to the upper and lower transparent electrodes, the negatively charged ink particles migrate to the surface to display the color of the fine particles. In this case, however, clustering and coagulation of fine particles have become a serious obstacle to practical commercialization. Therefore, microcapsules (average particle diameter 70 μm) in which white fine particles having a positive charge in a transparent fluid and black fine particles having a negatively charged particles are dispersed as shown in Figure 7.40. The black and white are displayed in a manner that is more advanced than the above method.

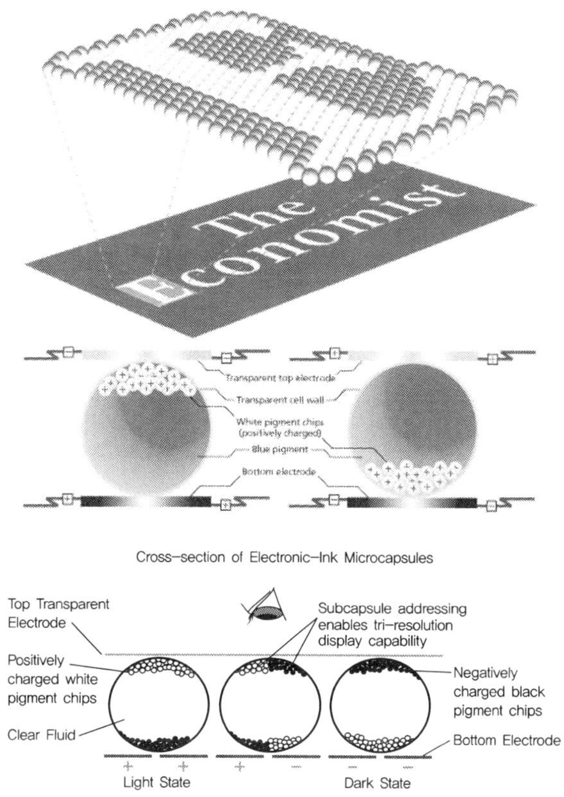

Figure 7.40
Operation principle of encapsulated electrophoresis display.

The electronic ink that is composed of microcapsules, is a new material invented by the combination of chemical, physical and electronic technologies. Each microcapsule contains a white powder (mainly TiO_2) and a black powder (−) charged with mainly carbon black) that are filled with liquid. When (−) electricity is applied to it, the white powder is raised above the microcapsule and visible to the human eye, and at the same time the black powder falls to the bottom and becomes hidden. In the opposite case, the black powder will rise up and the area will appear black. For an actual E–ink electronic paper display, after the electronic ink is coated on the plastic film, it must be laminated with the circuit layer, and the circuit layer forms a pattern of pixels. The microcapsules are mixed with a liquid carrier medium to enable screen printing and ultimately to any surface, like glasses, plastics, fibers and even paper. In recent years, electrophoretic displays with a reflectance of 45% and a contrast ratio of 15: 1 have been developed. As shown in Figure 7.41, a color electrophoretic display has also been developed. Examples of the implementation method include a method using a microcapsule having charged particles such as red, green, and blue, and a method using a shutter mode in which red, green, and blue reflectors are disposed under capsules containing black and white charged particles. However, although the response speed has recently been getting a little slower, it is still ~ several tens of milliseconds, and many improvements are required to implement the video quality. It is based on the passive driving principle, so there is no threshold voltage, although the resolution is limited. Therefore, passive driving is efficient for low information content applications, but there is a disadvantage that active driving must be applied to realize a high–resolution image.

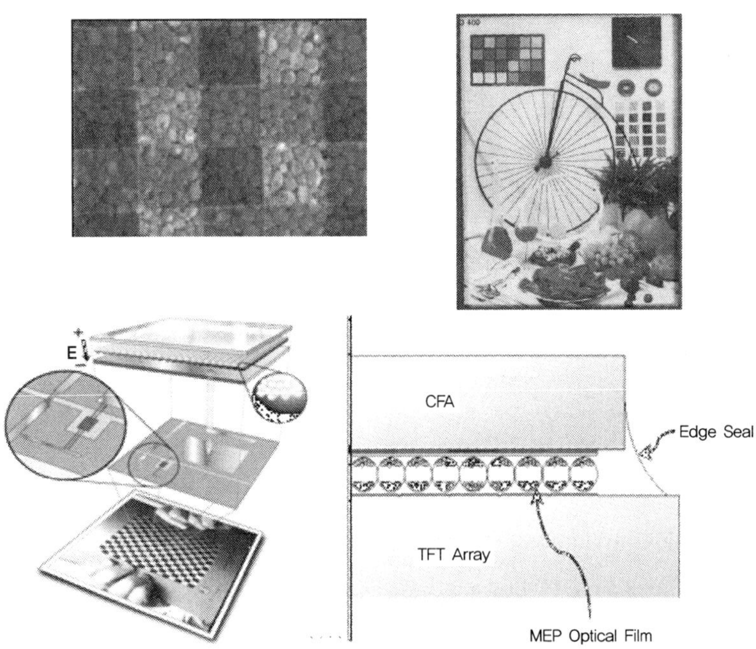

Figure 7.41
Color electrophoresis display.

In the future, E-ink's encapsulated electrophoretic display is expected to launch various models for commercialization along with various applications for market expansion. E-ink, including France's largest economic newspaper, Les Echos, is developing a new paradigm for newspaper distribution, including mobile phones, fashion watches and e-books. In particular, Les Echos' electronic paper newspaper will showcase new technology that combines a huge amount of information and speed of the Internet in the form of downloading the Internet site though the wireless LAN and downloading it to the portable display. E-ink, which shows the most advanced application possibility in commercialization, is cooperating with the leading companies of application development to secure the market preemptively and to supply low-priced electronic ink through cheap materials and simple manufacturing process.

(2) microcup EPD

Founded in Silicon Valley, California in 1999, SiPix Imaging is an electronic paper specialist that provides customized production of raw materials for electronic paper display modules and technology transfer for successful market entry. Fremont, Calif. is responsible for R&D and roll-to-roll production, as well as module production and application development at Chung Li in Taiwan. In addition, it has succeeded in forming electrodes by printing technology in cooperation with Toppan printing. It is possible to make electronic paper thinner than usual paper thickness, thinner than 200 μm including an electrode, and printing pattern of XGA (1024 × 768) resolution became possible. Based on these technologies, SiPix Imaging recently developed an electronic credit card that generates a new password every time the electronic paper is loaded on a credit card. In 2005, a total of 60,000 counterfeit credit card victims occurred in the United States, with damages of over $ 1.4 billion. The new credit card is a way that microchip randomly creates a new number on the LCD screen by pressing a button on the card each time you trade it, giving the same effect as changing the serial number. Figure 7.42 shows the application of SiPix Imaging to a real credit card. The smart function allows you to modify information freely. The advantages are low power, flexible, strong shock and moisture resistant.

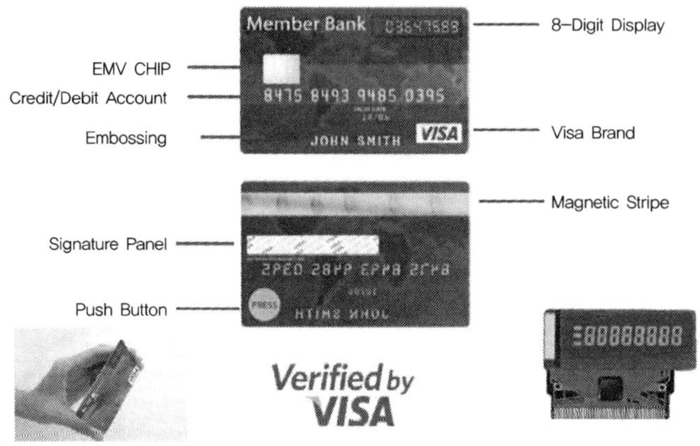

Figure 7.42
Example of applying SiPix Imaging on a credit card.

SiPix Imaging uses particles as electrophoretic driving type display device. Among them, as shown in Figure 7.43, SiPix Imaging has a function of displaying black and white by applying a voltage to the particles injected into the microcup. That is, a the microcup electrophoretic display is an electronic paper that injects electrophoretic charged particles into a microcup and implements characters or images by moving particles under an electric field.

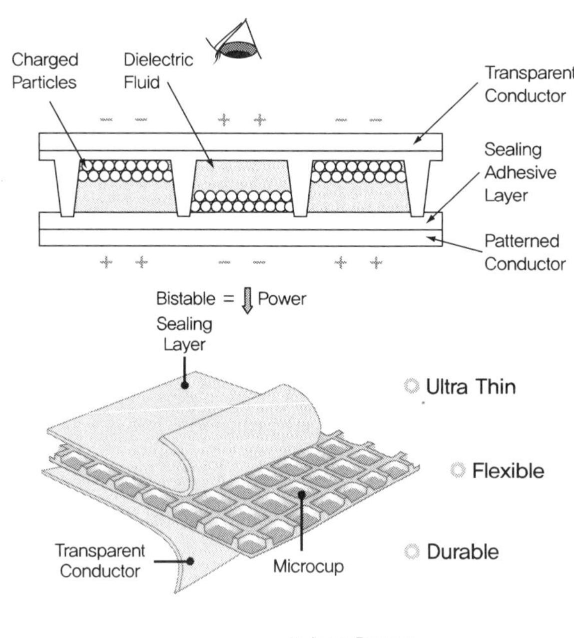

Figure 7.43
Microcup electrophoresis display driving principle.

This method is basically capable of mass production and exhibits excellent bistability. Microcups can be said to be microcontainers containing precise materials such as liquids, particles. The structure of the microcups is protected by a flexible PET plastic layer which is generally 150 μm thick and can withstand impact and pressure. Figure 7.44 shows the manufacturing method of microcups. This method is capable of high-speed microembossing process, seamless filling, and sealing process in mass production.

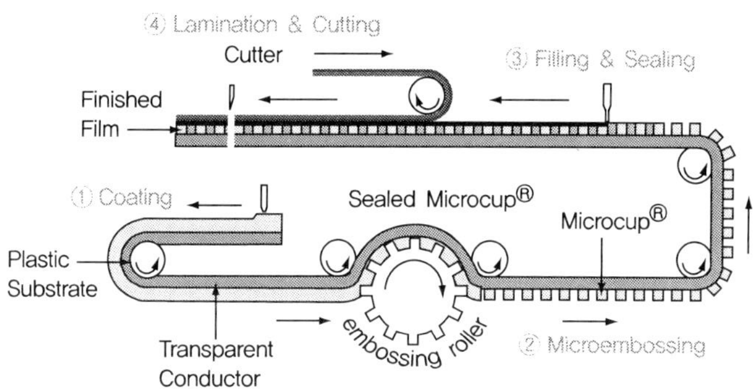

Figure 7.44
Microcup electrophoresis display manufacturing process

The microembossing process is a core technology for forming microcups with the shape shown in Figure 7.44. It is carried out by the stamp method which is very simple compared with the photolithographic process and the sandblast process. In addition to the microembossing process, the injection and sealing process, which is an important process, is a completely different method from the conventional display manufacturing process and is characterized in that charging and sealing of the charged particles are simultaneously carried out. In addition to the microembossing process, the injection and sealing process, it is a completely different method from the conventional display manufacturing process and is characterized in that the injection and sealing of the charged particles are simultaneously carried out as shown in the figure. When the sealant is mixed with the charged particles and filled in the microcups, the sealing material is phase-separated to the upper surface of the microcups and brought up to the surface, and then the sealing is completed by curing. A PET plastic substrate coated with a transparent conductor such as ITO is held on the surface of a specific resin through a coater and then molded through a mold having a three-dimensional structure for micro-embossing. This makes it possiblc to stably employ submicron and non-aqueous density-matched particles. This roll-to-roll process enables the molding process to be customized using various embossed mold shapes. For example, square and hexagonal structures can be fabricated with microcups as shown in Figure 7.45. The ITO coated PET plastic microcups showed excellent properties without weight loss even after storage for 5 days in an 80°C oven before laminating. The cell gap is kept constant at 15 to 30 μm. Figure 7.46 is a schematic diagram to illustrate the color technology of a microcup electrophoretic display. Three basic functions must be activated to implement color. 1. The microcups are filled with red, green, and blue fluids having white charged particles, which must act as a color filter. 2. The microcup wall has conductivity and acts as an electrode. 3. The black absorbent layer should be introduced at the bottom of the microcup to absorb the light from outside. For example, in the blue and green microcups, white charged particles are moved to the wall, so that the light that is absorbed from the outside is absorbed by the black absorbing layer and thus can not realize a specific color. On the other hand, since the white charged particles present in the red microcups are moved to the bottom, light can be scattered from the outside, and scattered light passes through the red fluid and only the red component is emitted to the outside of the panel.

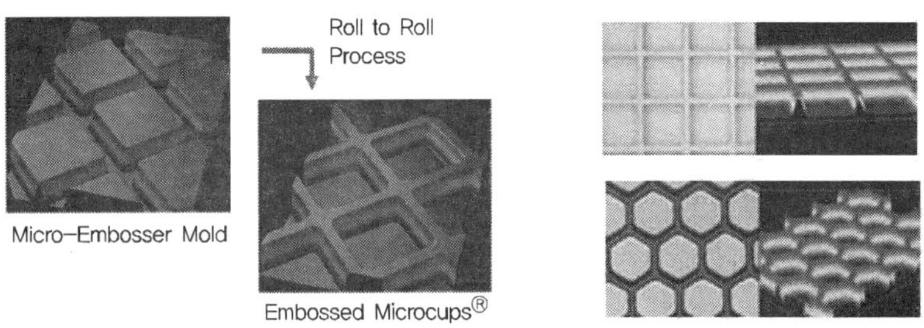

Figure 7.45
Microcup manufacture through microembossing process

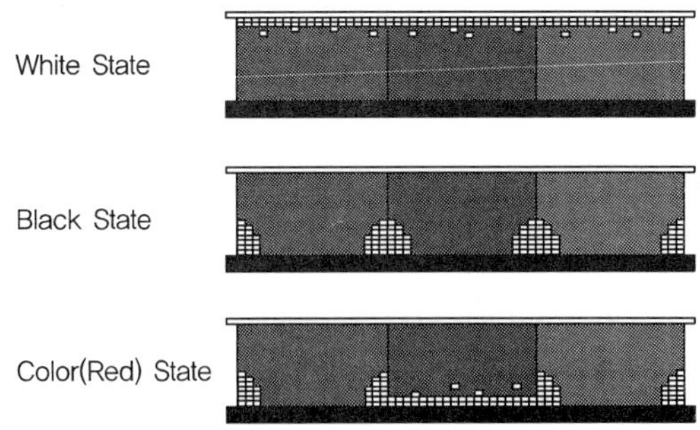

Figure 7.46
Principle of microcup electrophoresis display color

SiPix Imaging announced the image stability of an active microcup electrophoretic display (AMEPD) with passive driven microcup electrophoretic display (PMEPD) and a-Si TFT based on microcups and top sealing technology. This display product means that it is possible to realize low cost and high efficiency passive and active type microcup electrophoretic display by a roll-to-roll method. In 2003, SID realized the implementation of microcup efficient electrophoretic display based on a-Si TFT.

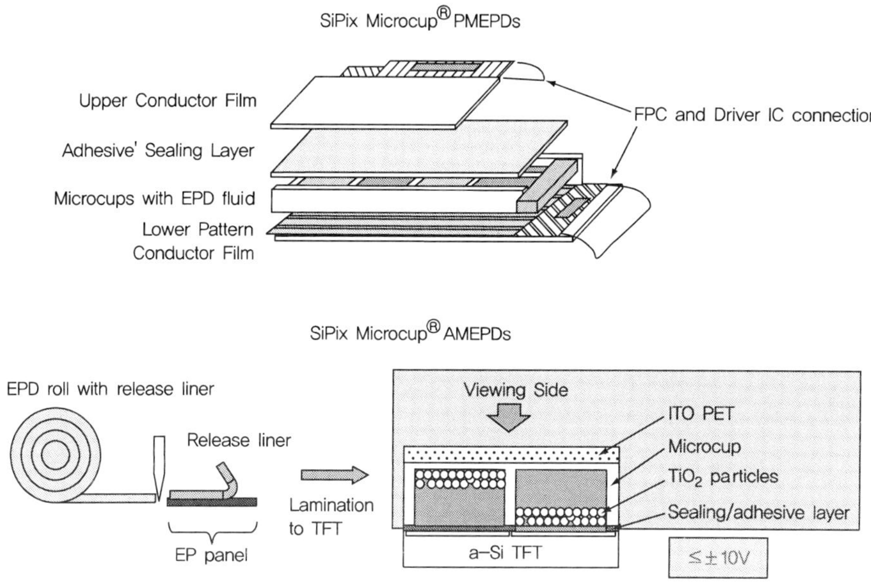

Figure 7.47

SiPix's passive, active-drive electrophoresis display schematic.

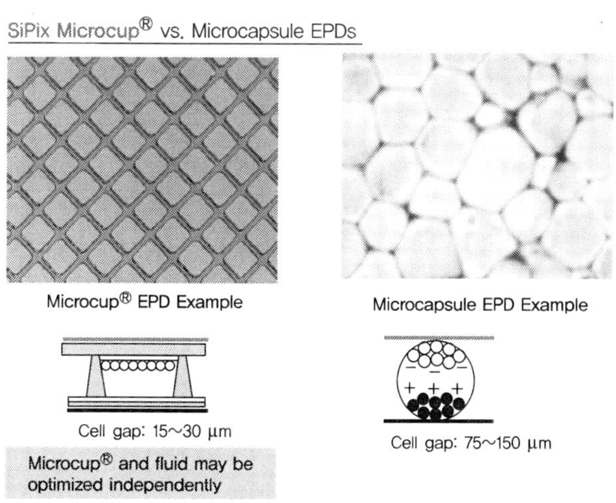

Figure 7.48

Comparison of microcup V·s microcapsule electrophoresis display.

Among the electrophoretic driving methods, SiPix's microcup electrophoretic display has a significantly smaller cell spacing than microcapsulated electrophoretic displays, enabling individual optimization between microcups and fluids. The main advantages are high productivity, high throughput. Continuous injection and sealing is possible due to its excellent physicomechanical properties. Submicron particles are adopted through narrow cell spacing, and response mobility of more than 1,000 $\mu m^2/V \cdot s$ can be expected even at low voltages.

(3) In-plain EPD

In Canon, a separate in-plain electrophoretic e-paper was proposed as shown in Figure 7.49. The transparent fluid was filled between the upper and lower substrates, and the black electrophoretic particles were dispersed. At this time, the lower substrate has a structure in which electrodes and a white scattering insulating layer are stacked. When a voltage is applied to a pattern electrode that is formed on a lower substrate to concentrate black particles, a white scattering layer is exposed and the white display becomes possible.

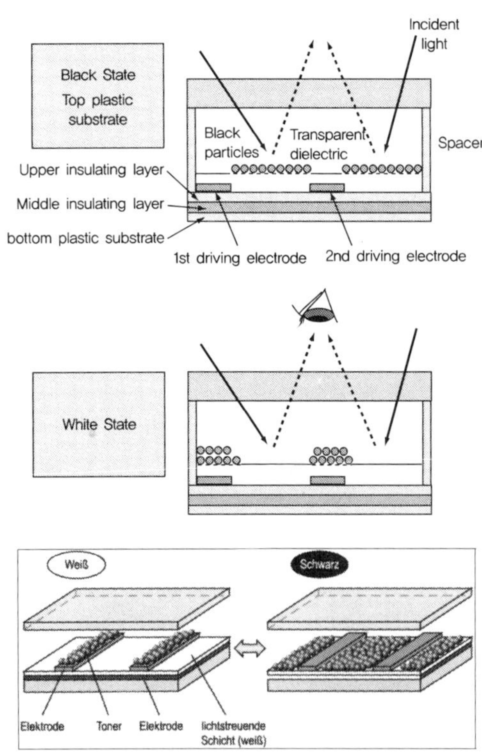

Figure 7.49
In-plain EPD driving principle.

The electronic paper panel structure for the full-color implementation of the in-plain EPD is shown in Figure 7.50. It consists of red, green, and blue color filters formed on the charged black particles, the transparent fluid and the bottom part. In the case of green and blue pixels, the black particles move to the bottom, so that all the incident light is absorbed. On the other hand, in the red pixel, the black particles move to both spacers so that the light can reach the color filter, thereby enabling red display. Therefore, it is possible to display a color with one pixel, which is advantageous for a high resolution, but problems such as improvement of display stability remain.

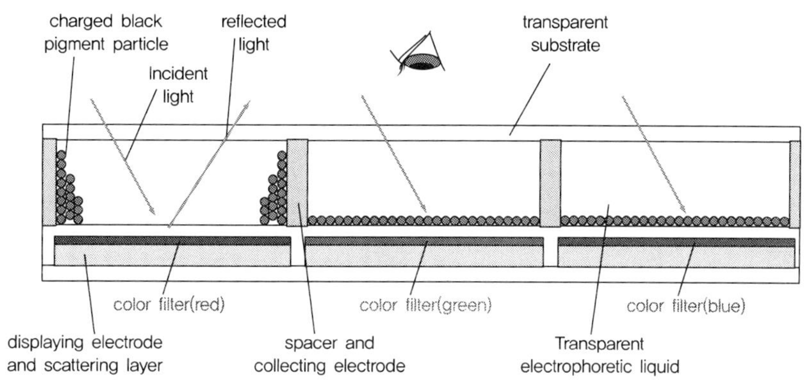

Figure 7.50
In-plain EPD color implementation principle.

(4) Quick-response liquid powder display (QR-LPD)

Reiji Hattori, the professor at Kyushu University in Japan, developed a passive-driven QR-LPD that is commercialized by Bridgestone Corporation, which corresponds to the dry transfer method. The particles are charged differently

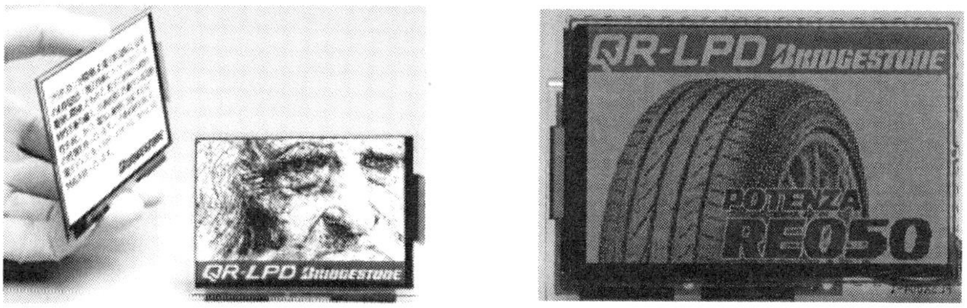

Figure 7.51
Bridgestone's QR-LPD

charged and the colored particles in the aerosol form are injected between substrates with ribs. An image is displayed by moving the substrate along an electric field. Bridgestone, known to us as a tire manufacturer, is a supplier of plastic products, conveyor belts, etc. In particular, Bridgestone in Japan was the first company to mass–produce film–type filters for PDPs.

In 2002, 'liquid powder', a polymer with fluidity like liquid and electronic paper technology using QR–LPD (Quick Response–Liquid Powder Display) were developed. The greatest strengths of the QR–LPD are its fast response time of fewer than 0.2 μs, bright images that show white surfaces of paper, the reflectance of more than 42%, infinite interactivity, wide viewing angle with ideal diffuse reflectance that provides a large matrix addressing drive capability with half–tone images. In particular, QR–LPD has a characteristic that it can maintain the picture by a kind of memory function even when the power supply is cut off and the viewing angle is close to 180°. The manufacturing process is simpler than liquid crystal too. In the case of manufacturing, it can be made to have a thickness of about 0.25 mm, so that an ultra thin display such as paper thickness is possible. However, in order for QR–LPD technology to be used, the driving voltage of 200V or more must be significantly lowered under 10V. Figure 7.52 shows the driving principle of QR–LPD.

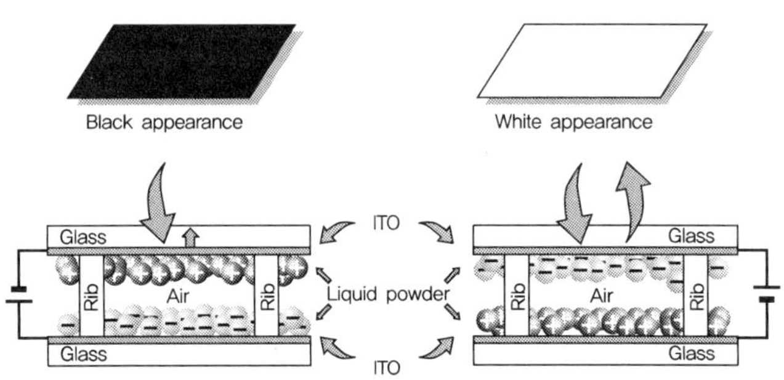

Figure 7.52
How QR–LPD works.

Like conventional electrophoresis, an electronic paper that uses particles as color display elements must overcome a system with a slow response time that particles dispersed in colloid have as a fluid resistance. Bridgestone's QR–LPD uses liquid powders, which are particles characterized by a fluid–like fluidity that is very dispersive, as the ink material. These liquid powders can overcome the problem of slow response time and long sticking image as in E–ink's commercial e–book. In addition, the resistance of the fluid can be greatly reduced by using a gas other than liquid. This is because the response speed is inversely proportional to the fluid viscosity. This dry transfer method is a method to overcome the limitations of the response speed in the conventional electrophoresis method (wet type), and it has achieved a response speed of 0.2 ms/line by using a liquid powder having fluidity like liquid, The electronic paper using the moving type and the particle rotating type has superiority over the visual characteristic with a response speed of about 100 ms.

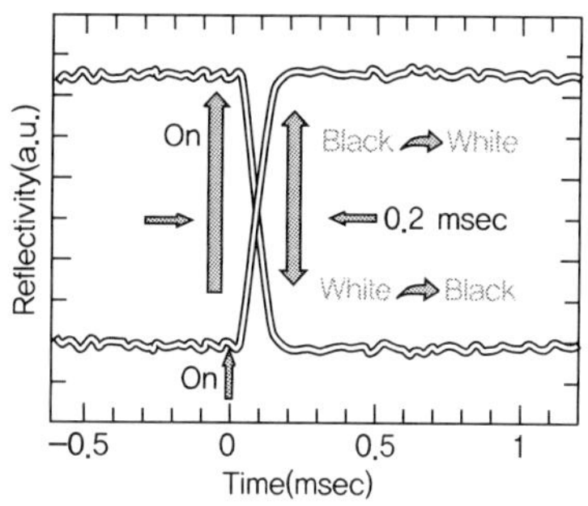

Figure 7.53

Response of reflectivity.

Figure 7.53 shows the contrast ratio corresponding to the fast response time. It is a short time, but it can be seen that the conversion from black to white or black to white occurs well. Also, the contrast ratio is the clearest among electronic paper types. The wide viewing angle of the QR-LPD is shown in Figure 7.54.

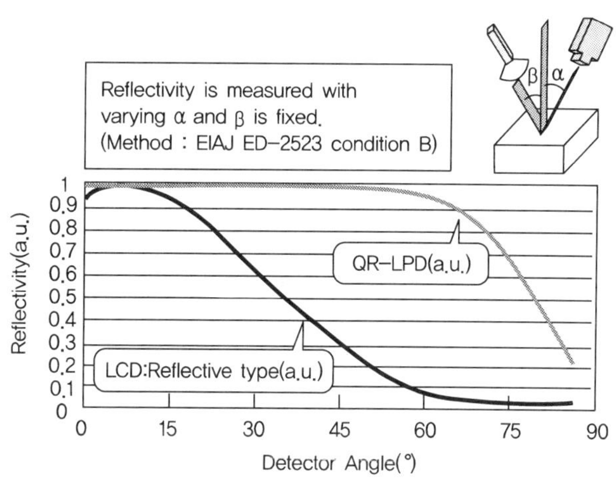

Figure 7.54

QR-LPD's Detector angle

The conventional LCD has a reflectivity of 30% at a 45 degree field of view, but the QR-LPD method maintains the best state. It is seen that the viewing angle decreases from 60 degrees, but it has a wider viewing angle than the conventional reflection type LCD. At SID 2006, LG Electronics made a 5.7-inch panel with 320 × 240 resolution as shown in Figure 7.55. The bulkhead of the pixel is formed in a honeycomb-like hexagonal shape, and the limit of warping is possible only to a small curvature due to the presence of the partition wall.

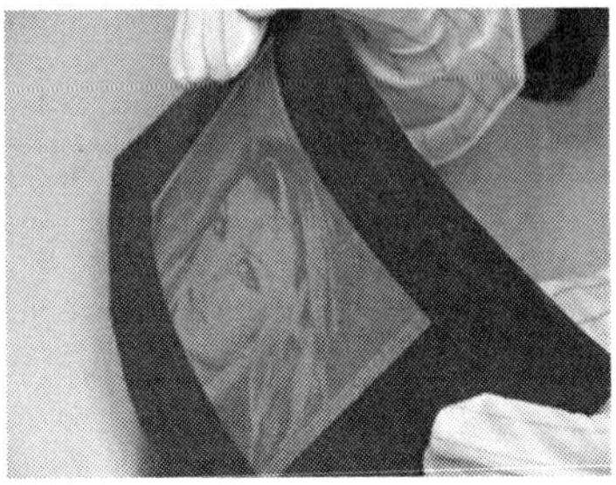

Figure 7.55
LG Electronics' QR-LPD.

As mentioned earlier, QR-LPD has a fast response speed to implement moving pictures, but it is difficult to apply it to the electronic paper that can realize moving images because of its large driving voltage. It is difficult to implement a gray scale with a large contrast ratio. The QR-LPD has excellent bending characteristics, and can be controlled by passive matrix driving without a thin film transistor. There is almost no change in the temperature of the liquid. On the other hand, it is pointed out that the driving voltage is higher than 200 V, but recently it has been realized that the driving voltage of 100 V or less. Bridgestone recently demonstrated a full-color QR-LPD with RGB and W-type color filters to show an 8.1-inch 4,096 color implementation with 480 × 384 pixels.

Figure 7.56
Color QR-LPD.

(5) Gyricon display

The twisted–ball display (Gyricon display) was developed by Dr. Sheridon of the Xerox Palo Alto Research Center (PARC) in the United States in 1975. In 2000, Xerox founded a company called Gyricon Media to promote commercialization. In recent years, Xerox has been working with 3M for research and development aimed at the mass production of display materials.

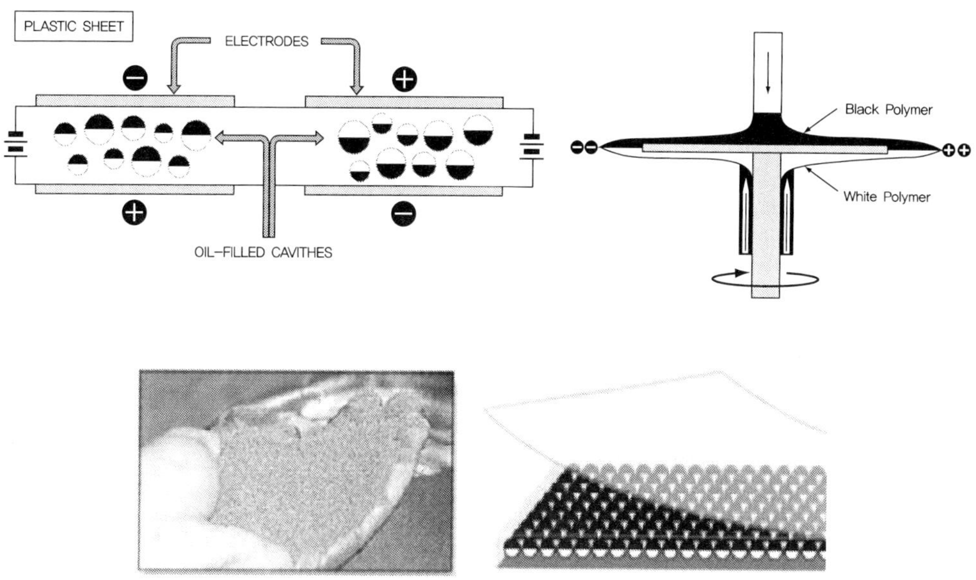

Figure 7.57
Gyricon Display Structure and Materials.

As shown in Figure 7.57, millions of small balls are dispersed in an elastomer matrix cavity filled with oil between transparent plastic sheets with transparent electrodes such as ITO, where the balls are strongly contrasted hemispheres. One is black which absorbs light and the other is white which reflects light. White and black areas of the ball have opposite charge. Gyricon Ball is a wax–like plastic made by spray molten process, which is a simple process that can be mass–produced. It has an advantage in realizing a low price and it is said to be stable. Since the white and black regions of the ball have opposite charges, it is possible to rotate not only in the electric field but to move up and down also in the cavity, thereby realizing characters and images.

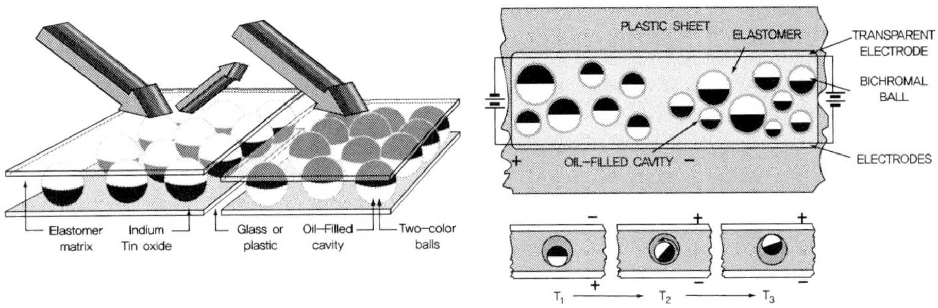

Figure 7.58

Gyricon Display Principle.

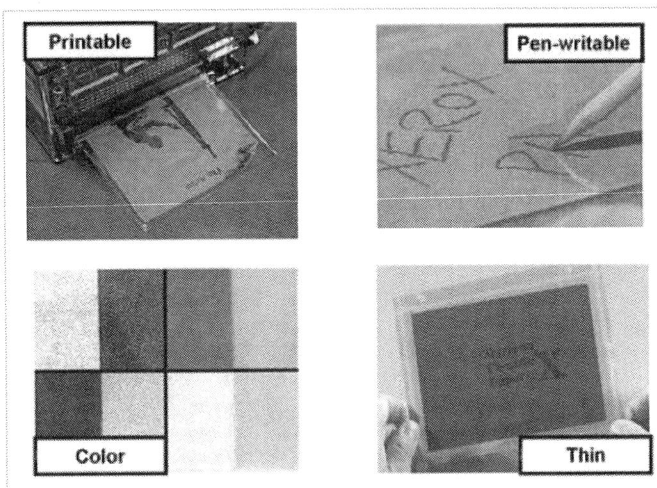

Figure 7.59

Various implementations of Gyricon displays.

Because the specific gravity of the ball and the oil are close to each other, if the ball is positioned by the electric field, the ball does not move easily, and therefore, even if the field is removed, the Gyricon display image shows bistability that can last for hours or days. Gyricon displays are typically made using balls with a diameter of about 100 μm and exhibit a control ratio of at least 6:1. The driving voltage and switching time are ranged about 50~150V and 80~100ms demanding high voltage and long switching time. These values are dependant on the electrical characteristic, size of the ball, and amount of damping values. A light reflectance of white light is about 20% which is about the one half of a newspaper value. Gray scale image could be observed by driving voltage of medium level. The thin Gyricon display is stable and can be used for a long time without an aging problem. However, the major disadvantage of Gyricon display is that there is no threshold voltage for a phase shift of the ball. In other word, the phase of the ball can be changed at an arbitrary voltage, thereby limiting the higher resolution. When a large number of pixel display is needed, the more complex control of the electric field is required in the Gyricon display. The color of the Gyricon display is realized either by controlling the color of the ball or by introducing the color filter.

6) Electrowetting display (EWD)

In September 2003, an electronic paper using electrowetting was introduced on the cover of Nature proposed by Hayes and Feenstra of the Philips Research Laboratory. What is noteworthy is that conventional electronic paper has been researching and developing in the direction of increasing the contrast ratio of bistability and black and white rather than emphasizing the fast response speed of moving images. Philips' model focuses on video implementation. It was a brilliant idea that proposed the classic electro-wetting principle as an ON/OFF switch of the display. Liquavista, spearheaded by Philips at the Society for Information Display (SID) in 2006, presented animated cartoons based on ideas proposed to Nature. Liquavista suggested a reflective type, but the University of Cincinnati Hikenfeld presented an emissive type electrowetting-based electronic paper. In the case of the reflection type, the implementation of Red, Green and Blue is regarded as the color of the droplet, and in the case of the emission type, there is a difference in using a luminous body which emits visible light in response to ultraviolet rays.

Figure 7.60
Image implemented by electrowetting method (SID 2006).

Both methods, however, are driven by electrowetting, which regulates the movement of the droplet under an electric field. As shown in Figure 6.61, a transparent electrode (ITO) is disposed on a white substrate, and a hydrophobic insulating material such as a fluoro−polymer, a colored oil (nonpolar dye + C10 to C16 alkane). Electro wet display in which electrodes (ITO) are sequentially laminated is a technique in which a potential difference is generated between a hydrophobic insulator and a liquid when an electric field is applied, and the water pushes out the oil (oil shrinkage). If the electric field is not applied, the color of the oil will be displayed. Under the direct electric field, the white color and the oil color (visually recognized as white) appear at the same time due to the microfluidic movement.

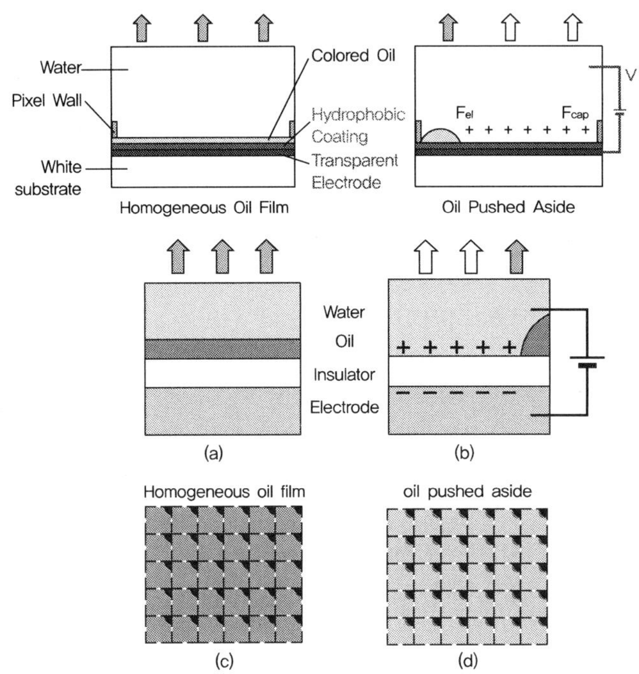

Figure 7.61
Electric wet display drive principle.

The implementation principle and image of a reflective electro pneumatic display using droplets of actual color are shown in Figure 7.62. (A) shows that the red droplet oil spreads evenly on the substrate without the electric field applied, and the reflected light shows the color of the droplet oil, and (c) is a picture of (a) seen from above. When the electric field is applied, the states of (b) and (d) are shown, and the liquid droplets are seen to converge to one side. The reason for this is that the water droplets drift to one side due to the electric field distribution above and below. If there is a hydrophobic insulator layer on the electrode and no voltage is applied, the oil will cover the surface. However, when the voltage is applied, the electric wetting causes the water to push the oil to one side.

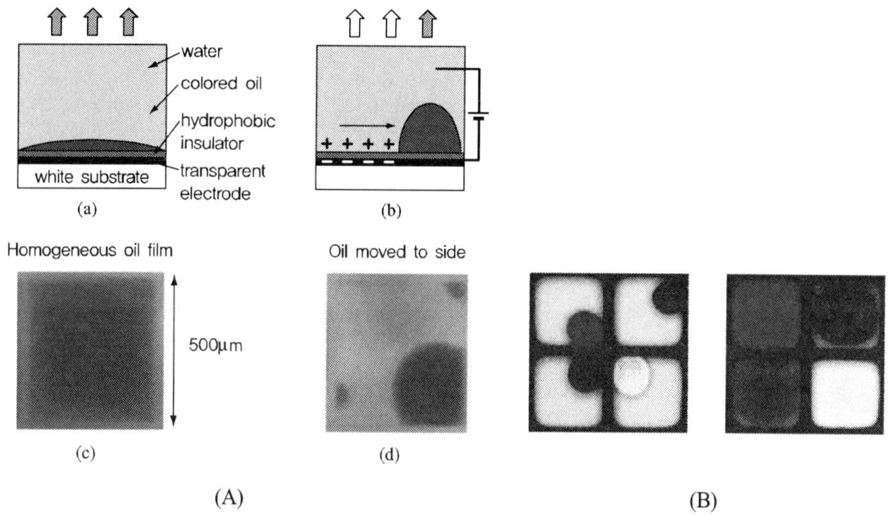

Figure 7.62

The driving principle (A) of the reflection type electrowetting method and the image (B) implemented in four pixels.

In Figure 7.62 (B), four pixels show different shapes of oil in different colors and scattered in the pixels. This pixel has the same function as the color filter in the LCD. The LCD exists as solid using the color pigment while droplets play the roles of both liquid crystals which act as a light ON/OFF switch and color pigment. The movement of droplets within a pixel is a function of the thickness of the oil, the pixel size, the viscosity of the oil, and so on.

The electroluminescent electrophoretic electronic paper was developed by the University of Cincinnati. It was introduced by Steckel and Dr. Heikenfeld. A light-wave coupled display using LEDs is a nonpolar oil that acts as a switch, just like the reflective mode. The theoretical efficiency is 10 ~ 40 times higher than the luminance of conventional LCD. The maximum luminance of Red, Green, and Blue is about 950, 5530, and 530 cd/m^2, respectively. The waveguide of Figure 7.63 is a UV-storage plate with a refractive index of ~1.5. Light in the ultraviolet region generated through the LED is propagated through the optical waveguide and stimulates a luminophore contained in the oil to emit visible light.

The droplet is playing the role of the pigment. The movement of droplets within a pixel is a function of the thickness of the oil, the pixel size, the viscosity of the oil, and so on.

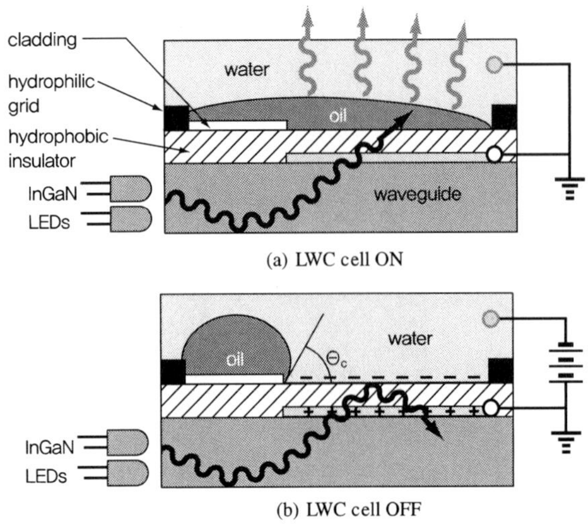

Figure 7.63
The driving principle of the electroluminescence type electrowetting system

At this time, the cladding layer has a refractive index of less than 1.3 and totally reflects ultraviolet light to return UV to the storage plate. When the voltage is applied and the oil collects on one side, the ultraviolet light returns to the UV storage plate due to the refractive index of water (~ 1.3). Figure 7.64 shows an LWC display implemented on a flexible substrate. Instead of using electrowetting, it is realized by filling a groove in a letter and filling it with oil containing a luminous body. The photovoltaic display implemented by applying the principle of electrowetting to date is not disclosed but is expected to be introduced soon.

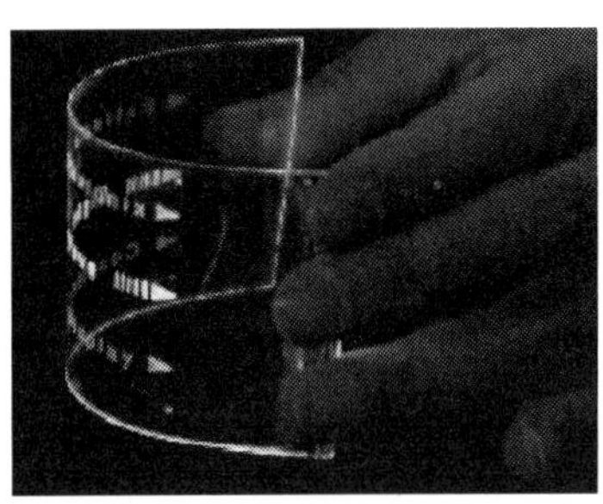

Figure 7.64
Electronic paper using electroluminescence type electrowetting.

A comparison of the driving power with the method using electrowetting and the other method is shown in figure 7.65. IMoD, which is a driving method of electronic paper using MEMS (Micro Electro Mechanical Systems), and the method using electrowetting show relatively much lower power consumption than OLED and conventional LCD. Also, it can be seen that the electro-wetting method has a great advantage in driving power. The reflectance and contrast ratio are also superior to other methods in comparison to other display systems.

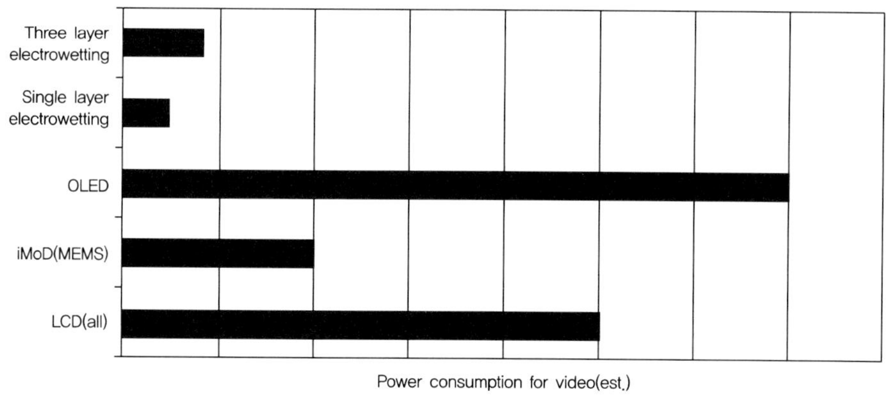

Figure 7.65

Power comparison with various display methods.

It's color expression is as good as that of a paper, which is a great advantage in color implementation. Similar to a general liquid crystal display driving, the opening ratio of the oil is increased in proportion to the applied voltage. Figure 7.67 shows a schematic view of an electro-wet display and a cross-sectional view of a panel employing a color filter. As shown in the figure, there is no dependency on cell spacing, which is advantageous for a flexible implementation.

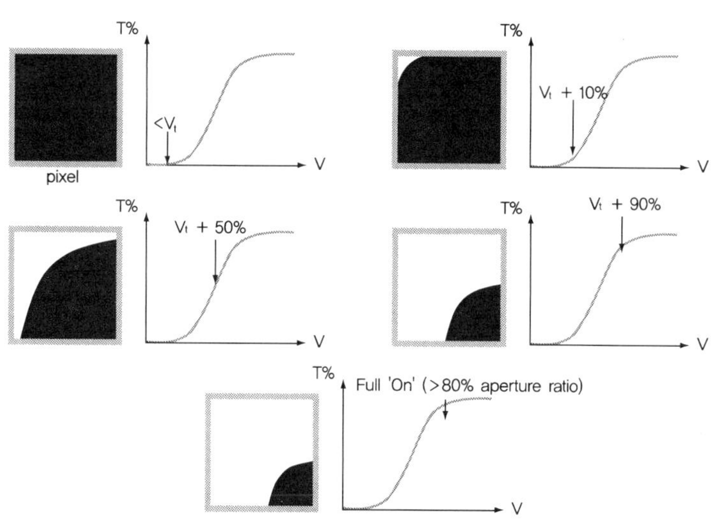

Figure 7.66

Electrowetting display drive scheme.

Similar to a general liquid crystal display driving, the opening ratio of the oil is increased in proportion to the applied voltage. Figure 7.67 shows a schematic view of an electro-wet display and a cross-sectional view of a panel employing a color filter. As shown in the figure, there is no dependency on cell spacing, which is advantageous for a flexible implementation. The implementation of a full-color electrowetting display is possible by mounting a color filter and using black oil. Another implementation of the full-color electro-wet display is to add a color oil layer and introduce cyan, magenta, and yellow color filters into the laminate structure as shown in Figure 7.68. The three subpixels are composed of cyan, magenta, and yellow in up, down, left, and right directions, and a wide range of color gamut can be realized according to the appropriate combination of these colors.

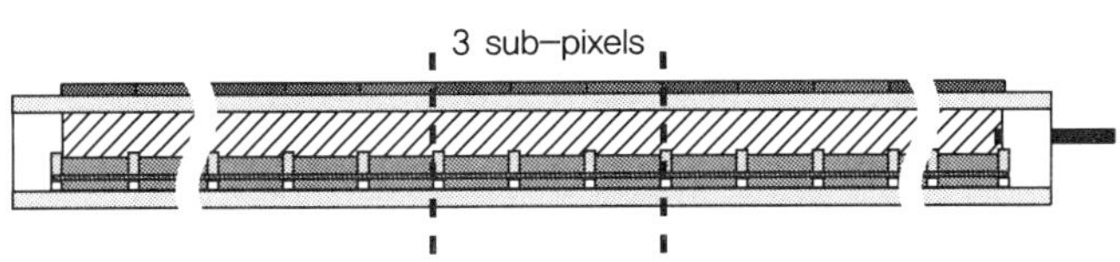

Figure 7.67
Color Implementation Using Color Filter

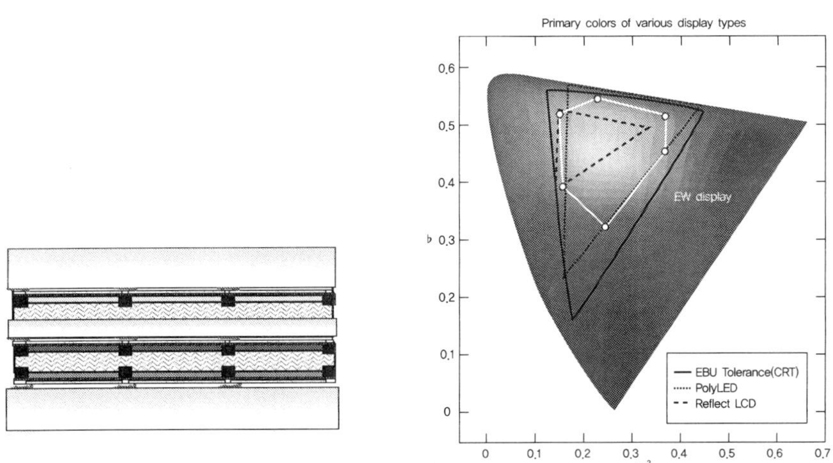

Figure 7.68
Electric wet color display structure and color range.

(7) Cholesteric LC Display

When a liquid crystal is applied to a flexible display, the conventional liquid crystal mode can be divided into a case where the liquid crystal mode is applied to a flexible substrate and a case where a liquid crystal mode having a memory characteristic is applied. Table 7.11 and Table 7.12 summarize the electronic paper type and performance comparison using liquid crystal mode. In 1993, Kent display developed a technology for implementing electronic paper using a cholesteric liquid crystal that can selectively reflect light of different wavelengths and express color.

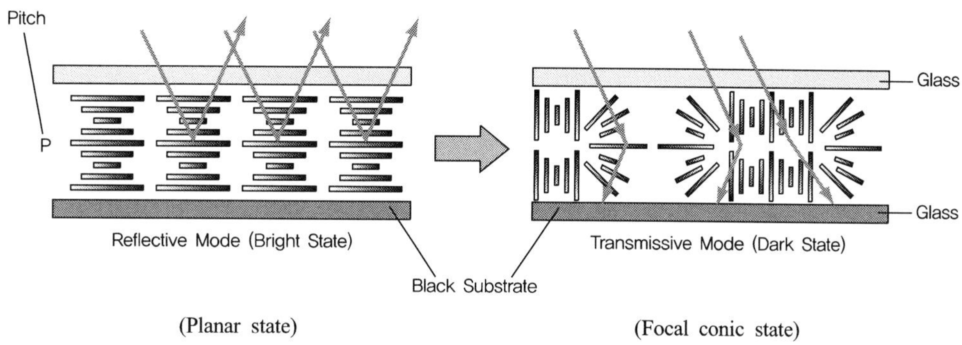

Figure 7.69
the state change of the cholesteric liquid crystal.

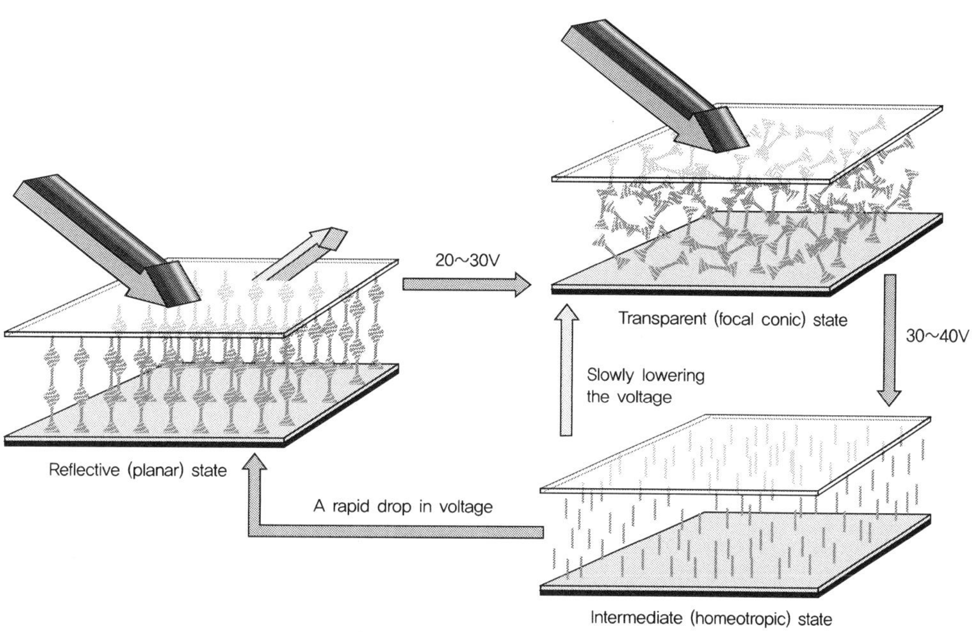

Figure 7.70
Driving principle of the cholesteric LCD.

In the cholesteric liquid crystal, the director of the liquid crystal is twisted along the rotation axis and the distance until the director turns 360 degrees is defined as the pitch. It is notable that the cholesteric liquid crystal determines the wavelength of the reflected light according to this pitch. When a voltage is applied, it travels in a planar state and a focal conic state, where reflection occurs in the planar state and transmission occurs in the focal conic state.

Table 7.11 Electronic paper using liquid crystal mode.

Mode	Technology	Related company
Normal liquid crystal mode	TN, FSTN, VA, IPS	Softpixel, Samsung
		Sharp/Seiko-Epson/Matsushita
Single liquid crystal or liquid crystal lamination method	BTN(Bistable Twisted Nematic)	Nemoptic
	ZBD(Zenithal Bistable Display)	ZBD
	SSFLC(Surface Stabilized FLC)	
	Ch-LC(Cholesteric LC)	Kent/Nanox/Matsushita/Fujitsu
Liquid crystal dispersion	PD-LC(Polymer Dispersed LC)	ASET

Table 7.12 Electronic paper using a liquid crystal mode.

Item	News paper	Ch-LC	PD-LC	BTN
Reflectivity	> 60%	30~40%	15~25%	30~40%
Color	Mono/Full	4,096	Yes	Need C/F
Response Time	–	30~100ms	< 100ms	30~100ms
Contrast	10:1	20~30:1	10:1	18:1
Driving	–	Passive	passive	Passive
Voltage	–	> 40V	> 10V	15V
Flexibility	Yes	Yes	Yes	–

The color filter is not required to implement the color, but the thickness of the module becomes thick because three layers of red, green, and blue must be stacked in order. Like the electrophoresis method, the response speed does not reach the level suitable for the implementation of the moving image. However, it is enough to be applied to a night vision device or map for a military use. When a voltage is applied between two electrodes containing a cholesteric liquid crystal, it is possible to switch between a transparent state in which light is not reflected and a state in which light is reflected. Also, as shown in the figure 7.71, in a structure in which red, green, and blue liquid crystals are laminated, each color is selectively reflected or combined to form a color. It has realization potential because it has a relatively high reflectance and selectivity, and Kent Display, Nanox, etc. have made prototypes. The cholesteric liquid crystal mode is characterized by excellent optical properties and mass productivity, which is promising for e-book and can significantly reduce power consumption and does not require a separate polarizing plate, unlike an LCD. In addition, since color can be implemented without a color filter, application to a portable flexible display is expected to in the future. However, it has drawbacks such as difficulty in realizing a high resolution, difficulty in full-color implementation, and high driving voltage. A complicated driving waveform is required due to the application of a reset pulse for the transition between a planar liquid crystal (planar) and a focal conic liquid crystal so as to be vertically aligned.

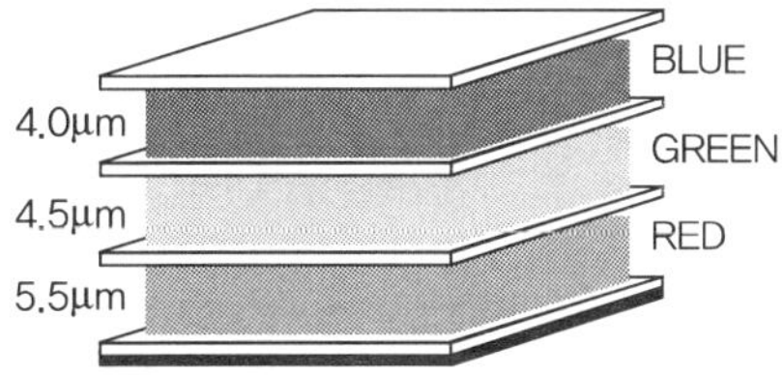

BLACK ABSORBER

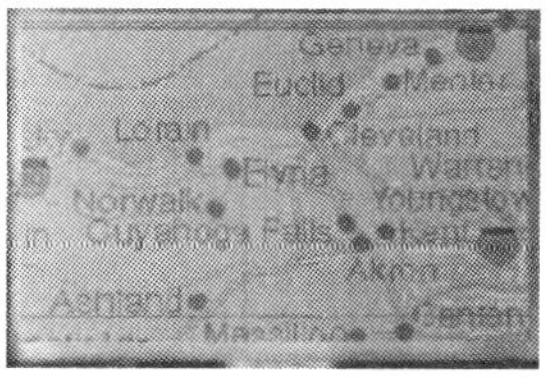

Ch-LCD, 1/8 VGA, Eight Color Image

Figure 7.71

Cholesteric color LCD structure and image of display.

In principle, a cholesteric liquid crystal has a characteristic that a relatively bright reflection type display is possible, and a backlight is not necessary and a high resolution can be realized. In the cholesteric liquid crystal, a quaternary liquid crystal molecule takes a helical structure by the action of a chiral agent, which is called a cholesteric phase. The liquid crystal molecules have different refractive indices in the major axis direction and the minor axis direction. The circularly polarized light of light can be defined as a combination of two slowly changing polarization components whose polarizing direction changes clockwise to the right and that changes to the left.

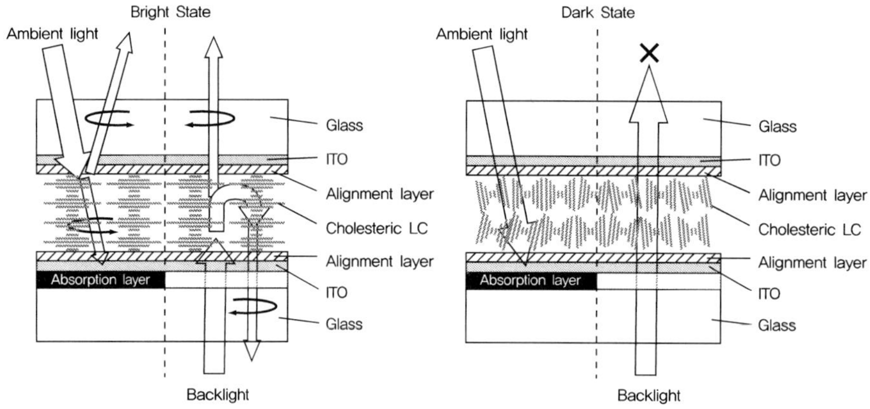

White and black(semi-transmission type)

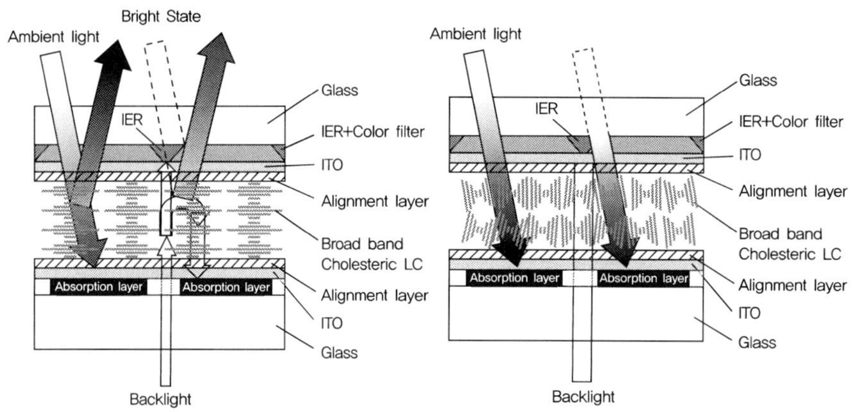

When implementing a color using a color fiter(semi-transmission type)

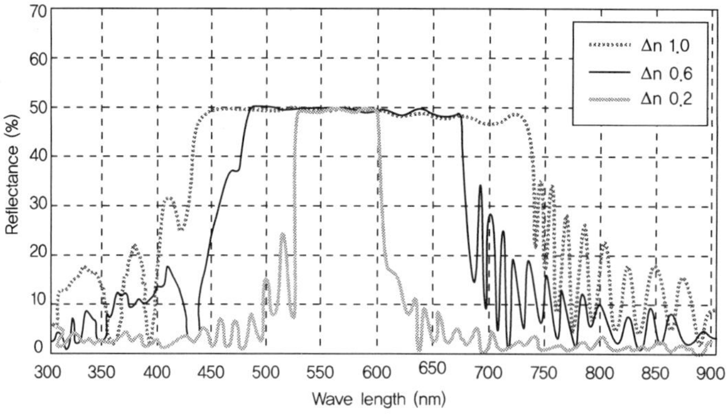

Figure 7.72

Principle of cholesteric liquid crystal device implementation and selective reflection wavelength according to refractive index anisotropy

The circularly polarized light in the same direction as the twist direction of the crystal is transmitted through that state by the optical rotation effect and becomes a film in which the refractive index is periodically changed with respect to the opposite circularly polarized light component and Bragg reflection occurs. Therefore, the principle of operation of a cholesteric liquid crystal is that the cholesteric liquid crystal can be made into two stable states, a planar state and a focal conic state, in a voltage unapplied state between narrow substrates, thereby exhibiting bistability. The helical axis of the cholesteric liquid crystal taking the planar arrangement state is almost perpendicular to the substrate surface. The light incident on the liquid crystal cell in this state is divided into left-handed and right-handed circularly polarized light, one of which is transmitted and the other is reflected. Theoretically, the maximum reflectance is 50%. The wavelength λ At which the reflection of the circularly polarized light becomes the largest is $\lambda=n \cdot P$. where n is the liquid crystal average refractive index in a plane perpendicular to the helical axis, and P is the helical pitch. The half width of the reflection spectrum is adjusted to a value proportional to the refractive index anisotropy Δn of the chiral nematic liquid crystal material. On the other hand, when the focal conic array is arranged, the helical axis is almost in equilibrium with respect to the substrate surface. When the helical pitch is shorter than the wavelength of the visible light, most of the visible light incident on the liquid crystal cell passes through the cell. When these two states are electrically replaced, a similar exchange is performed. Since the selective reflection mode element of the cholesteric liquid crystal has a memory property in which the display state is maintained even when the voltage is not applied, a simple matrix electrode structure can be used without requiring an active driving element represented by a thin film transistor. Therefore, it is possible to easily manufacture a display device with high accuracy at a low cost, and to use a plastic substrate, so that a thin and light display device can be obtained. The display state is determined by the presence or absence of selective reflection due to the difference in the arrangement of the cholesteric liquid crystals. Therefore, high brightness display can be obtained because there is no loss of incidence and outgoing light by the polarizing plate. Since a polarizing plate is not used, it is not necessary to deal with the liquid crystal molecules in the direction of the polarization axis, so that there is no need for a general nematic type or alignment treatment required for a smectic type device. When used as a reflective display element, a light absorbing layer is disposed on the side opposite to the light incidence side. The light absorbing layer absorbs unreflected one-side circularly polarized light components and light other than a selective reflection in a planar arrangement and absorbs the total light transmitted through the liquid crystal layer when the focus cone is arranged. By this effect, it is possible to display a high contrast of selective reflection. As a liquid crystal material, it is common to use a cholesteric liquid crystal material in which a chiral agent is appropriately added to a nematic liquid crystal material having a positive dielectric constant anisotropy to give a beating force to liquid crystal molecules. The cholesteric liquid crystal material can be used as a material for a reflective display device by adjusting a cholesteric phase in a practical use region so that selective reflection occurs in a visible light region. Adjustment of the reflection wavelength is carried out with the addition of a chiral agent, and when the addition amount is increased, the biting force of the nematic crystal is increased and the spiral pitch on the cholesteric phase is decreased, and the reflection wavelength is shifted toward shorter wavelength.

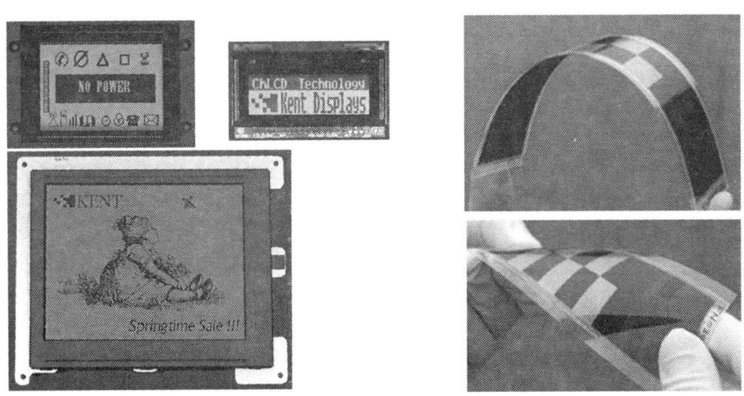

Figure 7.73
Various cholesteric display prototypes used.

When the conventional liquid crystal mode flexible display is classified into a-Si TFT, poly-Si TFT, and OTFT depending on the type of the driving device, it is considered most feasible to apply the existing liquid crystal method to a flexible substrate. The problems to be solved at this time are the maintenance of the liquid crystal gap, optical structure, and minimization of misalignment of the upper and lower plates due to misalignment. The liquid crystal mode using the memory characteristics is characterized in that it has a bistable property, which can not be implemented in LCDs, PDPs, and OLEDs. The PD-LC has the advantage that it has a reflection characteristic of paper due to the fine scattering method and has a simple process, and it can be colorized in the guest–host type.

Table 7.13 Comparison of electronic paper performance in various ways

Core Technology	Display type (Company)	Driving Voltage(V)	Contrast Ratio	Response Time(μs)	Reflectivity (%)
Paticle	Paper	–	15	–	70
	Microcapsule(E-ink)	15~90	10	100	40
	Microcup(SiPix)	30	~10	200	40
	Gyricon(Xerox)	50~150	~8	80~100	20
	QR-LPD(Bridgestone)	100	~9	0.2	40
Liquid crystal	ChLCD(Kent Display)	40	20~30	30~100	40
	STN(ZBD)	5	5	20	<5
Fluid	Electrowetting(Philips)	20~30	15	<10	60
MEMS	iMoD(Qualcomm)	5	10	10	50

Table 7.13 summarizes the performance comparisons for the implementation of the various types of electronic paper previously discussed. In contrast, the cholesteric liquid crystal dominates in the control ratio, but the QR–LPD and the electro–wetting method are excellent in the response rate and the reflection rate, respectively.

5 Research Trends and Forecasts Flexible Display

5.1 Technology Trends and Prospects

Organic thin film transistor materials for flexible displays can be broadly classified into organic semiconductors, organic insulators, electrodes, and substrate materials. Ultimately, they are flexible and capable of low cost, low power consumption and flexible display. In the early days of the flexible display market, a small–sized display with a slim–line and flexible characteristics such as a mobile phone, a PDA, an E–book entered the initial market. It is expected that the market will be developed in two directions with the small-sized mobile display based on ubiquitous, large-sized display such as advertisement signs and TV. Ultimately, high-speed mass production technology is expected to create a market for ultra-low-cost, disposable flexible displays and low-cost large-area displays. The development of organic semiconductor materials for flexible displays is a core material for electronic paper, display, solar cell, radio frequency identification (RFID), transparent conductive substrate and stealth technology. The market size of flexible display technology based on organic thin film transistors has not been formed yet, so it varies widely according to the research institute, and various forecasts are announced from pessimistic forecasts to optimistic forecasts. However, with the development of materials and device manufacturing technology, it is predicted that the market will expand rapidly in the future. However, the flexible display market, which is an important application field of organic thin film transistors, is expected to grow by 5–7% over the next two to five years and predicts growth in market size.

Flexible displays can be made from ultra-low cost, lightweight electronic paper, flat and thin, lightweight, robust displays, curved or conformed displays create new markets for flexible displays, foldable displays, rollable displays, wearable displays, and irregularly shaped displays. Therefore, the direction of the organic thin film transistor material market is expected to evolve into electronic species, flexible OLEDs, LCDs, and rolled organic thin film transistors. Therefore, the direction of the organic thin film transistor material market is expected to evolve into electronic paper, flexible OLED and LCD, and rolled organic thin film transistor as seen in Figure 7.74.

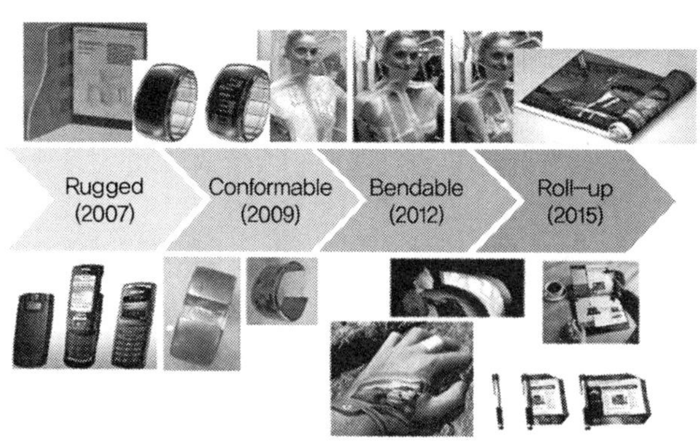

Figure 7.74
Flexible display evolution concept.

It is expected that the development of p-type organic semiconductors for roll-type solution process is important as well as the development of p-type organic semiconductor materials for low-grade (E-paper) solution processes, development of p-type organic semiconductors for a solution and vacuum deposition, development of p-type organic semiconductors for OLED and thin film transistor (TFT).

Since organic semiconductor materials do not have enough materials to dominate the market, it is believed that if only excellent organic semiconducting materials are developed, they will enjoy the same status as current Merck in liquid crystal materials. In order to develop exclusive materials for the market, such as liquid crystal, we are investing huge research and development expenses in the development of organic semiconductor materials. Organic semiconductor materials for the flexible displays are expected to develop organic thin film transistor materials for AMOLED or AM-LCD after organic thin film transistor materials for electronic paper are introduced first, rather than flexible display itself. Therefore, It is anticipated that possible low-molecular materials or oligomer materials will first come to market. Organic thin-film transistor materials for flexible displays with a large area of the low cost will be developed over the long term, so that polymer organic thin film transistor materials with the roll-to-roll method and stable characteristics are expected to come into the market.

In the case of organic insulator materials, although it is an essential material in addition to organic semiconductors for the application of organic thin film transistors to flexible displays, the development level is still insufficient globally. Since there is no material that can dominate the market yet, if an organic insulator material is developed, it will be possible to enjoy a monopolistic position such as the current Merck in the liquid crystal material as mentioned above. Interestingly, organic semiconductors are being actively developed in academia and industry in Europe and the United States, but organic insulators are considered to be a highly promising area in which research is limited. Organic thin film transistors using organic semiconductors are indispensable for the development of organic thin film transistors. Therefore, it is certain that if organic material development and technology are secured, organic thin film transistors and flexible display market will be secured in an exclusive position. In the case of organic insulators, it is expected that the organic thin film transistor will be developed in the same direction as the organic semiconductor development direction. In the case of the initial organic thin film transistor, the combination of the inorganic insulator and the organic semiconductor may be possible. However, development of organic insulators is essential for the ultimate roll-up solution process. Organic insulators for deposition type organic semiconductors are limited, but organic insulators for solution process type organic semiconductors are required to have an organic insulator that is chemically stable to organic solvents in the organic semiconductor thin film manufacturing process. In the case of plastic substrate materials, attempts to replace minerals and metals with organic materials in the information electronics industry are still in its infancy and it is expected that the market will grow only if many technical problems are overcome. However, research on materials should be given priority.

Table 7.14 The rode map of flexible display technology.

	Device/technology Performance			2006~2009	2009~2012	2012~2015	Current skill level			target
Flexible- Display	OTFT - backplane	Resolution		More than 100 ppi	More than 100 ppi	More than 150 ppi	o			High resolution
		Main performance		Conformable	Bendable	rollable	o			flexible
		Additional performance		Glass	Thin film		o			Improved reliability Long lifetime
				Lifetime 10,000 hours	Lifetime 10,000 hours	Lifetime 20,000 hours	o			
		Main technology	Design and modelling	Panel/long channel device	Panel/short channel device	Panel/3D design		o		Integration Low-cost standardization
			Vacuum deposition	Low-molecular semiconductor/metal electrode	Low-molecular semiconductor			o		
			printing	Gate insulator & S/D	High-molecular semiconductor	Roll to roll	o			
			Library	100ppi class device	100ppi class device	150ppi class device	o	o		
			module	100ppi class module	100ppi class module	150ppi class module				
		Market size		$0.8~2.4 billion	$1.9~5.7 billion	$3.1~9.5 billion	low	medium	high	Market formation and expansion

source: Information Display Technology Roadmap (2006)

In the case of display materials, about 20 kinds of display materials have been developed depending on the driving method from electronic paper to prototype or commercialized, and the display materials to be used also vary depending on the driving method. Representative electronic paper display materials include the micro-capsule type of E-ink, rotating ball type of Xerox, and cholesteric LC of Kent

Display. Since electronic paper is a display that has not yet been established in the market, the display material for electronic paper is being produced and sold for research and development, and the market for display materials alone may not be meaningful. On January 3, 2007, Plastic Logic invested US $ 1 million to develop a portable electronic paper production line using E–ink, which is the electrophoretic microcapsule type. In Dresden, Germany. It announced that it will produce 41.6 million products in 2010. Sony produces and sells an E–book viewer under the brand name LIBRie. It is expected that various other methods will take more time to apply to specific products and form a market.

According to Display Search data in 2006, the electronic paper market is expected to grow to about 320 billion won in 2010 and about 600 billion won in 2015. Among them, the display materials are estimated to account for about 5% of total assets 16 billion won in 2010 and 30 billion won in 2015.

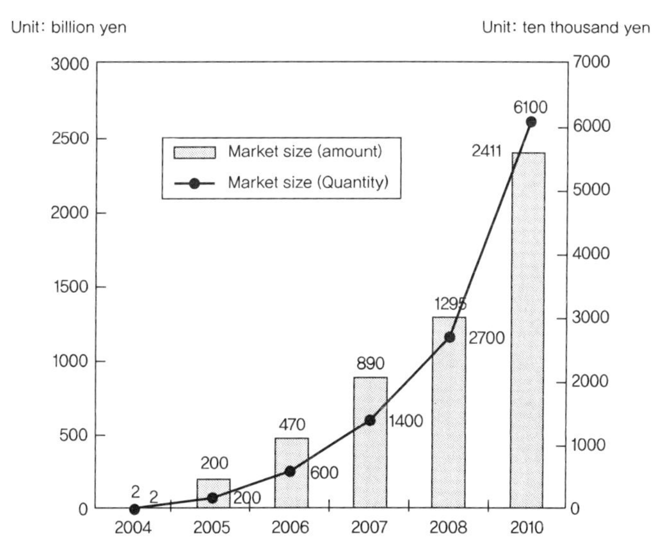

Figure 7.75

Electronic paper (E–paper) World market scale.

The Fuji Chimera Research Institute expects the global market for the electronic paper to grow from 2 million units, 20 billion yen in 2005 to 61 million units, 241 billion yen in 2010 as shown in Figure 7.75. The average annual growth rate of the electronic paper market is expected to reach a high growth rate of at least 50% to 125%, but the price per capita has been steadily declining. The expected price per capita in 2010 is $ 0.7 (iSupply data, 2004) and $ 0.16 (STEMI, 2003), and the market growth rate of the display materials is expected to be smaller than the market growth rate of electronic paper due to unit price competition.

Table 7.15 Average annual growth rate and unit price of electronic paper market.

Market research firm	Average annual growth rate (%)	Unit price ($)	
		2004	2010
iSupply	124.8	7	0.7
STEMI	86.9	–	0.16
Fuji Chimera	64.5	100	40

Source: Electronic Paper, KISTI R&D Strategic Task Pre-feasibility Analysis, 2005.

Since Japanese and UK companies including US E-ink have patents on the display materials and driving methods for almost all electron papers, it would be difficult to enter the market unless it is possible to implement color, has quick response characteristics that can realize moving images, and is a breakthrough display material with price competitiveness. It seems to be realistic in securing market competitiveness by licensing and using other methods that are already building or planning the production line for simple purpose electronic paper such as signage, price tag, billboard, E-book, etc.

5.2 R & D Trend

5.2.1 Flexible Display Research Trend

The next generation of flexible display technology is currently being studied extensively by panel makers, companies, national research institutes, academic college and so on. As the speed of technology development is accelerating, it is expected that the market of the flexible display will rapidly develop in the next decade. In the case of LCD, OLED and the electronic paper type flexible display, about 30 ~ 40 companies are researching and developing in the world. In the case of including research institutes and academics, hundreds of research institutes are currently conducting research on flexible displays. In Europe, Philips in the Netherland has been developing flexible displays using polymeric materials. Polymer Vision is established and invested in order to develop and commercialize flexible displays. Meanwhile, Plastic Logic in the UK is developing an ink-jet-based organic thin film transistor and a backplane for flexible display. In the United States, research on flexible LCDs, OLEDs, and electronic paper has been conducted mainly by corporate research institutes such as IBM, Sarnoff Corporation, Universal Display Corporation, E-Ink, Bell Labs and Palo

Alto Research Center (PARC). It is expected to occupy a very important position in the field of the flexible display in the future as it has secured many original technologies, core technologies and patents for displays, and government agencies and companies have invested heavily. In Japan, researches on organic thin-film transistors, poly-Si TFTs, and flexible OLEDs have been widely carried out by Hitachi, Sony, AIST, Seiko-Epson and other research institutes and national institutes. Recently, Bridgestone announced QR-LPD and it is expected to be commercialized. Sony and Seiko-Epson have developed a prototype of a poly-Si thin film transistor with excellent characteristics on a polymer substrate based on excellent poly-Si fabrication and transcription technology. Hitachi and AIST recently developed a flexible LCD (OTFT-LCD) using OTFT as a switching device and demonstrated prototype products. NHK Broadcasting Research Laboratories is currently studying active-matrix OLEDs using organic thin film transistors. In Korea, KETI, ETRI, Kyunghee University and Donga University are conducting research on flexible LCD, OLED and electronic paper including corporate research institutes such as Samsung Electronics, LG Display and Samsung Display. Korea's flexible display technology is approaching world-class in some fields, but overall it is somewhat lagging behind other advanced countries. In particular, the technology and patents related to flexible displays, core parts, equipment and materials are far behind compared to the US, Japan and Europe. However, as LG Display recently showcase the world's best technologies and the technology for flexible displays is gradually accumulating, it is highly likely to become a technological leader in flexible displays in the future.

5.2.2 Development trend of E-ink application technology

In accordance with this tendency, various applications have been disclosed in which electrophoretic displays are applied. Motorola's "MONOTINE F3" was selected as "Best Ultra-Low Cost Handset 2007". The mobile phone uses an E-ink display to reduce power consumption significantly. It combines a cutting-edge and low-cost combination to create an ultra slim bar (2.1 mm thinner 8.8 mm thick than the Samsung X820 mobile phone). It adopts E-ink's display, which not only lowers power consumption but also has a good contrast ratio and looks well under direct sunlight. The product is currently targeting sales strategies for low-cost products. These advantages provide low power consumption in the display portion of the mobile phone and can be achieved by using a low-cost E-ink electrophoretic display.

In addition, Polymer Vision, a specialist in the development of rollable displays as a venture company from Phillips, unveiled the "Readius", a collapsible e-book with built-in mobile communication capabilities, at the 3GSM event in Barcelona. It is called by its nickname, "Celluar Book" rather than "Readius" because it supports various mobile communication networks including 2G, 2.5G and 3G including GSM, GPRS, EDGE and UMTS.

Figure 7.76
World's First Mobile Phone with EPDs "MOTOFONE".

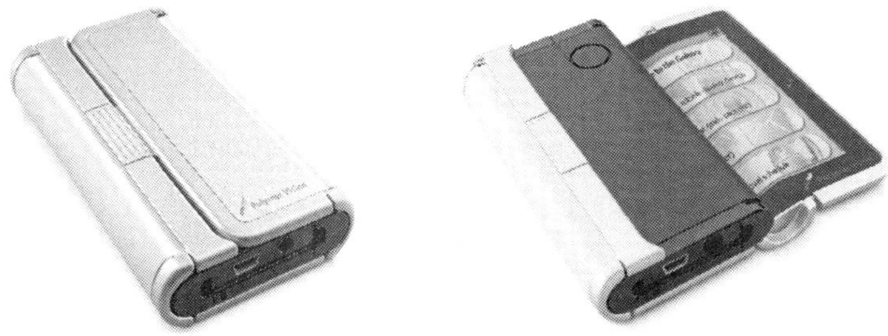

Figure 7.77
Polymer Vision's "Cellular-Book" with rollable display.

The display uses only E-ink's electrophoretic display to provide black and white gradation, but it has the advantage that it can be used for up to 10 days by charging once the built-in battery because of low power consumption. As such, E-ink's Encapsulated EPD is a leading developer of electronic paper using electrophoresis. E-ink's business strategy is to supply microcapsules, display modules and application system. In November 2000, electronic paper is developed using organic thin film transistor using printing method by Lucent's Bell lab and it shows that it is possible to realize a flexible process at low cost. In 2004, a plastic logic company in the UK published an electronic paper with a curvature radius of 5 mm and a resolution of 50 dpi. E–ink has continued to expand its business strategy and succeeded in commercializing Sony's Librie (E–book). Toppan Printing developed a front substrate containing ink and Phillips developed an active back substrate. In particular, E–ink developed Color Ink in cooperation with Toppan Printing in 2001. In July 2002, E–ink announced 4,096 color electronic paper of 5–inch and 80–dpi grade in July 2002 and is under development for commercialization. In addition, Seiko Epson has produced products that are applied to fashion watches using performance that can be freely designed and implemented.

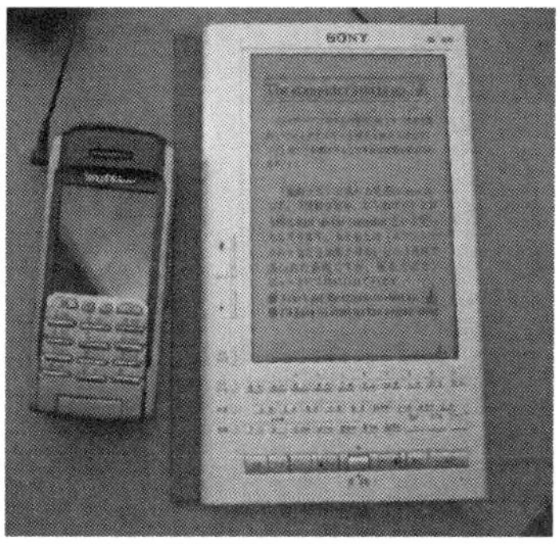

Figure 7.78
Sony's eBook Librie.

Figure 7.79
Seiko watches.

Future E–ink's encapsulated EPDs are expected to launch various types of models for commercialization while presenting various applications for market expansion. In addition to the example presented earlier, E–ink is in the process of being developed for the publication of an electronic paper newspaper with the French economic daily Les Echos in 2008. In particular, Les Echos' electronic paper newspaper will showcase new technology that combines a huge amount of information and speed of the internet in the form of downloading the internet site to the wireless LAN and downloading it to the portable display. E-ink, which introduced the most advanced application for commercialization, is developing the application field and securing the market preemption through major display system companies and cooperating with various companies to supply low-priced electronic ink through cheap materials and simple manufacturing process.

Figure 7.80

Future e-paper newspaper.

In Korea, LG Electronics, Samsung Electronics, and LG Display have been conducting various research and development for commercialization, and they are presenting prototype products using E-ink as described below.

1) LG Display: In October 2005, LG Philips LCD and E-ink jointly announced the development of a 10.1-inch electronic paper display with SVGA resolution of about 100 pixels per inch, a contrast ratio of 10:1 and four gray levels. In May 2006, a 14.1-inch VGA-grade black and white flexible electronic paper is implemented. In May 2007 the color flexible electronic paper is released. It adopts electronic ink of US E-ink company and can display up to 4,096 colors. In order to restore the original state of bendable flexible electronic paper, the a-Si thin film transistor was arranged on a substrate made of metal foil instead of glass, and a color filter was coated on a plastic substrate. A large amount of metal foil is supplied to LG Philips LCD and E-ink from Sumitomo and it is refined at Nippon Steel Industries.

Figure 7.81
Color electronic paper jointly developed by LG Philips LCD and E-ink.

2) Samsung Electronics: In FPD International held in Japan in 2006, we have been presenting a variety of prototypes starting from the presentation of A4 size plastic monochrome electronic paper display through FPD International. In 2007, SID unveiled the world's largest 40-inch monochrome electronic paper and A4-size plastic color electronic paper for the first time. The 40-inch E-paper is an electrophoretic display with a very large DID (Digital Information Display). It has HD (1,366 × 768) resolution and has 300 mW (1 frame/min) of ultra-low power consumption. The 14.3-inch color e-paper is an upgraded version of A4-size black and white electronic paper announced by Samsung Electronics in 2006 and implemented a 130°C low-temperature a-Si process technology. It shows that Samsung Electronics will lead the flexible display market in the future.

Figure 7.82
The electronic paper that Samsung Electronics announced at SID2007 (Left: 40-inch E-paper, right: 14.3-inch color E-paper).

3) Soft Pixels: Soft Pixels is a venture company specializing in flexible displays, which was established in April 2000. It is the first company to commercialize monochrome LCD flexible displays with a thickness of 0.7 mm and supply them to Fossil, a fashion watch company. It participated in the development of 5" flexible TFT LCD of Samsung Electronics. Samsung illustrated a product with 100 ppi resolution qsVGA, which combines color filter and liquid crystal made by Soft Pixel. Flexible products by Soft Pixel that weigh less than of less than 1.2 g/cm^3 and less than 0.76 mm standard are now expanding their application to digital clocks, smart cards, stereoscopic goggles, cell phone stereoscopic filters and liquid crystals.

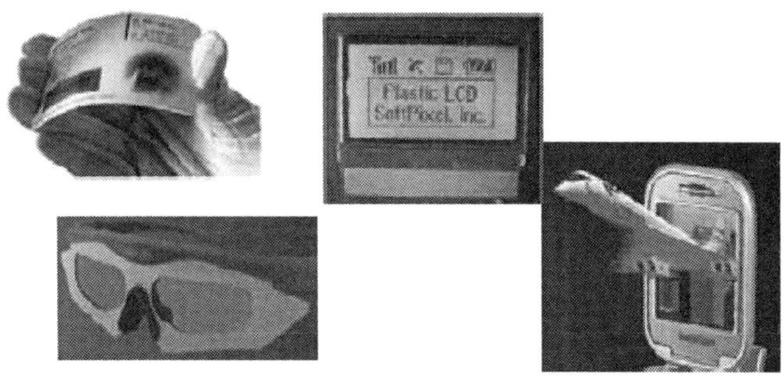

Figure 7.83
Soft Pixel's Flexible Display Products.

4) Neo-Lux: In cooperation with E-ink, they launched "NUUT", the first e-book terminal in Korea using electrophoretic display technology, and used it for a POP (point of purchase) advertisement for X-box which got a great response. NUUT is the first in the world to adopt the Vizplex Imaging Film electronic paper display, which is equipped with the latest technology, unlike the existing e-book exclusive terminals such as Sony Reader and iLiad which are based on E-Ink's electronic paper display. Vizplex Imaging Film is about 25% brighter than conventional electronic paper displays and has a reaction rate as fast as 1.8 times. Also, depending on the memory capacity, it is possible to store up to 1,000 books of e-book contents on the handset, and it is also a great advantage to be able to carry all the books desired by oneself.

5.2.4 Overseas R & D Trend

1) E-ink & Toppan Printing: As a venture company that was sprayed by MIT Media Lab in 1997, it possesses the technology for the electronic ink field and

manufactures E–ink Imaging Film in cooperation with Toppan Printing. In addition, it has been participating directly or indirectly in prototype and product development by supplying electronic ink or E–ink imaging film in partnership with many display module companies such as Polymer Vision, Plastic Logic Ltd, Seiko Epson Corp. and LG Philips LCD. In 2006, E–ink made a supply chain of active–driven electronic paper products as shown in IDW (International Display Workshops) 2006 in Japan. The supplied E–ink imaging film is combined with thin film transistor backplane in Prime View International or a display module manufacturer to implement electronic paper panels and modules and integrate them into products such as e–book from Sony and Librie. In Toppan Printing, a new concept of electronic paper was realized by assembling a color filter on the upper plate formed on the transparent amorphous oxide thin film transistor and the lower plate of the E–ink imaging film.

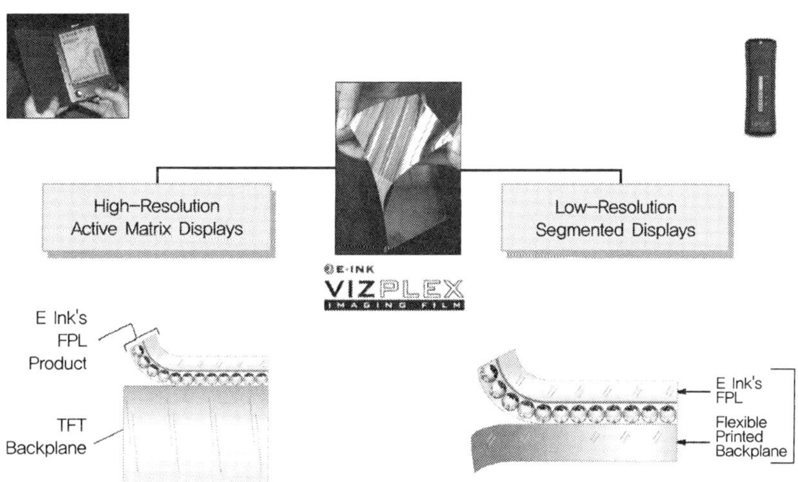

Figure 7.84
E–ink's electronic ink supply chain.

Figure 7.85
Electronic paper newspaper implemented by Toppan Printing

2) Hitachi: In order to provide a stable supply of electronic paper modules, Hitachi has partnered with Bridgestone to establish a strategy to pioneer and commercialize electronic paper solution market and to carry out various marketing. Bridgestone has decided to supply electronic paper modules and install products such as drivers, wireless communication circuits, and memories with Hitachi. Since 2006, they gradually expand and develop color and large size products from black and white A4 size paper. The company plans to expand its business to public and distribution areas by taking the transportation-related products such as electric trains as its main market and aims at annual sales of 10 billion yen by 2010. In May 2006, Hitachi introduced the "Albirey", a multipurpose display using electronic paper, which is a 13.1-inch monochrome electronic device with a resolution of XGA (1024 × 768). Hitachi announced that it had developed 4,096 color electronic paper in August 2006, and RGBW-type color filters were used. The resolution was 512×384 and the diagonal size was 13.1 inches. With the use of color filters, both vertical and horizontal resolutions were halved compared to the black and white Albirey launched in May 2006. Hitachi said it plans to release color displays in 2007.

Figure 7.86

Color electronic paper developed by Hitachi.

5.3 Flexible display issue analysis

The advantage of electronic paper technology is that it has characteristics of the electronic display, but it has similar characteristics to existing paper, so it is very likely to be used in e-book, tag, billboard etc. Until now, it is difficult to realize perfect color reproducibility and fast response speed like LCD or

OLED. However, it is expected to accelerate the development of products using this technology in the future because it is superior in price, thinness, and stability compared to other display technologies. The electronic paper should be easy to read like existing paper, fast response speed for video implementation, low power consumption, and so on. However, even if these technical requirements are satisfied, it is difficult to expect a large market growth if the flexible display can not have a specific application. ISupply, a market research firm, estimates that the flexible display market will reach $18 million in 2010. Many experts are still skeptical or hesitant to judge the possibility or timing of the market for flexible displays. Finding an application that has a value that can bring about flexibility is a key factor in enhancing the marketability of flexible displays. However, there are still no groundbreaking applications that require bending or folding displays. Typical applications of flexible displays, which have been mentioned so far, are electronic papers, newspapers, magazines, and newspapers made by IBM's electronic paper. It consists of 16 pages of electronic paper, which is smaller than conventional newspapers. The bottom corner of the newspaper is equipped with a battery, memory, wireless data port, buttons, and an electronic pencil. After going to work, you can read the e-paper placed on the desk in the morning. Going home you can read the changed news from the same e-paper. It is expected that newspapers will be able to display videos instead of related pictures.

A completely new application area can be imagined, which is to attach a transparent flexible display to the windshield of a car and use it as a navigator, such as on-screen navigation when needed. When the intelligent transportation system is realized and the car is surrounded by tens of thousands of sensors, it will be possible to see the position of the inner car by the display attached to the windshield while traveling with the view of the third person. In addition, when the flexible display is fully equipped, it is possible to push the large-screen TV, which is at the center of the living room, around the home, and fill the space with beautiful interior accessories. A blind-type rolled-up TV may be installed close to the ceiling and it may be displayed only at the time of watching, or the display may be attached to a large-sized window such as a car. However, they do not yet fully explain the marketability of flexible displays, as they appear in science fiction and movies. Early product launches are expected to include various smart cards with displays, credit card electronic devices that perform PDA functions, and electronic books. Going one step further, it is possible to display an electronic map of a military map or a display attached to work clothes at a construction site. However, they are only a niche market, and they are not enough to change the competitive paradigm of the flexible display industry. So far, the flexibility of flexible displays is focused and finding new core applications are not clearly visible. However, considering the viewpoint of replacing flat-panel displays such as TFT-LCDs, PDPs, and OLEDs, which replace glass substrates in the main applications of PCs and TV·s with flexible displays, the market potential of flexible displays should be different. The advantages of flexible displays can be found not only in flexibility but also in light weight and high durability. The thickness of the glass substrate is limited to 0.4 to 0.5 mm even if the glass substrate is as thin as possible. On the other hand, since the plastic substrate is generally in the range of 0.1 to 0.2 mm, it is possible to reduce the thickness from 1/2 to 1/5 of the glass substrate and the weight is less than 1/5 of that of the glass substrate. In addition, the plastic substrate has the high durability that does not break even when dropped. However, if only the panel is thinner and lighter without all the peripheral circuits being integrated on a flexible substrate to make the display itself thinner and lighter, it will not be possible to create a great response in the market as well as a restriction in a future application. Unbreakable displays can be an important selling strategy because of the characteristics of small mobile devices, which often fall off frequently. The advantages of the flexible display devices are flexibility, thinness, durability, and cost reduction. Until now, there is not much demand for plastic substrates and few companies participating in production are inferior in quality and the price is 2 ~ 3 times more expensive than glass. However, if the market grows and economies of scale are achieved on plastic substrates, the cost is expected to drop significantly. It is worth noting that by using thinner plastic substrates, you can take advantage of innovative low-cost processes. Currently, TFT-LCDs and PDPs employing glass substrates are basically difficult to escape from the batch process, in which substrates are sequentially processed one by one. However, when a flexible plastic substrate is used, it can be mass-produced by a continuous roll-to-roll process similar to a newspaper printing method and can be manufactured at a low cost. The technology of forming electrodes by the roll-to-roll process has already been advanced considerably by companies that produce flexible LCDs using plastic substrates. Benchmarking the know-how accumulated here will probably make the process stabilization of the flexible display much earlier.

With the existing advancement of flat panel display technology, light weight, device thickness, power consumptions are no longer the key differentiations of display. The source of future display competition is low cost structure. From this point of view, we must overcome the limitations of the current batch process somehow, and it is considered that flexible display is the best alternative. Here, new form factors such as ergonomically designed curved display monitors in panoramic form are possible. Flexible displays are sure to be a leader in next-generation displays that will follow TFT-LCDs and PDPs. At present, the growth of the market is delayed because the physiology of large companies trying to concentrate on profitable areas in the early stage makes active participation in flexible display difficult. However, research and development centered on academia, venture, and government research institute are steadily progressing, and it is expected that organic thin film transistor will be applied in 2010 and the development of flexible display is also expected to take a turn. As flexible displays are not limited to new core applications, they have the potential to replace the existing flat panel display market, and display companies are expected to reexamine their value as a next-generation revenue source.

5.4 History of electronic paper product development

The development history of the company-specific products related to the electronic paper developed so far is briefly summarized as follows.

- 1975: Sheridon developed a rudimentary Gyricon prototype to implement Xerox's "X" image. Nick Sheridon of Xerox PARC Institute first developed electronic ink using small spinning spheres to develop Gyricon system.
- 1976-77: Hideki Shirakawa, Alan MacDiarmid and Alan Heeger, three chemists first announced conductive plastics.
- 1977: Xerox abandons Gyricon and excludes Sheridon from e-paper projects.
- 1996: Joseph Jacobson joined the MIT Institute of Media Labs to develop electronic paper versions, and in April, Jacobson and two students established E-ink to commercialize electronic paper.

* 1997: Bell Labs from Lucent Technology announced transistor manufacturing from organic polymers.
* 1999: Xerox and 3M announced joint development of electronic paper.
* 1999: E-ink and Lucent announce partnership to create flexible electronic paper using E-ink and organic thin film transistors.
* 2000: E-ink / Lucent jointly demonstrated the first electronic paper using E-ink and flexible organic thin film transistor.
* 2000: Xerox spun off Gyricon to establish Gyricon Media.
* 2002: At SID, Philips exhibited flexible electronic paper models capable of displaying 16 grayscales of text and images. Sony provided a paper type that operated by electrochemical deposition and dissolution of dissolved metal ions (Ag) in solid electrolytes. Iridgim display (US) presented reflective color display using the structure of MEMS, E-ink announced image stability of active microcapsule electrophoretic display and 5 inch color display. Gyricon Media announced Smart PaperTM, which displayed text and images by electrically charging beads made of white and black and rotating them by the electric field in solution. In Stanley Electric (Japan) suggested small powder dispersed in liquid crystal, called MFPD (mobile fine particle display). Zenithal used Bistable LCD and Nemoptic (France) E-book based on reflective LCD. In addition, Seiko-Epson introduced an active matrix electrophoresis device based on ink-jet printing technology. Reflective LCD for E-book, Kent State University (USA) surface-stable cholesteric LCD and IVF (USA) screen-printed electrochemical display.
* 2003: SiPix Imaging Inc. (USA) manufactured passive matrix electrophoretic displays (PMEPDs) and active matrix electrophoretic displays (AMEPDs) using microcups and top sealing technology. E-ink and Philips Research Lab. jointly announced the drive technology for AMEPD. Kyushu and Bridgestone announced 3.1-inch QR-LPD (quick response liquid powder display) with 0.2 ms response speed.
* 2003: Philips (Netherlands) developed reflective display using electrowetting technique.
* 2004: Philips built a prototype production line by 2005 for the production of electronic paper based on E-ink's technology and announced a production plan of 1 million units annually. Matsushita has developed Sigma book called electronic paper

product “1st paper-like display conference” with 7.2-inch diagonal size, 1024 × 768 (XGA), and 16 gray scales using cholesteric LCD licensed from Kent Display and e-paper as its main content Held in Florida, USA.

- 2004: Neo-Lux Co., Ltd. in partnership with E-ink, announced the plan to mass produce flexible display (electronic newspaper). E-ink's electronic ink and film will be supplied.
- 2005: “E-paper” E-ink, a developer of flat panel displays in the US, developed color display using screen technology and exhibited at FPD International held in Japan in October 2005.
- 2005: Ilex Technologies, a subsidiary of Philips Electronics in the Netherlands, developed the E-reader, a wireless capable tablet device based on E-ink's “electronic paper” display technology.
- 2005: Hitachi, JR East Japan Planning Department announced that it will conduct demonstration test of advertising display device using electronic paper for two weeks in December 2005. During the experiment period, news and weather forecasts, a notice of JR same bureau, instructions of Tokyo station are displayed at intervals of about 5 minutes. This display terminal is equipped with 13.1 inch electronic paper, drive, a communication circuit, memory. The electronic paper is a toner electrophoretic (QR-LPD) panel. The total number of pixels is 1024 × 768 (XGA), and the color is white and black. The reflectance is about 40% and the contrast ratio is 10: 1.
- 2006: The specification of E-leader, a wireless connection tablet device announced by iRex Technologies in 2005, named as iliad, is ER0100 and announced its launch in April 2006, It can be used for 3 hours a day for 3 hours a day. It can support various formats such as PDF, XHTML, TXT and MP3. It can display 16 levels of grayscale as 160 dpi, 1024 × 768 pixels, 8.1 inch black and white electronic paper.
- 2006: National Cheng Kung University in Tainan, Taiwan, succeeded in switching E7 liquid crystals containing 1 wt% of methyl red dye using different polarized laser light, liquid crystal and dye between 6 μm spaced ITO glass substrates. A polyimide alignment film was formed on one ITO glass substrate, a polarizer was attached to the other ITO glass substrate, λ/4 wave plate and a partially transparent mirror were attached to the UV cured polymer side. When the linearly polarized light of 532 nm is irradiated on the coating side, the long axis of the dye molecule is oriented perpendicular to the polarization of the laser

beam, which induces the orientation of the liquid crystal molecules, while the other side of the cell keeps the liquid crystals fixed by the orientation film. A 45 degree twisted nematic (TN) phase is created, and the LCD takes on a stable ON state in the light irradiation area. When the circular polarization is irradiated, the OFF state The principles embodied in back.

- 2006: Fujitsu exhibited electronic paper made in the form of an A-5 paper-size e-book reader capable of 7.8 inches, over 4,096 colors, on a combined exhibition of advanced technologies (SEATEC) presenting application to the field of information in curved surface form.
- 2007: Softbank Telecom announced the development of electronic paper terminals using wireless LAN. It is a system linked with Softbank's wireless LAN service "BB mobile point" in cooperation with Hitachi and Fujitsu, which are already developing the electronic paper. This product has already been installed in McDonald's nationwide and is being used for store advertising.

Table 7.16 The representative electronic paper technology in the commercialization stage

method	electrophoresis method		
developer	E-ink	Cyphix Imaging Particle-Based Type	Bridgestone
device example			
Basic structure (All diagrams except Asahi Glass are written by Nikkei Electronics based on manufacturer's data)	As the black particles in the microcapsules rise and fall indicating black & white or contrast microcapsule contains the projection liquid in addition to the particle.	Using the fact that the photosensitive film is placed under the cholesterol film substrate and the resistance changes in the photosensitive film when the light is irradiated, the liquid crystal is connected in series and the input voltage is adjusted accordingly	Black and white particles show up in black and white images or gradients as they move up and down in the air. Bridgestone calls these black-and-white particles a "liquid power", which is different from the particles used in other particle-based electronic paper
reflectivity	35%	40%	about 30%

contrast ratio	10:1	10:1	about 10:1
response ratio	400 ms (when switching to monochrome) 800 ms (midtones)	300 ms	0.2 ms
operating voltage	15 V	two types: 40 V and 15 V	70~80 V
Characteristics and trends	Commercialization, and after-image, as well as devices with improved reflectivity and contrast ratio. The new electronic paper products were first released in 2006. Dot display based on active driving method.	Use plastic substrates for roll to roll manufacturing. In October 2005, it merged with Dopan Holmes. Dodan Holmes sells modules in Japan.	As the particles move in the air to realize a high-resolution dot display, the reaction speed is as fast as about 0.2 ms. Bridgestone will first commercialize monochrome products using glass substrates, and then develop color, ductility and other product models in the future.
developer	Asahi Glass	Fuji Xerox	Fujitsu
device example			
Basic structure (All diagrams except Asahi Glass are written by Nikkei Electronics based on manufacturer's data)	A general method using the properties of cholesterol liquid crystals that selectively reflect light of a certain frequency band. Two liquid crystal films are laminated	A general method using the properties of cholesterol liquid crystals that selectively reflect light of a certain frequency band. Two liquid crystal films are laminated	Three cholesterol films are stacked and each film reflects red, blue, and green to provide a full color display. Use plastic board to secure ductility.
reflectivity	40~50%	30%	30%
contrast ratio	10:1~12:1	4:1 (7:1 under development)	6:1
response ratio	about 20 sec (panel writing hour)	about 1 sec (panel writing hour)	3 sec (QVGA writing hour)
operating voltage	Unpublished	400 V	Unpublished

Characteristics and trends	The entire process from liquid crystal development to final products (display devices used by the public) is handled by itself. Tokyo Metro Yotsuya Station and several other places are currently running tests on the product.	Unlike most electronic paper, this product can be rewritten with light. A final product, including a dedicated rewriter, has already been developed.	Developing applications that can provide full color and flexibility. Plan to supply modules in the form of final products.

Source: Nikkei Electronics Asia, e-paper market grows, 2006.

Examples

1. Describe definition of flexible display and a core technologies for implementing perfect flexible display.

2. Describe basic structure and principle of various modes for implementing E-paper.

3. A flexible substrate is a key factor in implementing a flexible display. Please describe the required properties.

4. The ultimate goal of flexible displays is to implement a roll-to-roll process based on all-organic materials. To do this, print-based processes should be developed. Explain the types and characteristics of the printing process.

5. Think about the unique distinctiveness of flexible display in comparison with existing flat display.

6. Find an idea about the new type of operation mode differentiated from current ones with full awareness of principles of an existing E-paper.

Reference

[1] Lee, Jin Ho et al., "Flexible Display Device", Electronic Communications Trend Analysis Vol. 20, No. 6, 2005.

[2] Koo, J. et al., "Trends of Plastic Electronics Technology", Electronic Communications Trends Vol 22, Issue 5, 2007.

[3] "Flexible Display Industry Technology Roadmap", Korea Technology Foundation Technology Policy Research Center, 2006.

[4] Kwon, Jae-Hong et al., "Trends in Printed Electronics Technology", Weekly Technology Trend, Vol. 1291, 2007.

[5] Jeong, HW et al., "Flexible Organic Device Technology and Printing Continuous Process", Polymer Science and Technology, Vol. 18, No. 3, June 2007.

[6] Weekly Electronic Information "Printing Electronic Device Technology and Industry Trend", EIC, 2007.

[7] Kim, Ki Hyun et al., "The Status of Plastic Substrate for Display", Electronic Communication Trend Analysis, Vol.21, No.5, 2006.

[8] Analysis of the status of promising electronic components - E-paper, Korea Electronics Technology Institute 2007.

[9] Kim, Sung Hyun et al., "Organic Electronic Devices", Electronic Communication Trend Analysis Volume 20, Issue 5, 2005.

[10] JG Kim, "High Performance Organic Thin Film Transistor Using Interfacial Engineering", News & Information for Chemical Engineering, Vol. 22, No. 6, 2004.

[11] Weekly electronic information "E-paper technology trend based on particle technology", EIC, 2005.

[12] B. Comiskey, J. D. Albert, H. Yoshizawa, and J. jacobson, "An electrophoretic ink for all-printed reflective electronic display", Nature 394, 253-255, 1988.

[13] E. Nakamura, H. Kawa, N. Kanae, and H. Yamamoto, "Development of Electrohporetic Display Using Microcapsulated Suspension", SID98 Digest, 1014-1017, 1998

[14] Hwangyo Seon et al., "Status and Prospect of Electronic Paper", Electronic Components, 2006.

[15] Hong Sung Hwa et al., 2005 Future promising commercialization item issues analysis - Flexible Display, Korea Institute of Science and Technology Information, 2005.

[16] Hayes RA, Feenstra BJ, "Video speed electronic paper based on electrowetting", Nature, 425, 383, 2003.

[17] Roques-Carmes T, Hayes RA, Feenstra BJ, Schlangen LJM, "Reflective display pixel based on electrowetting". Journal of Applied Physics 95, 4389, 2004.

[18] Heikenfeld J, Steckl AJ, "Intense switchable fluorescence in light wave coupled electrowetting devices", Applied Physics Letters 86, 011105, 2005.

[19] Park, Eun Soon et al., "E-paper Technology Roadmap", EIC, 2005.

[20] Korea University Display and Nano Systems Laboratory (http://diana.korea.ac.kr).

[21] Electronic Information Center (http://www.eic.re.kr).

[22] Next Generation Information Display Technology Development Project (http://www.display.re.kr).

[23] Korea Institute of Science and Technology Information (http://www.kisti.re.kr).

[24] http://www.liquavista.com

[25] http://www.eink.com

[26] http://www.polymervision.nl

[27] http://www.bridgestone.co.jp

[28] http://www.plasticlogic.com

[29] http://www.sipix.com

[30] http://www.kentdisplays.com

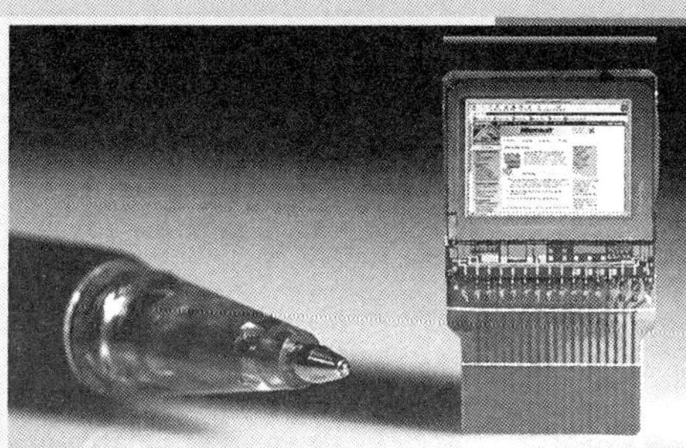

Chapter 07 → 08

Microdisplay

1. What is microdisplay?
2. Types and Operating Principle of microdisplay
3. Technical Trend of microdisplay

1 What is microdisplay?

1.1 Changes of the Information Display Environment

With the rapid development of the information age, a large amount of information is needed and there is a growing demand for electronic displays that display such information. It has evolved to the level of flat panel displays including LCDs and PDPs and flexible, curved displays based on OLEDs from the level of early information displays based on CRTs. Before the large-sized flat display was released in the market, there was a demand for a large-sized screen. Such a demand provided a chance to develop the large-sized flat display called the 'Projection Display'. Such a type of screen is based on the method of enlarging a small display screen by using the optical system. Through such a process, the initial type of microdisplay has been realized. The microdisplay technology, which is combined with the optical system that can enlarge images of such a small-sized display in order to show large-sized images, has been applied to the initial type of large-sized projection TVs and projectors. In recent years, attention has been focused on ultra-small microdisplays such as a head mounted display (HMD) and a micro-projector capable of providing a large screen on a mobile phone.

1.2 Definition & Characteristics of the microdisplay

(1) Definition & Method of the microdisplay

Generally, the microdisplay is a small-sized display with the screen whose size is less than 2 inches. Also, it can be defined as the display whose screen needs to be enlarged through an optical system since the size of the screen is extremely small.

The microdisplay can be used for a rear-projection TV, a front projector, an HMD and the view finder of a camcorder or a digital camera. The typical microdisplay is shown in Figure 8.1. The microdisplay can be generally distinguished in two types, the projection microdisplay and the virtual microdisplay. As shown in Figure 8.2 (a), in the case of the projection-type method, an image or an image of the microdisplay can be enlarged to a large screen of 100 inches for a large number of users. Such a projection type microdisplay panel has a size ranging from 0.7 to 2 inches, and a projection type such as a rear projection TV or a projector has a magnification of 100 to 200 based on the degree of the optical system.

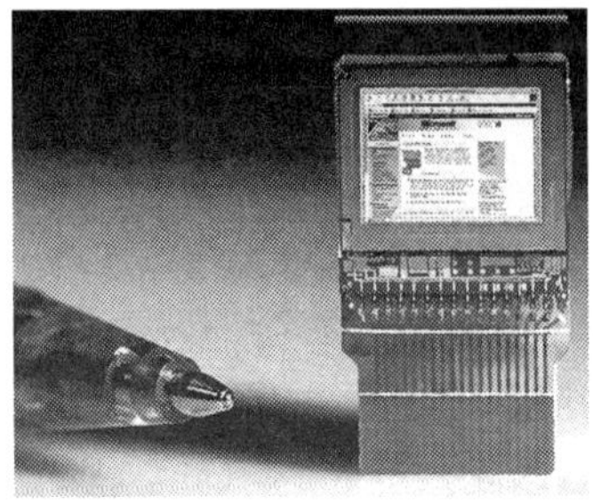

Figure 8.1

Flexible display.

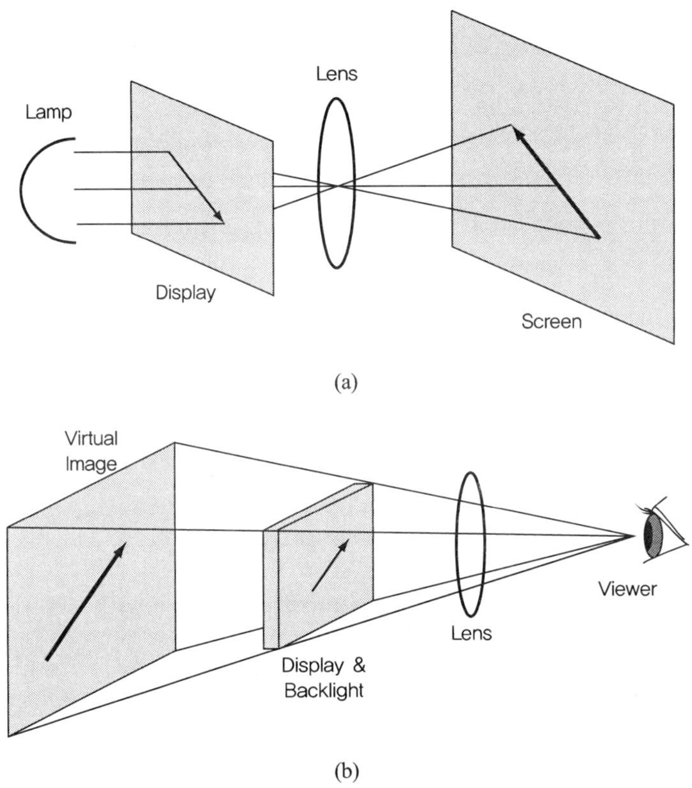

Figure 8.2

Types of the microdisplay (a) projection, (b) virtual.

Also, the virtual microdisplay shown in Figure 8.2 (b) is the method of directly looking at the image or video clip enlarged through the optical system. It is possible to realize the virtual size of the display panel up to 50 inches based on the magnification of the optical system. In virtual microdisplay, the size of the panel is usually between

0.4 inch and 1 inch. In general, the panel size is limited, so it is necessary to implement a high resolution pixel density in a small area. The size of the panel will also vary depending on the application. The projection microdisplay used for a projection TV or a projector can realize HD–class high resolution, while the virtual microdisplay used for an HMD or an Eye Glasses Display (EGD) can currently realize QHD–class resolution.

Table 8.1 Comparison between types of the projection and virtual.

Type	Projection microdisplay	Virtual microdisplay
Size of screen(image)	~ more than 100 inch	~ 50 inch
Size of panel	0.7 - 2 inch	0.4 - 1 inch
Resolution	SVGA, XGA, SXGA, UXGA, HD	QVGA, VGA, SVGA, HD, QHD
Application	Projection TV, Projector	HMD, Digital camera, Camcorder, EGD, etc.

(2) Characteristics of the microdisplay

As mentioned above, microdisplay has been developed in an effort to solve the problems that arise in the process of developing a direct view flat display, especially the very sensitive factors such as screen size, volume and price. The demand for a low–cost large–screen display or a small–sized display for a large–screen display can be solved by reflecting two important key features of a microdisplay. It is possible to substitute high–priced direct–view type flat panel display such as LCD and PDP with low–cost projection type of microdisplay system. Light–weight, extremely small sized and portable micro projector can also offer high quality of video information.

2 Types and Operating Principle of microdisplay

2.1 Types of the microdisplay

Similar to direct–view type flat panel displays such as LCD and PDP, micro–displays can be classified into reflective, transmissive, and emissive according to their operation modes. In the case of the reflection type microdisplay, the light source is present in front of the microdisplay panel in a manner of modulating the light incident from the outside while changing the characteristics of the reflection surface, thereby reflecting the image. In the case of the transmission type microdisplay, The light is modulated while passing through the panel where the light source is behind the display. In the case of a self–emission type microdisplay, a light source is generated by itself and a separate light source is unnecessary.

(1) Reflective microdisplay

The representative technologies used for the reflective microdisplay include the one of the MEMS (Micro Electro Mechanical Systems)–type and the other of the LCoS (Liquid Crystal on Silicon)–type. The LCoS type microdisplay is composed of liquid crystal thin film on the silicon CMOS chip.

1) MEMS microdisplay

MEMS microdisplay refers to a display based on MEMS devices fabricated from ultra–small electrical and mechanical components using semiconductor processes. A small–sized MEMS device is a part of a pixel in a display, and a reflection type display is realized by modulating the size or phase of light incident from the outside by adjusting pixels by an electric and mechanical method. The MEMS microdisplays based on such technologies include DMD (Digital Micromirror Device) of TI (Texas Instruments) and GLV (Grating Light Valve) of SLM (Silicon Light Machines).

a) DMD Technology of TI

TI has combined DMD technology and DLP (Digital Light Processing), which converts digital images into digital light pulses to recognize them as colored analog images, resulting in a high–resolution, large–screen display. The structure of the DMD pixel fabricated by the MEMS process is shown in Figure 8.3.The DMD pixels are manufactured with the minute 4–level structure made on the top of the memory cell (CMOS SRAM) part through the micro semiconductor process. In other words, the structure is made up of the mirror layer, the yoke and hinge layer, the metal 3–layer and the CMOS SRAM layer. The mirror layer plays the role of reflecting the light and is strongly connected to the yoke layer located below. The yoke layer is connected to two thin torsion hinge layers to support the posts formed in the lower layer. The york and hinge layer plays the role of changing the signals from the SRAM as the slope of the mirror. By creating the electrostatic torque based on the voltage (ON/OFF) from the SRAM, the torsion hinge part can be twisted with the york part tilted by +/−10 degrees. During the tilting process, the movement shows the superhigh speed of mere 15 μsec. When the yoke is tilted, the 3-layer of metal becomes the landing point of the yoke and supports the torsion hinge. As shown in Figure 8.4, the two DMD pixels including the memory cell can be indicated in three dimensions.

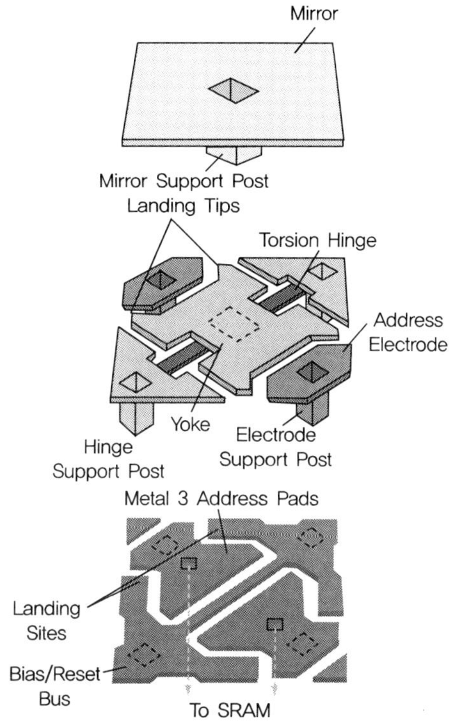

Figure 8.3

Structure of the DMD pixel.

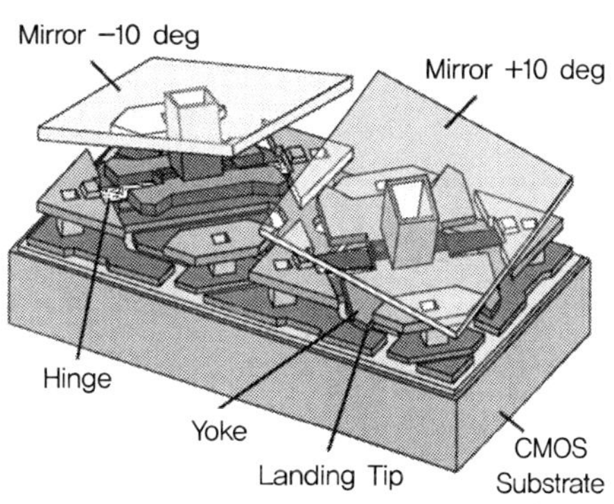

Figure 8.4

Two DMD pixels.

In Figure 8.4, the mirror made of aluminum film plays the role of the light switch which is operated according to the signals from the memory cell. When the memory cell is in the state of '1', the mirror is rotated to the direction of +10 degrees. When the memory cell is in the state of '0', it is rotated to the direction of −10 degrees. As a result, by combining the subminiature DMD pixel array which has the function of the optical switch with the optical source and the optical system, a large-screen display can be materialized. By rotating the mirror at the superhigh speed and switching the light in order to reflect it towards the inside or outside of the projection lens, the light can be changed to the signals of the digital optical pulse which can be recognized as an analog image by the eyes of the viewer. Such a technology which can make the light recognized by the eyes of the viewer is called the binary pulse width modulation. It is one of the tone-indicating technologies widely used for displays. In order to materialize the DLP projection system based on the DMD, it is possible to consider three methods using the single-chip, the two-chip and the three-chip. Such three methods are closely related to such factors as the price, the optical efficiency, the amount of electricity used and the weight and volume of the system, so the choice can be made based on the use. Figure 8.5 shows one example of the single-chip DLP projection system.

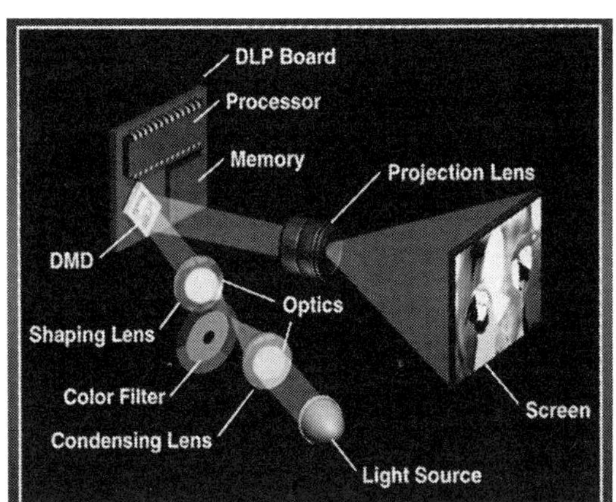

Figure 8.5
Projection system of the Single-chip DLP (Reference source: T1 Co.).

Unlike the LCD-based method, the DMD-based method is the reflective type whose panel is composed of millions of DMD (Digital Micromirror Device). The light coming from the optical source is changed to the one of RGB through the color filter wheel, which is reflected on the lens by one DMD in order to materialize images or video clips. Therefore, there is only one color for the immediately-formed images.

However, since the responsive speed of the eyes is slow, three images are combined and indicated as one full-color image.

Figure 8.6
Samsung World's largest DLP TV - 71 inch 1080p in 2005.

For the materialization of the projection system, the DMD-based projection display which uses one chip is small and light. However, the resolution, the DLP projection system which uses three chips is the best. Because of such a reason, it is quite expensive. In commercialization, the color breakup which was a problem of the initial DLP technology has been solved with the frame-rate high-speed technology. According to the development of the process technology, the mirror pitch of 13.6 μm has been reduced to 10.8 μm, which is 1080p HD level. Figure 8.6 shows the 71-inch 1080p DLP TV released by Samsung Electronics, which showed the biggest size in the world in 2005.

(b) GLV Technology of SLM

The GLV (Grating Light Valve) technology was designed by Stanford University in the early 1990s and provided the basis for the establishment of SLM (Silicon Light Machines) in 1994. SLM was absorbed as a subsidiary by Cypress Semiconductor in 2000. Since then, it has developed various elements based on the GLV technology. As shown in Figure 8.7, the GLV technology starts from one pixel composed of six ribbons.

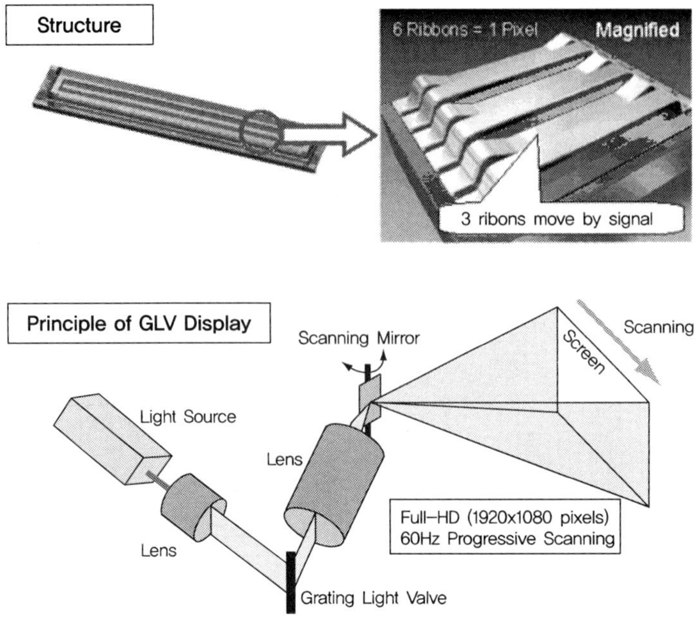

Figure 8.7
Principle and pixel structure in GLV projection displays.

An aluminum film having excellent reflectivity is coated on a silicon nitride film to make a ribbon. The size of the ribbon is 2~4 μm in width, 100~300 μm in length, and 100 nm in thickness. By supporting both ends of the ribbon, the 650 nm air gap is made. By arranging six ribbons made in such a way in parallel, one pixel is completed as shown in the upper picture of Figure 8.7 .The principle of the operation of the pixel is that by applying an electric field between the ribbon and the electrode at the lower part of the air gap, the ribbon is pulled towards the lower part at the high speed of 20 nsec by the electrostatic force. The GLV projection system which is based on the pixels composed of the six ribbons shows the linear array structure, not the two-dimensional pixel array. It is possible to inject images due to the rapid switching speed. As shown in Figure 8.7, SLM demonstrated the full HD-class projection TV system by using red, green and blue laser optical sources.

(c) LCoS microdisplay

Other than the transmissive LCoS methods of Canon and Kopin, most LCoS microdisplays show the reflective method. The basic structure of the LCoS microdisplay is shown in Figure 8.8.

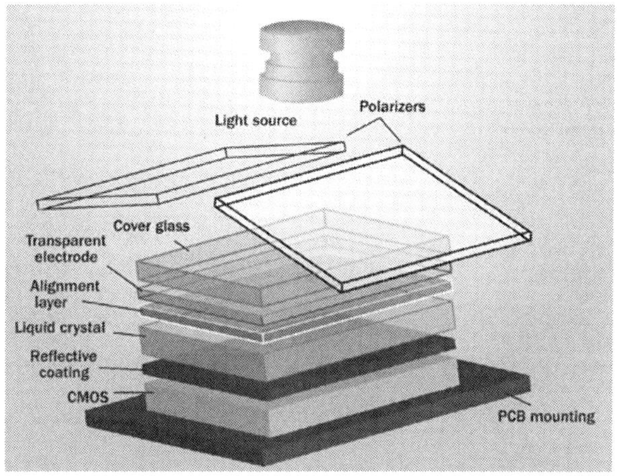

Figure 8.8
Structure of the LCoS microdisplay.

The LCoS microdisplay is similar to the TFT−LCD, which is the direct−viewing display since it uses the light−modulating characteristic of the liquid crystal. In the TFT−LCD, there is a liquid crystal layer between two glass plates, which is controlled by the TFT (Thin Film Transistor) formed in each pixel. In the LCoS microdisplay, a silicon plate is used instead of a glass plate on one end where the TFT is formed. For the purpose of controlling pixels, the CMOS chip with the single−crystal TFT element plays the role of the operating element.

The LCoS microdisplay is similar to the TFT−LCD, which is the direct−viewing display since it uses the light−modulating characteristic of the liquid crystal. In the TFT−LCD, there is a liquid crystal layer between two glass plates, which is controlled by the TFT (Thin Film Transistor) formed in each pixel. In the LCoS microdisplay, a silicon plate is used instead of a glass plate on one end where the TFT is formed. For the purpose of controlling pixels, the CMOS chip with the single−crystal TFT element plays the role of the operating element. The reflective film, the liquid crystal, the alignment film, the glass plate and the polarizing plate are located on the CMOS chip. The LCoS display made of such components is operated based on the modulation of the retardation of the entering polarized light. By simply looking at the modulation process in the TN (Twisted Nematic) liquid crystal structure, the entering polarized light is twisted and made to penetrate the display from the front side to the back side. Then, the light is reflected on the electrode located on the back side and then it is twisted again and made to penetrate the display. The second polarizer serves to allow twisted light to pass through the polarizer at right angles. By applying a voltage to each pixel location, it is possible to control the twisting angle of the liquid crystal. When the maximum voltage is applied, the polarization of the light is rotated and located on the analyzing polarizing plate at a right angle, making the passage of any light impossible. By applying a medium level of voltage, it is possible to rotate only one part of the light and enable the passage of the light with a medium level of brightness. In other words, in the LCoS structure including the liquid crystal, the liquid crystal plays the role of a valve which is used to control the amount of light reaching the surface of the reflective film. Thus, when more voltage is applied to the liquid crystal of a certain pixel, a larger amount of light is allowed to pass through. Meanwhile, as a method of materializing the color LCoS microdisplay, it is possible to use three LCoS chips and mirrors in order to produce Red, Green and Blue (refer to Figure 8.9) or use one LCoS chip and one color filter wheel in order to continuously materialize colors.

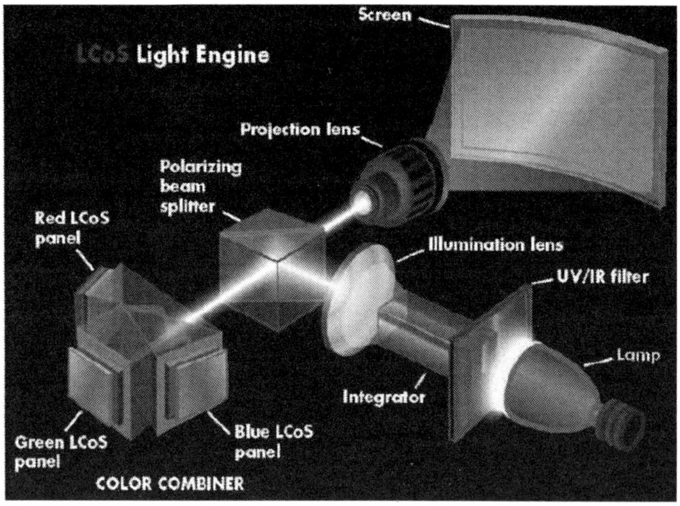

Figure 8.9
Structure of the LCoS microdisplay.

As a type of the LCoS technology, Sony has developed the SXRD (Silicon X-tal Reflective Display). The main characteristic is that the contrast ratio is 3000:1, which is quite high with the pixel density of about 2.4 times compared to the high-temperature polycrystal Si liquid-crystal panel. In addition, the planarization technology of the silicon driving device and the uniform orientation film-forming technology for a uniform orientation film have been developed to materialize the liquid crystal with the cell thickness of less than 2 μm. The specific structure and characteristics of the SXRD technology are shown in Figure 8.10. Also, examples of the chip with the application of the SXRD technology and the SXRD rear projection TV are shown in Figure 8.11.

JVC (Victor Company of Japan, Ltd.) has developed the LCoS technology called the D-ILA (Direct-Drive Image Light Amplifier). The company has developed the 1.75-inch 8k x 4k D-ILA device for the front projector with the resolution of 8192 x 4320 by using the D-ILA technology (refer to Figure 8.12). The screen ratio is 17:9 and the pixel pitch is 4.8 μm, while the inter-pixel gap is 0.24 μm and the aperture ratio is more than 90%. By adopting a confusion reduction technique by the orientation of the liquid crystal, scattering and diffraction of light on the reflection surface is suppressed to achieve a contrast of 20000:1, and the response speed is about 4.5 msec. It completely corresponds to 7,680×4,320 of "Super High-vision" which is pushed ahead with as the next-generation broadcasting technology of NHK. Also, it is possible to materialize high-definition video images which are 17 times better than the full HD images and four times better than the 4k digital cinema. The 1.75-inch 8k×4k D-ILA device is the first one of its kind in the world in terms of the realization of the 8k/4k resolution with a single plate.

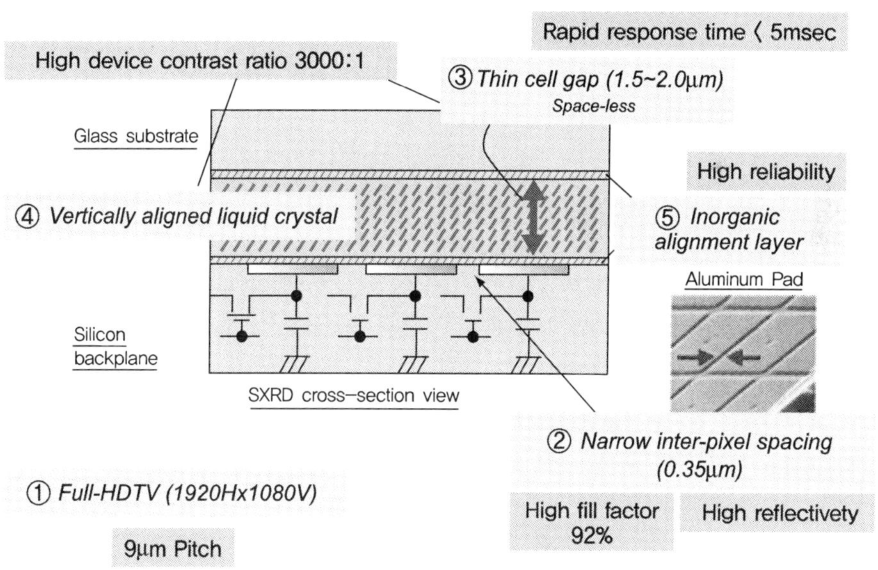

Figure 8.10
Cross section and main characteristics of the Sony's SXED.

(a) (b)

Figure 8.11
Sony's (a) 0.78 inch SXRD chip, (b) 70 inch SXRD 1080p rear projection TV.

Figure 8.12
JVC's 1.75 inch 8k×4k D-ILA device.

(2) Penetrative microdisplay

Transmissive microdisplays are mostly based on liquid crystal displays (LCDs). An AM (Active Matrix)–LCD using TFT (Thin Film Transistor) is the most widely used in order to satisfy the high performance requirement of the display defined by the number of pixels, color purity, and response speed. Such a TFT is fabricated on polycrystalline polysilicon or monocrystalline silicon.

1) Poly-crystal Silicon TFT-LCD microdisplay

The microdisplay based on the poly-crystal silicon TFT-LCD, the TFT is manufactured through a high-temperature process. Therefore, the quartz which can endure a high temperature is used as a plate. Compared to the amorphous silicon TFT which is used as the driving element of the TFT-LCD, the poly-crystal silicon TFT makes it possible to obtain a high level of mobility and a high level of pixel density. In the penetrative LCD microdisplay, the optical source is located in the back side of the display. Light is made to penetrate the LCD microdisplay and modulated in each pixel for the materialization of pictures or video images. In Japan, there are major projector companies such as Seiko-Epson and Sony that produce microdisplays based on the poly-crystal silicon TFT. For a projector, the display size is at least 1 inch, and for virtual types it is between 0.5 inch and 1 inch. In Figure 8.13, the structure of the projector using the poly-crystal silicon TFT-LCD microdisplay is shown. As shown in the figure, the LCD microdisplay either enables the passage of light separated in R, G and B away from the optical source or blocks light, modulating light and creating pictures or video images on the screen.

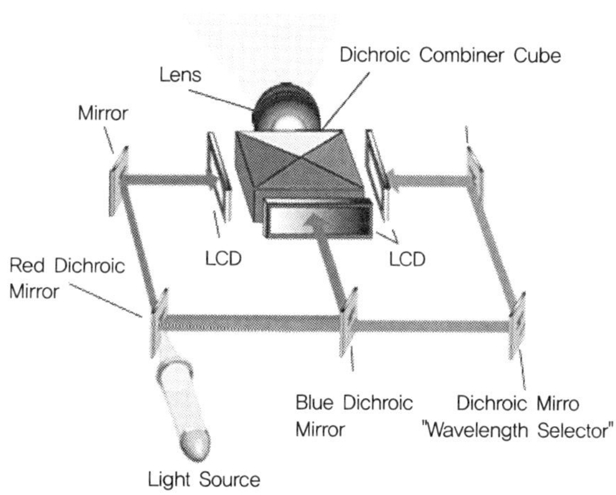

Figure 8.13
Projector structure using the LCD microdisplay.

(a) Single-crystal Silicon TFT-LCD microdisplay

Compared to the poly-crystal silicon TFT, the single-crystal silicon TFT has a relatively high level of mobility and pixel density, and high-quality integrated circuits can be easily configured on the panel. This has advantages in terms of reliability and price competitiveness. Also, since circuits are made on an opaque plate such as silicon, it is general to materialize the display mainly according to the reflective method. In the Cyberdisplay LCD, the patented lift-off process for transferring a very thin IC layer to a glass substrate is used. In this way, a monocrystalline silicon circuit on a silicon substrate is transferred onto a transparent glass substrate to enable a penetrative microdisplay. The cross-sectional view of Cyberdisplay LCD using a color filter is shown in Figure 8.14. The basic structure is similar to the one of the TFT-LCD.

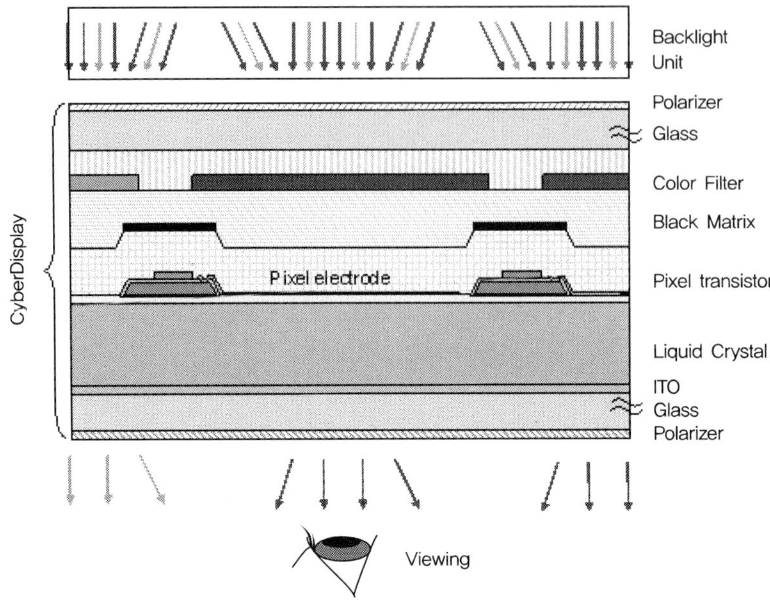

Figure 8.14
LCD cross section of the Kopin Co's Cyberdisplay

In the penetrative microdisplay based on Cyberdisplay LCD, the single-crystal silicon TFT is used. Therefore, the pixel size of the penetrative microdisplay is much smaller than the one of an LCD TV or a cellular phone using the amorphous or poly-crystal silicon TFT. As a result, it has an advantage like resolution and is mainly used for such virtual mobile displays as the HMD and the wearable display. The products of Kopin are produced in various kinds ranging from the 0.29-inch WQVGA class to the 0.97-inch SXGA class, and mainly applied to camcorder viewfinder, personal computers, games and mobile GPS devices.

(3) Emissive microdisplay

In the emissive microdisplay, it is unnecessary to have an external optical source. Therefore, it is possible to establish the simplest microdisplay. Until now, various emissive elements have been tried for microdisplays.

The displays with emissive characteristics include the inorganic EL, OLED, FED, VFD, LED and QLED. The displays which have been commercialized until now include the OLED-based, the VFD-based and the LED-based displays. The displays which have been grafted on microdisplays include the low-molecular and the high−molecular OLED displays. eMagin of the United States and MED (MicroEmissive Displays) of the Britain have developed the low−molecular and the high−molecular OLED microdisplays, respectively, and they have been carrying out the commercialization process for each type of microdisplay.

The major component of the P(Polymer)−OLED microdisplay of MED is the P−OLED structure shown in Figure 8.15.

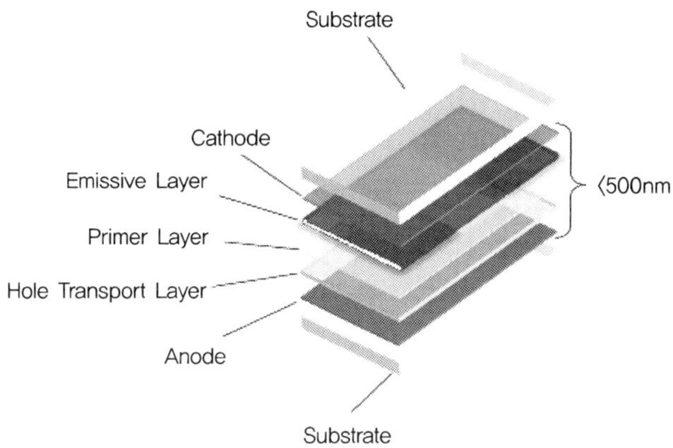

Figure 8.15
Basic structure of the P−OLED.

The basic structure of the P−OLED consists of two electrodes, the HTL (Hole Transport Layer), the emissive layer, the primer layer and the plate. Such a structure is same as the one of the high−molecular OLED display.

In order to fabricate a microdisplay, an anode electrode, a polymer EL layer, and a transparent cathode electrode are deposited on a CMOS chip having a built−in circuit by a semiconductor process with a thickness of about 500 nm. A protective P−OLED microdisplay as shown in Figure 8.16 is produced when the thin film encapsulation process is performed on the front side and the glass substrate with the color filter is attached. The "eye screen" ME3024 of MED shown in Figure 8.16 reflects the picture of the 0.24−inch commercialized product sold with the QVGA resolution. By combining the product with the optical system including a lens, it is possible to complete a virtual microdisplay like the HMD.

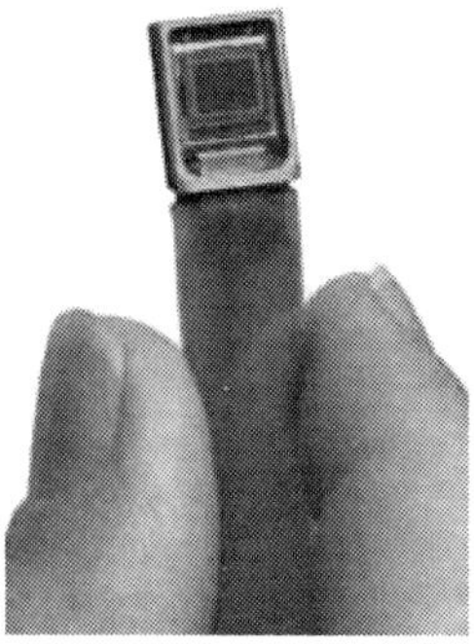

Figure 8.16
MED's P–OLED microdisplay– "eye screen" ME3024.

Meanwhile, eMagin which is using the low–molecular organic EL for the microdisplay panel has been developing the technology to manufacture a microdisplay panel after receiving the patent of the low–molecular organic EL from Eastman Kodak. As shown in Figure 8.17, the manufacturing process consists of the flattened insulator film, the anode, the white low–molecular EL, the transparent cathode, the transparent seal layer, the black matrix, the color filter, the transparent protective film and the reflection preventative film, all of which are metalized or beamed on the single–crystal silicon CMOS circuit plate in order.

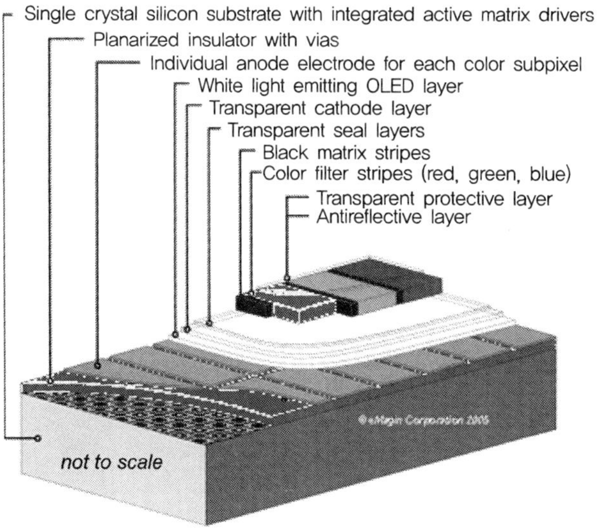

Figure 8.17
eMagin's low molecular weight OLED structure.

As shown in Figure 8.18, the OLED microdisplay made in such a way is combined with the optical system containing a non-linear lens and made into a virtual microdisplay like the HMD or the NTE display.

eMagin is manufacturing and selling the 0.59-inch SVGA(800×600) and the 0.61-inch SVGA+(852×600) microdisplays.

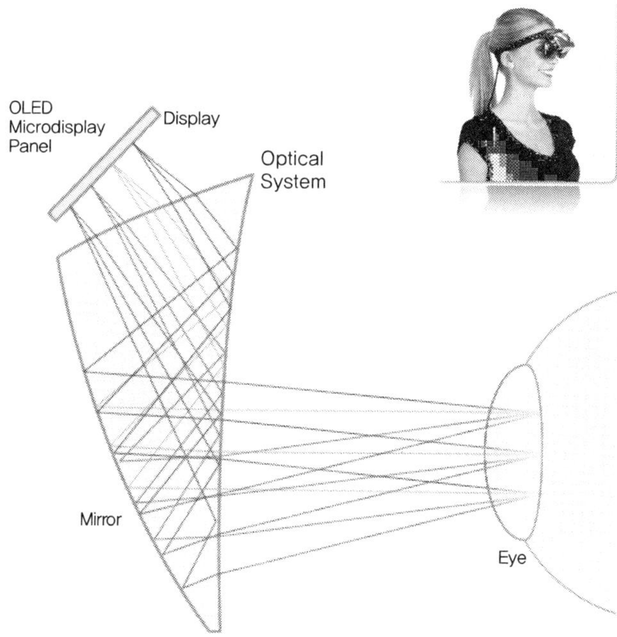

Figure 8.18
One example of microdisplay and 0.5 inch reflective LCD display with screen resolution 800×600.

2.2 Applications of the microdisplay

The applications of the microdisplay can be classified into various categories based on how to use the display. According to different applications, there can be two large categories. The first one is the application based on the projection method, while the second one is the application based on the virtual image method. The projection method is currently being used in the widest range of applications. It is possible to see the method being mainly used for such electronic devices as displays and projectors. Meanwhile, the virtual image method is being used in such limited applications as the camcorder viewfinder and the HMD (Head Mounted Display).

The technologies to materialize the projection display range from the initial penetrative CRT (Cathode Ray Tube) technology to the current penetrative LCD,

DMD and LCoS. The projection display using this technique has attracted a great deal of popularity due to its advantage in large screen, weight, and volume while the conventional CRT had the limitation of the weight, volume, and screen size. As the OLED and QLED which are light-emitting diode are released, the projection microdisplay had to compete with the current light-emitting diode. As a result, it is now expected that it will be more difficult for the projection microdisplay to occupy the market. Figure 8.19 shows the projection TV product based on various technologies.

The high-temperature poly-crystal silicon-based projection LCD has been mainly used in the projector until now. However, the DMD-based DLP (Digital Light Processing) projector has recently increased its share in the market.

(a) (b) (c)

Figure 8.19

Projection TV (a) LCD method, (b) DLP method, (c) LCoS method.

As the applications of the virtual microdisplay, the HMD (Head Mounted Display) and the EGD (Eye Glass Mounted Display) have stood out the media of providing various kinds of information together with the rapid development of the data communication. Since the sizes of the video images shown on such mobile devices as cellular phones and PDA phones are extremely limited, there are often various situations in which the amount of information required by the user is provided in a limited way. As a solution to such a problem, the HMD providing virtual reality has been developed and commercialized. Also, the EGD which is used for the wearable PC can be regarded as one of the major applications attracting popularity among young people including teenagers. Figure 8.20 shows the examples of various applications for the virtual microdisplay.

Figure 8.20
Variety of applications of the virtual microdisplay.

It is expected that there will be much more applications based on the aspect that the virtual microdisplay is to be gradually used for the wearable display which is one of the future displays. In particular, there will be preferential applications in such entertainment fields as the mobile phone, the portable game console, the PMP and the DVD. Also, the virtual microdisplay will make a lot of contributions to the effective performances in such fields as the military, the education, the industry and the medical treatment by transferring and expressing information in real time.

3 Technical Trend of microdisplay

3.1 Domestic and Foreign Trends for the Development of the microdisplay Technology

Several companies including Sony and Epson are dominating the market for the high-temperature poly-crystal silicon-based LCD display. In the future, it is expected that there will be great changes in the market due to the rapid development of the MEMS-based and the LCoS-based microdisplays. The DMD-based microdisplay has been greatly adapted to products such as the projection TV and the projector. Also, researches on the subminiature projector have been continuously carricd out. Meanwhile, LCoS-based micro-displays such as SXRD and D-ILA are being applied to products such as projection TVs and projectors, and much research is underway to develop technologies for implementing very large displays. Recently, as the demand for the mini projector has increased, many types of research have been conducted by various companies and research institutes, and they are continuously receiving interest such as exhibiting prototypes at various exhibitions and conferences.

(1) Domestic Current State for the microdisplay

In Korea, large enterprises such as Samsung Electronics, Samsung Electrics and LG Electronics and public research institutions such as ETRI and KETI have carried out research. In particular, the level of interest for the development of the mini projector has increased (refer to Table 8.2). The development of such products as the green laser and the MEMS scanner, which are the core elements for the development of the mini projector, has shown a great progress. Meanwhile, researches regarding HMD technology show considerable improvements. HMD technology has been improved with Samsung's Gear VR system which works with smartphones.

Table 8.2 Microdisplay research development trend.

Corporation	Product	Characteristic	Note
Samsung Electro-mechanics	Pico Projector (S beam)	• VWGA 800x480, RGB LED	
Samsung Electro-mechanics	Gear VR	• QHD 2560x1440, Utility of Smartphone	
LG Electronics	Mini Projector	• HD 1280x720, LED	

(2) Domestic and Foreign Current State for the Development of the microdisplay

In the foreign countries, the companies possessing their own technologies have initiatively led R&D. In particular, those companies or research institutions in the United State, Japan and Germany have actively carried out researches consistently. In Japan, large enterprises have carried out a lot of research on the mini projector and the large-sized projection display. In the United States, R&D has been carried out for the DMD-based technology of TI, the virtual low-molecular OLED microdisplay technology of eMagin and the MEMS scanning mirror technology of

Microvision. Also, in Europe, researches have been continuously carried out for the MEMS scanner technology of Fraunhofer Institute, the high-molecular OLED microdisplay technology of MED and the mini projector technology of Simens.

Table 8.3 Japan's microdisplay research development trend.

Corporation	Application	Content	Note
Canon	Mini Projector	• Development of the resolution SVGA small size device • Development of the MEMS scanner	
Sony	GLV TV	• Using the laser optical sources • 3000:1 high contrast	
Epson	Mini Projector	• Application of the transmissive LCD panel • high brightness LED	
Mitsubishi	Mini Projector	• SVGA DMD panel • Lumileds LED	
NEC	Mini Projector	• Rehearsal of the Future work environment connected Mini projector and computer	
JVC	Projection Display	• Technical development of D-ILA	

Table 8.4 USA's microdisplay research development trend.

Corporation	Application	Content	Note
Texas Instruments	DMD	• Development of the DMD/DLP system-based MEMS • Feature size 12 μm pixel	
Symbol	Mini projector	• Application the Barcode scanner to mini projector	
Microvision	Mini Projector Engine	• Development of the MEMS scanner • Commercialization of the HMD/HUD • PicoP: full color prototype exhibit (2008 SID)	
eMagin	HMD	• Development of the SVGA and SVGA+ low molecular weight OLED microdisplay panel	
Displaytech	Pico-projector	• Development of the FLCoS (Ferroelectric Liquid Crystal on Silicon) technology • Development of the VGA, WVGA, and SVGA FLCoS panel	

Table 8.5 Europe's microdisplay research development trend.

Corporation	Application	Content	Note
Fraunhofer Institute	Mini projector	• Development of the MEMS scanner • Development of the OLED projector • Development of the OLED on CMOS microdisplay • Development of the OLED microdisplay embed the Photosensor	
Simens	Mini projector	• Cyan monochrome • Application of the UV LD and phosphor screen	
MicroEmissive Displays	HMD	• Development of the polymer OLED microdisplay panel • Development and commercialization of the 0.6-inch QVGA OLED panel	
Fourth Dimension Displays	NTE	• Development and commercialization of the 0.88-inch XGA panel-based FLCoS • Fast switching liquid crystal (40μsec) • 24-bit Native (8-bit per color) • High fill factor >93%	
Holoeye Photonics AG	Hologram projector	• Development of the real time hologram projector • LCoS panel + laser optical source + hologram S/W combination • LCoS microdisplay size from 0.55" to 0.7" • Resolution:1280×768 (WXGA), 1920×1200 (WUXGA) • Frame rate: 60 Hz to 180 Hz	
Visitech	Micro-projector	• Development of the micro projector-based DLP	

3.2 Future Prospect for the microdisplay

It was expected that the projection TV would be less competitive in the large-sized TV market due to the rapid growth of the LCD and the PDP, which are the leaders of the flat display. However, it has continuously grown mainly in the North American market. The projection TV which is subject to the application of the microdisplay technology did not depend on the performance much in the early period since its screen was greater than that of the CRT TV. However, as the performance of the projection TV has been gradually improved with more competitiveness, its share in the market has increased. Also, as the digital broadcasting has become regularized, it is expected that the consumer's demand for the high-quality microdisplay-based HD projection TV will increase. Compared to the LCD-based projection display, the DMD-based projection TV showing a greater level of resolution and accessibility towards the digital broadcasting has gradually attracted a lot of attention in the market. It is expected that the DMD-based projection display will continue to grow in the future due to its relatively cheaper price compared to the LCD TV or the PDP TV as well as its excellent resolution and small size. Currently, the DMD-based and the LCD-based methods are fiercely competing with the projection TV in the projector field. It is expected that the trend will progressively move towards the DMD-based method in the future. Samsung Electronics and LG Electronics are expanding their dominance as market leaders in North America and Europe. The digital DMD-based projection TV has made consumers strongly recognize it as the display with an excellent kind of resolution through various products of many companies together with the aggressive marketing of TV. These market changes have led to the early introduction of three-plate LCD-based projection TVs to LCD-based projection display makers such as EPSON, Fujitsu, Hitachi, Matsushita, Sanyo and Sony. In Japan, many companies have shown their effort to release the 3-LCD-based projection TV. While the DMD-based projection display and the LCD-based projection display are fiercely competing with each other in the projection TV market, the LCoS-based projection display products have been gradually released. As the LCoS-based projection display has entered the projection TV market, it seems that the trend for high quality and low prices will be accelerated more. Also, due to the investment competition carried out by the LCD and PDP TV companies, the price of the direct-viewing flat display has rapidly decreased. In addition to the rapid decrease of the price of the flat display, the performance of such a display has continuously increased, which could definitely become a threat to the projection display in the future. In order to survive in the fierce future competition, the projection display must be equipped with high resolution and a low price. In addition to the rapid technological development, it is required to establish a competitive product group by finding difficult application fields which cannot be easily materialized. The product group subject to such competitiveness is the subminiature nano or pico projector. Second, it is expected to have a super-sized high-resolution projection display. In addition to the technological development of the microdisplay, it is expected that we could see the subminiature and the super-sized projection displays which cannot be materialized with the current display technology in the near future.

Also, the market for the HMD and the EGD, which are the application fields of the virtual microdisplay, is still in the early stage, but it is expected that the market will rapidly become diverse based on the trend of the rapid information society. The level of demand regarding mobile displays for personal purposes has increased. Due to the expansion of leisure time, it seems that there will be a lot of fields including entertainment fields such as games and movies, medical treatment, industry and military in which it is possible to increase additional values for special purposes. It seems that the virtual microdisplay will be essentially applied to the wearable PC or the virtual reality system which could lead to the realization of the virtual reality in the future.

Exercise

1. Discuss and compare the advantages and disadvantages of projection microdisplay and virtual microdisplay.

2. Compare and discuss the differences between characteristics of Kopin Co.'s virtual LCD microdisplay and Reflective LCD microdisplay.

3. List up the major components of microdisplay. Explain each component.

4. Explain the principle of FLCoS.

5. Discuss the techniques of JVC's D-ILA and the possibility of application of large projection displays.

6. Compare and explain the differences between LCD, LCoS and DLP about basic technology of the projection microdisplay.

7. Explain reasons why the commercialization of microdisplay-based LCoS was late compared to microdisplay-based LCD or DMD.

Reference

[1] David Armitage, microdisplays move outside the box, SPIE's OEMagazine, PP. 28-30, 2002.11

[2] David Armitage, Ian Underwood, Shin-Tson Wu, Intoduction to microdisplay, Wiley, 2006. 11

[3] Byung Seong Bae, microdisplay : High temperature polymer TFT-LCD Iljin Display Co., Ltd

[4] Hyeon Chan Mun, Micro-Projection evolutionary technological & market tendency(I), KETI research report, electronic parts researcher, 2007. 7

[5] Byeong-Gwon Ju, MPD(Micro Projection Display) industrial trend, KETI research report, electronic parts researcher, 2006. 3

[6] microdisplay Web Page(http://tfcg.elis.ugent.be/microdis/)

[7] Kopin Co. Web Page(http://www.kopin.com)

[8] TI Co. Web Page(http://www.dlp.com)

[9] eMagin Co. Web Page(http://www.emagin.com)

[10] SLM Co. Web Page(http://www.siliconlight.com)

[11] MED Co. Web Page(http://www.microemissive.com)

[12] SLM Co. Web Page(http://www.siliconlight.com)

[13] Sony Co.'s SXRD page(http://pro.sony.com/bbsccms/ext/SXRD/index.shtml)

[14] JVC Co.'s D-ILA page(http://www.victor.co.jp/english/pro/dila/feature.html)

Plasma Display

Chapter 08 → 09

1. Definition of Plasma
2. Formation of the Plasma
3. Electron Emission
4. Dielectric breakdown of the gas
5. Overview and history of the PDP
6. Principle and type of PDP
7. PDP manufacturing technique
8. Driving principle of PDP
9. Technology development trend of PDP

1 Definition of Plasma

1.1 Basic Concept of Plasma

When the energy is applied to the gas particles (energy transfer due to the impact of the accelerated electrons, thermal transfer or microwave), the valence electron leaves the outermost electron orbit becoming a free electron and makes an atom to have a positive charge. These positively charged ions and negatively charged electrons recombine to become neutral, it emits a unique light from ionized gas or plasma. Depending upon the way the electrons are accelerated or the condition such as electron density or temperature at the time when the plasma is being formed, various kinds of plasma are formed. Figure 9.1 shows the status of plasma according to electron density and temperature. A glow discharge area located at the center of the picture is mainly utilized for the display applications.

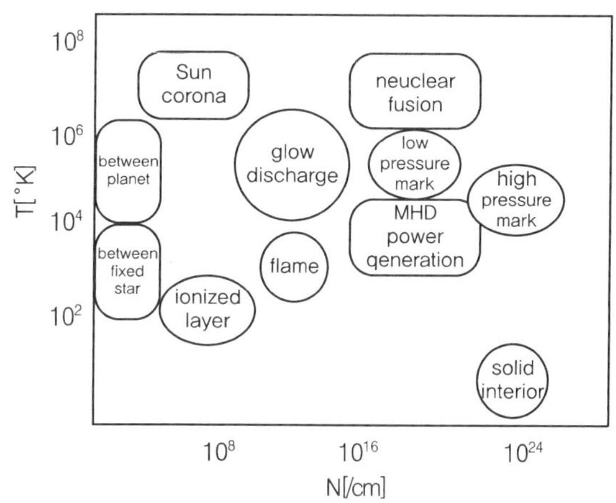

Figure 9.1
States of plasma according to the density of electrons and temperature.

1.2 Formation and Structure of Plasma

Plasma can be distinguished by three states as shown in Figure 9.2. (a) The electrons have a weak kinetic energy and this energy does not change when colliding with an atom or a molecule (repulsion). However, as shown in (b), accelerated free electrons collide with atoms and repel the outermost electron so an ion and a free electron can be created (ionization). This newly formed electron and the accelerated electron are accelerated again and increase the ionization by multiple−collisions. Meanwhile, as shown in (c) the free electrons or ions which don't have enough energy to ionize other atoms, but they energize the collided atoms to excite the outermost electrons. The excited electron comes back to the stable orbit after emitting the light energy which is the same as the interval between the specific energy gap (excitation).

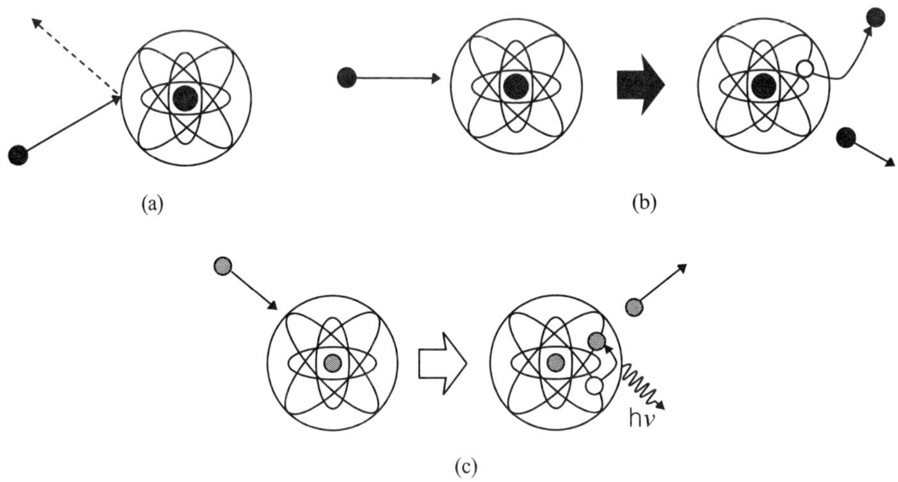

Figure 9.2
Formation of Plasma (a) Repulsion, (b) Ionization, (c) Excitation.

The other typical plasma generating mechanisms are Penning ionization and Penning excitation. When it contains an inert gas, the gas particles collide with stable atoms or metastable atoms and makes them ionized or excited.

$$X^* + Y \rightarrow X + Y^+ + e \qquad \text{(Ionization)}$$

$$X^* + Y \rightarrow X + Y^* \qquad \text{(Excitation)}$$

As stated above, a precondition for generating plasma is that the particle should be accelerated more than the threshold value during the collision between particles. To maintain the certain state of plasma, the appropriate number of collision are needed. Therefore, the external voltage and pressure become important process parameters. However, when the pressure is too high, the distance between the collision becomes too short to make the electron to have a sufficient energy which results in interference of ionization. In short, plasma is a condition that a cation (+, positive ion) and an anion (−, negative electron) exist in the ocean that is full of neutral atoms [Figure 9.3].

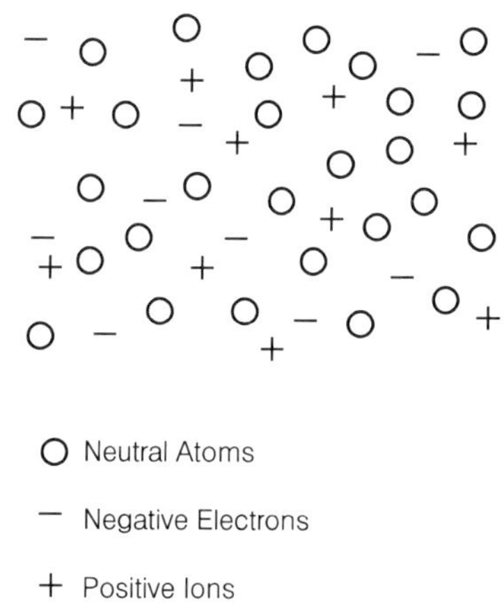

Figure 9.3

General formation of Plasma.

1.3 Density and Temperature of Plasma

The density of the plasma is represented by the number of particles per unit volume. The plasma temperature uses an energy unit called eV to avoid constant dependent on the dimension. 1 eV = 1.6×10^{-19} J, and the temperature corresponding to 1 eV is 11,600 °K.

1.4 Characteristics of Plasma

(1) Plasma is electrically neutral as a whole, however, it consists of a number of ions and free electrons, so it can allow flow of current by an external electric field.

(2) Active electrons and ions inside of plasma can ionize or excite the other materials so that it can promote a chemical reaction with other substances.

(3) The mass difference between the electrons and ions inside of plasma is very large. That is why the speed of movement of each particle shows a big difference which causes physical phenomena such as sheath[7] and Debye shielding[8].

---

7) When plasma is inside of conductor surface such as probe or wall of vacuum vessel which is confining the plasma, the electron is much faster than the ion. Therefore it arrives at the surface of conductor faster. A remote area of the conductor surface has a positive potential as they have the relatively large amount of ions. In this case, the area that has most of the voltage drop is called sheath and the potential formed by the plasma ion at a distance from the conductor surface is called the plasma potential.

8) The entire plasma is electrically neutral, but when a floating object is inserted, the local neutrality is broken and the particles inside the plasma are redistributed by themselves. This kind of phenomenon is called debye shielding and the range that has the influence Is called debye length.

(4) When the electrons and ions of the plasma get into a magnetic field, they move in circular motion whose direction is perpendicular to the magnetic field. In this way, it can have a magnetic property that can confine the plasma in one place or concentrate the density of the plasma as desired. The electrical resistance of high-density plasma becomes lower.

* 2 Formation of the Plasma

2.1 Energy for Ionization

Ionization is required to turn to plasma state from a gaseous state. The energy needed to break electrons that were bound to the nucleus, from the bondage is considered through a simple hydrogen model. Hydrogen is composed of a nucleus which consists of one proton and surrounded by electrons as shown in Figure 9.4.

Ionization energy is the energy needed to move an electron from the atom to the infinite distance. According to Bohr's (N. Bohr, 1985–1962) quantum theory,

(1) Electrons trapped in an atom normally move in a certain orbit according to the laws of classical mechanics and electrostatics.

(2) Movement trajectory of the electrons must satisfy the following quantum conditions.

$$\int p\, dq = nh \qquad \text{(Equation 9-1)}$$

Where p is momentum, q is coordinate and n is a natural number (n= 1,2,3,...), h is Planck's constant.

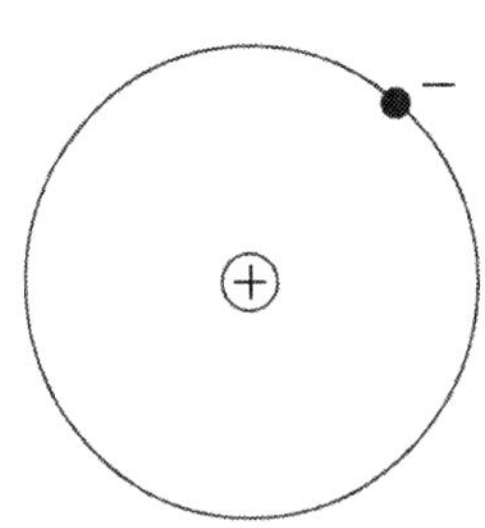

Figure 9.4

Atomic Structure of Hydrogen(+ : proton, – : electron)

According to the theory, electron's orbital radius r in the coordinate system which set the nucleus as an origin becomes

$$r = \frac{n^2 h^2 \epsilon_0}{\pi m_e e^2} \qquad \text{(Equation 9-2)}$$

where ω is angular frequency and $-e$ the charges of electrons and $+e$ the charge of nucleus as shown in Figure 9.5. When n=1,

$$r_1 = \frac{h^2 \epsilon_0}{\pi m_e e^2} = 0.53 \times 10^{-10} \equiv a_0$$

The generalized form of r becomes $r_n = n^2 a_0$

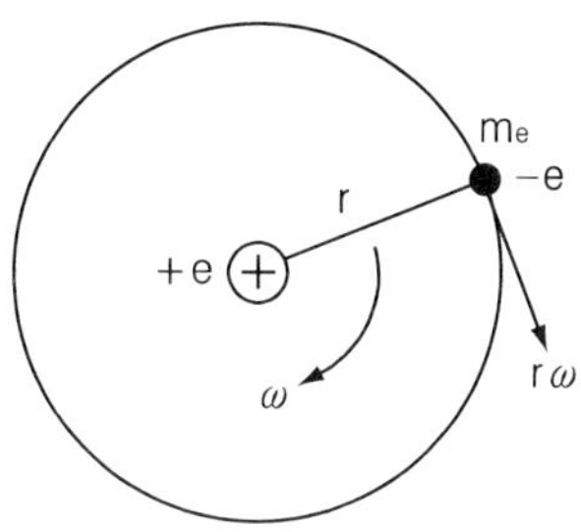

Figure 9.5

a_0 is Bohr radius and $\pi a_0^2 = 0.88 \times 10^{-20} m^2$ is called Bohr surface which becomes a reference in calculating the surface area of atoms. If the potential energy of electron is zero when $r \to \infty$, electron's total energy, which is the sum of potential energy and kinetic energy becomes,

$$E = \frac{1}{2} m_e (r\omega)^2 - \frac{e^2}{4\pi\epsilon_0 r} \qquad \text{(Equation 9-3)}$$

and E_n (energy with r_n) becomes

$$E_n = -\frac{m_e e^4}{8\epsilon_0^2 h^2} \frac{1}{n^2} \qquad \text{(Equation 9-4)}$$

and when n=1,

$$E_1 = -\frac{m_e e^4}{8\epsilon_0^2 h^2} = -2.168 \times 10^{-18} J \qquad \text{(Equation 9-5)}$$

Thus, it becomes as $E_n = \frac{E_1}{n^2}$

The total energy level of each electron orbit becomes as (a) in Figure 9.6.

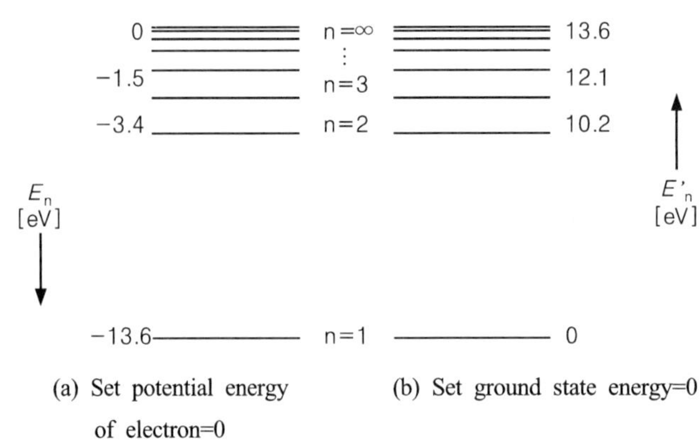

Figure 9.6

Energy level of Hydrogen

From the above theoretical discussions, the bound electron cannot have an arbitrary trajectory r and energy E and has a quantized value which is determined by n. The energy state determined by n is called energy level. Especially the state when the n=1 is called ground state and the state that has the value of n of more than 1 is called excited state. If an external energy which is the same amount of energy with the n=1 energy level is produced, the electron can be free from the confinement of nucleus although the potential energy of the electron is 0 when $r=\infty$ as shown in Figure 9.6. Thus, this 13.6 eV corresponds to the ionization energy of a hydrogen atom. The ionization energy can also be called ionization potential and expressed with eV unit. Ionization energy of other atoms also can be calculated in the same way. Examples of ionization potential are shown in Table 9.1.

From Table 9.1, it can be seen that the ionization potential corresponds well to the layout of the periodic table. In the case of inert gas, the outermost orbit is totally filled with electrons so their electrons which have a great bondage with the nucleus needs a large ionization energy.

In the case of inactive atoms (He, Ne, Ar, Kr, Xe, etc.), electrons are filled up to the outermost layer of each layer and their electrons are strongly bonded to the nucleus. In an atom having only one electron (hydrogen, alkali metal, etc.), the outermost electron can be relatively easily removed. This can also explain why the inert gas that has a lower number of electrons has higher ionization potential. An inert gas such as He, Ne, Ar, Xe is mostly used in the display application.

Table 9.1 Ionization potential of atoms.

Atom	H	He	Li	Be	B	C	N	O	F	Ne
Ionization Potential [eV]	13.60	24.59	5.39	9.32	8.30	11.27	14.53	13.62	17.42	21.57
Atom	Na	Mg	Al	Si	P	S	Cl	Ar	K	Ca
Ionization Potential [eV]	5.14	7.64	5.99	8.15	10.49	10.36	12.97	15.76	4.34	6.11
Atom	Sc	Ti	V	Cr	Mn	Fe	Co	Ni	Cu	Zn
Ionization Potential [eV]	6.54	6.82	6.74	6.77	7.43	7.87	7.86	7.64	7.73	9.39

Until now, the case of the atom has been considered so far, there is a molecule that consists of more than 2 atoms in normal temperature and pressure. In order to ionize them, they often go through a single dissociation process. To dissociate the molecule, energy is needed to break the bondage, just like the ionization of atoms. The energy needed for dissociation is called dissociation energy and the examples are shown in Table 9.2.

Table 9.2 Dissociation energy

Type of Dissociation	Dissociation Energy [eV]
$H_2 \rightarrow H + H$	4.4
$H_2O \rightarrow OH + H$	4.7
$O_2 \rightarrow O + O$	5.1
$NO \rightarrow N + O$	6.1
$N_2 \rightarrow N + N$	9.1
$CO \rightarrow C + O$	10.0

2.2 Ionization by collision of particles

The energy for ionization has been discussed and how the energy can be produced to the electron inside of an atom will be discussed. The simplest case is caused by the collision of particles. Not only the collisions between atoms but also the collisions between ionized atoms with charged particles (electrons and ions) should be considered in a collision between particles. There are 6 kinds of collision as shown in Figure 9.7. It can be divided into two cases whether the collision happens between particles that have similar size and weight or not.

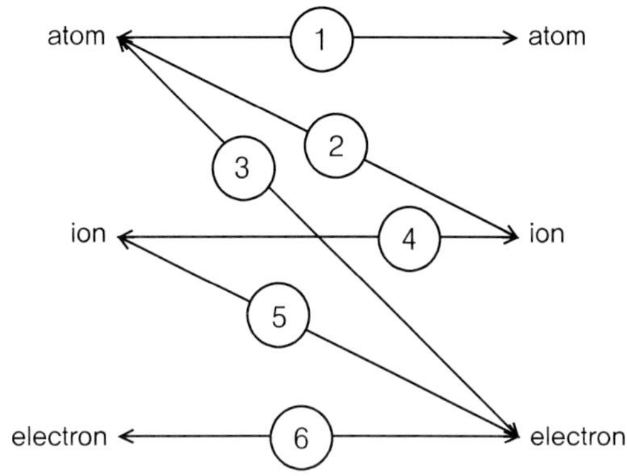

Figure 9.7
Various type of collision of particles

When the kinetic energy of the other particle is lower, the kinetic energy is preserved in any case of collisions. Like billiard balls striking at each other, the collision occurs as much as the energy is transferred. This kind of collision is called elastic collision. Even though the relative kinetic energy is large during the crash, a collision between the electrons (⑥) and collision between completely ionized ions (④) always become the elastic collision. However, in case of ①, ②, ③, ⑤, and ④ when the ion has not been ionized completely, all of or part of kinetic energy can be transferred to electron potential energy between ions and atoms. This kind of collision that can not preserve the kinetic energy after the collision is called inelastic collision.

When the mass of the particles that makes a collision are m_1, m_2 and speed of each particle are v_{11}, v_{12}, v_{21}, v_{22}, a necessary condition for inelastic collisions to occur would be as follows.

$$\frac{1}{2}m_1v_{11}^2+\frac{1}{2}m_2v_{21}^2=\frac{1}{2}m_1v_{12}^2+\frac{1}{2}m_2v_{22}^2+\phi \qquad \text{(Equation 9-6)}$$

In (Equation 9–6), ϕ should be greater than the energy required to make an ionization or to go up from ground level to excitation level. A collision that makes an excitation is called exciting collision and a collision causing ionization is called ionizing collision.

An electron that has been transferred to excitation level can stay in the level only for a certain lifetime. After that, it is transferred to the lower level spontaneously and finally transits to stable ground state. Normally the lifetime is really short which is 10^{-8}~10^{-9} s but some electrons have a lifetime more than 10^{-4} s or more than 1 s in some cases. As their meta–stable state has a long lifespan, it has a higher possibility to have a collision with other particles. It has high probability that the atoms in a meta-stable state give their transition energy to the other particle and can ionize it when the particle has lower ionization potential. In addition, the meta-stable atom itself plays an important role in the ionization process since the ionization energy of the atom itself only requires lower energy than the ground state.

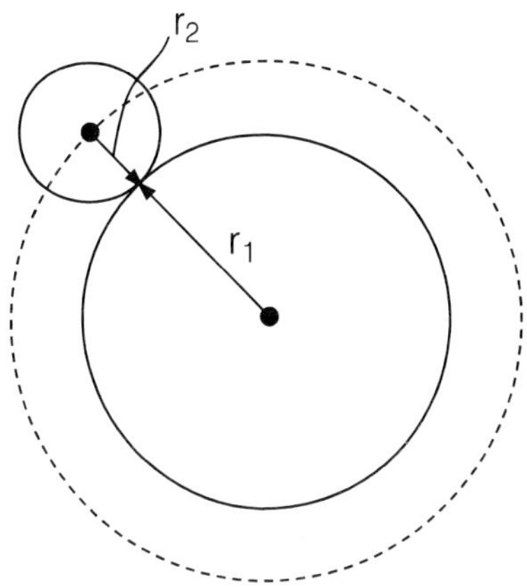

Figure 9.8
Cross section of collision

The concept of "cross section of collision" can be used when considering the collision between particles. As shown in Figure 9.8 if we use the cross section of collision of billiard balls which have radius r_1, r_2, it would be easier to explain when we expand the concept to the collision of atoms.

$$\pi(r_1+r_2)^2 \qquad \text{(Equation 9-7)}$$

Using this radius value, the impact cross-sectional area at the time of collision between the same or different atoms can be obtained from Eq. (9-7). When the velocity of the electron is fast enough, it does not need to be considered as a wave. The radius of the electron which is too small compared to the radius of atom could be ignored. The value of cross section of collision can be obtained from [Table 9.3]

Table 9.3 Radius of collision between atoms.

Atoms	Radius r[m] $\times 10^{-10}$	$\pi r^2(\pi a_0^2)$ [m^2] $\times 10^{-20}$
He	1.09×10^{-10}	4.3
Ne	1.30	6.1
Ar	1.82	11.9
Kr	2.08	15.5
Xe	2.42	21.0
H	0.53	1.0
Hg	1.08	4.2

When the speed of the electron is low and the De Broglie wavelength$\left(\lambda = \dfrac{h}{m_e v_e}\right)$ of the electron wave is as short as the radius of an atom, electron should be treated as a wave. This can be shown in Figure 9.9 when the energy of electron collision is used as a variable.

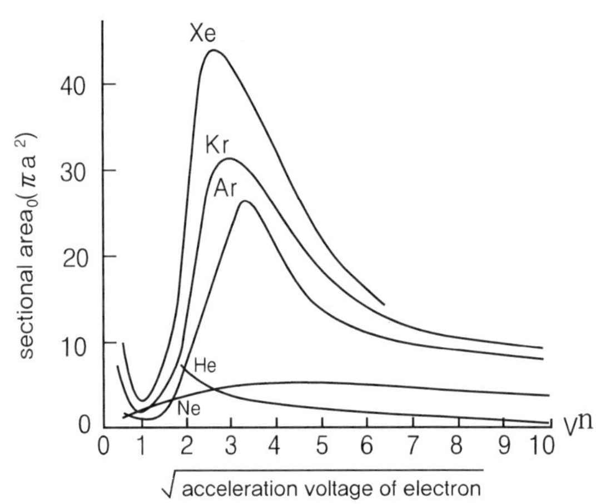

Figure 9.9
Cross section of collision between electrons.

First, Let's think about the distance (mean free path, m) that each molecule travels from one collision to the next collision and the number of collisions during the unit time (collision per second).

To compute these, consider the case where the molecular beam with the velocity of I_0 and the radius of r_1 is incident on the target of the stationary molecular group with radius r_2 (particle density n) as seen in Figure 9.10. The velocity of incident beam decreases as the beam collides with the target and the rate can be calculated assuming only dI as it proceeds from an arbitrary position x to $x + dx$.

It becomes as (Equation 9–8)

$$dI = -\sigma n I dx \quad \text{(Equation 9–8)}$$

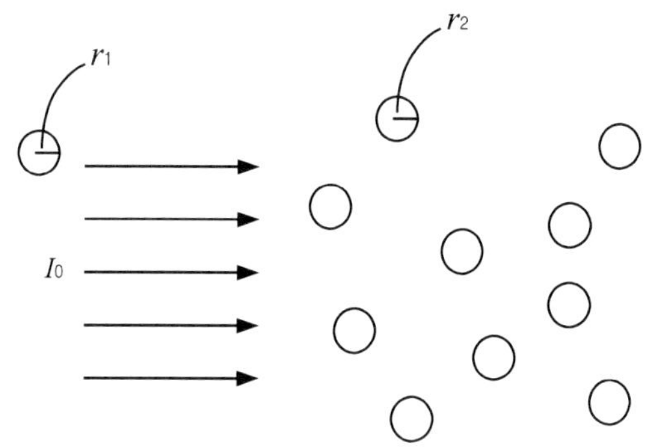

Figure 9.10

Model of beam(atomic radius r_1) colliding with stopped target(atomic radius r_2)

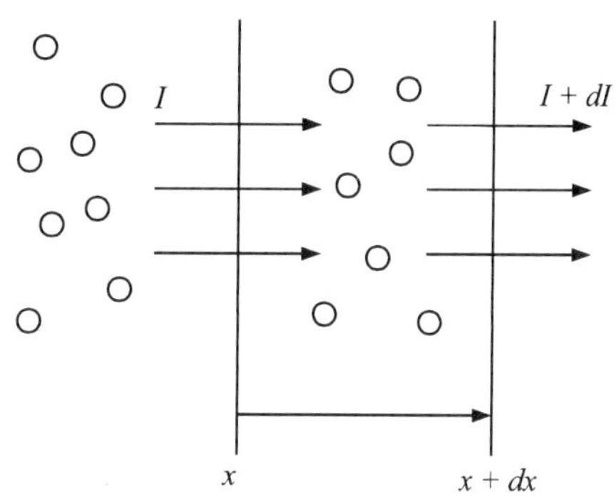

Figure 9.11

Attenuation of incident beam

σ is the cross section of collision between the molecules, which becomes $\pi(r_1 + r_2)^2$ in Figure 9.9 and (Equation 9–7). If we integrate (Equation 9–8), where I_0 is the velocity of the beam at x=0, it becomes (Equation 9–9)

$$I = I_0 \exp(-\sigma n x) \quad \text{(Equation 9–9)}$$

In formula $\frac{I}{I_0} = \frac{1}{e} = \frac{1}{2.718}$, most of the atoms collide only once and the value of x is called the mean free path length expressed with λ as (9–10)

$$\lambda \cong \frac{1}{\sqrt{2} n \sigma} \quad \text{(Equation 9–10)}$$

3 Electron Emission

3.1 The electron Emission in solid

So far, we have discussed the generation of charged particles by ionizing electrons in atoms and molecules, or by attaching electrons. Electron emission from solids for plasma is applied in the fields of plasma displays, field emission displays, and vacuum tube displays. The electron of metal in a vacuum has an energy state that is lower than vacuum by the work function (ϕ_m). The way to give the electron more energy than ϕ_m is secondary electron emission by the electron impact, thermionic emission, photo–emission, field emission, etc. These ways will be described briefly.

3.2 Secondary electron emission by electron impact

Secondary electron emission by electron impact can emit electrons on a solid surface by injecting electrons of energy greater than ϕ_m onto surface of the material. However, the detail contents of this emission process are not known yet. The number of a secondary electrons produced by impact from one electron is called secondary electron emission coefficient and is denoted by δ.

3.3 Thermionic emission

Thermionic emission occurs when electrons are emitted at a higher temperature of the material is increased. The electrons that have more energy than ϕ_m is emitted from the material. Current density, j, current per unit area of the material, is expressed by Richardson's equation.

$$j = AT^2 \exp\left(\frac{\phi_m}{kT}\right) \qquad \text{(equation 9–11)}$$

Where A is emission constant by material and kT is the temperature (11,600K converted to 1eV). A material with large A and small ϕ_m is suitable as cathode materials because it can have high j. Such a material can be used for various discharge tubes.

3.4 Photo-emission

In photo-emission the electrons are emitted in solid by photon energy of light. Such particles can be detected by Einstein's photoelectric effect experiment. The condition of photo-emission is that the photon energy of the electromagnetic wave, hc/λ, has to be larger than ϕ_m. The wavelength of electromagnetic wave $\left(\lambda \leq \frac{1239.8[nm]}{\phi_j[eV]}\right)$ can be obtained from the condition that $\phi_j = \phi_m$. The number of the electrons emitted by the incident photons is called quantum efficiency. It is denoted by η_q. A representative example of the value of η_q is shown in Figure 9.12. Figure 9.12 shows that photo-emission can occur starting from the infrared radiation from the various compounds. Materials with large η_q are used as a phototube and cathode material on the surface of photo multiplier.

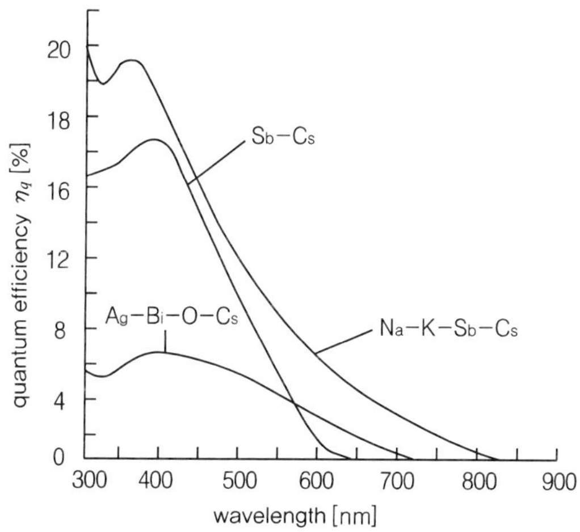

Figure 9.12

Quantum efficiency of front light of optoelectronics bilge pipes.

D i s p l a y E n g i n e e r i n g

* 4 Dielectric breakdown of the gas

The method of generating a plasma through ionization process at the molecular groups in the gaseous state is by applying an electric field. When the gas becomes a plasma, it gets the properties of a conductor, so this is called dielectric breakdown or simply breakdown. Let's consider the circuit shown in Figure 9.13 to examine the dielectric breakdown process. First, a container at vacuum state is filled at a gas of suitable pressure. Next, a counter electrode in the container is connected to the electric circuit and a voltage V is applied. These charged particles are accelerated by the electric field and change the gas to electrons and ions by colliding with molecules and atoms. The ionization effect by the electron is important because light electrons are accelerated much faster than ions by electric field $E=V/d$ (d: distance between C and A) applied between C and A. If the electron and ion pairs are gradually proliferated by repeating ionization of molecule and atom, the degree of ionization is increased and finally the gas in the container comes to the dielectric breakdown state.

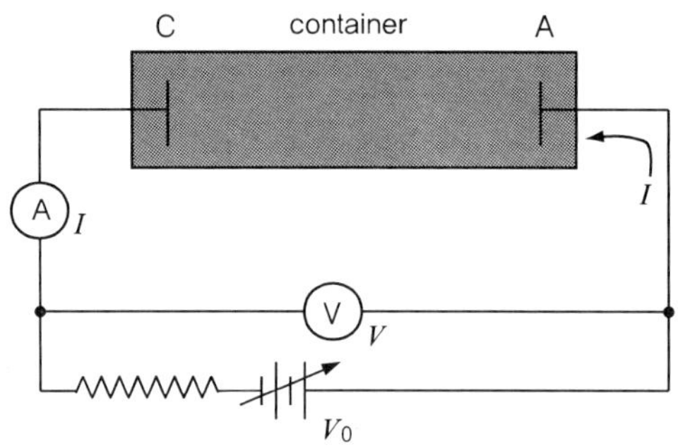

Figure 9.13
Discharge of gas

When one electron proceeds causing collision and ionization, the number of times ions and electrons are generated in addition to the original electrons is called alpha (α) and it is called the coefficient of ionization by the collision of electrons. While one electron moves from χ to $\chi+d\chi$, the number of electrons is increased by $\alpha d\chi$.

Thus, a number of dN for N electrons is shown in (Equation 9-12).

$$dN=\alpha N dx \quad \text{(Equation 9-12)}$$

Assuming that the number of the electrons from the cathode surface (χ=0) per hour is N_0 (initial electron), following equation can be obtained.

$$N=N_0\exp(\alpha x) \quad \text{(Equation 9-13)}$$

The number of electrons increases to $N_0\exp(\alpha d)-N_0=N_0\{\exp(\alpha d)-1\}$ and the same number of cations are generated in the space between the electrodes. If the number of electrons emitted for one cation is γ, $\gamma N_0\{\exp(\alpha d)-1\}$ of secondary electrons are emitted. The current by electrons entering the anode can be found as

$$j = e[N_0\exp(\alpha d) + \gamma N_0\{\exp(\alpha d) - 1\}\exp(\alpha d) + \gamma^2 N_0\{\exp(\alpha d) - 1\}^2\exp(\alpha d) + \cdots\cdots$$

$$= eN_0\frac{\exp(\alpha d)}{1 - \gamma\{\exp(\alpha d) - 1\}} \quad \text{(Equation 9-14)}$$

Even though N_0, the initial electron is not supplied, the condition that j can be maintained is when the denominator of (equation 9–14) is zero,

$$\gamma\{\exp(\alpha d) - 1\} = 1 \quad \text{(Equation 9-15)}$$

The voltage needed to satisfy (Equation 9–15) is called discharge starting voltage. Paschen's law for determining the discharge starting voltage can be obtained as follows using Equation 9–15 and equation expressing a with electric fields E and pressure p.

$$\frac{\alpha}{p} = Aexp\left\{\frac{-B}{\left(\frac{E}{p}\right)}\right\} \quad \text{(Equation 9-16)}$$

$$V_s = \frac{Bpd}{\ln(Apd) - \ln\left\{\ln\left(1 + \frac{1}{\gamma}\right)\right\}} \quad \text{(Equation 9-17)}$$

The influence of the gas having various cross–sectional areas on the macroscopic quantity a is expressed by an approximate expression including constants A and B (Equation 9–16).

(Equation 9–17) denotes that the number of ionization $\alpha\lambda_e = \left(\frac{\alpha}{p}\right)\lambda_{el}$ (unit of p is Torr, λ_{el} is mean free path when p=1 Torr) can be decreased by the energy $eE\lambda_e = e\lambda_{el}\left(\frac{E}{p}\right)$ obtained from the field during the electrons process in the mean free path λ_e. The result of Paschen's law is shown in Figure 9.14. It can be seen that the minimum of V_s is present on any value of pd.

For the experiment shown in Figure 9.13, the voltage–current characteristics are shown in Figure 9.15 when the terminal voltage V between C and A is increased gradually. Even if the discharge conditions (kind of gas, pressure, the distance between electrodes, etc.) are changed, distinctive appearance of the change of o–a–b–c–d–e shown in the figure is not changed. First, o–a part is that gas in Figure 9.13 acts as an insulator so, the applied voltage becomes the inter–terminal voltage. However, if the voltage is increased to increase current around self-sustained discharge starting voltage V_s given in (Equation 9–17), the number of electrons and ions from ionization increase gradually and the plasma region is gradually formed at both the sides. In this plasma region, the electric field of the direct current is hardly applied because of the conductivity of the plasma, so that the anode extends to the region where the self-discharge is maintained in the region between the virtual anode and the cathode.

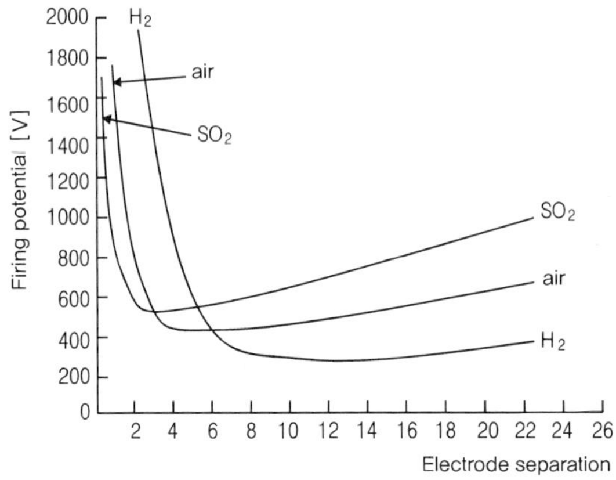

Figure 9.14

Discharge starting voltage curve between the parallel flat electrode

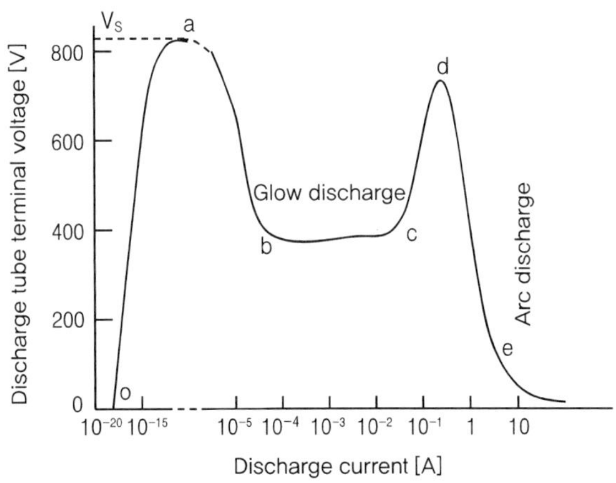

Figure 9.15

Therefore, if d is reduced and (pd) is small as seen in Figure 9.14, the voltage needed to maintain the self-sustained is lowered. Once self-sustained is started, it leads to dielectric breakdown by increasing the discharge current rapidly in $(pd) > (pd)_{min}$ by Paschen's law [Figure 9.15(a–b)]. If it reaches near $(pd)_{min}$, self-sustain maintains the value of nearly close to V_{smin} until self-sustain is expanded to the entire cathode [Figure 9.15 (b–c)]. If voltage V_0 increases, (pd) tries to be smaller than $(pd)_{min}$, so the voltage required for discharge must increase [Figure 9.15 (c–d)]. If discharge current is increased more, another secondary electron's emission mechanism such as thermionic emission becomes important than the secondary electron emission on the cathode surface by ion, and then it changes to arc discharge [Figure 9.15 (d–e)]. Until now, the simplest way of generating plasma by DC voltage is shown as an example. In reality, to maintain the plasma state, the generation of charge carriers due to the ionization action must be balanced with the loss due to diffusion or recombination, or the generation of charged carriers should be larger. After setting the type of gas, pressure, etc., it is the electric field E that governs the amount of charged particles (Equation 9–16), but there is a large change in space charge after the gas is converted to plasma. Therefore, it can be expected that, in the state where the plasma which is more right side than b in Figure 9.15 fills the space between A and C in [Figure 9.13], the density and the electric field of the charged particles are not uniform.

5 Overview and history of the PDP

5.1 History of PDP

Light emitted from the plasma, which is naturally observed, is representative of light emitted from the sun, which is a hydrogen plasma mass.

Observation of the light–emitting phenomenon from artificial plasma is started by being able to make artificial gas discharge by the battery invented by Volta in the early 18th century. The monochrome gas–discharge display device was invented by Gray and others of Bell Labs in 1927 based on extensive researches on gas discharge phenomenon by researchers and engineers in the middle 19th century and this was the beginning of the first plasma display device in the world. At the same time Nicholson invented gas discharge TV display device using row drive system in the beginning. However, with the rapid development of CRT in the early 1930s, the interest in using gas emitting device for TV display was lost. So utilization was limited as outdoor advertising displays and alpha–numeric displays.

One of the most successful gas emitting display device is Dekatron Tube invented by Bacon and Pollad in the 1940s. It was character display device that a lot of cathodes were placed as an annular shape around the common anode. After that, Nixie discharge display device was the first practical display using a gas discharge by Burroughs Corporation in the 1950s. Also "mercury sealing method" developed at that time is the important technology used for stabilization and long lifetime of current DC type plasma display.

A gas discharge with a matrix structure to reduce the cathode drive terminal significantly started to be reported from late 1956. It had shown a possibility that a gas discharge display device could be used as a display device of high capacity and high-resolution by the development of these new structures. However, when one cell discharges by addressing, the other cells on the same line can not discharge due to lowering the potentials of the same line by the characteristics of the discharge phenomenon. It was impossible to solve this problem with manufacturing technology at that time and the plasma display with the way of dot matrix could not reach the commercialized stage. This problem could be resolved by using AC-type structure by Bitzer and Slottow at Illinois University in 1964. Through this structure, a problem when one cell discharges, the other cell next to it does not discharge, was resolved. Furthermore, very useful memory function was also obtained. The invention of this AC-type plasma display got practical with development by Owen and others at Illinois University in 1968 and this inspired Holz and Ogle at Burrough Corporation to invent Self-scan™-type and became practical in 1970. Self-scan™-type display is driven by the technology that uses shift addressing with the diffusion of a charged particle, and had an advantage that could reduce the number of the driven elements significantly. Meanwhile, NHK, Hitachi Corporation, Sony Corporation and others independently invented PDP enlargement and high purification technologies in the early 1970s in Japan, and they steadily developed the technology for implementing Work Station Monitor and HDTV by selecting each unique electrode structure. A very large-scale full-color PDP having a size of 40 inches or more with high quality as good as CRT was actively put into practical use. However, from a consumer perspective, it needs to be inexpensive and capabilities including brightness, efficiency, power consumption and contrast ratio should be improved.

6 Principle and type of PDP

6.1 Principle of PDP

PDP (Plasma Display Panel) is electronic display using neon light as display light by putting gas such as Ne+Ar, Ne+Xe between the front glass and rear glass and by applying the voltage to the anode and cathode for neon light emission.

Generally, PDP is driven by the continuous pulse with constant voltage and gas discharge usually requires relatively high voltage as much as several hundred volts, so it is driven by amplifying the video signal.

As shown in Fig. 9.16, the driving of the PDP is an active light emission method in which ultraviolet rays generated by a gas discharge excites a fluorescent film to realize an image.

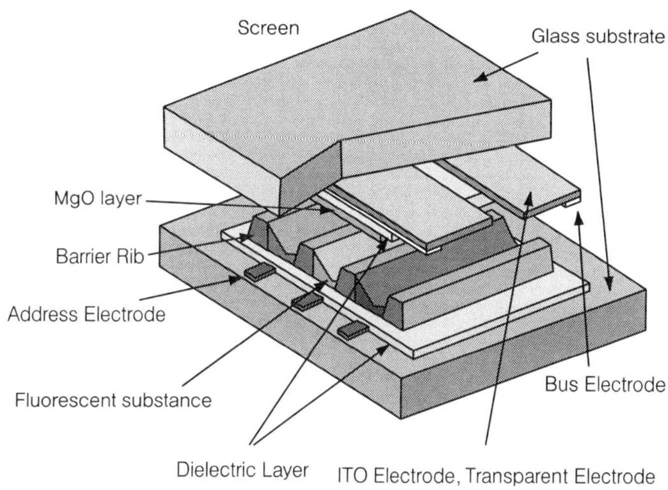

Figure 9.16
Structure and Principle of PDP

Basically, a voltage is applied to the gas sealed by the front and back glass and the barrier rib by a cathode/anode to cause neon light emission, and this light emission is used for display. At this time, a tiny gap of 0.1 mm is maintained between two sheets of glass substrate and (R, G, B) phosphors are applied to a discharge space composed of a large number of barrier ribs filled with a gas such as neon and xenon, and when a voltage is applied to an electrode provided on a glass substrate, a discharge occurs. The generated UV hits the phosphor (the opposite side of the glass substrate and the side of the barrier rib), combines the light, and implements the image.

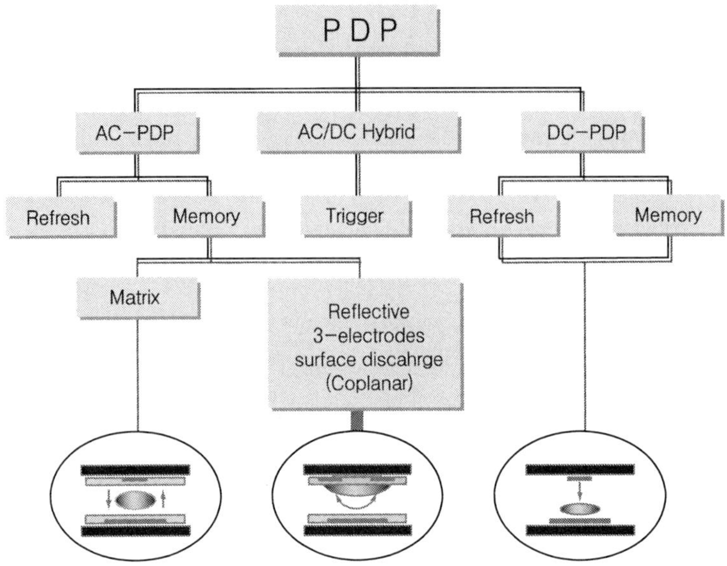

Figure 9.17
Classification according to driving method of PDP

PDP is divided into AC driving type (indirect discharge type) and DC driving type (direct discharge type) according to discharge structure. There are memory type and refresh type depending on whether the memory function is used or not.

(1) DC driving type (direct discharge type)

In this method, the electrode is exposed to the discharge space and driven by the DC voltage. Since the DC type is driven by the refresh method, the driving circuit is simple, but the panel structure is somewhat complicated by providing a partition for separating the cells to prevent discharge. In the structure of a DC type PDP, each cell is separated by a partition wall, and an anode and a cathode are formed on the front and rear panels. Also, in the case of DC, there is a current limiting resistor and the discharge gas is charged and the DC power is applied while sealing the front and rear panels.

On the front substrate, an anode is formed in a direction perpendicular to the cathode, that is, parallel to the partition, and the partition is not required on the anode. Next, the front substrate and the back substrate are bonded, the edge is sealed in vacuum, and the gas is finally injected. At this time, when a discharge protection resistor is inserted between the anode and the cathode and a DC voltage is applied, the light is emitted at the intersection.

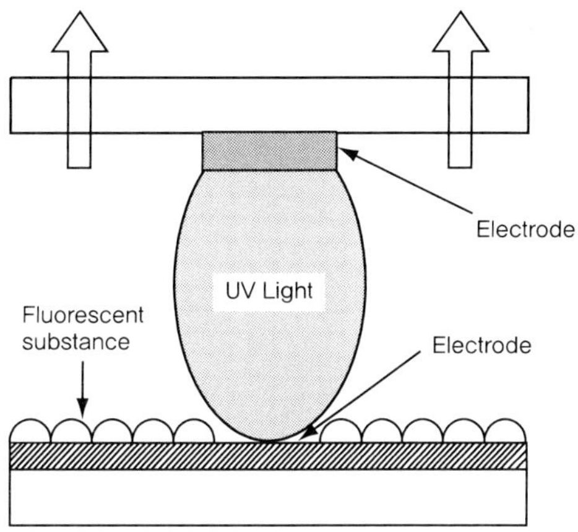

Figure 9.18
Structure of DC-type PDP

The major difference between the AC type and the DC type is that the DC type is a simple structure while the electrode is exposed to the discharge space to make an external resistance for the current limitation. On the other hand, the electrode is covered with a dielectric in AC type and the natural current limit can be made so that the electrode is protected from ion impact during discharging, which makes it possible to have a longer life than that of the DC type.

(2) AC driving type(indirect discharge type)

The electrode is covered with a thin glass-made insulator and is driven by a pulse voltage of about 200 kHz. In the AC type driving method, there is a refresh method in which a discharge is sustained when a low discharge sustaining voltage is applied after discharging once, and a discharge method is performed by applying a pulse voltage having a high voltage between electrodes to be displayed. In the AC PDP, each cell is separated by barrier ribs and the electrodes are formed on the front and rear panels. A sinusoidal AC voltage or pulse voltage is applied to each of the electrodes to cause discharge so that light can be seen through the front glass substrate. It is essential to insulate the vertical electrodes arranged in parallel on the rear substrate and the electrodes are insulated with a dielectric material. MgO is deposited on the surface of the dielectric to improve discharge characteristics.

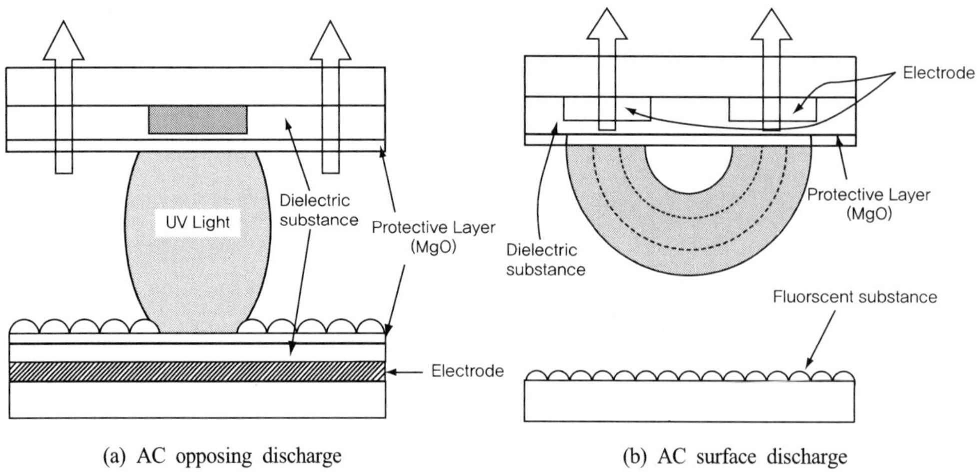

Figure 9.19
Structure of AC type PDP

The AC type is divided into an opposite discharge type and a surface discharge type. The opposite discharge type has a problem that the lifespan of the phosphor is deteriorated due to the ion impact, while the surface discharge forms the discharge to the opposite side of the phosphor to minimize the phosphor deterioration. It overcomes the problem of the opposite structure, and it is used in most PDP structures at present

6.2 Features and advantages of PDP

PDP has many merits that are suitable for large size flat panel displays of various fields such as LCD (Liquid Crystal Display), FED (Field Emission Display) and ELD (Electroluminescence Display). The reason why the PDP can be enlarged as a flat panel is because two glass substrates, each having a thickness of about 3 mm are used, and a suitable electrode and a fluorescent material are applied on the respective substrates. A gap of 0.1 to 0.2 mm is maintained between the substrates in which plasma is formed.

(1) Very strong non-linearity

Even if a voltage is applied between electrodes, the PDP has strong non-linearity such that no discharge occurs with respect to an applied voltage lower than the discharge starting voltage.

(2) Memory function

PDP has the memory function where the following state is determined by the conditions of the previous state.

(3) Long life

In the case of AC type PDP, the lifetime is extended by using a protective film such as MgO having a good sputtering property on the dielectric film. In the case of DC type, by using a resistor for current control and sealing a small amount of mercury or raising the gas pressure, the sputtering damage of the electrode material due to the ion bombardment is reduced, and thus the life is prolonged.

(4) High brightness and an efficiency of light emission

PDP is capable of high luminance of over 500 cd/m^2 and high efficiency of 4 lm/W or more in the theory of gas discharge phenomenon.

(5) Wide viewing angle

PDP has a wide viewing angle as a self–emissive display. AC type and DC type have a wide viewing angle over 160° both left/right and upper/lower and it is almost the same level as the CRT.

(6) Ease of full colorization

A color implementation of PDP uses photoluminescence mechanism that UV created from the discharge cells stimulates phosphor film and emits visible light, and it is possible to make the color reproduction of the CRT level.

(7) Low manufacturing price

Soda–lime glass is generally used as the substrate glass of PDP, and materials of the electrode, dielectric film, partition and so on can be commonly made cheaply.

(8) Heat resistance·Cold resistance Properties

The plasma generated in each pixel is hardly affected by the temperature of the barrier rib or electrode in the range of about −100 °C to 100 °C.

(9) Lightweight

The weight of TV set is over 100 kilograms if CRT with 40 inches is made. However, the weight is just about 20 kg if PDP with the same size is made. It is expected for the weight to be reduced more with lighter parts.

(10) Seismic characteristic

PDP has a good seismic characteristic or earthquake proof, because it does not use a filament, unlike CRT or VFD.

(11) High resolution

The PDP is ultimately a display suitable for HDTV. However, in order to produce a PDP having a high resolution while maintaining a high luminance, a technique of making a high-temperature and high-density plasma more effectively is required.

*7 PDP manufacturing technique

7.1 PDP manufacturing process

The common manufacturing process of PDP is shown in Figure 9.20. First, after ITO or SnO_2 is coated on the front substrate and the transparent display electrode is patterned by the photolithographic method. More fine BUS electrode is formed by Cr/Cu/Cr or Cr/Al/Cr on display electrode to prevent voltage drop. Transparent glass dielectric layer is formed by printing powder glass paste on the display electrode. About 5 mm wide frit glass sealing layer with low melting point is formed on the substrate and then it is annealed. After evaporating the protective layer of MgO on an inner marked zone of sealing layer,

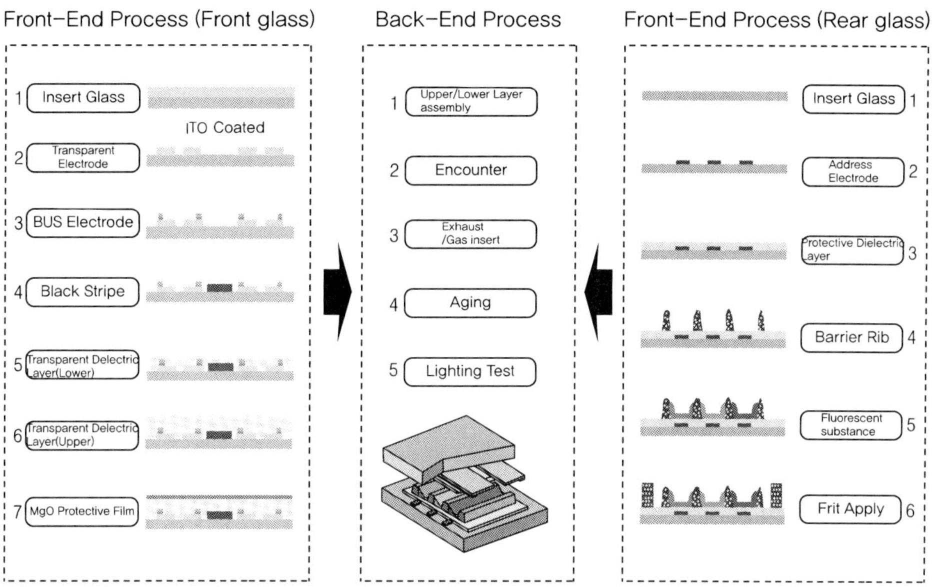

Figure 9.20
PDP manufacturing process is divided to pre–step producing front substrate and back substrate and post step.

a hole with about 1 mm in diameter is pierced in the back substrate for exhausting and filling the gas and the address electrode is printed over it. Address electrode is coated with a dielectric layer and the bulkhead between address electrodes is formed by the low melting point frit glass. For the mass production of a large screen, sand blasting method and a photo etching method in which an ultraviolet ray hardening resin is added to a barrier material are mainly used.

Colorization of PDP uses the photoluminescence phenomenon caused by vacuum ultraviolet like a fluorescent lamp and uses the R, G, B fluorescent substances for CRT. Since the improvement of brightness of each substance by a few percents leads to improvement of overall brightness by 20~30 cd/m^2, it is important to improve fluorescent substances. Methods of building fluorescent substance layer include photosensitivity paste method and ink jet method for which the ink already contains the fluorescent substance. Then, the complete glass vessel is formed on the exhaust pipe, and the two front and rear substrates and the exhaust pipe are fused. The exhaust pipe is connected to a vacuum device, and the entire PDP is installed in an electric furnace in an atmosphere of 400°C to remove impurities from the MgO surface and the surface of the phosphor. After exhaust, discharged gas such as Xe/Ne/He is filled in a pipe by 670 hPa and then the substrate is adhered to the bulkhead by force of external atmospheric pressure. After gas pole is completely made, glass vessel is plugged and cut. When the appearance of PDP is completely formed, the PDP goes through a long time aging with voltage exceeding the initially discharged voltage applied to PDP. Final PDP is completed after passing some characteristics tests.

7.2 Material of PDP

(1) Glass substrate

In the PDP manufacturing process, the frit or paste material, which is a pixel forming material, is fired and formed at a high temperature of about 550 to 600°C, so that the substrate glass is also subjected to the repeated heat treatment within the temperature range. Thus, for the glass substrates of PDP, obtaining the stability against heat treatment is important.

(2) Transparent dielectric material

Dielectric acts as a condenser for AC type PDP when it is discharged. For the material, glass paste which restricts the current and gives the memory function is used. The requirement for the transparent dielectric is over 80% of permeability in 530~580°C. The most important issue is how to increase the transmittance of visible light. In addition, even if there is only one foreign substance, it becomes defective, so strict management is required for manufacturing conditions.

(3) Bulkhead material

Glass paste is used as the barrier rib forming material, but since the line width and height of the barrier rib are high and the smoothness of the top surface of the barrier rib is high, the development of the paste material for satisfying these requirements is proceeding.

(4) Fluorescent substance

Fluorescent is the substance that generates visible light by resonance radiation light of Xe. Three kinds of fluorescent substances generating RGB light are needed for full colorization. The minute regions (cells) coated with these three kinds of phosphors are combined to form one pixel.

(5) Protective film material

Without the protective film, the surface of the dielectric layer is sputtered by the plasma ions generated between the electrodes, so that the discharge voltage rises and the electrode life deteriorates. The protective layer also has a function of maintaining the discharge state or suppressing the discharge current by acting as a cathode when discharging. The characteristics required for the protective film material include high permeability, resistance to sputtering, low discharge voltage, wide memory margin, and driving voltage stability. Currently, MgO is used as a material satisfying the requirements.

(6) Near-infrared and electromagnetic waves cutoff filter for PDP

It is pointed out that a near-infrared ray emitted from an enclosed gas used in a PDP is likely to malfunction due to interference with an infrared ray of a remote controller of various home appliances. Therefore, how to absorb electromagnetic waves and near-infrared rays emitted from a PDP is necessary development task.

Currently, near-infrared and electromagnetic wave-shielding filters for PDPs include transparent conductive films formed on tempered glass and products made of PMMA resin coated with absorbing dye or polyester resin for optical use. Manufacturers for tempered glass are Asahi Glass in Japan and Viratec in the USA. Resin film suppliers are Sumitomo and Kuraray in Japan.

(7) Panel sealing, Seal material

The low melting point glass frit is used for sealing the substrate of the PDP glass panel. The required characteristics for the glass powder for sealing are a matching linear expansion coefficient with other materials, dielectric properties, insulation characteristics, strength, airtightness and chemical characteristics. In particular, substrate glass and sealing frit are closely related to each other. The glass frit is a material which satisfies most of the characteristics required in panel manufacturing, so it can cope with existing material performance. However, in order to shorten the manufacturing process of the panel and to prevent cracking of the MgO protective film in the future, it is necessary to develop a sealing material which can be sealed at a lower temperature.

(8) Filling gas

Currently, Ne–Xe gas is commonly used in AC type and He–Xe in DC type of PDP panel. There are requirements for the filling gas of the PDP, which is high UV emission intensity of UV, low visible luminescence of gas, low change on standing, good stability for discharge and low discharge.

Generally, mercury, which is used in a fluorescent lamp, is much used for high efficient gas generating UV in low voltage discharge. The increased pressure of sealing gas is necessary to extend the life of both AC type and DC type panel. The luminescence from discharged gas and electric light from fluorescence substances affect the color reproducibility.

*8 Driving principle of PDP

8.1 Realization of Gray scale

In order to realize a moving picture on a screen of a certain display device, at least 60 picture frames per second must be continuously displayed. In order to display an image having a color of 24 bits, the amount of light should be adjusted to 256 scales (8 bits) on a gray scale for all cells. When a moving picture having 60 picture frames per second and a gray scale of 256 tones is displayed on the PDP, each frame is completely separated in time, so it is necessary to know how to display an image of one frame in 1/60 second. This one frame is represented by a time–field based on time. Before discussing how to display the entire screen, let's look at the operation of only one sub–pixel.

In order for a sub–pixel to emit its assigned light, it must generate a large number of discharges. Rather than discharging by counting each number of times of 256, it is more efficient to divide the time–field into 8 subfields and then divide the ratio of the number of discharges of each subfield to 1: 2: 4: 8: 16: 32: 64: 128, and apply the data to the sub–pixel only 8 times to cause discharge. The amount of light perceived by the human eye is superimposed in an amount averaged over time to obtain the desired 256 gradations.

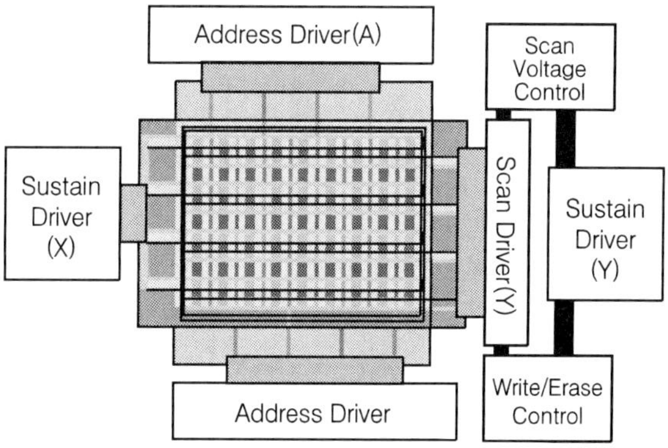

Figure 9.21

General arrangement of electrode of PDP.

All current PDPs implement a gray scale based on this method. This is because the amount of visible light that the PDP can emit in one discharge is small and the response time required for the discharge is very fast, that is in μs. Also, it is very difficult to adjust the amount of light emitted here with a single discharge in several steps, and it is almost impossible to constantly adjust the steps for each pixel on the entire screen. Therefore, the PDP can be regarded as an element that displays the image in the most digitized manner of the digitally processed image data.

Driving multiple cells has a more complex in time. The address electrodes allow one column to be selected and the scan electrodes to select one row.

As a first step, consider driving cells that are adjacent to each other (one row in the matrix). Cells in a row have one or two common electrode or electrodes. A common electrode is used to select the row, and the electrodes connected to the cells in a row are in the column direction. At the moment when this row is selected for data input, only the voltage of the address electrodes is different according to each data, and all the remaining waveforms are shared. That is, cells in a row behave as if they were acting together.

Considering driving several rows, it is very important to select a certain row and perform a process of inserting data at a certain time interval and to maintain the discharge according to the data bit at the respective time interval. This is organically combined with the structure and discharge characteristics of the PDP, which greatly affects each other and determines the driving method.

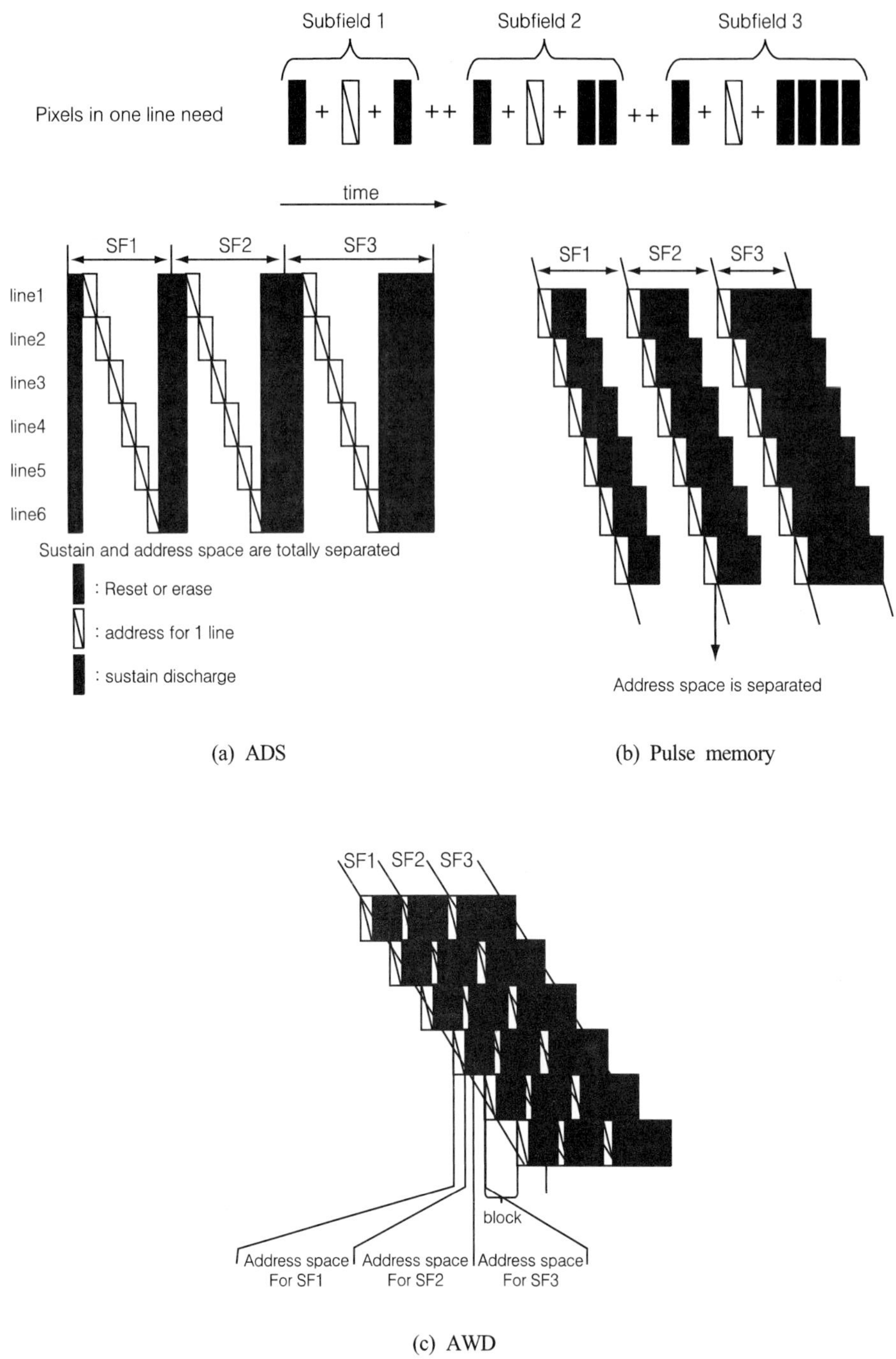

Figure 9.22

Classification of operation method by difference of timing.

Three subfields are needed for the pixels on one line. There are subfield reset process, addressing process, and sustain process. Reset process deletes the data of front subfield and prepares for addressing discharge.

Addressing process builds the addressing discharge for each time when the specific row is selected. Sustained discharge process maintains the proper number of discharge according to the weight of the data from the previous process. Depending on how they are allocated, they are divided into three types.

(1) Address-Display Separation(ADS) Method

ADS method conducts addressing while each line is scanned sequentially. When this process occurs on overall screen and data 1 or 0 is inputted, the voltage of maintaining discharge is applied on overall screen and only the cells with data of 1 keep the discharge continuously. Voltage maintaining discharge is able to emit the light according to the importance of that data bit.

At this time, the wall charge effect plays a role of data storage. However, there is a difference of data storage between earlier scanned row and later one. Since the effect of the wall charge in AC PDP maintains while at least more than a few msec, even if all the lines are scanned at a time, the data can be written without problems and the data can be maintained. At the end of this process, a scan is performed to insert the next bit of data. It can not be used to drive the DC PDP because in the DC PDP, the wall charge effect is necessary.

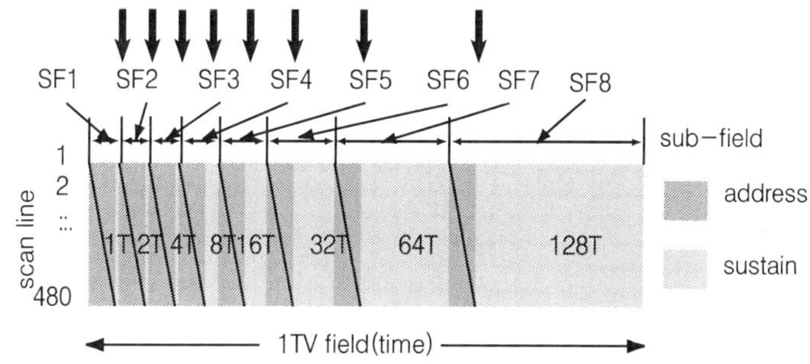

Figure 9.23
Constitutional diagram for each line according to time took by ADS method.

(2) Pulse Memory Operation Method

It looks like that other line is sustained at the moment when one line is addressed, but addressing is conducted between sustained discharges. A time-gap sustaining is used for addressing. The reason why pulse memory operation method is only used when operating DC PDP is that DC PDP has following properties.

First, addressing discharge can be applied quickly by narrowing the width of the applied pulse.

Second, the rise time or fall time of the pulse is short since the capacitance between the electrodes is low.

Third, the memory effect of the DC PDP utilizes the metastable species of charged particles, so its duration is short. It is impossible to use the ADS method which does not discharge all the time during the addressing period. A sustained discharge should be generated as soon as possible so that the memory effect persists.

(3) Address while Display(AWD) Method

The process that makes waveform by AWD method is as follows.

First, 4 pairs of sustained discharge pulse is applied to sustain electrode and scan electrode are bonded as one set.

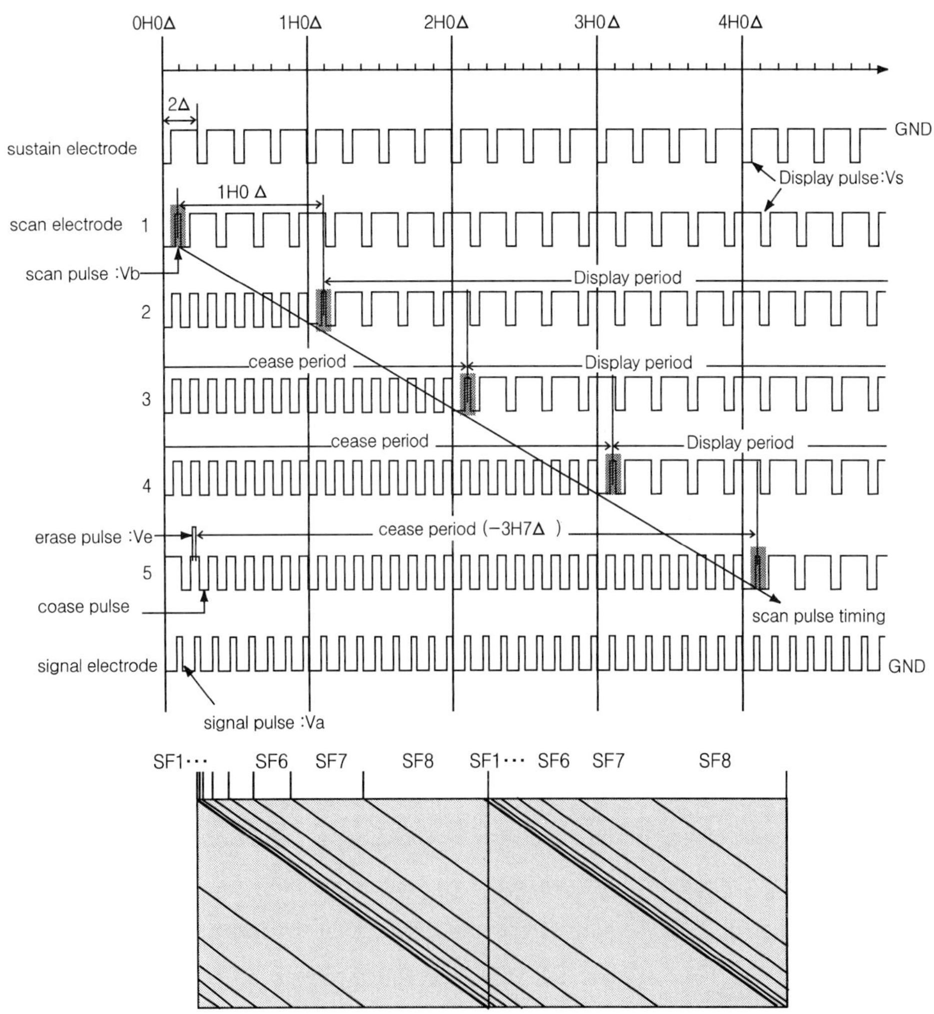

Figure 9.24

Practical example of AWD operation method.

Second, there are 8 spaces where the voltages are the same between the sustain-electrode and the scan electrode in the sustain-discharge pulse. The first space in these spaces is only used for addressing subfield 1. The second space is used to address subfield 2 in the same way.

Third, when a scan is done in this way, the data at the address electrode does not overlap every subfield.

Fourth, if one line fills the sustain discharge number of one subfield, then a cease pulse is applied to the scan electrode during the next two to four sustain batches to erase the wall charges written in the previous data and prepare for the next addressing. When these contents are summarized, a waveform as shown in Figure 9.24 is created.

*9 D i s p l a y E n g i n e e r i n g

Technology development trend of PDP

9.1 High-definition technology

Unlike conventional CRTs and LCDs, PDPs are attracting attention as a leading digital display device that can control the brightness by digital signals and lead the digital satellite broadcasting era. Nowadays, it is aimed to realize the high definition technology. In order for a PDP to have a market competitiveness as a display device, the display performance must be superior to other devices. For a display device, the first thing to be achieved is to realize a beautiful image.

Criteria for evaluating image quality include brightness, color temperature, contrast, and clarity of pseudo contour. The trends are as follows.

(1) High peak brightness technology

The reason for a lower the peak luminance of the PDP than that of the CRT is due to the gray level indication method of the PDP, although the difference in the luminous efficiency is a major cause. In order to solve this problems, an adaptive luminance enhancement technique has been developed. This method is a signal processing driving technique that dynamically changes the number of subfields and the number of pulses according to the characteristics of an image. By developing this method, the average luminance is lowered and the peak luminance is raised without harming the gray level of the low luminance portion so that the peak luminance can be improved by more than 40% compared to the conventional one, without supplying power separately. In the AI (adaptive intensifier) method, the peak luminance can be improved without increasing the power. In addition, NEC is developing dynamic PLE (Peak Luminance Enhancement) technology. This method also raises the average luminance at a low ASL (average signal level) and lowers the average luminance at a high ASL, thereby suppressing peak power consumption.

(2) color temperature development technology

The reason why it is hard to improve both brightness and color temperature for traditional PDP is that light emitting efficiency of blue fluorescent substance is low. The signal levels of green and red were reduced to take a white balance for low-intensity blue light. However, in the asymmetric cell structure, it is also possible to balance the luminescence sensitivity of each cell of RGB, and when driven by the maximum pulse, the color temperature of white can be increased by 10,000K or more.

(3) High contrast technology

In order to realize a clear image, it is necessary to improve the image contrast. To do so, it is necessary to reduce the background light of the panel or to increase the emission of the discharge cell and to reduce the reflected light from surrounding. Due to the change in driving technology, it is now possible to achieve a high contrast ratio of about 3000:1 in the dark room.

Among the ADS driving methods, background light can be reduced by introducing a ramp waveform during reset, that is, during the initialization period to induce a weak discharge and perform initialization. In particular, as discussed above, in the asymmetric cell structure proposed for improving the color temperature, the difference in the discharge characteristics was remarkable due to the difference in the shape of cell, which posed a serious problem. However, by introducing the ramp voltage waveform in the initialization, this problem can be solved. When the ramp voltage waveform is initialized, a wall voltage corresponding to the discharge starting voltage of each cell can be formed.

Accordingly, it is possible to absorb the deviation of the discharge voltage due to non-uniformity in the panel regardless of the difference in discharge characteristics due to the difference in the cell width, thereby achieving the two goals of improving the color temperature and improving the contrast ratio.

(4) Dynamic image quality improvement technology

In operating method representing 256 grays as subfield method like PDP, a distorted image may be seen when observer's line of sight changes. It is known that the direction of integration of light emitted is inconsistent with the direction of integration of light actually recognized by a person. Therefore, in order to realize a high-quality image, studies for minimizing the false contour of a moving image have been actively conducted.

The most typical pseudo contour resolution method is to divide sub-field bits to increase the number of sub-fields. However, as the number of subfields increases, the address period increases and the sustain-period of plasma decreases, resulting in a decrease in luminance. Therefore, methods for minimizing the pseudo contour while appropriately arranging the number of subfields have been sought.

The picture quality of PDP has surpassed that of the early development stage and it shows great potential as a representative of the next generation HDTV class large-screen display device. However, in order for the PDP to compete with CRT HDTV or projection TV type large-sized display device, it is necessary to achieve higher efficiency and lower price with higher picture quality.

Figure 9.25 shows the required PDP picture quality and other target specifications as HD class. It is anticipated that it will be able to achieve its target soon as research and development are actively carried out with a lot of investments both at home and abroad.

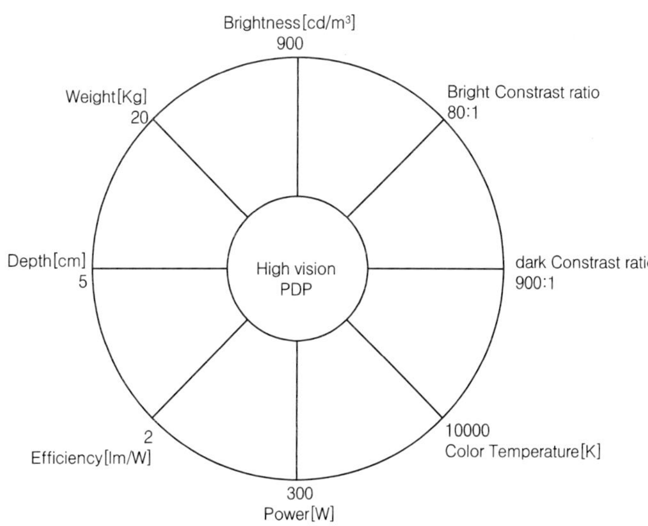

Figure 9.25

Objective specification of HD level PDP.

References

[1] Hwang Ki-woong, Seok-hyun Lee, Tae Hong, "Fundamentals of Plasma Engineering"

[2] Brian Chapman, "Grow Discharge", John Wiley & Sons, 1980.

[3] Alfred Grill, "Cold Plasma in Materials Fabrication From Fundamentals to Applications", 1993.

[4] Alexander Roh, "Vacuum Technology", third, updated and enlarged edition, 1990.

[5] Giovanni Bruno, Pio Capezzuto, Arun Madan, "Plasma Deposition of Amorphous Silicon-Based Materials", ACADEMIC PRESS, 1995.

[6] Brian Chapman, "Glow Discharge Processes", SPUTTERING AND PLASMA ETCHING, 1980.

[7] Lee Chang-hwan and 4 others, National Strategic Industry Analysis PDP

[8] Lee Doo-hee, Intellectual Property 21, Technology Trends of Plasma Display Panel (PDP)

[9] Joo Byeong-kwon, the latest technology analysis of information display industry

[10] Shin Sung-tae, Display Device Technology Trend

[11] Jang Myoungsoo and others, Trend of PDP material technology

[12] Hwang Ki-woong and 1 other, high-efficiency PDP technology trend

[13] Kwon Sang-jik, Display Equipment and Vacuum Packaging Technology

[14] Image processing laboratory of Seoul National University, Principle of operation of AC Plasma Display Panel

[15] Korea Display Research Institute, Process Technology of Plasma Display Panel

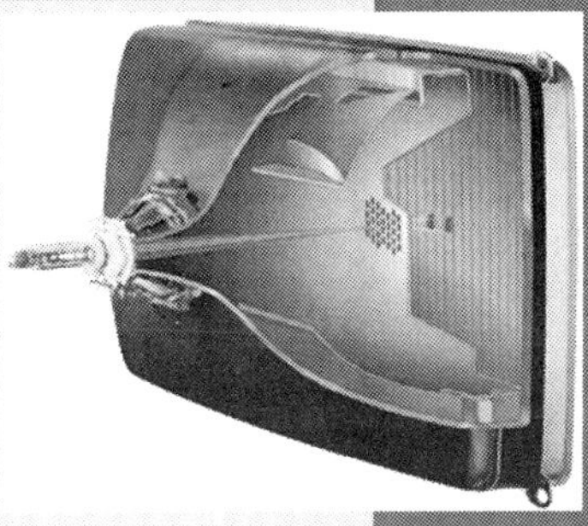

Chapter 10

Touch Screen Panel

1 Overview of the touch screen panel

2 Actual structure and operating principle of typical touch screen technology

3 Technology development trend of the touch screen panel

* 1 Overview of the touch screen panel

1.1 Concepts and components of the touch screen panel

(1) The concept of the touch screen panel

The touch screen panel is an interfacing devices that enables user to operate the machine easily with interactive and intuitive manner by simply touching the button on the display. In addition to smart−phones that we see around, touch screen panel has been applied to ATM machines, navigation and touch pads in a notebook computers. There is some confusion with touch screen and touch pad. In a typical touch screen, a touch panel is attached onto monitor screen. When the touch panel is used as a replacement of mouse, the structure is called touch pad. The touch panel is the most basic structure of touch screen.

(2) The components of the touch screen panel

1) Touch Panel

A touch panel is fixed with glass(as lower plate) and film(as upper plate). TCO(Transparent Conductive Oxide) is deposited on the glass (typical TCO is Indium tin oxide(ITO)). If the upper plate of touched point is in contact with the lower plate, the electrical analog signals along with its X, Y coordinates are transmitted to a controller. Although typical glass and film are non−conductive, but the glasses used in touch panel are coated with transparent conducting oxide films onto them enabling electrically conductive property.

2) Controller

As a device for controlling the input and output signals of the touch panel, the controller converts analog signals transmitted from the touch panel into digital signals through the A/D converter and has a function of printing the digital signals on the display at a specific rectangular coordinate. The controller connected to the main body of the PC exchanges data.

3) Driver

The S/W driver and control program for operating the touch screen on the PC maintains compatibility with the mouse driver, so that software can set all types of mouse–based applications directly to the touch screen driver. There are various forms of S/W driver depending on the operating system environment. The driver is software optimized to implement the touch panel for each operating system (DOS, windows, etc.).

1.2 Classification and features of the touch screen panel

Touch panels exhibit different characteristics along with different types and, therefore, it has been widely used in various environments to suit a certain application.

(1) Resistive overlay method

Two substrates coated with a transparent conductive film (ITO) are bonded together so that the transparent conductive films face each other with a dot spacer interposed between the two substrates. Since the transparent conductive film is coated on the inner side of the special film, a signal for position detection is generated when the upper substrate is touched with a finger or a pen. By the pressure of pen or finger, the upper substrate is contacted with the ITO of the lower substrate. The electrical signal from upper ITO is delivered to lower ITO thereby the touched position can be recognized.

(2) Capacitance method

A transparent electrode is formed in touch screen sensor by coating a special conductive material on the top glass and bottom glass. The glass plate where we have conductive layer, can flow a certain amount of current. The principal of capacitance method is to detect touched position using a capacitance of the human body(Parasitic Capacitance) by recognizing the position where current has changed and also by estimating the current.

(3) Ultrasonic reflection method

Piezoelectric elements are used to convert structural change into electrical signal.

- When external voltage is applied to piezoelectric elements its volume may change.

- Conversely, if external force is applied, the piezoelectric elements it can generate a voltage.

Using these features, when the user touches a glass surface, the surface wave is generated and some of the reflected waves arrive to the piezoelectric element to generate a voltage. Using this voltage, it is possible to calculate the time difference between generation of surface wave and the reflected wave. The principal of ultrasonic reflection method is to calculate the distance of each input point by generating a surface wave alternately to the X and Y directions.

(4) Infrared rays (IR) method

Infrared rays (IR) method is to use straight line of IR light which is blocked or interrupted by obstacles. The structure of IR method is an arrangement of many light elements around the panel. Elements receiving or emitting the light are arranged to face each other like a matrix. When the user touches the screen, the light from the emitting element is interrupted and the amount of light arriving at the light receiving element decreases. By obtaining X, Y coordinates of the decreasing point, the input coordinates can be determined.

Table 10.1 The main features of each touch screen panel method.

Method	Resistive overlay	Capacitive	SAW (Ultrasonic reflection)	IR (Infrared rays)
light penetration ratio	75~85%	90%	More than 92%	100%
durability	Low	Excellent	Excellent	Excellent
resolution	Excellent	Good	Good	Not Good
multi-touch	possible (difficult)	possible	impossible	possible (difficult)
size	1~20"	2~22"	3~8", 10~40"	20~150"
input	Anything	Special pen, finger	Special pen, finger	Special pen, finger
merits	Low cost high-Resolution Various input methods	Multi-touch Excellent sense of touch high-Resolution available	Durable High light transmittance high-Resolution	Durable High light transmittance Wide screen available

demerits	Low light transmittance Wide screen difficult Low durability	Limited input methods Expensive Wide screen difficult	Vulnerable to noise Vulnerable to moisture Pollution malfunction	Expensive Slimmer difficult low-Resolution
Main Application	Mobile phones , game consoles, navigation, etc.	Smartphone, tablet PC, laptop, etc.	ATM, Kiosk, POS etc.	ATM, Kiosk, POS etc.

1.3 Application Area of the touch screen panel

Application fields of touch screen panels are being extended from the public sector unit (ATM, Kiosk, etc.) to personal information devices (mobile phone, PDA, game consoles, etc.). Touch screen panel market has been formed based on some electronic devices (PDA, PMP, Navigation, etc.) and public sectors (ATM, Kiosk, etc.). However, in recent years, higher growth is expected in the market of personal information device sector like mobile phones, games, VR and others. In particular, after the big hit of the Apple's I–Phone with touch screen panel, major mobile phone manufacturers launch making smart–phones with touch screen panel as strategic models. A variety of mobile phones with the touch panel has also been launched in Korea.

* 2 Actual structure and operating principle of typical touch screen technology

The touch screen is classified into various types (resistance film / capacitance / ultrasonic / pressure, vision) along with the implementation principle and operation method. Here, the touch screen mounted on the familiar smart phones or tablet PCs, is mainly applied with a resistive film type and a capacitive type.

2.1 Resistive overlay method (analog)

(1) Principle & pros and cons

If a user touches the substrate using a pen or a finger, a transparent electrode of the upper and lower substrate makes contact with each other generating electrical signals and detects the location. Resistive overlay method is to determine the touched position by detecting the signals and its input durability is about one to two million times. The reaction speed of resistive overlay method is slower than capacitance method and malfunctions due to a pressure-drop but it has some advantages, that are the possibility of fine touch and the inexpensive manufacturing cost.

(2) Structure

Touch screen panel of resistive overlay method is composed of two ITO film(PET printed with ITO) layers. ITO films are coated with ITO transparent electrodes facing each other. By using DAT(Double-sided Adhesive Tape), it can create some separation between two layers of ITO films. By placing the Dot Spacer at regular intervals, it is possible for the two layers of ITO film not to contact each other. After flowing a certain current to both terminals of ITO film, if a screen is touched by a finger or stylus pen, X, Y coordinates of a touched point is detected by contacting the upper ITO film and lower ITO film. Depending on the number of wires used in detecting a signal, resistive overlay methods are classified as 4, 5, 8-wire resistive overlay method.

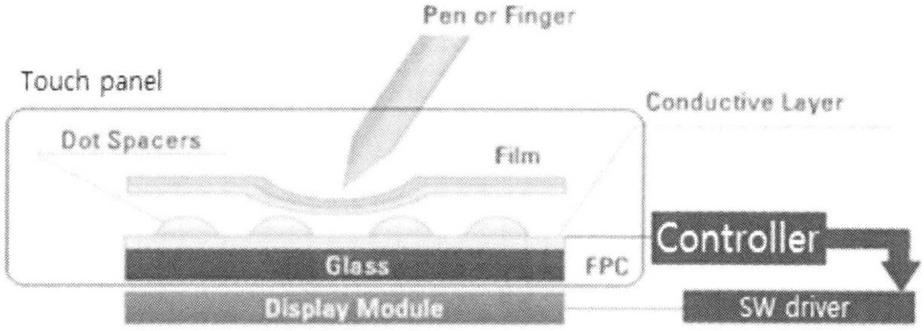

Figure 10.1

Touch screen panel structure of general resistive overlay method

(3) Difference between 4 wire resistive overlay and 5 wire resistive overlay

1) 4 wire resistive overlay method

4–wire resistive overlay method consists of the film coated with the conductive material(ITO) and glass. Film and glass are printed with a silver pattern. If the film and glass are touched by the pressure of a finger or stylus pen, the voltage of touched point is transferred to the controller. The controller converts the voltage into a certain signal and transfers to the computer. The computer recognizes the touched point. 4–wire resistive overlay method is less durable than 5–wire but it has some advantage like it is inexpensive and that it can be designed to a smaller size. Because of the wide application range of 4–wire method, it can be applied everywhere such as mobile phone, PDA, industrial equipment, POS, KIOSK, touch monitors, etc.

2) 5 wire resistive overlay method

Like a 4–wire resistive method, 5–wire resistive overlay method consists of the film coated with a conductive material(ITO) and glass. However, the silver pattern is coated only on the glass because of the weak durability of the film. When a user touches, the film just reads the voltage from the glass. Thus, a 5–wire resistive method is more durable than the 4–wire method. However, the price is slightly higher and the specification is larger because the structural outer margin range should be wider.

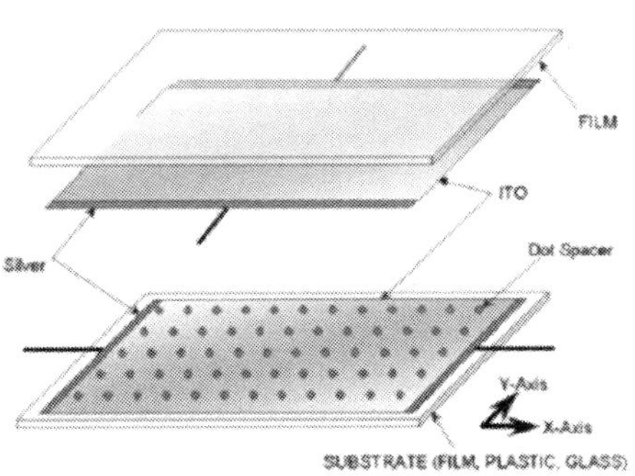

(4-wire resistive overlay method)

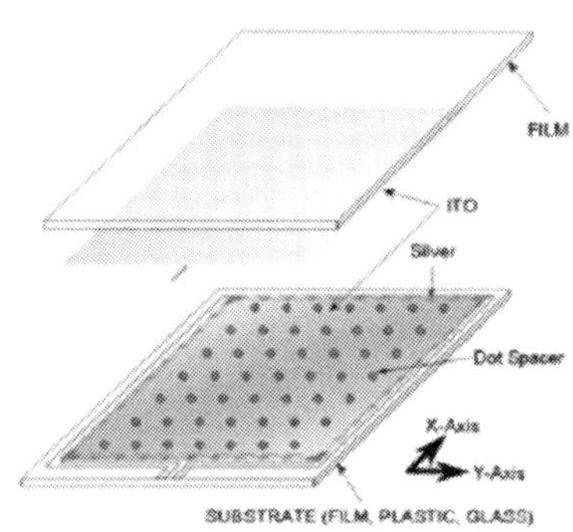

(5-wire reresistive overlay method)

Figure 10.2
Principle of 4, 5–wire resistive overlay method

2.2 Capacitance method (digital)

(1) Principle & pros and cons

Capacitance touch panel is made of special transparent conductive material (Indium tin oxide: ITO, Antimony Tin Oxide: ATO) that is coated on both sides of the glass and uses capacitance of the human body. It has better durability and has higher light transmittance than the resistive touch panel for consisting of only one glass. However, a nonconductive material like a gloved hand can not be used at capacitance touch panel.

(2) Structure

Generally, capacitance touch panel is made of two ITO films. One is patterned with X-axis electrode, the other is patterned with Y-axis electrode. The two films are combined by using OCA(Optical Clear Adhesive) film. The combined film is attached to a window type upper plate(tempered glass or PMMA). Generally, the pattern of a touch sensor is of diamond shape, however, it can be made of many other shapes. Finally, control IC is attached to the touch panel and it is possible to detect fine change in generated charge by touching the panel.

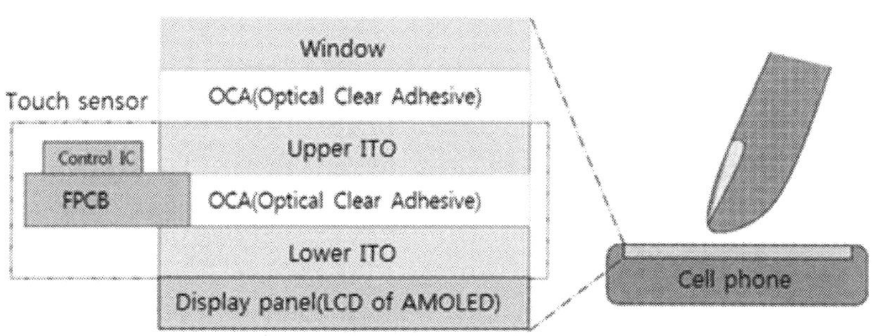

Figure 10.3 Typical touch screen panel sensor of capacitance method

The touch screen panels of capacitance method are classified as following three categories.

① Add-on type

② Touch-integrated with cover window type

③ Touch-integrated with display (LCD/OLED) type

First, add-on type is classified as film method and glass method. In the film method is the ITO layer that is patterned on the film. The GFF and GF2 belong to the film method. The glass method is the patterning of an ITO on a glass. GG and GG2 belong to the glass method. The touch–integrated cover window type is a method of making sensor by patterning ITO on the cover window. G1F, G2, and G1 are touch–integrated with cover window type. Finally, the Touch–integrated display is classified as top of the display or inside of the display, or by the so called On–cell or In–cell, respectively.

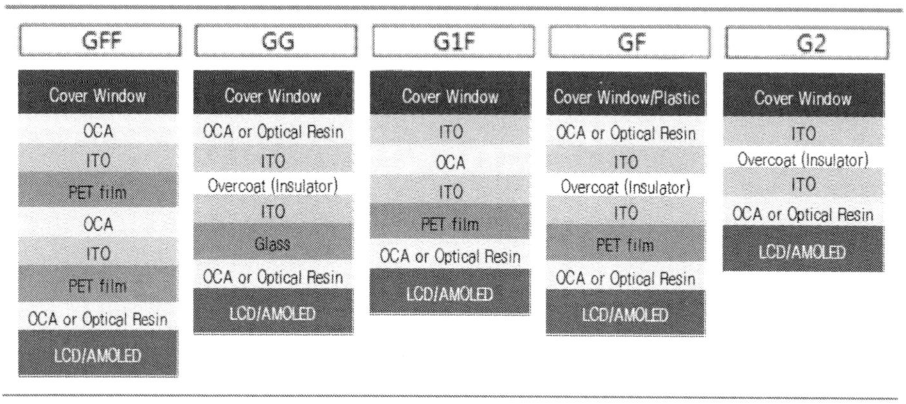

Figure 10.4
Technologies of capacitance method touch panel

(3) The principle of capacitive sensor

The basic principle of capacitive touch sensors is to measure the capacitance change between electrodes. Since the change in the capacitance is attributed to the delay of stabilization time of the voltage, any contact or touch can be detected by the time delay. As shown in [Figure 10.5] (a), an electric field is formed between the electrodes A and B, as a response to the voltage applied between A and B. This field gives the capacitance of C_a. When the object like a finger approaches to the conductors, electrons are grounded through the object, as shown in [Figure 10.5] (b). For the ground, the electric field between A and B decreases, thereby giving C_b.

The difference between C_a and C_b is too small to be detected. [Figure 10.5] (c) and [Figure 10.5] (d) are practical models that add insulator by complementing the limits of [Figure 10.5] (a) and [Figure 10.5] (b). For inserting an insulator, an electric field is produced only on the upper part of the screen giving C_a. When a finger comes close to the screen, the gap between insulator and conductor is detected as C_c and C_b. It can be a criterion of checking the touch point.

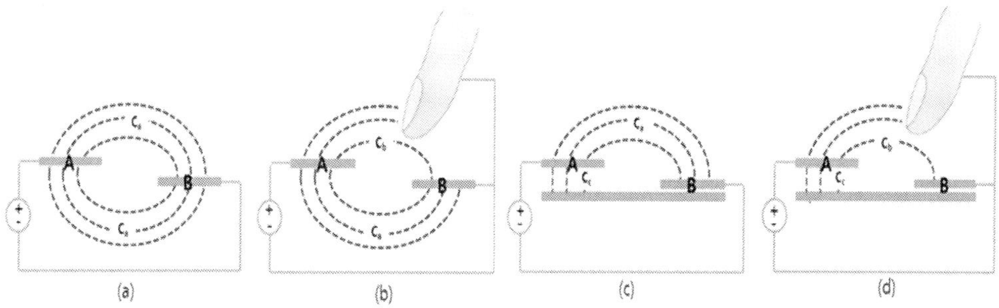

Figure 10.5

The principle of capacitance sensor

After detecting a touch signal as above, the signal is handled by the following process.

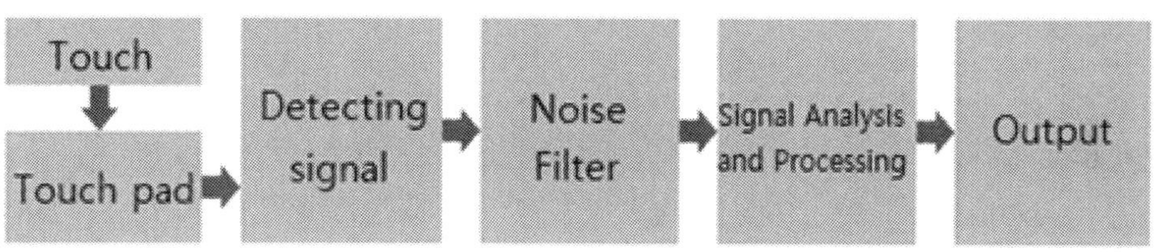

Figure 10.6

How to detect the touch signal

2.3 Process of the touch screen panel

Touch screen panel(TSP) method is divided broadly into capacitive, resistive, infrared and ultrasonic. Among many TSP production methods, capacitance method is the top-trend and this trend is expected to continue. Therefore, let's find out the structure and process of the capacitive method(add-on type and integral window cover type) as introduced earlier.

(1) Add-On Type

The add-on type is to pattern the touch sensor on a film and a glass, then laminate the cover glass and produce the TSP module, and laminate the TSP module on the TFT-LCD again. Following GFF and GF2 structures are all film method.

The GFF is a standard structure of touch-screen panel. It is made by laminating a glass and two ITO films patterned with X/Y-axis sensor. The GF2 structure needs a glass and an ITO film. It could be made by patterning the X/Y axis sensor on both faces of the film and sticking it to the cover glass. It is a relatively simple method and various decoration on the window is possible. Even though it has a lower optical characteristic, it has an advantage of fine property and low production cost. It is most widely used from a low price product to a high price product.

GF2 has advantages of inexpensive unit price and excellent transparency by reducing one film sheet. However, its mass production is difficult because of limitations in obtaining double-sided ITO film and its difficult pattern etching process.

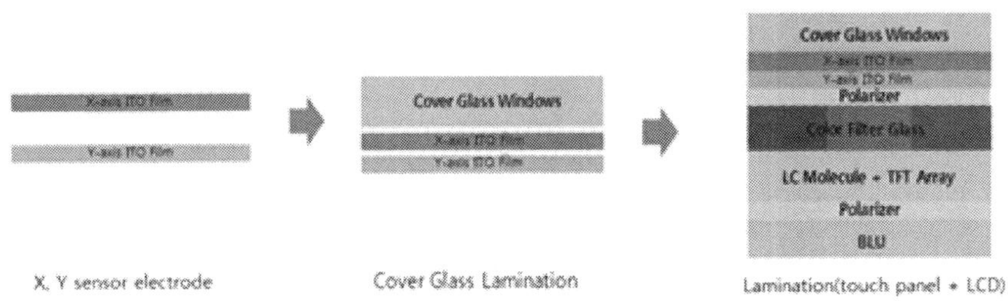

Figure 10.7
Module process of GFF touch panel display

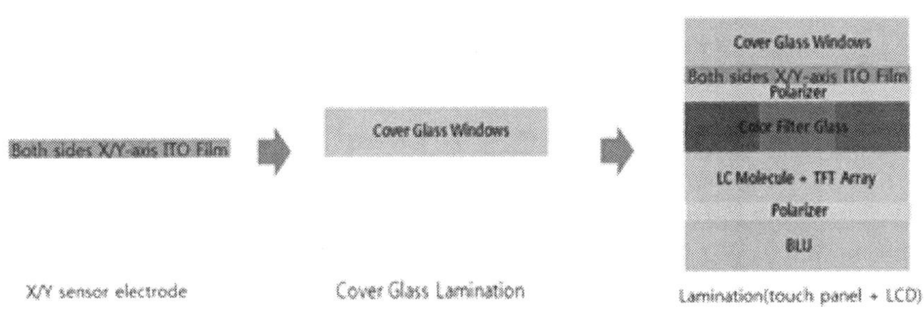

Figure 10.8
Module process of GF2 touch panel display

Next, the GG structure using glass is fabricated by depositing ITO on the top and bottom of the sensor glass, patterning the X and Y axis sensors on ITO, and then attaching a cover glass. The GG2 structure consists of two glasses. Two layers of ITO are deposited on one side of the sensor glass, and the X and Y axis sensors are patterned on ITO, respectively, followed by cover glass. Such a glass method has advantages in that it can realize a fine line width by utilizing a high transparency and a photolithography process, but it is disadvantageous in that it is thick and heavy.

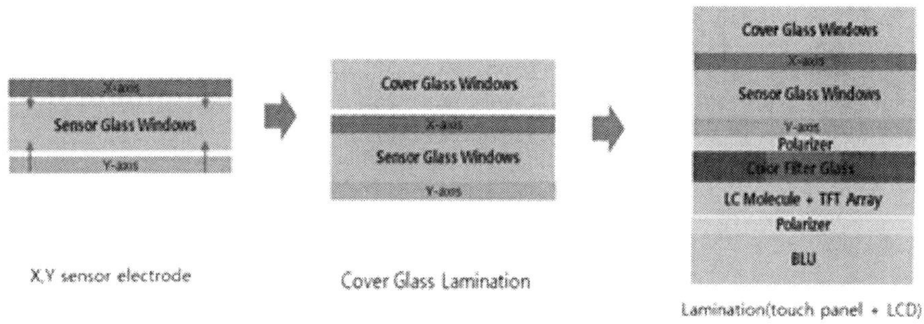

Figure 10.9
Module process of GG touch panel display

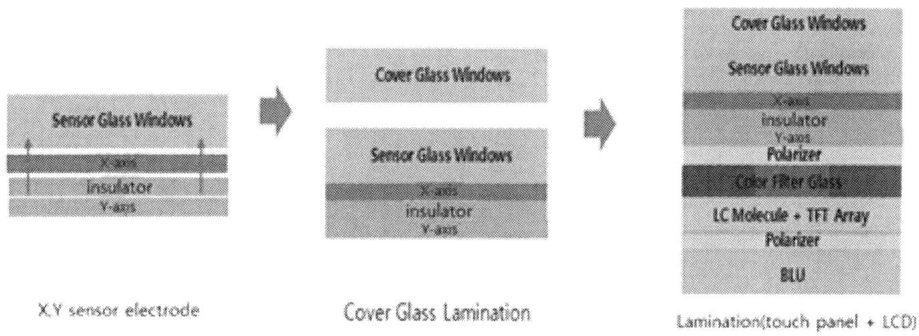

Figure 10.10
Module process of GG2 touch panel display

(2) Integrated cover window type

Unlike GFF method which laminates by depositing ITO onto a film, glass integrated method deposits ITO onto a tempered glass. One sheet of glass and a film are used in the G1F method. For the G1F method, ITO is deposited on the backside of the cover window. The X–axis and Y–axis sensors are fabricated by patterning ITO and ITO films deposited on the cover window, respectively.

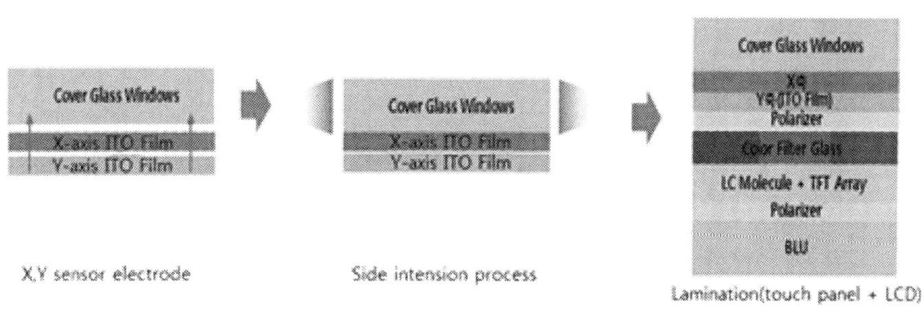

Figure 10.11
Module process of G1F touch panel display

Since only one ITO film is used, glass integrated method is thinner, lighter and optically superior to GFF method. However, it is hard to obtain the yield and has expensive capital investment costs due to the difficulty of ITO deposition and patterning process.

The G2 method is to deposit two layers of ITO on the back side of a cover window. On the deposited ITO, X–axis sensor is patterned and an insulation shield is deposited on the sensor. Then, after depositing another ITO over the insulation shield, Y–axis sensor is patterned on it. This production process of the G2 method is largely divided into sheet method and cell method. Sheet method is to put a sensor on a firstly strengthened glass and implement cutting and side strengthening. Advantages are that the required investment cost is less than the cell method and it is easy to transform into a lot of different forms. However, it has a high rate of defective production, because the strengthened glass cutting process is basically too complicated and sometimes, it must be reproduced because the strengthened part becomes loose after cutting so the defect rate is high. On the other hand, the cell method is to cut a glass board in accordance with the device, strengthen, and put a sensor on the board. The durability of the products is the same as the original method, like GFF. However, it is disadvantageous in that the productivity is lower and the initial investment cost is higher than the sheet method due to the process of the cell unit.

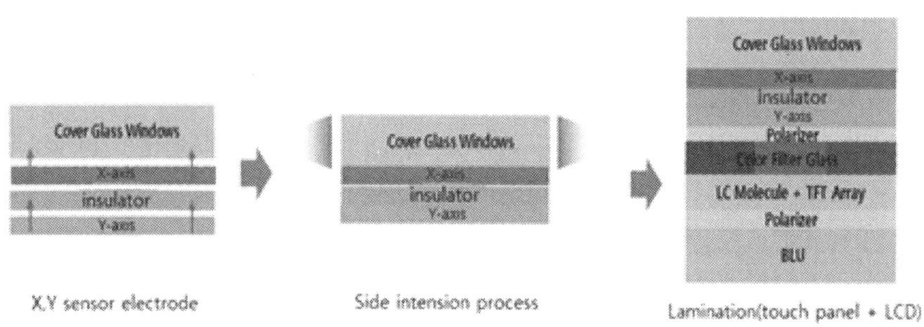

Figure 10.12

Module process of G2 touch panel display

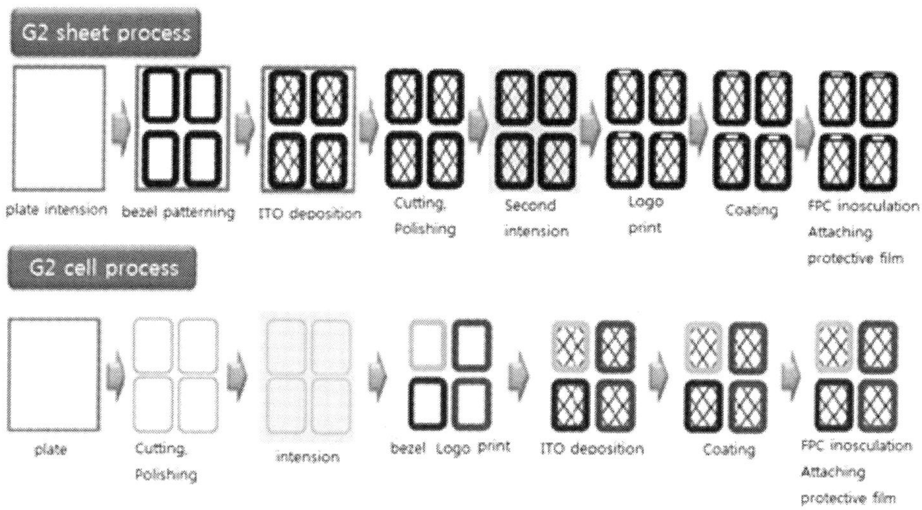

Figure 10.13
Production process of G2 Sheet and Cell method

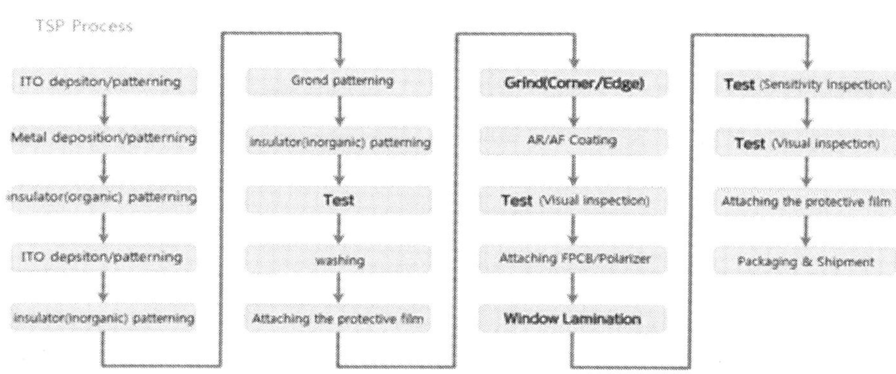

Figure 10.14
Production process of touch screen panel

Display Engineering

* 3 Technology development trend of the touch screen panel

3.1 Multi-point touch screen panel

(1) The kind of touch screen and multi-point system

The user interface has been changed in various forms with the development of techniques. The recent user interface is called NUI (Natural User Interface) which receives input of the gesture and voice. In the future, we will use our hand as input device according to changes in demand for electronics from PC to smart phone and smart watch. They will use 3D hologram display, not like the current 2D liquid crystal. According to this demand, 'Multi–touch' technology is essential to operate the smart devices freely. Among many touch screen systems such as a capacitive system, analog resistive system, ultrasonic waves system and infrared lines system. The capacitive system is most notable touch screen because of excellent sensitivity, high transmissivity, high resolution and excellent durability. Capacitive touch screen is divided into two types. One is the surface capacitive type and the other is projected capacitive type. The surface capacitive type is made up of a cover–layer upon the surface, a layer coated with ITO above board–layer and 4 electrodes. Unlike projected capacitive type, the electrode pattern is not necessary and surface capacitive type can be produced with a lower cost. However, it has difficulty to detect two or more touch points. The projected capacitive type is divided into self–capacitance touch screen and mutual–capacitance touch screen. The self–capacitance touch screen is based on measuring the capacitance of a single electrode with respect to the ground. A transparent conductor is patterned between the electrodes that are separated by one or two layers. When the electrodes are in a single layer, each electrode shows a different pair of touch coordinate and connects individually to the controller. The two layers normally consist of a column of layers and a row of layers. Column and row intersections represent a unique combination of touch coordinates. However, controller of self–capacitance touch–screen can not estimate each intersection because they only measure each row and column, one finger works best when you touch the screen. If your two fingers touch X2, Y0 and X1, Y3 as seen in Figure 10.15, four touch points will be recognized. Since the controller can recognize only two rows and columns, "ghost points" are generated and interfere with accurate touch location measurements. So, multi–touch is difficult in self–capacitance. Mutual capacitance is less sensitivity to EMI (Electromagnetic Interference) and more efficient to use space sensor. Mutual capacitance uses the fact that when most of the conductive objects are close to each other, they can hold an electric charge. In mutual–capacitance touch screen, transparent conductors are usually patterned in electrodes that are spatially separated from the two layers that make up the row and column since each intersection of a column and row forms an intrinsic pair of touch coordinates. By allowing each intersection to be recognized individually, mutual–capacitance touch screen can recognize multi–touch recognition. The ability to recognize every intersection of the electrode is the main advantage with the mutual–capacitance touch screens. In the mutual projected capacitive touch screen, the value of mutual capacitance is changed after applying input on touch panels. The touch detection circuit converts the change in the amount of charge of the mutual capacitance into an electrical signal. The controller processes the information generated by the touch detection circuit and converts the position into coordinates. The coordinates are transmitted to the host processor. At this time, we need a process that changes panel coordinates to display coordinates. A linear transform function changes panel coordinates to display coordinates. The host processor receives the coordinates data, analyzes and then orders the task.

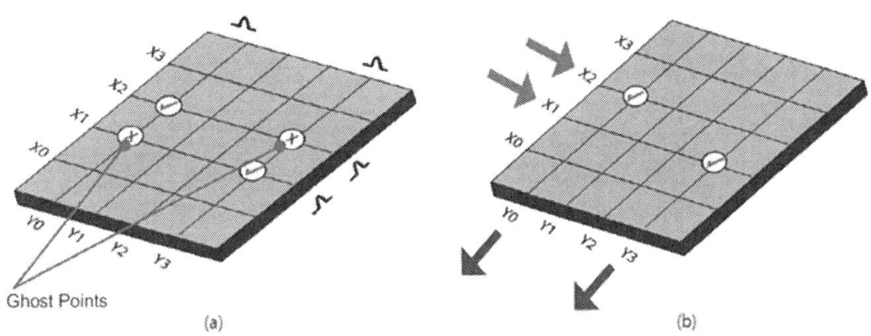

Figure 10.15

multi touch point recognition (a) self–capacitance, (b) mutual–capacitance

(2) Multi–touch point principle

Let's look at how the patterned electrodes in two layers are scanned. Typical projected capacitive touch screens are made of glass top layer, X sensor, an insulating layer, and Y sensor above glass substrate. Panels have a wire for X, Y sensor. For this, the 5×6 panel has 11 connections.(Figure 10.16)

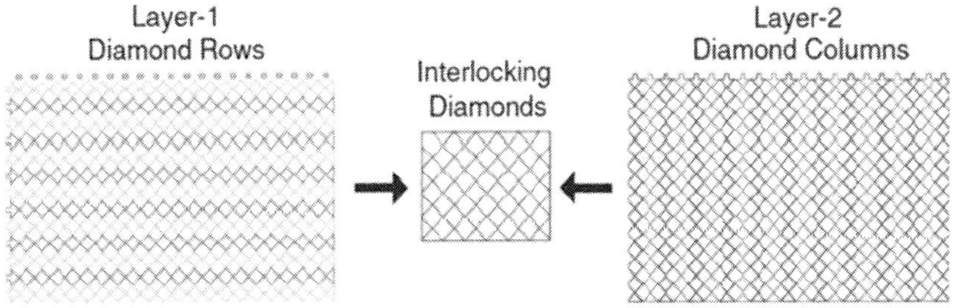

Figure 10.16

Arrangement of touch screen sensor of capacitive type

X, Y sensor is made by a column formed by one layer and a row layer formed by another layer. Glass substrates attached to each of the layers are made lattice–type by connecting to each other. At one layer, there are sensing lines above top plate and a driving line above the bottom plate. The interval of each line is 4~6 mm because the interval recognized on fingers is 4~6 mm.

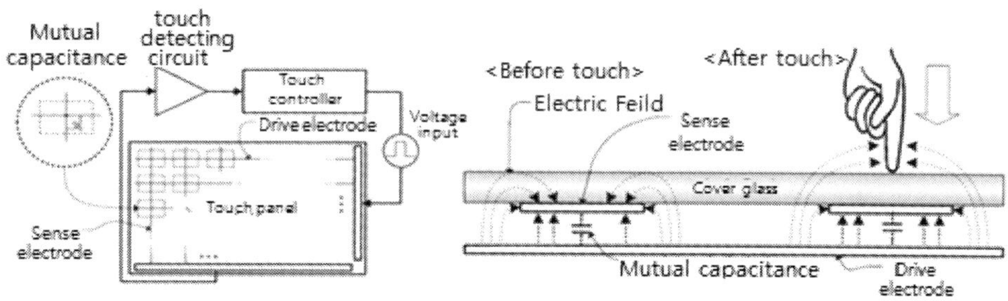

(a) (b)

Figure 10.17

Mutual capacitance touch screen (a) placement, (b) signal processing

In a mutual capacitive touch screen, a drive line for applying a driving voltage and a sense line for outputting are arranged so as to cross each other as shown in Fig. 10–17 (a). These two line's each coordinate point makes mutual capacitance. A mutual capacitance is formed at each intersection of the two lines. When a voltage is applied to the drive line, the mutual capacitance is charged by the electric field formed between the drive line and the sense line. The charge in mutual capacitance is delivered to a detection circuit by the sense line. Figure 10.17 (b) shows changing of touch panels mutual capacitance by touch. When you touch your finger on the touch panel, the body capacitance absorbs a part of an electric field that was formed between the drive line and the sense line. Thus, the amount of charge recharged at mutual capacitance is decreased. The pattern is transmitted to a controller and the signal is processed.

3.2 Internal (On Cell/In Cell)

(1) Structure, advantage and disadvantage of internal display

Conventionally, the TSP has been used as an add–on (out–cell) method, that is, a touch panel is mounted on a display area of LCD panel that is to be used as an input device. In–cell type of display have been investigated and manufactured while TSPs are integrated into display panels between the glasses, as shown in Figure 10.18.

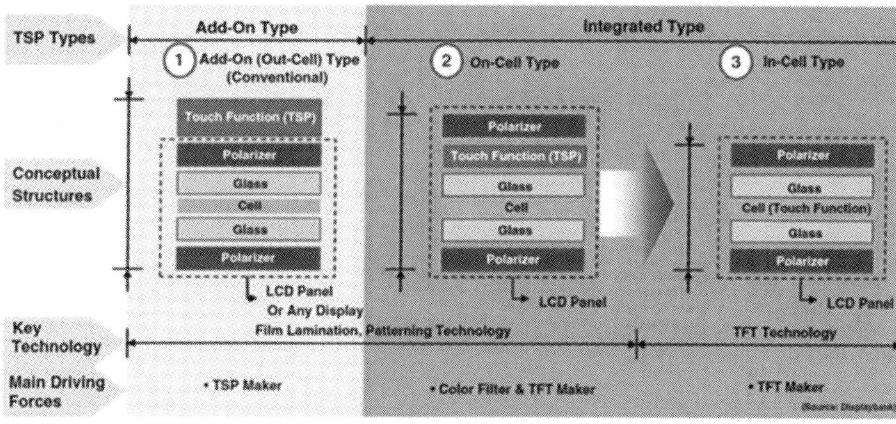

Figure 10.18

Structure by touch screen form

In the existing method of attaching the touch panel on LCD externally (add–on type), the external light is reflected on the upper surface of the liquid crystal panel and the lower part of the touch panel. This leads to a poor visibility in a bright environment such as outdoor. On the other hand, the system–integrated touch screen can be divided into an In–cell technology and an On–cell technology. In addition, the system–integrated touch screen can be made thinner and lighter than the conventional one. At the same time, the light reflection on the surface of the touch screen panel is reduced, and the power consumption required for backlighting is also reduced. In addition, touch screen panel module process disappears, and the production of major parts such as the window glass and the ITO film electrode disappears, which can reduce the cost, which can quickly reduce the proportion of existing add–ons. The in–cell technology, unlike the existing touch method, which is attached to the outside in the form of chips or parts, the touch panel is attached to the display itself, so that users can use thinner, lighter and brighter devices. It is a technology that fully supports touch screen and can be applied to various types of devices. In–cell technology has been studied for a long time, but there are some difficulties. First, a complicated semiconductor manufacturing process is used by integrating a touch sensor function in a pixel on a TFT array substrate, which has a difficulty in a manufacturing process. By integrating the touch sensor in the pixel, the effective display area is reduced, which causes detcrioration of image quality. On the other hand, in terms of cost structure and product development, an on–cell method is known to be dominant. In the on–cell method, a resistive film type touch panel is adhered between an upper polarizing film (polarizer) and an upper glass substrate (color filter substrate) with an OCA (optical transparent adhesive) or an electrostatic touch panel is vacuum-deposited on a glass substrate of a panel. The cost is lower and the yield is higher than that of the in-cell method. In addition, since the area of the effective display area in the pixel is not reduced, image deterioration hardly occurs. As a result, it has attracted attention as an on-cell and conventional add-on technology.

3.3 Pressing sense on touch screen panel

The main disadvantage of the touch screen is the absence of feeling. Because there is no physical force like a keyboard or a mouse, the reaction force is also 0. However, studies are being carried out to give users a more realistic feeling. The principle is that when you touch the touch screen, the minute vibration of the human hand is transmitted. It can realize the feeling of rubber or metal. Consumers of the future will be able to touch the goods before buying them on the internet.

(1) Inertia type Actuator

An inertia type actuator is a vibrating motor type commonly used in mobile phones because of its small size and relatively low power consumption. It refers to an eccentric motor and a linear resonance actuator, and an object with a mass rotates between the two magnets to provide a touch to the user. The eccentric motor suspends an object having a mass asymmetrically on a motor rotary shaft and produces a vibration effect by the eccentric force generated when the motor rotates. The response time of the eccentric motor is 50 ms, which is larger than other actuators, which is inappropriate as a tactile touch for virtual keyboard clicks in a touch device. However, it has excellent durability. Coin-type eccentric motors are very thin and economical. However, it is not used much recently because of slow response time and low durability. Linear resonant actuators have low mechanical wear, it is possible to maximize the intensity of vibration by using the resonance frequency. Currently, most smartphones have internal linear resonant actuator and response time is about 30 ms, which is faster and more durable than eccentric motors.

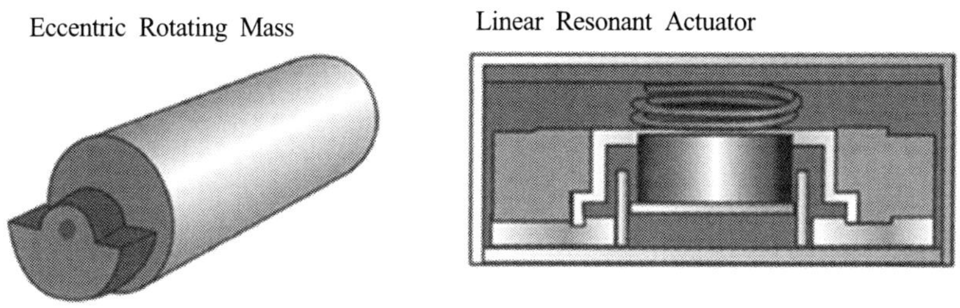

Figure 10.19
(a) Eccentric motor, (b) Linear Resonant Actuator

(2) Piezoelectric Actuator

The piezoelectric actuator is a new type of actuator that operates without magnets or coils. It utilizes a piezoelectric element that has the characteristics of instantaneous change in size or shape by an electric field. Since the response speed is usually less than 5 ms, immediate tactile representation is possible, just as when typing a keyboard. Since the operating frequency is also wide ranging from 150 Hz to 300 Hz, various tactile effects can be obtained using the frequency change. In addition, there are no limitations on size and thickness, so customized production can be done for specific applications or packaging. There are many advantages compared to Inertia type actuator. First, it has no defined shape. Second, it has a fast response speed. Last, it has an advantage of using accurate vibration control and various driving frequencies. However, there is a characteristic that a relatively high current consumption is required for a tactile effect of sufficient strength. The shape is mainly formed into a beam shape or a disc shape. Despite the characteristics of piezoelectric devices using ceramic materials that are fragile, it is expected to be embedded in various portable devices in the future after an attempt to design them durable.

(3) Electroactive polymer actuator

Actuators using electroactive polymers are relatively recent emerging technologies. It is possible to replace the conventional ceramic materials. As shown in Fig 10–20, the electroactive polymer instantaneously changes the chemical bonding structure inside the polymer by the electric field and attaches a mass on the EAP film to produce a vibration effect through the movement of the mass. In general, a separate mass is used, but in the case of a cell–phone, the internal battery itself is often used. Polymers are suitable for use in tablets and smartphones that require cutting-edge technology because it is durable and can be freely designed. The response speed is within 5 ms similar to that of the piezoelectric actuator. Since the reaction speed is fast, the input voltage can be adjusted to realize various irregular waveform vibrations. However, because of the very large potential difference required to drive an EAP actuator, the challenges of high-voltage amplifier design follow.

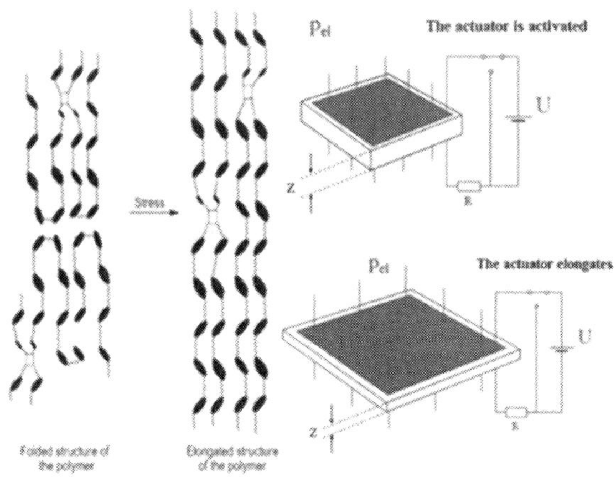

Figure 10.20
Operating principle of the Electroactive Polymer actuator

(4) Electrostatic actuator

Electrostatic actuators are actuators whose surfaces are directly vibrated by touch, unlike other actuators. Instantaneous charging with different electric charges or charging with the same electric charges causes the repulsion and attraction to be generated on the two glass surfaces, thereby generating vibration. There are few mechanical components except for the spacer which acts as a buffer between the two glass surfaces. It is difficult to provide partial touch because the entire film is vibrated. However, it provides a very realistic button click feeling.

3.4 Flexible Touch Screen Panel

Touch screens using conventional ITO films are not suitable for flexible touch screen materials due to their property of being easily broken in case of severe or repeated bending. To replace ITO film, there are CNT, metal mesh, graphene and metal nanowire. However, the CNT has disadvantages that their electrical resistivities are too high and their electrical conductivity is weak.

The graphene is a two-dimensional planar carbon molecule with a very low thickness, which is more than 100 times more electrically conductive than copper and can move electrons more than 100 times faster than monocrystalline silicon used in semiconductors. However, the technology level is still too low and its commercialization takes a lot of time. The metal nanowires are now being studied extensively. Silver nanowires are used to decrease the resistance. The silver nanowire is relatively easy to manufacture and has a long and thin shape, so it has excellent transparency and bending properties. In order to utilize silver nanowires for large-area displays, a coating which can withstand oxidation and physical stress should be used. When using the polymer coating, the physical thickness becomes thicker and transparency and the electrical conductivity is lowered. However, In January 2013, a university research team improved the stability, uniformity, and transmittance of the transparent electrodes by using oxidized graphene. The oxidized graphene is a carbon nanomaterial composed of a single layer. It has high transmittance and durability to organic solvent, heat, light, gas (H_2S) and so on. It is not only suitable for use as a protective layer but also environmentally friendly because it can be mixed with water due to its hydrophilic property. The oxidized graphene is twice as resistant as a transparent electrode with silver nanowires. It also has improved haze (blurred phenomenon) and transmittance. The copper, silver nano-ink, and silver nanowire have attracted attention as materials for a metal mesh. However, a copper metal mesh has a visibility problem and for the silver nanowire, a haze phenomenon occurs in the process of continuously growing a wire. It is also a disadvantage that there is a possibility that a short circuit will occur due to the dendrite phenomenon(growth like branches). The silver nano ink is arranged in a lattice pattern (orthogonal form) to form an electrode. Electricity passes through a thin white wire (silver nano ink), and light is passed through to the black vacant area to secure visibility. In the meantime, since the line width of the nano ink cannot be reduced, the visibility is lowered and it is difficult to commercialize. The problem to be solved is to make the line width as fine as possible to allow light to pass through or to increase the light transmittance of the material. Even if the line width is only 3 μm, users can identify it on devices such as smartphones. The visible array is called the Moire phenomenon that is a visible phenomenon of the lattice pattern on the screen when using a smartphone under the sun. It is related blackening but discussion on this topic is omitted here. The silver nano ink developed by 'I' company is blackened and does not reflect sunlight, so Moire phenomenon does not occur.

3.5 R&D Status of Transparent Conducting electrode

Nowadays, utilization of smartphone has increased and so the demand of TSP, which is essentially applied to smartphones is increasing exponentially. Currently, most transparent electrode materials used in TSP adopt ITO, and conventional ITO transparent electrodes are not applicable to flexible display touch panels due to their low mechanical properties. Therefore, new applicable transparent electrodes should be developed. In recent days, graphene, carbon nanotube (CNT), and Ag nano wire have been actively studied to replace ITO. On Feb. 16, 2014, Sungkyunkwan Univ. the team of Prof. Hyo yeong Lee has developed a transparent electrode that "stretches" beyond the "bending" phase using Ag nano wires.

(1) Graphene

The graphite, which is used in the pencil lead and is familiar to us, is a structure in which carbon layers are piled up with layers arranged like honeycomb hexagonal nets. One layer of graphite is called graphene. The graphene has the thickness of 0.2 nm and very high physical and chemical stability. In 2004, researchers from Britain's Andre Geim and Konstantin Novoselov succeeded in separating graphene from graphite using transparent tape at room temperature, which in turn won the Nobel Prize in Physics in 2010. Graphene has more than 100 times more electricity than copper and has more than 100 times faster electrons than silicon, which is mainly used as a semiconductor. Strength is 200 times higher than steel and has twice the thermal conductivity than diamonds with the highest thermal conductivity. Also, since most of the incident light passes through, it is transparent and has excellent stretchability. Because of these characteristics, the graphene is evaluated as a material that surpasses the carbon nano tube, in the spotlight as the next-generation's new material, and called 'the dream nano-particles'. The graphene and carbon nanotubes have very similar chemical properties and can have separate metallic and semiconducting properties through post-processing. However, graphene have more uniform metallic properties than carbon nanotubes. There are many applications of graphene such as high-speed semiconductors that utilize high electrical properties, bendable displays that use transparent electrodes, computers that operate with displays only, and high-efficiency solar cells with high conductivity. In particular, it can be used for bendable displays, wrist-mounted computers and electronic paper and it is attracting attention as a new material of the future.

(2) CNT(carbon nano tube)

Dr. Ijima Sumio of the Institute of Electrical Engineers of Japan (NEC) discovered CNT during the analysis of the carbon mass formed on the cathode of graphite using the electric discharge method in 1991. The hexagonal shapes of six carbon are connected to each other to form a tubular shape. The diameter of the tube can be from a few to several tens of nanometers and is called carbon nanotubes. A nanometer is one-billionth of a meter, a hundredth of a thousand times the thickness of a hair. Electrical conductivity of CNT is similar to that of copper, its thermal conductivity is equal to that of natural diamond, and its strength is 100 times better than steel. The carbon fibers can be broken even when only deformed by 1%, while carbon nanotubes can withstand 15% of deformation. Since the discovery of this material, scientists have focused on synthesis and applications, and there have been numerous developments in devices using carbon nanotubes, including semiconductors and flat panel displays, batteries, super fibers, biosensors, and television cathode ray tubes. It is also used as a nano forceps that can transfer nano-sized materials.

(3) Ag nano wire

The technology to make this with a microscopic wire with a cross-sectional diameter of about 20 to 40 nm is considered as one of the 10 new technologies to change the world and is now considered as one of the most efficient fields in nanotechnology. Nanowires are used in various fields such as lasers, transistors, memories, and sensors for chemical detection (detectors). The materials used here are semiconductor silicon, chemically sensitive tin oxide, and gallium nitride, which is a light emitting semiconductor. A nanowire, based on silver, is a substitute to indium. It has relatively simple production process, thin and long shapes, resulting in good transparency and flexibility. However, in order to utilize silver nanowires in an area display, it is necessary to coat them to withstand oxidation and physical stress. When the nanowire is coated with a polymer in a conventional method, the surface becomes thick, and transparency and electrical conductivity deteriorate.

(4) Organic transparent electrode

A conductive polymer, which an organic material, can be coated on a large sized flexible substrate using a simple method like spraying. The conductive polymer, which is a π−conjugated polymer, is an ICP(intrinsically conductive polymer) material like poly−acetylene, poly−pyrrole, poly−aniline, PSS (poly−phenylene sulfide), poly−thiophene and PEDOT. Among them, PEDOT:PSS, which has excellent processability, excellent thermal stability and high electric conductivity, is the most widely used conductive polymer and has a conductivity of 500 to 1,200 S/cm. In case of PEDOT group polymer, by using atmosphere pressure plasma method, a film can be made to have about 300 Ω/sq and T=80% at room temperature or through combination of single wall CNT or Au nano particles, it is possible to fabricate a transparent conductive film having a sheet resistance of 100 to 150 Ω/sq and T = 80%.

(5) ITO ink−jet printing

Indium, the main ingredient of ITO (Indium tin oxide), needs to be replaced due to its high price and resource depletion. From this point of view, attention has been paid to a technique of using a conductive polymer as a transparent electrode. The conducting polymers have been actively studied due to their applicability in various fields including fuel cells, displays, actuators, and electro−static/electro−magnetic shielding conductive coatings. Particularly, a conductive polymer patterning technique for forming a thin film transistor or a wiring electrode as a conductive polymer in a flexible display, is considered as a next generation display device. It has been studied in academic and industrial fields. With such a conductive polymer patterning technique, a solution process for printing on a flexible support using an ink−jet printer has recently been developed. Such a solution process using ink−jet printing has the advantage that the desired pattern can be freely formed without a photomask in a short time and no waste is generated during the process. However, in the existing technologies, there is a problem of different optical transmittance between the region where a conductive polymer pattern is created and the other region. In this case, the visual quality of the display or the touch sensor, deteriorates.

(6) ITO nano powder and ink

ITO powder is essential material for a transparent electrode that drives information displays like TFT-LCD, PDP, Organic EL. Since the specific surface area of nanoparticles is much larger than that of conventional particles, the density of the evaporation material increases as the size of the particles decreases. When the electrode is deposited using the nanoparticles, its characteristics can be improved. However, the existing process technology for manufacturing ITO particles is the wet process using melting and precipitation reaction. Since these raw materials contain a large amount of chlorine and nitric acid components, the treatment process is very difficult. This is followed by wastewater treatment for the removal of chlorine and nitric acid to remove the adverse effects on the particles and devices resulting in the enormous consumption of water resources. Recently, a low-temperature synthesis method is applied to improve the environmentally hazardous and high-temperature heat treatment process. In other words, studies have been actively conducted to prepare oxide powders by excluding existing chlorine and nitric acid from raw materials.

3.6 Connection to transparent OLED

Organic electroluminescent displays (OLEDs) are gaining popularity to display lighter, thinner, and faster-moving images than liquid crystal displays (LCDs). Furthermore, efforts are being made to extend the visual constraints of existing displays by implementing them as transparent pixels and transparent driving elements. If this transparent OLED and touch screen are connected, it can be applied to various forms such as automobile navigation system, transparent advertisement window, transparent screen mat, transparent digital table, and goggles.

(1) Transparent OLED and touch screen interface process method

When a touch screen is applied to a transparent OLED panel, the process is as follows.

(1) Step of providing a transparent substrate having two surfaces

(2) Step of forming flat-panel organic light emitting diode display on one side of the substrate

(3) Step of forming resistive film on the other side of the substrate using a

low-temperature technology

(4) Step of forming a resistive touch screen on the resistive film

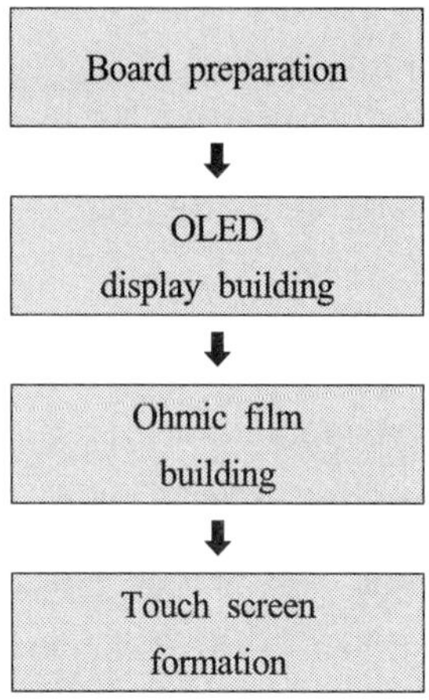

Figure 10.21
Transparent OLED and touch screen interface process method

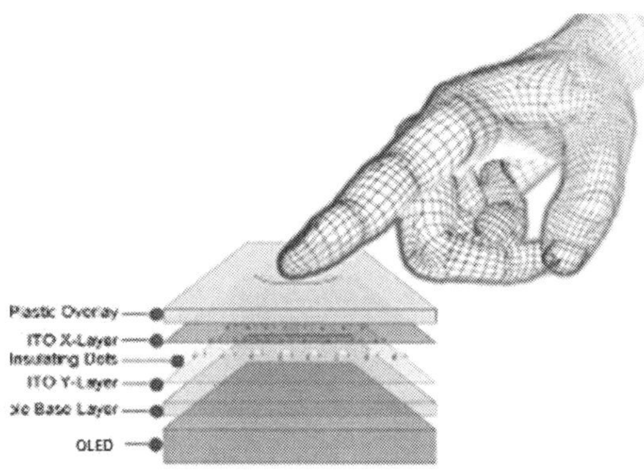

Figure 10.22
Schematic of transparent OLED and touch screen

When the panel is pressed, the ITO X-layer and the ITO Y-layer are brought into contact with each other, so that the electrical signal is transferred to the controller. In the driver, like the PC, the operation value is transferred to the OLED through operation, and the OLED of the pressed part or the commanded part is operated.

(2) Problems and development trend of transparent OLED and touch screen connection

When a touch screen is used in a flat panel display, the touch screen is simply placed on the flat panel display and coupled by a mechanical mounting means such as an enclosure. However, this prior art arrangement of combining a touch screen and an

OLED display has several disadvantages. Also, the use of a frame increases the number of parts, weight, and cost of the device. The distance between touch screen and the display increases thickness. Excess components found in displays and touch screens increase costs and reduce performance compared to more integrated solutions. In addition, the need for separate cabling for the touch screen increases manufacturing costs. Thus, there is a need for an improved touch screen flat panel display system in which device weight is reduced, excess materials are removed, the cost is reduced, no special mechanical mounting design is required, reliability is increased, and degradation of image quality is minimized.

The most common use of transparent OLEDs is to connect with wearable devices or flexible displays, such as attaching them to automotive glass, as mentioned above. The most important thing is the flexibility of the touch panel. It is also important that the flexible transparent display substrate is bent and the touch panel should have the same flexibility or similar specifications as the display substrate, but the biggest problem is that such a panel is still in its early stages of development.

The second problem is the transmittance of the touch panel. One of the biggest revolutions that transparent OLEDs can bring is the design revolution that allows you to change the design of an electronic device or any object at will. The flexible transparent OLED is needed to accelerate and commercialize the design revolution and the transmittance of the touch panel will emerge as a major issue. If the transmittance of the transparent OLED and the touch panel are both 90%, The transmittance of the display will significantly fall to 81%. To avoid this, transparent materials such as graphene, carbon nanotubes, etc. should be used or more transparent touch panels should be developed through rearrangement of materials.

3.7 Non-contact vision-based system

The non-contact vision system is an early stage of the market, which is beginning to be applied mainly to large-sized displays such as electronic signboards and TVs, which are difficult to implement in a capacitive manner because the inputs are transmitted by means other than touch or buttons.

The non-contact vision system refers to the way in which the input device interprets the user's actions and accepts the commands, rather than in direct contact with the input device. The input device interprets the motion of the user's hand, iris, and the like and accepts the command through it. Because there is no need to learn to use the input device, the vision-based input device provides a more intuitive interface and convenience to the user. Although the touch screen also provides an intuitive interface, the intuitive aspect of vision is beyond the touch screen that requires minimal contact with the device, and unlike the disadvantages of the touch screen, which becomes more difficult to use as the size of the display increases, the size of the display is not affected. In addition, the vision method gives the user some kind of pleasure by issuing the commands, so 'Kinect' using the vision method became the fastest selling home appliance.

Recently, as the device has become slimmer, the need to simplify various operations such as wireless internet use, video, and multi-touch on one screen has become more important, and the touch screen has become an optimal UI for user convenience. The noncontact vision system can be seen as an extended concept for this purpose. The noncontact vision system has various meanings, including voice recognition method, camera image recognition method, input method by motion change, and proximity sensor input method. Usually, a user inputs an image or an image of a user-object using a video input device such as a camera and derives a user's instruction that can be deduced from the processor. So basically it depends heavily on the algorithm of the image analysis technology, the development of the processor, and the camera must necessarily be accompanied.

Siri of iPhone and Galaxy series Svoice are voice recognition methods. These applications are evolving into a new service type in combination with an intelligent system, and user's gesture recognition through the camera can be realized by object recognition. The application range is large and the research is actively performed. The noncontact vision system can be divided into three methods. There are advantages and disadvantages for each method.

(1) Ultrasonic method

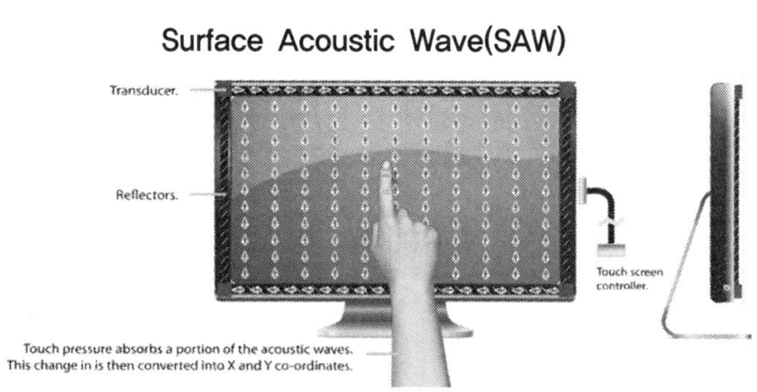

Figure 10.23
Elastic surface wave

The ultrasonic method is a method in which the reduced size of the emitted ultrasonic wave due to the obstacle is sensed. It consists of a glass substrate arranged as a sound wave reflector on the panel rim. The ultrasonic waves emitted from the panel corners are reflected by the sonic reflector and travel to the opposite side of the panel. When a finger touches the panel, a portion of the ultrasonic waves is absorbed and the receiver senses the input by detecting the ultrasonic waves having weaker intensity at the receiver. The ultrasonic method is advantageous in large size, has a light transmittance of 92% and has high durability. However, it is vulnerable to sensor contamination and liquid, has high entry barriers, has a very high manufacturing cost, and has a technical limitation that multi-touch is not possible. The ultrasound method has been implemented only for monitors of 10 inches or more. It has many advantages, but deficient technology has emerged in present application areas.

(2) Infrared(IR) method

The infrared method utilizes the characteristics of the straightness of infrared wave while it is not visible to the human eye and it works by recognizing obstacles and sending appropriate signals. The IR method can be considered to be performed by using the surface acoustic wave converted from the bulk wave generated from the electrostrictive element, and the position detection is performed by converting the receiving time difference of the energy. These IR systems have a 100% light transmittance and excellent durability, but they are expensive to manufacture, and the resolution is low and the position of the touch frame is on the screen so that before the finger or pen actually touches the surface, it is susceptible to initial activation. In addition, since the recognition rate of the IR light is low, it is impossible to recognize one point, unlike the case where the user directly touches the display, and only motion such as movement, rotation, and enlargement can be recognized. Infrared is being applied to 15 inches or more TV, 10 inch or more monitor, ATM, KIOSK and medical devices. In addition, it is gradually evolving through various improvement directions such as improvement of a device of a noncontact touch switch, recording of physical contacts with a screen, and positioning reliability.

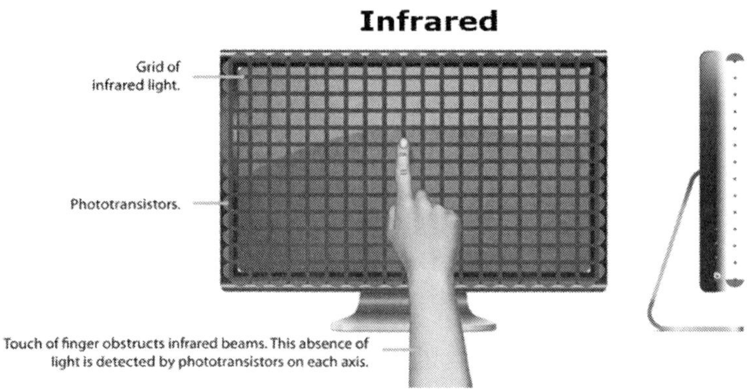

Figure 10.24
Infrared ray method

(3) Optical method

The non–contact vision system uses a video input device such as a camera to store an image or an image of a user–object and derive a user's command that can infer from the processor. So basically it depends heavily on the algorithm of the image analysis technology, the development of the processor, and the camera must necessarily be accompanied. Therefore, there are inevitable disadvantages in the case of the contactless vision system. First, if you use only one camera, the recognition rate is significantly low even for just motion recognition. It is impossible to plot background and subjects, in other words, background and user behavior is recognized only by color information. In a color environment that is easy to be confused with the user, the accuracy of vision–based on a single camera is extremely poor. In addition, because there is no depth information, only two – dimensional motion can be recognized and there is a limit to the recognizable motion. For example, Vega's motion recognition function and Galaxy S4's iris recognition function are very low in accuracy, and the kinds of information that can be accepted are limited. Above all, vision systems that rely on cameras become obsolete in low light situations.

To overcome this, a vision system using two cameras was introduced. It is installed by crossing two cameras, analyzing images that came in with each camera using a processor, and creating a virtual 3D space to calculate depth information. This can solve the problem that only information that can be received by a single camera is RGB information. It is now possible to recognize depth information. Therefore, it is possible to recognize the distance from the user, the three–dimensional motion, and the like, rather than reading only the color value of the user. Therefore, it is possible to solve the problem that the motion and command of the user may not be recognized due to the surrounding color, the problem that the recognizable operation does not vary, and the problem that the accuracy of motion recognition is not high. However, this also has not solved the disadvantage that the accuracy of the distance is limited and it is useless in low light conditions. This is Kinect 1.0. Kinect 1.0 succeeded in measuring the depth of the space and the user by using two cameras, but it did not solve the difficulty in the situation where the accuracy of the distance value was limited and the light was low. The disadvantage of this method itself is that it is impossible to measure the distance value again if the two cameras are out of the overlapping space. Therefore, it is possible to measure the distance value and the depth of the user in this manner only in a predetermined space, and there is a disadvantage that the user is forced to use in a narrow space. To solve this problem, there is a method of using one camera for obtaining RGB values and an infrared projector for supporting the camera. It recognizes RGB color information according to a single camera, then emits a number of infrared beams to an infrared projector, receives and analyzes the reflected values, and obtains distance values. After this information is combined, the RGB value and the distance value can be obtained accurately. In this method, the distance can be obtained more accurately than the conventional method using two cameras. Even if the RGB value is not obtained in the light-reduced state, it is possible to obtain the distance. Therefore, even if there are some incomplete parts, There is an advantage that it is possible to recognize the command. In addition, since the range of recognition by two cameras is narrow, the usable space of the user is limited. Also, since the common recognition range of the infrared projector and the RGB recognition camera is very wide, the usable space can be relatively widened. The product that uses it is Kinect 2.0.

The most advanced noncontact vision methodology reported in March 2014. It uses two cameras, one infrared projector, and a high-level processor. In real time, the object and other spaces are read into a single device, and the virtual 3D space is reconstructed. It is aimed at enhancing reality by linking with reality. This is under the name "Project Tango" by Google, and it is still only available at the prototype stage and has limited use.

(4) Non-contact vision-based system development trends

1) Retro-reflection

The retroreflective (infrared image sensor, camera) system consists of two infrared ray emitters and infrared image sensors combined. The emitted infrared rays are reflected from the reflection frame placed on the three sides of the touch area and come back to the image sensor. As the reflector attached to the bicycle, the light comes back in the direction of incidence, and the infrared light does not come back when touched, so the part becomes a shadow on the image sensor. You can see the angle of the touch position at the shadow position. If you measure this at two different locations, you can calculate the location touched by the triangulation. It is mechanically simple and don't break down, is easy to make, light and it can be installed regardless of subject.

It is also advantageous that it is easy to make in a large-size and the area, and it can be made thin. It is advantageous in that a plurality of locations can be detected by a suitable algorithm, the light transmittance is 90 to 100%, the external light reflection/visibility is excellent, and the environmental durability is also excellent. However, it is difficult to miniaturize, it is easily affected by dust or insects, and the image sensor is expensive. In order to overcome these shortcomings, we have been developing a chip capable of processing large data or by reducing the resistance of glass.

Such ultrasonic, infrared, and optical methods represent touch panels that can be used regardless of the input method such as stylus pen, durability and a drag of capacitance resistive method and glove or bare hand. At CEATEC JAPAN in 2010, Alps Electric has exhibited a TV moving system with gestures. When a human being gestures in front of a TV, the machine responds to the signal and can perform channel change and volume control functions according to various commands. In recent years, mobile phones manufactured by LG and SAMSUNG have recognized the motion of people, and have put in a technique to process the operation in a non-contact format that does not touch the screen. Also, in order to solve disadvantages of ultrasound and disadvantages of infrared and optical methods, these companies have been researching and filing patent applications to advance the development of contactless vision method.

2) Organic photodetector sensor

Organic optical devices are made of semiconductor materials obtained from the organic chemical industry. Printed electro-optical elements are made by accumulating organic materials on a plastic-based material by engraving patterns. Typically, an optical diode converts a photon into an electron current. The major advantage of an optical element is that it can detect a large area with a sensor. In addition, it is possible to detect from a wide spectral range, that is, a visible light range to an invisible light range, consumes less power than conventional sensors and has a wide dynamic scanning range. When installed on the display, it does not leave afterimages on the image displayed on the actual screen, and the noise is also small. Prices are more competitive than amorphous silicon and CMOS technology. In the real world, this encompasses the area of the amorphous silicon and shows better performance. For example, an organic optical device does not use a camera lens, and there is no additional current consumption, space, or power to be consumed by the device when the camera is used in this area. It is thin, light and flexible, so it can be applied to the next generation flexible LCD market. Also, there are few restrictions on the first recognition distance problem, which is a disadvantage pointed out in CMOS. For example, a CMOS sensor needs to be separated by a certain distance, but an organic optical element has no such inconvenience.

Exercise

1. Describe key features of each type of touch screen panel.

2. Provide a structural diagram of a resistive touch screen panel sensor and explain how it works.

3. Describe the structure of the capacitive touch screen panel sensor and explain the operation method.

4. Explain the various structural diagrams of the capacitive touch panel.

5. Present the process diagram of the touch screen panel and describe each process.

6. Describe the built-in touch screen in terms of in-cell technology and on-cell technology.

7. Describe the configuration and operation principle of contactless touch screen panel.

9. Derive the response rate formula from the formula for the response speed of the LCD. Describe the properties of the liquid crystal material and the liquid crystal cell parameters.

10. Discuss what points to consider when designing an automotive LCD

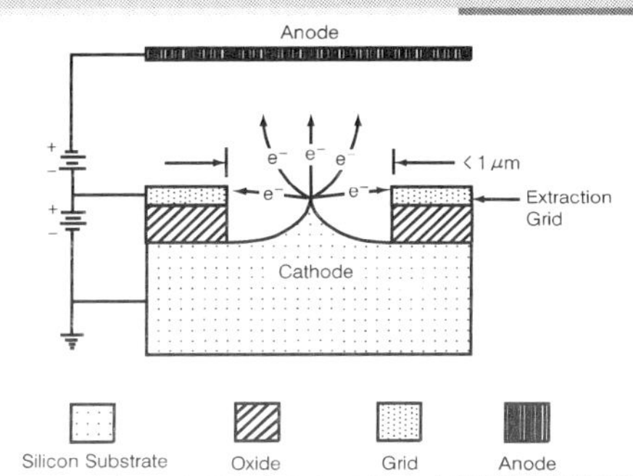

Display using electron beam

Chapter 10 → 11

1 ntroduction and History of Display using electron bear

2 Electron motion and field emission principle in electric field

3 Principle of luminescence of electron beam utilization display

4 Structure and Operation Principle of Electron Beam Display

5 Types of Electron beam display

6 Status of Electron beam Display

Introduction and History of Display using electron beam

1.1 Display using electron beam

CRT (Cathode Ray Tube) was invented in 1897 and has been steadily used by many users. It is withdrawn from the market due to the remarkable development of a flat panel display such as LCD and LED displays. However, it is obvious that CRT has been used for the longest period of time. In addition, in the past when the insufficient performance of the FPD could not replace the CRT, those who are familiar with the excellent performance of the CRT focused on a FED (Field Emission Display) which had a similar principle to that of the CRT. The concept of FED first appeared in 1970. It has evolved with intensive research since 1985. The FED is a cold cathode electron source and uses field emission array (FEA) in a matrix form. Its working principle is similar to the CRT; it uses phosphor where bombarding electron beam generates image by cathodoluminescence. Like this, CRT and FED have the similar principle. These are called displays using the electron beam. Display using electron beam has high color reproducibility and high reactive display having a wide viewing angle. Although it is not currently used, it is clear that it is a display that has been used for a long time due to its outstanding advantages.

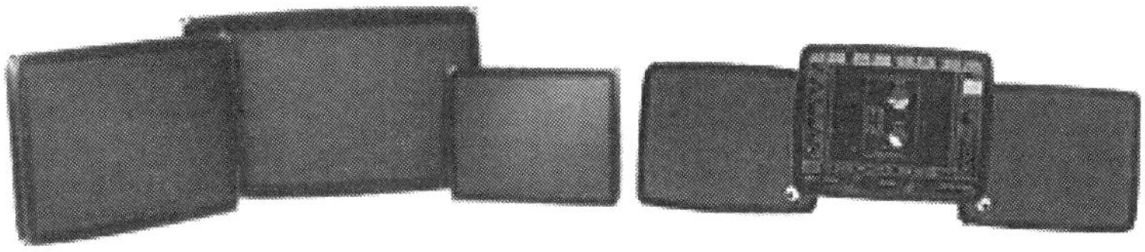

Figure 11.1
Various types of CRT (cathode ray tube)

1.2 History of the FED

FED was introduced in 1970 for the first time. It has evolved by "Silicon Video Corporation1" since 1991. FED technology started from the creation of a cold cathode array capable of emitting electrons at a low voltage by Dr. Spindt of SRI in 1979. It was possible to have a 25,000 hour lifetime due to the development of the Spindt electrodes. A full-fledged research was conducted in accordance with the creation of Monochrome display in 1985. Microwave tube was created in 1986 for the first time. LETI in France made a new change by successful commercialization of flat panel display, driven as the matrix, using FEA as a unit pixel. In Korea, although there were developments like the creation of 4" color FED (Spindt-type) in 1996, they disappeared eventually because of failure of enlargement.

1.3 History of the CRT

The history of the CRT goes back to 1603. The first artificial phosphors were discovered by an Italian alchemist "Vincenzo Cascariolo". Since then, CRT has been developed steadily by the effort of many scientists. Early 1897, "Karl Ferdinand Braun" (1850–1918) invented the basic CRT including all the functional concepts. It is the basic feature of all the current CRT that contains electron source, focusing, deflection and acceleration components, phosphors, screen, etc. In the same year, Braun announced all five components of CRT, the origin of cathode ray (cathode), an engine providing a centralized point(diaphragm), an engine to accelerate a beam of a cathode ray (anode), a deflection system to indicate several points on the screen and a phosphor screen to make the cathode ray beam visible. In the United States in 1921 CRT was first commercialized by "H. By J. van der Bijl" and "John B. Johnson". In 1950, a shadow mask method, the most important technology of CRT, was completed. This advanced CRT dominated the display market for about 50 years. However, due to the steady development of the FPD, it started to disappear from the market

Figure 11.2

Initial electrostatic deflection CRT of the first inventor Dr. Karl Braun and Von Ardenne

Display Engineering

* 2 Electron motion and field emission principle in electric field

2.1 Motion of particles having electricity

Electrically charged particles are influenced by the electric field and magnetic field. In this chapter, the motion of the charged particles under the influence of the external force will be considered. The effect of electric and magnetic fields on particles having charge q is given by the Lorentz[1)] force and is shown in equation (11–1) below.

$$F = q(E + v \times B) \quad \text{(Equation 11–1)}$$

In this equation, v is the speed of charged particle and B is the magnetic field. $v \times B$ is the cross product of v and B expressed as equation (11–2) below.

$$(v \times B) = vB\sin\theta \quad \text{(Equation 11–2)}$$

This means that the direction of charged particle and electric field have the direction of 'right–handed screw motion' during rotation. Equation (11–1) determines the motion of all the particles described in this chapter. When a current flows through the conducting wire in the magnetic field B, the conducting wire receives the force from the magnetic field. The current is the flow of electric charge, so the electric charge that moves in the magnetic field receives the force. When a current flows through the conducting wire in a uniform magnetic field B, this conducting wire receives the electromagnetic force expressed as $F = BIl$, and its direction is perpendicular to the directions of B and I. At this time, free electrons move opposite to the direction of current in conducting wire and this electron receives the force f toward the direction of the force F. If the current I flows through the conducting wire with length l that lies perpendicular to magnetic field B for time t, the force F applied to the conductor becomes $F=BIl$. Since the current I flowing through the wire is the amount of charge passing through the section of the wire in unit time, $F=BeNl/t$. The e is the charge amount of free electrons and N is the number of free electrons passing through the cross–section of the wire, so the electromagnetic force is expressed as $F=BeNv$, and v is the speed of free electrons passing through the conducting wire. Therefore, the force that one free electron receives becomes $f=F/N=evB$. In general, the force applied to the charged particles with charge q, which moves at a velocity v in a direction perpendicular to the magnetic field B, is $F=qv$ B(N). This force is called electromagnetic force or Lorentz force. On the directional side, when the (–) charge moves, current flows in the direction opposite to (–) charge's motion and When the (+) charge moves, current flows along the direction of (+) charge's motion. Then, we can know the direction of the electromagnetic force.

---

1) Lorentz, Hendrik Antoon, 1853~1928 : he focused on Maxwell 's theory of electromagnetic phenomena and clarification of electronic phenomena and optical phenomena. In particular, he has studied the electromagnetic properties of materials and interaction with light expressed by various material constants.

2.2 Movement of electrons under a uniform electric field

When the electron with mass m and charge $q = -e$ (-1.6×10^{-19}C)moves along the x-direction under a uniform electric field with no magnetic field, the relation between Newton's equation[2)] and equation (11-1) expressed by equation (11-3).

$$m\frac{dv_x}{dt} = -eE_x \quad \text{(Equation 11-3)}$$

The electric field E is the x component and the acceleration a is the derivative of time respect to velocity. Thc velocity is the derivative of the distance respect to time, so the acceleration can be expressed as a second derivative of the distance respect to time. Therefore equation (11-3) is:

$$\frac{dv_x}{dt} = \frac{d^2x}{dt^2} = -\frac{e}{m}E_x \quad \text{(Equation 11-4)}$$

Equation (11-4) is the differential equation of the particle motion. When equation (11-4) is integrated, the speed can be obtained.

$$\frac{dx}{dt} = v_x = v_0 - \frac{e}{m}E_x t \quad \text{(Equation 11-5)}$$

When equation (11-5) is integrated again, it becomes:

$$x = x_0 + v_0 t - \frac{e}{2m}E_x t^2 \quad \text{(Equation 11-6)}$$

x_0 and v_0 are the initial point and speed at time $t = 0$, respectively. If the electrons do not move at the initiatory stage, $x_0 = v_0 = 0$. Then the position is the secondary function of time which is given by

$$x = -\frac{e}{2m}E_x t^2 \quad \text{(Equation 11-7)}$$

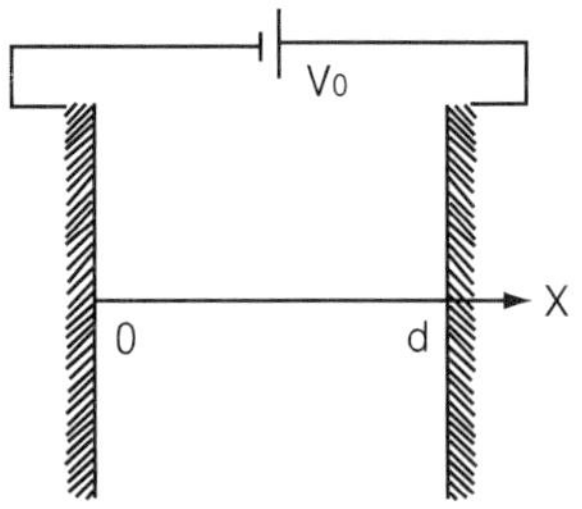

Figure 11.3

E between two parallel plates

2) An object is moved by pushing with force. If given more force, it moves faster. This relationship is expressed by Newton's equation $F = ma$ where the force is F, the mass is, and the acceleration is a.

Example 11-1 An Example

Consider the acceleration of electrons between two parallel plates as shown in Figure 11-1. Initially, the electron has speed 0 and the position x is 0. Then, a potential is applied between two parallel plates placed at $x = 0$ and $x = d$. The electric potential is V_0 volt. The electric field is uniform between two plates and expressed by equation (11-8)

$$E = -\frac{V_0}{d} \quad \text{(Equation 11-8)}$$

(−) is due to the direction of the electric field. To get the time for the electrons to reach the plate of $x = d$, x is changed to d in Eq. (11-7), and E_x is rewritten using Eq. (11-8),

$$d = \frac{eV_0}{2md} t_d^2 \quad \text{(Equation 11-9)}$$

$$(x = -\frac{e}{2m} E_x t^2,\ E = -\frac{V_0}{d})$$

or

$$t_d = \left(\frac{2m}{eV_0}\right)^{1/2} d \quad \text{(Equation 11-10)}$$

Substitute the equation (11-10) in the equation (11-5) and the speed v_d at $x = d$ is defined as the equation (11-11).

$$v_d = \frac{eV_0}{md}\left(\frac{2m}{eV_0}\right)^{1/2} d = \left(\frac{2eV_0}{m}\right)^{1/2} \quad \text{(Equation 11-11)}$$

If the acceleration potential V_0 is 1kV, then

$$v_d = \sqrt{\frac{2 \times 1.6 \times 10^{-19} \times 10^3}{9.1 \times 10^{-31}}} = 1.87 \times 10^7 m/s$$

$$v_x = v_0 - \frac{e}{m} E_x t,\ E_x = -\frac{V_0}{d},\ t = \left(\frac{2m}{eV_0}\right)^{\frac{1}{2}} d$$

Equation (11-11) can be predicted from the energy relationship because the energy of the electron is $\varepsilon_d = eV_0$ and $v_d = \left(\frac{2\varepsilon}{m}\right)^{1/2}$. Therefore, when the acceleration potential V_0 is 1 kV, the final velocity of free electrons is $1.87 \times 10^{17} m/sec$.

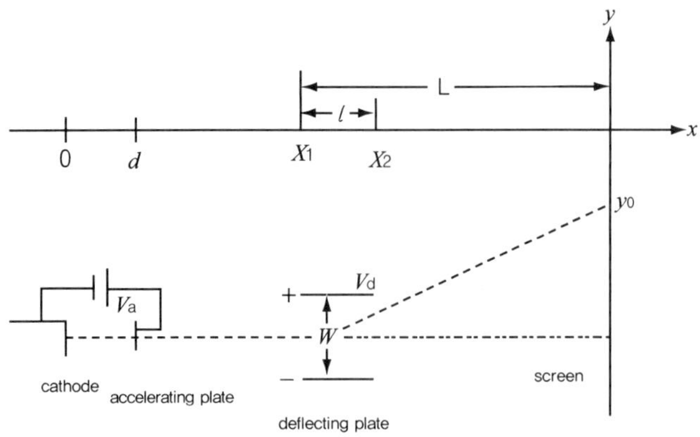

Figure 11.4
Cathode ray tube having an electrostatic deflection

Accelerating or deflecting electrons using an electric field is relatively simple. It is commonly used in cathode ray tubes and uses an electric field for acceleration and deflection. Consider a two-dimensional tube as shown in Figure 11.4. The electrons are accelerated by the potential V_a and emitted from the electron gun with velocity v_0

$$v_0 = \left(\frac{2e\,V_a}{m}\right)^{1/2} \qquad \text{(Equation 11-12)}$$

Since the deflection potential and the electric field act only on the y-axis, the x-velocity is constant as v_0 after the electron passes through the point $x = d$. This deflecting field acts on the electrons during the time the electrons are between the deflecting plates. This time is given by equation (11-13).

$$t_d = \frac{x_2 - x_1}{v_0} = l\left(\frac{m}{2e\,V_a}\right)^{1/2} \quad (v_0 = (2e\,V_a t_d = \frac{x_2 - x_1}{v_0} = l\left(\frac{m}{2e\,V_a}\right)^{1/2}, \quad \left(v_0 = \left(\frac{2e\,V_a}{m}\right)^{1/2}\right)$$

(Equation 11-13)

In this equation, $l = x_2 - x_1$. The y position of the electron can be obtained by determining the movement of the electron. While the electron is between the deflecting plates, the motion of the y-direction is given by equation (11-14).

$$\frac{dv_y}{dt} = -\frac{e}{m}\left(-\frac{V_d}{w}\right), \left(\frac{dV_y}{dt} = -\frac{e}{m}E_y, \quad E_y = -\frac{V_d}{w}\right) \qquad \text{(Equation 11-14)}$$

So, v_y is given by equation (11-15)

$$v_y = \frac{e}{m}\left(\frac{V_d}{w}\right)t \qquad \text{(Equation 11-15)}$$

If t = 0, the electron velocity in the vertical direction is $v_{yo}=0$ in the y-axis and this represents particles at $x=x_1$. At x_2, v_y is expressed by equation (11–16).

$$v_y=\frac{e}{m}\left(\frac{V_d}{w}\right)l\left(\frac{m}{2e\,V_a}\right)^{1/2},\qquad \text{(Equation 11–16)}$$

$$\left(at\quad x=x_2,\ t_d=l\left(\frac{m}{2e\,V_a}\right)^{1/2}\right)$$

Also, equation (11–17) is obtained by integrating velocity of equation (11–15).

$$y=\frac{e}{2m}\left(\frac{V_d}{w}\right)t^2 \qquad \text{(Equation 11–17)}$$

The x-direction movement of the particles is given by

$$x-x_1=v_0t=\left(\frac{2e\,V_a}{m}\right)^{1/2}t \qquad \text{(Equation 11–18)}$$

$$\left(t_d=\frac{x_2-x_1}{V_0},\ v_0=\left(\frac{2e\,V_a}{m}\right)^{1/2}\right)$$

or

$$t=\left(\frac{m}{2e\,V_a}\right)^{1/2}(x-x_1),\ \left(t=\frac{x-x_1}{v_0}\right) \qquad \text{(Equation 11–19)}$$

By substituting the equation (11–19) to the equation (11–17), you can get y.

$$y=\frac{e}{2m}\left(\frac{V_d}{w}\right)t^2,\ \left(t=\left(\frac{m}{2e\,V_a}\right)^{1/2}(x-x_1)\right)$$

$$=\frac{e}{2m}\left(\frac{Vd}{w}\right)\left(\frac{m}{2e\,V_a}\right)(x-x_1)^2$$

$$=\frac{V_d}{4w\,V_a}(x-x_1)^2$$

$$\therefore y=\frac{V_d}{4w\,V_a}(x-x_1)^2 \qquad \text{(Equation 11–20)}$$

At x_2, y becomes

$$y_d=\frac{V_d}{4w\,V_a}l^2,\ (x_2-x_1=l) \qquad \text{(Equation 11–21)}$$

The accelerating electric field no longer acts on the electrons at the right side of the deflecting plate. The speed is constant and has the following value.

$$v_x=v_0 \qquad \text{(Equation 11–22)}$$

$$v_y = \frac{e}{m}\left(\frac{V_d}{w}\right)\frac{l}{v_0}$$ (Equation 11-23)

$$\left(v_y = \frac{e}{m}\left(\frac{V_d}{w}\right)t\ ,\ t = \frac{l}{v_0}\right)$$

The time to reach the screen from position x_2 is

$$t_s = \frac{L-l}{v_0}$$ (Equation 11-24)

Using equation (11–21) in v_0, deflection y becomes:

$$y_p = v_y t_s + y_d = \frac{e}{m}\left(\frac{V_d}{w}\right)\frac{l(L-l)}{v_0^2} + \frac{V_d}{4wV_a}l^2$$
$$= \frac{V_d}{2w\,V_a}\left(L - \frac{l}{2}\right)$$

$$v_0 = \left(\frac{2eV_a}{m}\right)^{1/2},$$

$$\frac{e}{m}\left(\frac{V_d}{w}\right)\frac{l(L-l)}{v_0^2} = \frac{e}{m}\left(\frac{V_d}{w}\right)\left(\frac{l(L-l)}{\frac{2eV_a}{m}}\right)$$

$$v_y t + y_d = \frac{V_d l(L-l)}{2wV_a} + \frac{v_d l^2}{4wV_a} = \frac{v_d(2lL - 2l^2 + l^2)}{4wV_a}$$
$$= \frac{2\,V_d l}{4w\,V_a}\left(L - \frac{l}{2}\right)$$ (Equation 11-25)

The deflection is a linear function of both the acceleration potential and the deflection potential. The deflection on the screen is a linear function of the deflection electric potential. So, If electric potential varying with time is applied to the deflection plate, a signal can be seen on the fluorescent screen. This electric potential changes the position where the electron beam reaches the screen. The position on the screen is a linear function of the amplitude of the applied deflection electric potential. This is the principle of the cathode–ray oscilloscope.

Example 11–2 An Example

Obtain the maximum deflection in a typical tube when V_d is changed to 100 V with L = 30 cm, l = 3 cm, w = 0.5 cm, and V_a = 1.5 kV.

Sol. Substituting the respective variables from the equation (11–25),

$$y_p = \frac{v_d l}{2wV_a}\left(L - \frac{l}{2}\right) = \frac{4\times100\times3(30-15)}{2\times0.5\times1.5\times10^3} = 5.7cm$$

The maximum deflection y_p is 5.7 cm.

2.3 Movement of electrons under a uniform magnetic field

Figure 11.5 is an index showing the direction of force along the direction of movement of electrons within the uniform magnetic flux density B. If the magnetic field is selected with reference to the x–direction, the force in equation (11–2) depends on $\sin\theta$ when the angle between the magnetic field and the velocity is θ. It can be seen that the velocity of the electrons moving parallel to the direction of B is zero. If the velocity of an arbitrary electron is decomposed into x, y, and z components, the velocity vx in the x–direction is not affected by the magnetic force. That is, since any motion along the x–axis is not affected by the magnetic force, only the motion in the yz plane is the subject of interest. In other words, when turning I toward B, the direction of the right hand screw is the direction of the force, so that the y–direction velocity acts on the electrons in the (+) z–direction with respect to the electrons (opposite direction of −e). The z–direction velocity produces a force in the (−) y–direction.

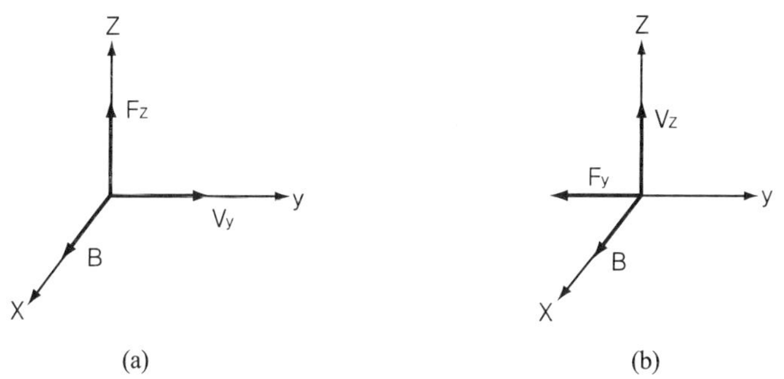

Figure 11.5

The magnetic force acting in accordance with the direction of electron movement (a) the electrons move in the + y–direction, (b) the electrons move in the + y–direction

Considering the sign of the electric charge without considering the right hand law for force and field effect in Equation 11–1, we can set up equations like equation (11–26) in the case of (a) in Figure 11.5 and equation (11–27) in the case of (b) for bidirectional motion.

$$\frac{dv_y}{dt} = -\frac{e}{m}v_z B \qquad \text{(Equation 11–26)}$$

$$\frac{d}{dt}\left(\frac{dv_y}{dt}\right) = -\frac{e}{m}\frac{dv_z}{dt}B, \quad \frac{d^2v_y}{dt^2} = -\frac{e}{m}\cdot\frac{e}{m}v_y B^2 = -\left(\frac{eB}{m}\right)^2 v_y$$

$$\frac{dv_z}{dt} = \frac{e}{m}v_y B \qquad \text{(Equation 11–27)}$$

Substitute the equation (11–27) to the differential equation of the equation (11–26) and also substitute the equation (11–26) to the differential equation of the equation (11–27). Then you will obtain two second order differential equations of motion,

$$\frac{d^2v_y}{dt^2}=-\left(\frac{eB}{m}\right)^2 v_y \qquad \text{(Equation 11–28)}$$

$$v_y'' + w_c^2 v_y = 0$$

$$v_y = ce^{\pm iw_c t} = A\cos w_c t + B\sin w_c t$$

$$\frac{d^2v_z}{dt^2}=-\left(\frac{eB}{m}\right)^2 v_z \qquad \text{(Equation 11–29)}$$

The values of these harmony equations are:

$$v_y = A_1\cos\omega_c t + A_2\sin\omega_c t \qquad \text{(Equation 11–30)}$$

$$v_z = A_3\cos\omega_c t + A_4\sin\omega_c t \qquad \text{(Equation 11–31)}$$

and ω_c is defined as:

$$\omega_c = \frac{eB}{m} \qquad \text{(Equation 11–32)}$$

The equation (11–32) represents frequency called as cyclotron frequency or synchrotron frequency. The electrons in the magnetic field of the 1000 gauss (0.1 W/m^2) has a ω_c of 1.76 $\times$ 10^{10} rad/sec.

$$w_c = \frac{1.6\times 10^{-19}\times 0.1}{9.1\times 10^{-31}} = 1.758\times 10^{10} rad/\mathrm{sec}$$

Select A_2=0 as a guide, let A_1 = v_0, then

$$v_y = v_0\cos\omega_c t \qquad \text{(Equation 11–33)}$$

v_0 is a constant. Likewise, when $A_3 = 0$ and $A_4 = v_0$, following equations can be obtained by using the equation (11–32).

$$\frac{dv_y}{dt}=-\frac{e}{m}v_z B$$

$$\frac{dv_y}{dt}=v_0\omega_c\sin\omega_c t=-\frac{e}{m}v_z B$$

$$v_z = v_0\sin\omega_c t = A_3\cos\omega_c t + A_4\sin\omega_c t$$

$$\therefore A_4 = v_0,\ A_3 = 0$$

$$v_z = v_0\sin\omega_c t \qquad \text{(Equation 11–34)}$$

For understanding the phenomenon under the uniform magnetic field, determination of the position of the particle is needed from the equation (11–33) and (11–34).

$$v_y = \frac{dy}{dt} = v_0 \cos \omega_c t \quad \text{(Equation 11–35)}$$

$$v_z = \frac{dz}{dt} = v_0 \sin \omega_c t \quad \text{(Equation 11–36)}$$

Then by integrating the equation (11–35) and (11–36),

$$\int v_y dt = y - y_0 = \frac{v_0}{\omega_c} \sin \omega_c t \quad \text{(Equation 11–37)}$$

$$\int v_z dt = z - z_0 = -\frac{v_0}{\omega_c} \cos \omega_c t \quad \text{(Equation 11–38)}$$

To simplify the equation, the origin is chosen to be $y_0 = z_0 = 0$. The speed vibrates in time, and the position on the yz plane also vibrates in time.

$$y = r \cos \phi \quad \text{(Equation 11–39)}$$

$$z = r \sin \phi \quad \text{(Equation 11–40)}$$

Converting to the polar coordinate system, the particles will move along a circle having the radius:

$$r^2 = z^2 + y^2 = \left(\frac{v_0}{w_c}\right)^2 \quad \text{(Equation 11–41)}$$

In addition, since

$$\phi = \tan^{-1}\frac{z}{y} = \tan^{-1}\left\{-\frac{\cos \omega t}{\sin \omega_c t}\right\} = \tan^{-1}(-\cot \omega_c t) \quad \text{(Equation 11–42)}$$

$$\tan \phi = \frac{z}{y} = -\frac{\cos \omega t}{\sin \omega_c t} = -\cot \omega_c t = \tan\left(\omega_0 t - \frac{\pi}{2}\right)$$

$$\therefore \phi = \omega_c t - \frac{\pi}{2}$$

this equation is obtained.

$$\phi = \omega_c t - \frac{\pi}{2} (-\pi < \phi < \pi) \quad \text{(Equation 11–43)}$$

$$ref)\ \cot x = \tan\left(\frac{\pi}{2} - x\right) = -\tan\left(x - \frac{\pi}{2}\right)$$

Then the particles move along a circle having the radius v_0/w_c and it is apparent that they rotate at a frequency of w_c. Thus, particles have a stable orbit rotation in the magnetic field. Then, the angular momentum is $\frac{mv_0^2}{w_c}$ and the energy is $\frac{1}{2}mv_0^2$, respectively. w_c increases as the magnetic field increases. This means that the radius, v_0/w_c, and angular momentum decrease for the particles having a certain energy.

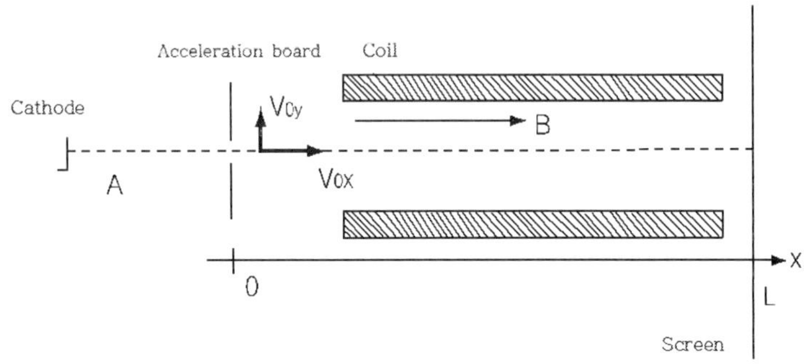

Figure 11.6
Focusing effect of the magnetic field

The particles rotate in a plane perpendicular to the magnetic field, which can be easily used for the focus effect of the magnetic field. To explain this, apply a magnetic field in a cathode ray as shown in Figure 11.6. Since an electron gun of the position A and a discharge opening have a finite size, the electrons are not fully focused. That is, when the electrons are emitted from the electron gun in the $x=0$ they have a small y and z components in addition to the x component of the velocity. Assume $v_{0z}=0$ to simplify. When the electrons are emitted from the gun at $t=0$ it has $v_0=v_{0y}$, a component of velocity, and enters the magnetic field at $x=y=z=0$.

The equation of the motion of the particles is:

$$y-y_0=\frac{v_0}{\omega_c}\sin\omega_c t \qquad \text{(Equation 11-44)}$$

$$z-z_0=-\frac{v_0}{\omega_c}\cos\omega_c t \qquad \text{(Equation 11-45)}$$

Since $y=z=0$ at $t=0$, y_0 and z_0 have values:

$$y=\frac{v_0}{\omega_c}\sin\omega_c t \qquad \text{(Equation 11-48)}$$

$$z=\frac{v_0}{\omega_c}(1-\cos\omega_c t) \qquad \text{(Equation 11-49)}$$

Then the motion is expressed as shown below:

$$y = \frac{v_0}{\omega_c}\sin\omega_c t \qquad \text{(Equation 11-48)}$$

$$z = \frac{v_0}{\omega_c}(1-\cos\omega_c t) \qquad \text{(Equation 11-49)}$$

Focusing on the beam, the magnetic field prevents dispersion of the beam by the y and z components of v_0. When the particle reaches the screen is multiplied by the rotational frequency, it is possible to obtain this effect easier. The transit time is

$$t_t = \frac{L}{v_{0x}} = \left(\frac{m}{2e\,V_a}\right)^{1/2} L,\ where \left(v_{0x}^2 = \frac{2e\,V_a}{m}\right) \qquad \text{(Equation 11-50)}$$

from Equation (11–10) where V_a is acceleration potential. It should meet the following conditions:

$$\omega_c t_t = 2n\pi \qquad \text{(Equation 11-51)}$$

$$\frac{eB}{m}t_t = 2n\pi \rightarrow B = \frac{2n\pi m}{et_t}$$

where n is an integer. If equation (11–51) is satisfied, the particle returns to $y = z = 0$ position when it reaches the screen. Therefore, the focus of the beam is adjusted. The value of the magnetic flux density is obtained by applying the equation (11–51) and equation (11–32).

$$B = \frac{2n\pi m}{et_t} = n\left(\frac{2\pi m}{e}\right)\left(\frac{2e\,V_a}{m}\right)^{1/2}\left(\frac{1}{L}\right) = \frac{2n\pi}{L}\left(\frac{2m\,V_a}{e}\right)^{1/2} \qquad \text{(Equation 11-52)}$$

The minimum (n = 1) value of B which has a focusing effect is 2.74×10^{-3} W/m^2 or 27.6 gausses[3].

$$w_c = \frac{eB}{m} = \frac{2\times1\times3.14159}{0.3}\times\left(2\times9.1\times10^{-31}\times1.5\times\frac{10^3}{1.6}\times10^{-19}\right)^{1/2} = 2.74\times10^{-3}\,Wb/m^2 \qquad \text{(Equation 11-53)}$$

3) 1 W/cm^2=10000 Gauss

2.4 Movement of electrons under electric and magnetic field

If both the electric and magnetic fields exist, the movement is the sum of each movement by electric and magnetic fields and the movement perpendicular to E and B. It is already noted that the velocity parallel to the magnetic field does not generate a magnetic force. There is no interaction if electric and magnetic fields are parallel to each other, so the movement can be understood by adding them after handling two components separately.

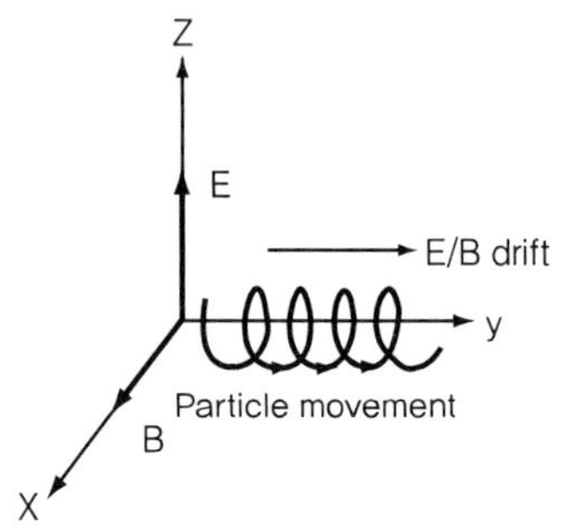

Figure 11.7

Drift motion of the electrons in the perpendicular electric and magnetic fields.

If the electric and magnetic fields are perpendicular to each other, something interesting happens. To make it simple, it is assumed that the electric field is towards the z-direction and the magnetic field is towards the x-direction. It is shown in Figure 11.7. The equation of force is

$$\frac{dv_y}{dt} = -\frac{e}{m}v_z B \quad \text{(Equation 11-54)}$$

$$\frac{dv_z}{dt} = -\frac{e}{m}E + \frac{e}{m}v_y B \quad \text{(Equation 11-55)}$$

$$\begin{aligned}\frac{d^2v_y}{dt^2} &= -\frac{e}{m}\frac{dv_z}{dz}B \\ &= \left(-\frac{eB}{m}\right)\left[\frac{-e}{m}E + \frac{e}{m}v_y B\right] \\ &= \left(\frac{e}{m}\right)^2 EB - \left(\frac{eB}{m}\right)^2 v_y\end{aligned}$$

from the equation (11–1). After differentiating each equation as in section 3 and substituting each equation, these following equations are obtained.

$$\frac{d^2v_y}{dt^2} = -\left(\frac{eB}{m}\right)^2 v_y + \frac{e^2}{m^2}EB \quad \text{(Equation 11-56)}$$

$$\frac{d^2 v_z}{dt^2} = -\left(\frac{eB}{m}\right)^2 v_z \quad \text{(Equation 11-57)}$$

With the exception of the second term in equation (11–55), this is the same as the equation derived in section 3. Both the electric and magnetic fields are constant, so added term is also constant. Applying this linear transformation,

$$v_y = v'_y + \frac{E}{B} \quad \text{(Equation 11-58)}$$

equation (11–55) becomes

$$\frac{d^2 v'_y}{dt^2} = -\left(\frac{eB}{m}\right)^2 v'_y \quad \text{(Equation 11-59)}$$

$$\begin{aligned}\frac{d^2 v'_y}{dt^2} &= -\left(\frac{eB}{m}\right)^2 \left(v'_y + \frac{E}{B}\right) + \frac{e^2}{m^2} EB \\ &= -\left(\frac{eB}{m}\right)^2 v'_y - \left(-\frac{e}{m}\right)^2 EB + \left(\frac{e}{m}\right)^2 EB \\ &= -\left(\frac{eB}{m}\right)^2 v'_y\end{aligned}$$

Equation (11–56) and equation (11–58) are the same as the equations shown in section 3 and those values are

$$v'_y = v_0 \cos \omega_c t, \qquad v_z = v_0 \sin \omega_c t \quad \text{(Equation 11-60)}$$

By equation (11–57), these equations are obtained.

$$v_y = \frac{E}{B} + v_0 \cos \omega_c t \quad \left(v_y = v'_y + \frac{E}{B},\ v'_y = v_o \cos \omega_c t\right) \quad \text{(Equation 11-61)}$$

$$v_z = v_0 \sin \omega_c t \quad \text{(Equation 11-62)}$$

The additional velocity term means that the particles drift by the combined effects of the electric and magnetic fields. The direction follows right hand rule when turning the screw from E to B. When expressed as a vector, this drift velocity term can be expressed as

$$v_{drift} = \frac{E \times B}{B^2} \quad \text{(Equation 11-63)}$$

Movement center of electrons is moving with the drift speed E/B. The electrons rotate with the angular velocity ω_c about the movement center. In section 2 of this chapter, it has been found that the direction of particle motion in the electric field depends on the sign of the electrons. In the cyclotron motion discussed in section 3, the direction of rotation depends on the sign of the charge (see equation about ω_c). However, the vertical electric field drift velocity E/B is independent of the charge. It only depends on the magnitude of the related electric field. Drift movement is often called the center induction movement. This movement overlaps with cyclotron movement discussed in section 3. In the case of the vertical electromagnetic field, if the electric field is small, it only serves to induce the vertical field drift and does not accelerate the particles. The electric field parallel to the magnetic field accelerates the particles along the electric field as discussed in section 1 of Chapter 11.

2.5 The principle of field emission

As shown in Figure 11.8, field emission phenomenon occurs when an electric field of 0.5V/Å or more is applied to a metal surface in a vacuum. The electrons in the metal are quantitatively tunneling[4] and are released into the vacuum.

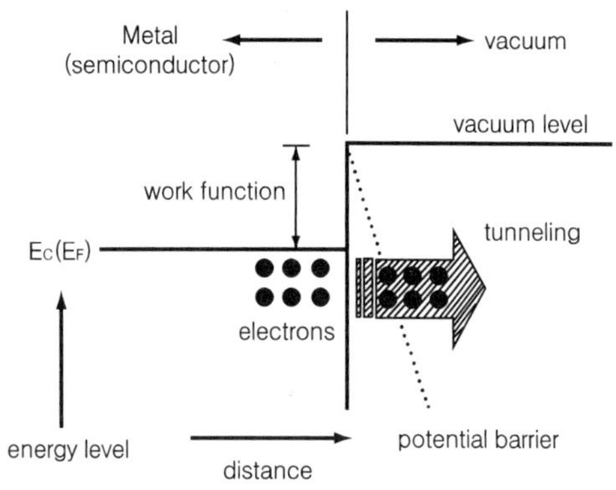

Figure 11.8
The phenomenon of field emission

The field emission current (equation 11-64) is expressed as a Fowler-Nordheim (F-N) expression. J corresponds to the density of emission current in A/cm^2, a and b are constants (which in fact is also slightly dependent on the electric field), E is the applied electric field in V/cm and ϕ is work function in eV.

$$J = aE^2\phi^{-1}\exp(-b\phi^{3/2}/E) \quad \text{(Equation 11-64)}$$

For F-N equation, current-voltage relationship is expressed as (equation

4) If the electron has a low kinetic energy inside the very narrow width of the potential barrier, the probability of finding an electron($|\Psi|^2$) refers to a non-zero. Therefore it means that electrons can exist beyond the potential barrier.

11-64) and (equation 11-65). In (equation 11-65), the relationship between *ln* (I/V^2) and (I/V) is of a straight line. The effective work function and emission area can be obtained while confirming the field emission current.

$$I = MV^2\exp(-N/V)\ \ (M \text{ and } N \text{ are constants}) \quad \text{(Equation 11-65)}$$

$$\ln(I/V^2) = -N/V + \text{constant} \quad \text{(Equation 11-66)}$$

In addition, if the F-N equation is formulated in the state of neglecting the small influence from (equation 11-63), it is the same as (equation 11-66). The predominant one in this equation is the exponential term. If the electric field is increased (E=1.74$\times 10^7$~8$\times 10^7$V/cm) and the work function is reduced (ϕ=0~5 eV), the current density (J) is increased very rapidly. For example, if the electric field applied to the emitter is increased by 25%, the emissive current is increased by 3.7$\times 10^4$, and when the work function of the emitter is increased by 25%, the emitter current is increased by 2$\times 10^6$.

$$J = 1.54\times 10^{-6}E^2\phi^{-1}\exp(-6.83\times 10^{-7}\phi^{3/2}/E) \quad \text{(Equation 11-67)}$$

Therefore, when designing electron emission source, the maximization of the electric field applied to the tip and the minimization of the work function of the emission source are important to get a large emission current from small applied voltage. Maximization of the electric field which is applied to the tip is related to the structure of electron emission source. Minimization of the work function of emission is related to material of electron emission source. One of the methods to increase electric field is externally applied voltage but it is limited by the matching with the peripheral circuit and the breakdown voltage. One of the methods of reducing the distance between the emitter and the electrode is limited by the limitations of lithography. Thus, the method of changing the shape of the discharge portion is most effective. A related factor is the electric field multiplication factor β. It is a proportionality constant of the voltage V applied to the electrode and the electric field E applied to the discharge portion as shown in equation (11-67).

$$E = \beta V \quad \text{(Equation 11-68)}$$

$$\beta = kh/r \quad \text{(Equation 11-69)}$$

The β has a value in the range of 50-500 and it is proportional to barrier height h by a proportionality constant (k), inversely proportional to radius r

through the (equation 11–69). That is, when the end of the tip is higher and sharper, β has a larger value. Therefore, a larger electric field is applied in emission part even if the applied voltage is the same. On the other hand, the work function is unique value of the material itself. It can be improved through the development of a material. The melting point and the work function for each material are shown in [Table 11.1]. The lower the work function, the higher the emission current.

However, to use as a field emission material, thermal–chemical–mechanical durability and temperature characteristics of the material (change in work function with temperature change) should be considered.

Table 11.1 The work function and melting point for each material

Material	Work function(eV)	Melting point(°C)	Material	Work function(eV)	Melting point(°C)
Ag	4.7	961	Nd	3.3	
Al	3.0	660	Ni	5.0	1455
Au	4.8	1063	Pb	4.0	327
Ba	2.5	850	Pt	6.0	1774
Bi	4.1	271	Rb	1.8	39
C	4.7	>3500	Sr	2.1	800
Ca	3.2	810	Ta	4.1	2850
Cd	4.1	321	Ti	4.1	
Cs	1.8	29	Th	3.4	1845
Cu	4.1	1083	W	4.5	3370
Fe	4.7	1535	Zn	3.3	420
Hf	3.6		Zr	4.1	1900
Hg	4.5	−39			
Ir	5.4		LaB_6	2.7	
K	1.8	62	NdB_6	4.6	
La	3.3		TaB	2.9	
Li	2.2	186	TaC	3.1	
Mg	2.4	651	ThO_2	2.6	
Mo	4.3	2620	TiC	3.4	
Na	1.9	98	ZrB	4.5	

3 Principle of luminescence of electron beam utilization display

3.1 Principle of phosphorescence

(1) Emitting cathode

In the middle of the 18th century, emission of light is discovered from the glass tube when a gas was put into a glass tube with a metal electrode at both ends and a high voltage was applied between the electrodes after lowering the pressure. The light emitted by the electrons produced from the cathode (cathode ray) collides with gas molecules to excite gas molecules and the excited molecules go down to the ground state. The wavelength of the light differs depending on the gas type. It was discovered at the end of 1800 that when a mercury vapor is put into a glass tube, mercury atoms emit ultraviolet rays and weak blue light is emitted by a cathode ray. It was used as a mercury lamp. When a glass tube wall is coated with a fluorescent material, the material is excited by ultraviolet rays emitted from mercury. It is used for a fluorescent lamp. The fluorescent agent used in the fluorescent lamp is a phosphor.

Strictly speaking, a fluorescent agent used in a fluorescent lamp is a phosphor that emits phosphorescence and the light emitted therefrom is phosphorescence. A phosphor is mainly used as transition metals or rare earth metal compounds. The fluorescent lamp has high energy efficiency compared to the incandescent lamp and it can have a variety of colors depending on the type of the phosphor. The lifetime is also long and it has been used for many years. A Braun tube TV(cathode–ray tube) uses the light emitted from the cathode by directly impinging electrons from the cathode against the phosphor coated on the glass wall.

(2) Phosphor for electron beam(CRT, FED)

As the CRT phosphors, sulfide or oxysulfide is used as the host crystal. $ZnS:Ag^{+}$, Al^{3+} as a blue phosphor, $ZnS:Cu^{+}$, Al^{3+} as a green phosphor, and $Y_2O_2S:Eu^{3+}$ as a red phosphor are used. The depth of electron penetration into the phosphor is several micrometers. Thus, the metal film is coated so as to suppress the transfer of the phosphor surface and deterioration.

FED has a low acceleration voltage of several kV or less, so electron beams are not allowed to enter the phosphor. Since the current density at a FED is higher than that of the CRT, the deterioration of the fluorescent material occurs more than the CRT. The recently developed materials are Al(N,O):Eu^{2+}.

It exhibits blue light emission of 470 nm with a high efficiency at an acceleration voltage of 5 kV and is lower in deterioration (1/3 or less) than the conventional oxide phosphor. In addition, the characteristics of carbon–based materials have been greatly improved as an electron emission source. Besides, FEL has been attempted to be developed as a special light source by combining phosphors.

3.2 CRT cathode–ray emission (Edison effect)

Edison thermionic emission (thermoelectric emission) is a phenomenon that electrons are emitted when a material is heated to a high temperature. In 1884, Edison put two filaments in a vacuum bulb and heated it. He found this phenomenon that the current easily flows between the two unconnected filaments.

For an example of a metal, the free electrons may move freely in metals but they cannot escape easily because of some sort of barriers. However, electrons having a high kinetic energy E can go beyond the wall. The height of the wall is called the work function of the metal. Electrons with more kinetic energy than work function can go out of the metal. The thermal energy required for thermionic emission is represented by W = kT J, using the Boltzmann constant k = 1.38062×10^{-23} J/K and the absolute temperature T K. When this thermal energy is larger than the work function, thermionic emission occurs. The higher the temperature of the metal and the smaller the work function, the greater the number of emitted hot electrons. The most direct application of thermionic emission is a vacuum tube.

3.3 FED Emitting Principle {FEA(Field Emitter Array)}

The FEA plays a role as an electron emission source that spatially ensures superior uniformity, low power consumption, low–cost driving circuit and low voltage matrix drive.

One of the FED is composed of hundreds of millions of Field Emitter Array which is μm size, and the number of the emitter is about 100 to 10,000. The technology level of the configuration of the FEA is the emitter–hole diameter of 0.15 μm, the emitter–emitter spacing of 0.32 μm, a tip radius of 20 ~ 40 μm, silicon tip of 1 ~ 10 μm, a current tip of 500 μm, and about 5 μm for a silicon tip. The maximum current tip is 500 μm for the metal tip, and about 5 μm for silicon tips. The current value is considerably larger than the value required to achieve the performance target of the FED presented in the following example. Therefore, 1000 lines of scanning lines, luminance of 342 cd/m^2 (100 fL), phosphor screen voltage of 500 V, 50% fill factor, phosphor efficiency of 5 lumens/watt, DC average current density of 84 μA/cm^2 and the ratio of the tips contributing to the field emission is about 50%. To obtain this specification, the spacing between tips is assumed to be 5 μm, the peak current per tip of 0.04 μA, and the average current per tip of 0.04 nA.

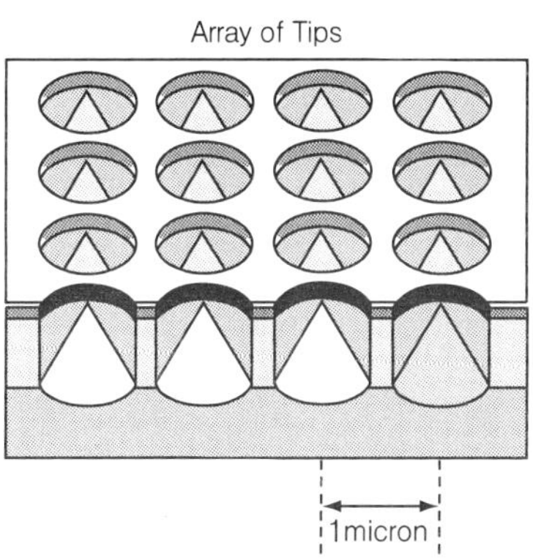

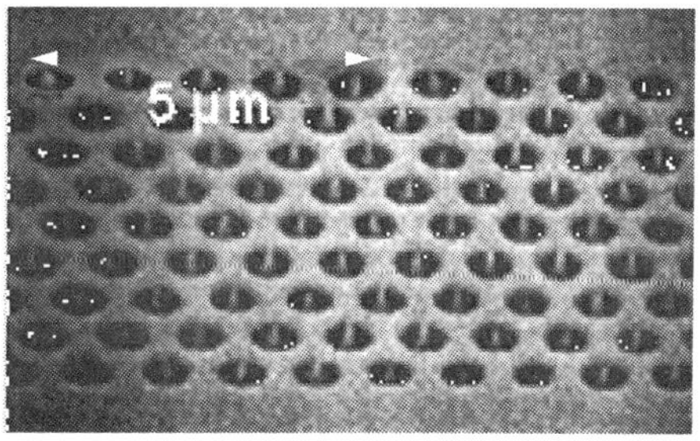

Figure 11.9
Field Emitter Array and real photograph

4 Structure and Operation Principle of Electron Beam Display

4.1 The operation of FED

(1) The basic principle of FED

An anode plate coated with a phosphor and a cathode plate composed of a cathode and a gate are arranged at the upper side with a small vacuum gap maintained by a spacer. There are row electrode and the column electrode in cathode plate, and FEA is driven for the matrix. The electrons are released to the gate and accelerated by anode voltage. These electrons go through the vacuum gap hitting the phosphor dot coated on the anode and emit light. A color display can also be achieved by using R (red), G (green), and B (blue) phosphor dots in one pixel.

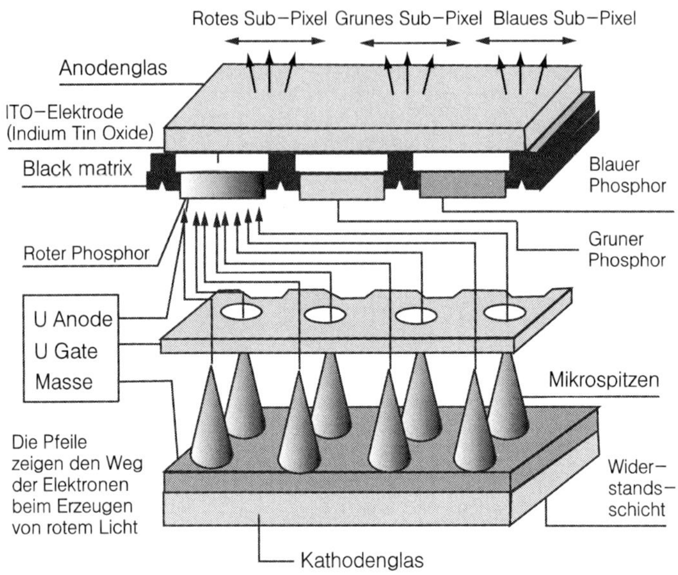

Figure 11.10
A cross sectional of FED(Siemens)

Figure 11.10 shows a cross–section of FED. The intensity of the electric field at the field emission tip can be obtained by adjusting the voltage of the row electrode or the column electrode. It controls the number of released electrons by tunneling to the surface of the emitter. The released electrons impinge on the phosphor without energy loss and emit bright light.

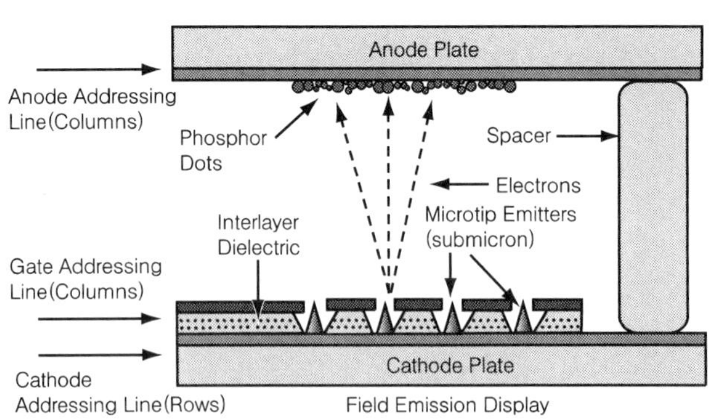

Figure 11.11
FED internal electronic schematic diagram of the release

(2) Configuration of the FED system

The basic components of FED are the FEA panel, the anode plate, and the driven circuit. First, FED panel assembled with field emitter source forms a pixel which is composed of two-dimensional form. The second anode plate is designed to apply a DC voltage of several hundred to several thousand volts to the FEA panel to accelerate and attract the emitted electrons. The accelerated electrons are transformed to visible light on the anode plate at the surface of the coated phosphor of the FEA panel. For this process, the anode plate is configured by forming a transparent electrode such as ITO on the transparent plate and coating fluorescence film thereon. Finally, it is composed of a driver circuit, control logic circuit, a column and row drive circuit.

(3) How to drive FED

The control uses the properties of a field emission device in which a cold electron is emitted only when the voltage that is higher than the threshold voltage is applied to the tip. The simplest method is to connect the gate and the cathode in a crossing manner and emit electrons by applying a voltage equal to or higher than the threshold voltage to the pixel. Gate lines and cathode lines are arranged to be perpendicular to control each pixel. The sealed package has been exposed to a high vacuum by the external end of each gate line and the cathode line. By applying an appropriate signal to each line of the pixel, it can be individually controlled and selected. The other way to drive FED is using four sides of the top, bottom, left and right sides to alternately form two electrodes. If a separate driving integrated circuit is connected, the driving frequency can be reduced by half and the driving circuit can be easily constructed. A video signal is applied in series in an actual panel drive. Like PM-LCD, the multiple signals are transposed by using the shift register. Then, pulse count will be the inverse of the number of pixels in the vertical direction.

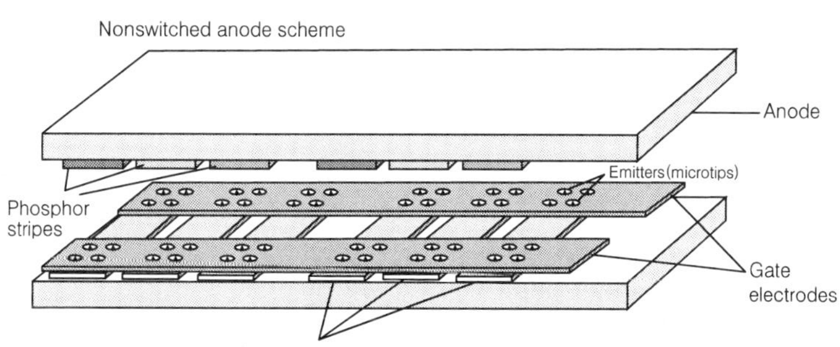

Figure 11.12

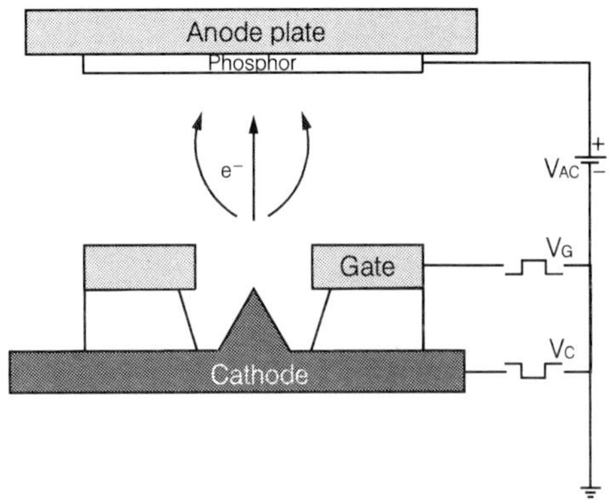

Figure 11.13

Cathode electrodes, Gate electrodes and the structure of Anode

※ Implementation of Gray Level

The goal is to allow a pixel to exhibit as much brightness as desired between a black state in which a minimum light emission state is achieved and a white state in which maximum light is emitted. Speaking about PAM and PWM in FED, pulse height modulation method is the easiest way to generate brightness proportional to a signal. However, the field emission source current is difficult to control because of its non–linearity, and it is not used widely for FED because of nonuniform emission characteristics of each pixel. The luminance step can be expressed by adjusting the ratio of the time that a constant light is emitted within a sufficiently short period of time, which is referred to as a pulse width modulation method in a display device. Only the combination of the reference clock signal and the counter can generate a precise analog signal by the digital signal. It is the essential method because precisely linear brightness steps can be obtained, regardless of discharge characteristics of the individual.

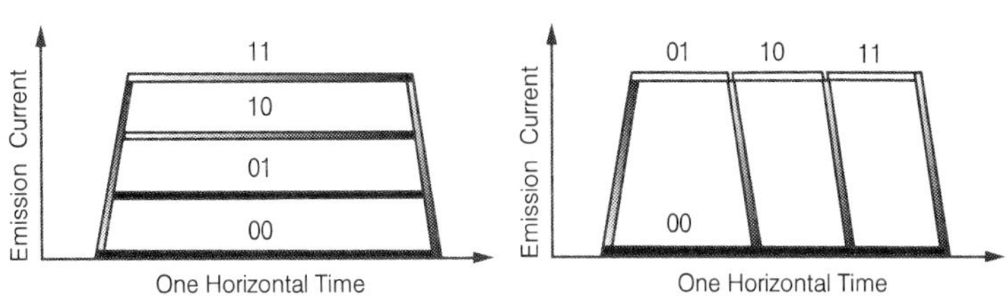

Figure 11.14

PAM(a) and PWN(b)

In the voltage driving method, a voltage equal to or greater than a certain voltage is applied between the cathode and the gate electrode to use the emitted electrons by the electric field effect. The advantage is that the configuration is simple. The disadvantage is that due to the nonlinearly increasing I–V characteristic, a large difference in current occurs when the input voltage changes slightly resulting in non–uniformity between pixels and noise is also amplified. The current driving method utilizes the inherent characteristic of the field emission device that the emitted light is proportional to the anode current.

(4) Implementation of color FED

1) The method of implementing color FED

In order to implement a color FED, it requires three or more color phosphors and needs a system that can be properly driven in accordance with that signal. There are close focusing pixel alignment method, the anode switching method, and a composite film pixel switching method with fluorescence.

2) Close focusing pixel alignment method

The simplest way to making color is to use a plurality of pixels as a basic unit, and the phosphor of the corresponding color is positioned at the anode. This is called a close focusing pixel alignment method. Square pixels are formed by using four unit pixels as a basic bundle. In this case, a so–called 4: 2: 2 pixel arrangement method of increasing the resolution of the luminance by placing more green than the red and blue colors, which have the greatest influence on the luminance signal, is generally used. A method of creating a pixel in which a square is vertically or horizontally trisected to correspond to three colors can reproduce the primary color without limitation. When the pixel is divided into red, green and blue, the blue region is arranged to be wider because the efficiency of the blue phosphor currently used is relatively low.

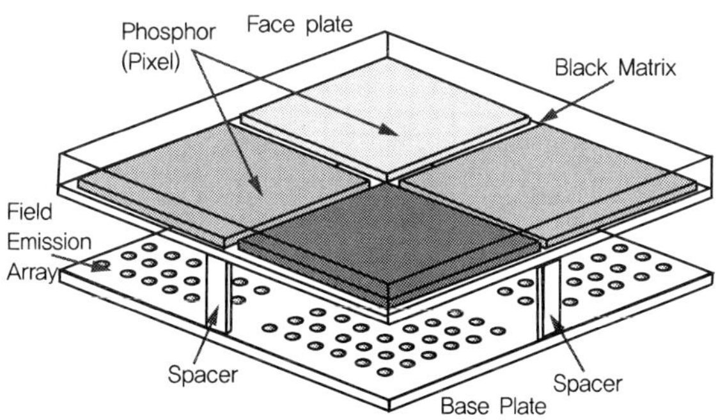

Figure 11.15
The method of implementing color FED by pixel to pixel

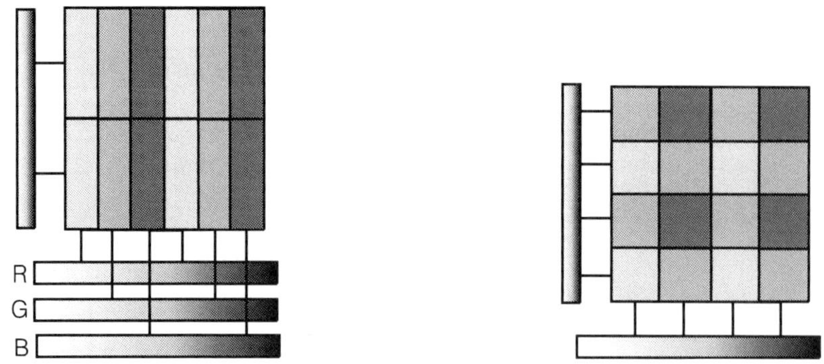

Figure 11.16
The example of implementing color FED by pixel to pixel

3) Anode Switching

In the case of the anode switching method, if the frame frequency is maintained at 60 Hz, the screen is updated every 1/20 seconds for one full screen. As a result, flicker noise appears. In order to prevent this, a double scanning method in which the screen is divided into upper and lower portions is used, or the frame frequency is increased. Therefore, when displaying an input signal with a frame frequency of 60 Hz, it is necessary to use a large capacity frame memory to adjust the scanning speed. This method can constitute a high-efficiency color FED in a relatively simple way. Also, this is particularly advantageous when a high-voltage phosphor is used because the distance between the anode and the cathode can be made sufficiently large.

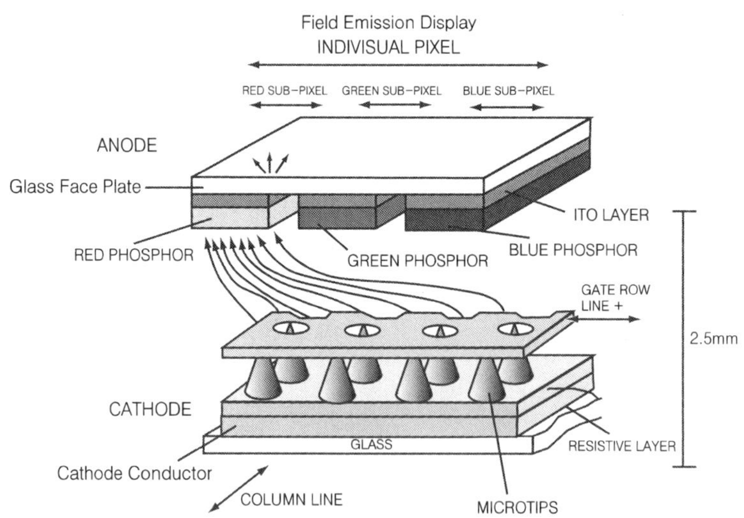

Figure 11.17
The configuration of the FED implemented using a switched anode method

4) Composite Switching

It is divided into two frames to apply a high voltage to the fluorescent film one line at a time. The pixel activates gates one after another to concentrate the electron emission by obtaining the plate focus electrode effect. In the pixel region, since the low voltage is applied to the gate of the inactive pixel to be used as the focus electrode, the area efficiency is improved. As for the fluorescent film, there are only two fluorescent film electrodes instead of switching to the three fluorescent films, so that the advantage is obtained for the pulse coefficient.

4.2 The structure of FED

(1) Emitter

Cone type emitters, which can be obtained through E–gun deposition of metal, are called Spindt cathodes and are most widely used. A lot of cone type emitters obtained by etching silicon have also been studied.

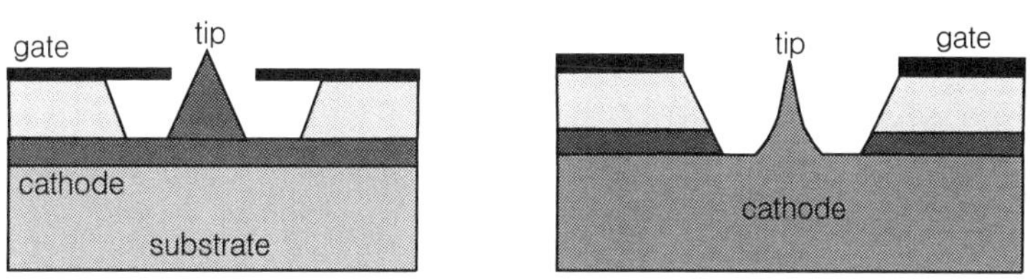

Figure 11.18
Cone Type Emitter

The edge emitters have also been studied because they can be easily fabricated with flat film and photomask work. In the figure below, an insulating film is obtained on a thin conductor plate, a metal edge emitter is formed thereon, and a metal electrode is formed on the next layer. The side with the edge emitter is used as the cathode and the side without the edge is used as the gate electrode.

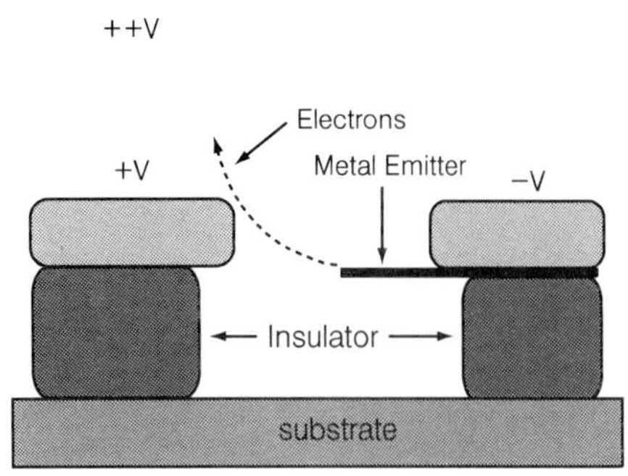

Figure 11.19
Structure of edge emitter

Figure 11.20 shows how to make another edge emitter that increases the intensity of the electric field at the emissive surface. However, the magnitude of the field enhancement factor, which increases the intensity of the electric field, is smaller than that of the cone type emitter. Therefore, The magnitude of the emission current at the same voltage is much smaller. As the structure of the emitter is determined, the uniformity of the spatial electron distribution becomes important. This is due to the human eyes sensitivity to the high spatial frequency patterns and sharp boundaries. One way to improve this uniformity is to maximize the number of emitters per pixel to level the distance between tips. By making the gate hole close to 1 μm and putting $10^5 \sim 10^6$ tips in one pixel, it is possible to increase the production yield and obtain high current. The other way is forming a resistive layer between the emitter tip and the metal conductor. This increases the overall uniformity by negative feedback.

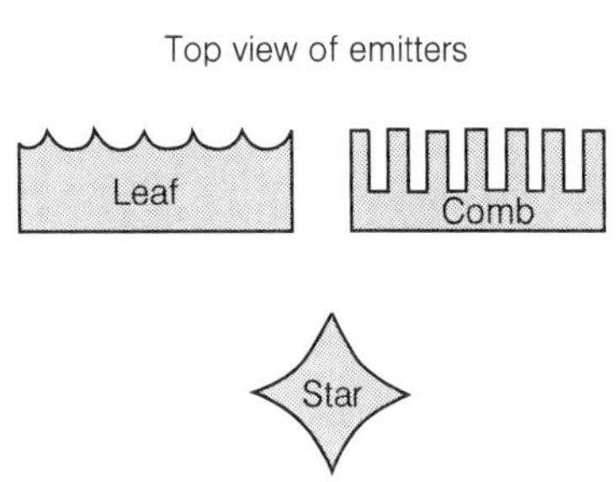

Figure 11.20
he various types of edge emitters to increase intensity of electric field

When driving with voltage, it is important to improve the structure of the emitter, to increase the value of the electric field strengthening factor, to improve material selection and process, and to lower the surface work function.

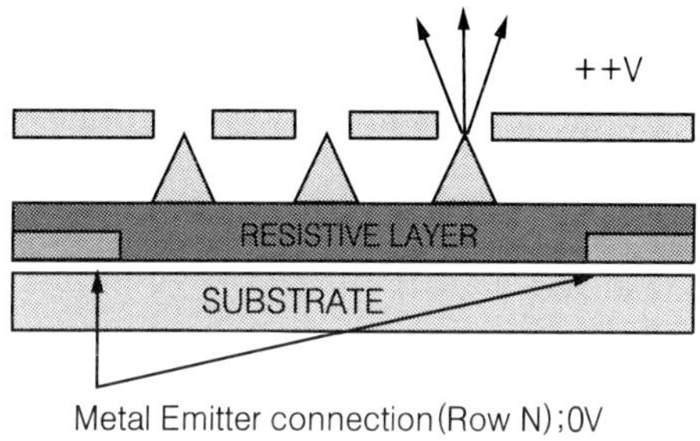

Figure 11.21

Resistance layer for improving spatial uniformity of field emission

(2) Spacer

The spacer is indispensable elements in the FED maintaining a constant gap between the front and back of the plate. The spacer acts to prevent the plate from breaking or bending by supporting the plate which forms a thin and light front and back at an external pressure of 1 atm and an internal pressure of 10^{-6} Torr. The spacer should be able to prevent the substrate from becoming thicker even if the display area is widened. To maintain proximity focusing of the FED resolution, the height of the spacer is similar to a pixel pitch because the emission angle of the electrons emitted from the FEA is large. The details required for the spacers are the mechanical strength, invisible small size, insulation material available in the high vacuum. Known materials include polyimide, which is used as an insulating material in IC process, and it is reported by SRI that FEA is coated and shaped as the last process of FEA fabrication. This material ensures mechanical strength and is made in a small size of $50 \times 50 \times 100$ μm^3 and has a very low vapor pressure. LETI uses a glass ball of 0.2 mm as a spacer, and SI Diamond Inc. sometimes uses Ni plating to create insulating spacer by electroplating.

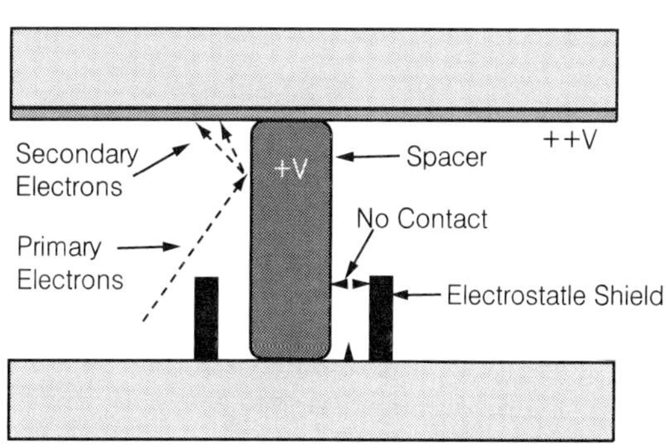

Figure 11.22

Structure of spacer to solve problems of high voltage breakdown and charging of electrons

Spacers in high-pressure(4~6 kV) displays are more challenging because they must be able to withstand the above three conditions and withstand high pressures.

※ The Vacuum Level of FED

All the electron-emitting devices require a high vacuum. This is because the cation generated by the ionization of the residual gas deteriorates the emitter tip of the FED and causes arc discharge between the emitter and the phosphor to deteriorate. Furthermore, the residual gas is gathered in the emitter, and this can change the work function. According to various research results including SRI, 1×10^{-7} Torr is appropriate. However, until the lifetime of the device is exhausted, it is desirable that the manual vacuum pump, depositing barium in the FED as a getter, is incompatible and mounted on the back side of the FED in a tube shape

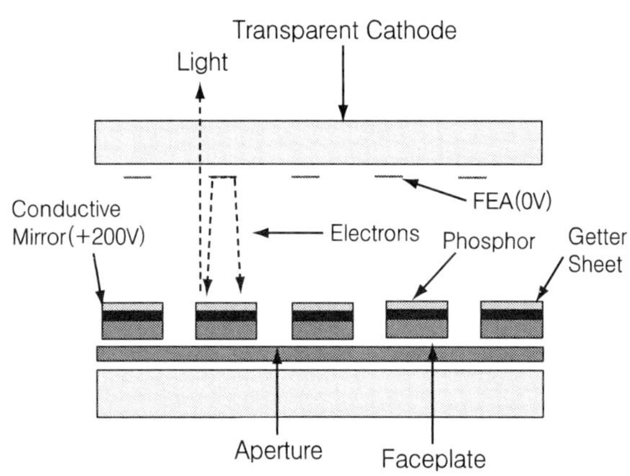

Figure 11.23
FED structure with a transparent cathode electrode with a large area getter in order to improve the vacuum level.

In an attempt to increase the amount of getter to improve the vacuum level in the FED, the getter area was increased by the display size by advancing the concept of the transparent cathode as shown in Figure 11.23. It is designed so that harmful substances generated by drilling a hole in the front can be taken by a getter. The size of this hole is less than 50 μm, so our eyes cannot detect it and it does not affect the image we see. Due to the structure of the element, it is difficult to measure the vacuum level inside, so the effect of vacuum level is measured by the emission current. Since the most stable current increases by about 100 times for use of a getter, the getter greatly reduces the vacuum level.

4.3 The operation of CRT

(1) Scanning System

The electric image created on the target surface of the imaging tube is made up of innumerable different points. To extract this as an electrical signal (image signal) of the image, it is disassembled in order from the left end to the right end, from top to bottom as shown in Figure 11.24. On the side of the receiver, the image signals sent on the tube fluorescent surface are assembled in order to reproduce the image. Scanning is the process of disassembling and assembling images. This scan proceeds at a very high speed, so it actually appears to be disassembling and assembling into a horizontal line.

Breaking the screen from the left end to the right end is referred to as a horizontal reference, and the horizontal line created by a horizontal scan is referred to as a scan line. In addition, the scan that lines the scan line from the top to the bottom is called a vertical scan, and the screen composed of the scan lines is called a raster. The distance from one scan line to the next scan line is called a flyback line, and this period should be as short as possible.

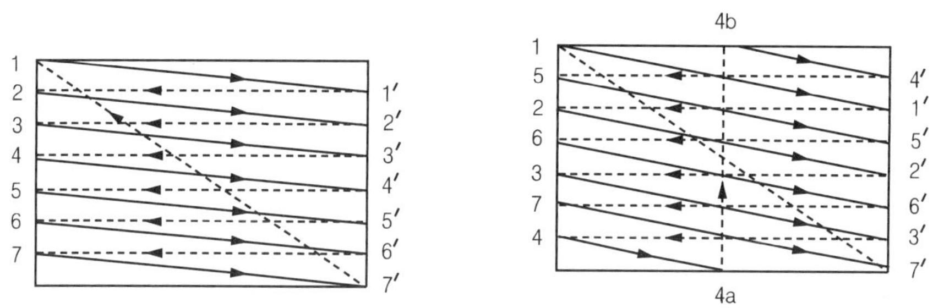

(a) progressive scanning method (b)inter-race scanning method

Figure 11.24
Scanning method, the real line is scanning line and the dotted line

In real TV, this scan line is set to 525, so it is very fast and takes 1/30 seconds to scan the screen. It is transmitted 30 images per second. Since there is a residual image on the fluorescent screen of the cathode-ray tube or the human eye, it is seen as a continuous screen. However, when assembling the image at the upper part of the fluorescent tube surface of the figure tube of the TV and assembling it downward, the upper part will erase the image, resulting in a slight flicker on the whole screen. In order to prevent this, we perform a horizontal scan at one interval. Creates a screen with 262.5 scan lines in the first vertical scan. A method is employed in which scanning is performed between the first scanning lines by the second vertical scanning to complete one screen.

(2) Brightness and color

In the case of black and white, there is only luminance control, but in the case of color, color control is also necessary, which is controlled by the current amount. The current control method is G1 drive method, but the cathode drive method is variously used. If a gray scale is displayed as shown in Figure 11.25, the voltage across the electron gun G1/K from the white level to the black level is applied. Although a different current flows through the cathode at the applied voltage, the voltage–current characteristic is not linear, and the current–voltage slope varies from about 2.52 to 3.02. Since the amount of current depends also on the setting of the cut-off voltage, the display monitor design uses the drive characteristic in the cut-off chart and cut-off voltage. In the case of CDT, the color display is performed by current control (voltage control) of each of the three electron beams. The number of display colors is determined by the product of the number of signal input gradations of each color, and 4,096 colors can be displayed in the case of 16 gradations.

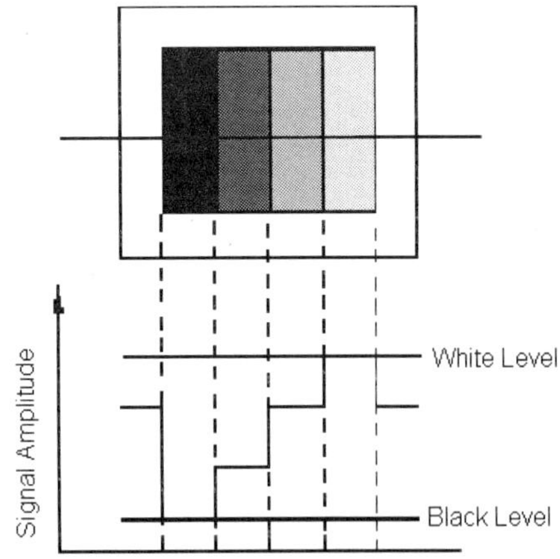

Figure 11.25
Relation between screen brightness and signal amplitude

(3) The circuit configuration of the CRT display monitor

The circuit configuration of the color CRT display monitor is composed of basic circuits such as a video circuit, the deflection circuit, a high−pressure circuit, the power supply circuit, a dynamic focus circuit and the horizontal deflection frequency conversion circuit. Since the CRT takes a sufficient light emission luminance of the phosphor, the anode voltage is high. Since an electric current flows in the form of an electron beam on the anode, it is necessary to consider that a stable high voltage can be obtained with respect to a change in current. The deflection circuit is a circuit in which a sawtooth current finally flows through the deflection yoke and is fundamentally different from the manner in which another flat panel display monitor sequentially applies a constant voltage to the matrix coordinates. The linearity of the sawtooth current is important. If the waveform has distortion, the linearity of the screen and the image distortion are affected. Recently advanced monitors are multimode compatible with a horizontal deflection frequency conversion circuit that automatically adjusts the horizontal deflection frequency to the input signal for use in various systems.

4.4 The structure of CRT

Braun Tube (CRT; Cathode Ray Tube) is a part of display equipment which changes electric signal to light signal. When an electrical gun installed on negative part of Braun Tube emits electrons, they are led to the strong positive electric field. They are then injected through the tube. This enables 2−D display by biasing electrons with deflection york.

Electrons run into the fluorescent substance positioned back of a phosphor−coated screen, displaying video of characters or figures. Electrons are emitted from heated cathode focused by electrical field lens. Then they become a minute beam moving toward the screen with high electricity. A part of the beam go through a shadow mask(S/M) and collided with phosphors, making them illuminate. Electrons return to the socket by the order of metal back, mask spring and panel pin. Electrons go into an inner conductive film through contact spring as shown in Figure 11.26. Secondary electrons are gathered in anode by the power of inner conductive film. The electrons go out again from anode to the outer circuit. The electrons are supplied to the cathode and launched as an electron beam.

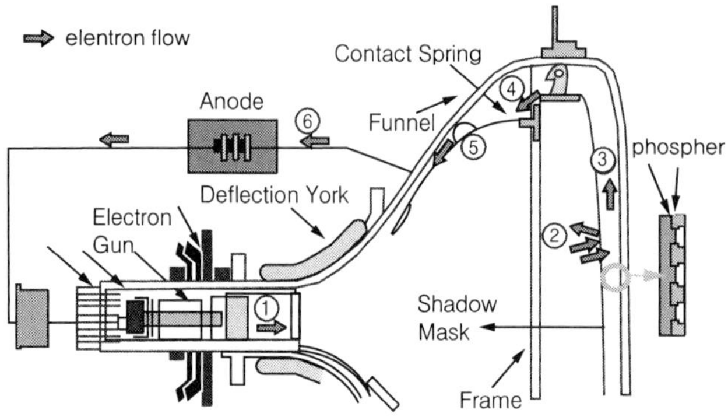

Figure 11.26

Principle of Movement CRT(Braun Tube)
① Electric Gun, ② Phosphor Collide, ③ Metal Back, ④ Mask Spring, ⑤ Funnel, ⑥ Both Poles

(1) Panel

As a part of the front screen of CRT, it is a place where video information from electric signal translates into final visual information as constituting some part of the vacuum tube. R(Red), G(Green), B(Blue) pixels are placed inside of the panel.

(2) Funnel

The funnel is a part of the vacuum tube with a panel. It consists of funnel-shaped cone and neck inserting electric gun. The cone portion is provided with a positive electrode voltage supply terminal and the inner conductive film is spread inside of the funnel. The inner conductive film includes shadow mask and the function that provide both poles to the final current of the electric gun.

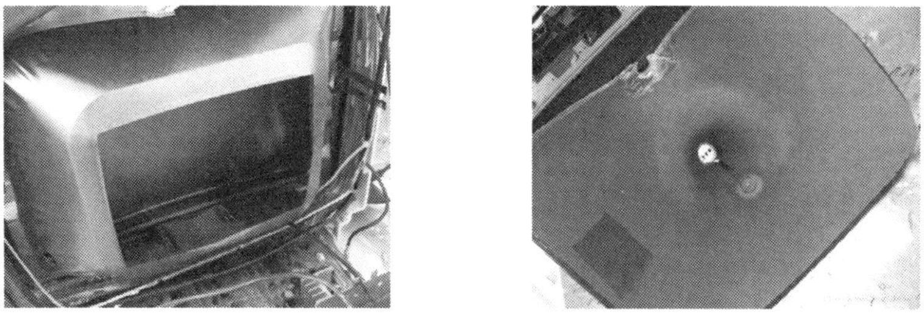

(a) (b)

Figure 11.27

Outside of the CRT's Funnel(a), Inside of the Funnel(b)

(3) Electron Gun

The electron gun is an apparatus for emitting an electron beam in accordance with an electric signal for displaying image information, and is inserted and mounted in a neck portion of a panel. Since the shadow mask image-receiving color tube needs to independently stimulate the pixels of the three primary color phosphors, the electron gun is composed of three elements.

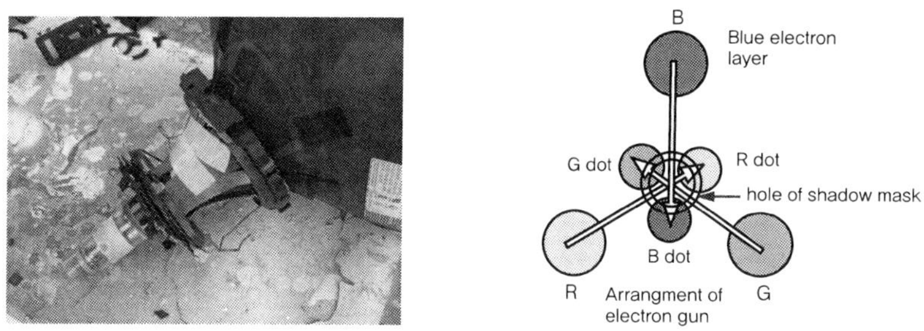

(a) (b)

Figure 11.28

Side of Electron Gun(a), Cross-Sectional Diagram(b)

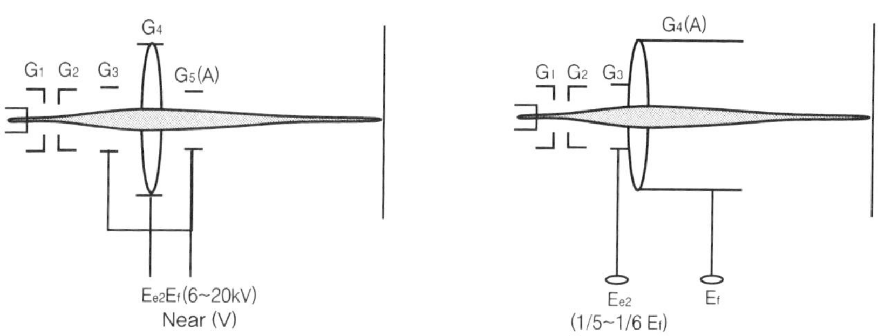

(a) unipotential form (b) bipotential form

Figure 11.29

Shape of Electron Gun According to the Position and Shape of a Grid

1) To make luminescence of R, G and B, 3 phosphor dots independently need 3 electron guns because it needs 3 electron beams.
2) 3 electron guns are placed at the point of the regular triangle as shown in Figure 11.28, the upper dot is B, right; R and left; G. The extension line of the axis of each electron gun is inclincd inward by about 1 degree on the surface of the shadow mask.
3) Emitted electron beam makes light by smashing each phosphor.
4) Inside the electron gun, heated cathode emits thermoelectron from the oxide at the tip.

5) This thermoelectron is accelerated toward axis through the small hole, attracting the positive charge of the second grid(G2).

6) As negative charge is occupied in the first grid(G1) comparing to the cathode, the movement of electrons is regulated by this charge.

7) The brightness of the screen is proportional to the current.

8) Electron flow which passed the second grid(G2) is accelerated more toward the third and fourth grid(G3, G4). After passing the fourth grid(G4), it rushes to the phosphor screen at constant speed.

9) Electron flow reaches phosphor screen as a beam-form because of an electrostatic lens which is composed of the third and fourth grid(G3,4).

Table 11.2 Compare Electron Gun to Grid's Location and Form.

Unipotential Electron Gun	Bipotential Electron Gun
① When focusing on the phosphor screen, the depth of focus is deep compared to the big spot and blur of the focus is less though electrical or mechanical error exists in phosphor front screen. The focus voltage is under a few percent of anode voltage. There is the advantage of the rareness of focusing out. ② It is used in black-white Braun tube.	① As mentioned before, the focus voltage is designed to be 16.8% ~ 20% of anode voltage so it needs high electric voltage for focus. As there is no narrow part of the electrode at the part of lens-formation, it is easy to correct the route of the beam and also the beam is less distorted. ② It is used in colored Braun tube.

(4) Deflection Yoke

One of the most important parts of CRT electrical devices is deflection yoke(DY). It makes magnetic field by flowing current to 2 coils placed perpendicular to each other inside of Braun tube and deflect electrical beam according to Flemings rule. In other words, to transfer the electrical signal in time series to video signal in CRT, the electron beam from the electron gun is replayed in the 2-dimensional image plane. The 2 coils are called a horizontally biased coil and a vertically biased coil as shown in Figure 11.30. The Electron beam is biased from side to side by giving 15.75 KHz current at the horizontally biased coil and the vertically deflected coil is fed with a 60 Hz sawtooth current [Fig. 11.31] to deflect the electron beam up and down.

The two coils are combined to sequentially scan the electron beam from the upper left to the lower right on the screen. The left end location becomes left end at Braun tube screen, middle parts become the center of the screen which electron beam goes straight and the right end location makes electron beam to the right end side of the Braun tube. It is then deflected sharply to the left side.

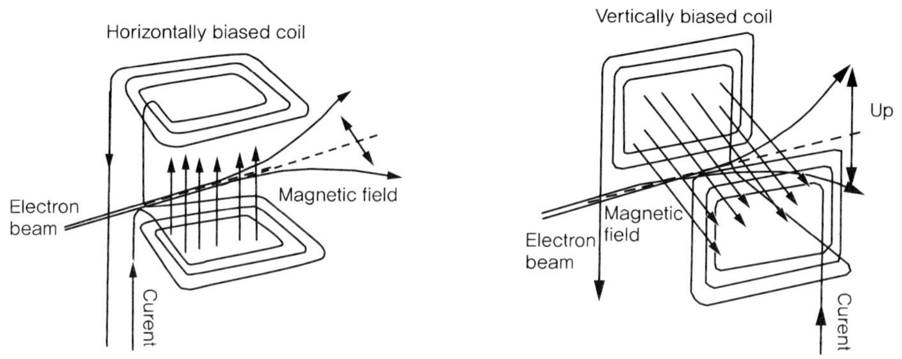

Figure 11.30

DY's form of horizontally and vertically biased coil.

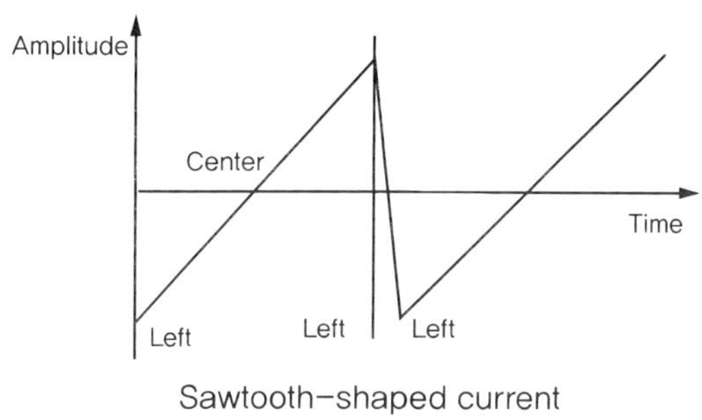

Figure 11.31

A sawtooth-shaped current wave inputted to deflect electron beam.

(5) Shadow Mask

Shadow masks play the most important role for 3 electron beams to reach the corresponding fluorescence points only. It is mounted in parallel with the fluorescent screen with a distance of about 10 mm from the fluorescent screen. The mask is regularly drilled with about 300,000 holes (depending on the size and type of the cathode ray tube) on a metal plate having a thickness of about 0.15 mm. Since the resolution of the screen is proportional to the square root of the number of holes, the better the resolution is, the smaller the pitch. However, if the pitch is smaller, the diameter of the fluorescent spot becomes smaller and the beam tends to cause color mixing. The shape of the hole is a conical hole (taper hole) that widens toward the fluorescent surfacc. So the passing electron beam does not scatter and collide against the wall of the hole. The diameter of the hole is about 0.25 mm in the center, slightly smaller as it goes to the end. The diameter of the hole at the end is about 0.22 mm. It varies from 0.2 to 0.3 mm depending on the type of pipe. The bigger the diameter is, the more the transparent of the electron beam, making the screen brighter. However, the color mixture tends to occur as in the case where the pitch is small. That is, when the mask pitch is determined, the diameter of the fluorescent spot is determined. The transmittance of the electron beam is 15 to 18% at the center of the screen. Only the transmitted electron beam acts to emit the fluorescent surface, and other electron beams collide with the mask, resulting in heat loss. For this reason, the back surface of the shadow mask is made black to improve the heat emission.

* 5 Types of Electron beam display

5.1 CRT

(1) Plate CRT

Monochrome Display Tube (MDT) type thin CRT

As shown in Figure 11.32, the line-shaped negative electrode is used to divide the beam on the first grating electrode through an aperture. The electron beam is deflected by the vertical deflection electrode and the horizontal deflecting electrode. As shown in Figure 11.32, a complicated process is required to create an image. The electron beam passes through the vertical and horizontal deflection -electrodes so that the beam can be focussed on to the screen.

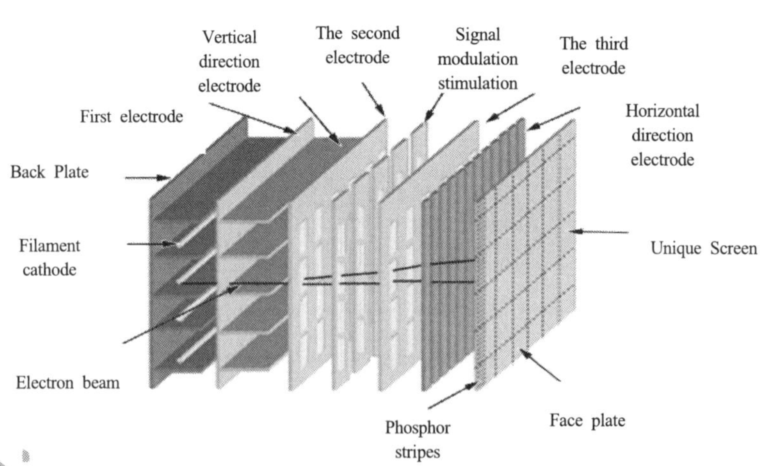

Figure 11.32
Structure conceptual diagram of MDT type thin CRT

(2) Cold cathode CRT

1) Sprindt type cold cathode CRT

Panel type display using Spindt type cold cathode, as shown in Figure 11.33, consists of a cathode made of a conically shaped electron emitting material, a gate electrode covering the conical cathode, and a cathode. A voltage is applied between the anode and the gate electrode to pick up the field emission current at the peak of the cone cathode. Each element is several μm in size, and the cold cathode element group need not be defect-free.

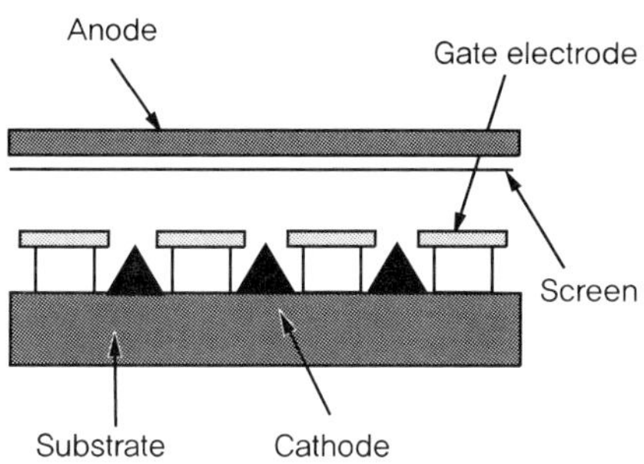

Figure 11.33
Sprindt type cold cathode CRT

2) Wedge typed cold cathode CRT

Since the Sprindt type cold cathode manufacturing method requires high technology, a wedge type cold cathode has been proposed. As shown in Figure 11.34, it is a structure in which a negative electrode material and a gate electrode disposed opposite to the negative electrode material are formed on an insulating material on a base substrate and the wedge-shaped electrode is arranged to face each other with a slight gap between the negative electrode and the gate electrode. Since the cathode is wedge-shaped, when a voltage is applied between the cathode and the gate electrode, the electric field is concentrated at the tip of the wedge-shaped cathode, and electrons are emitted. According to the three-dimensional simulation of the potential distribution at the tip of the wedge-shaped cathode, a maximum electric field of about 4.6×10^{-7} V/cm can be obtained at a distance of 5 nanometers from the surface of the cathode when the distance between the cathode and the gate is 1 micrometer, the distance between the gate and the cathode is 0.25 micrometer, the angle of the tip of the cathode is 30 degrees.

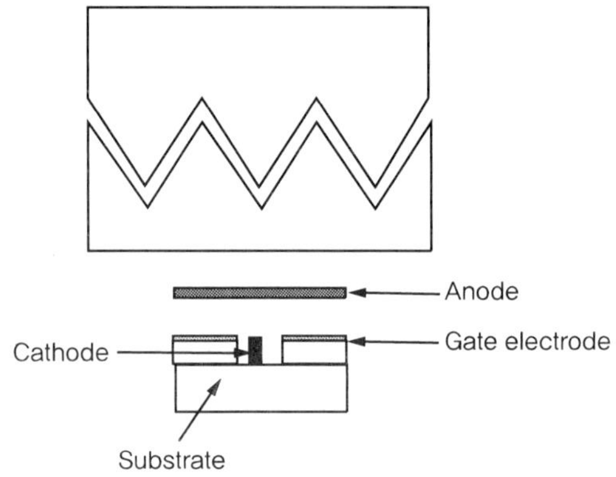

Figure 11.34
wedge typed cold cathode

(3) Others

1) Flatron

Flatron is used as a flat steel plate with applied tension without giving curvature to the shadow mask. The brightness is improved by 80% and the contrast is improved by 70% compared to the conventional CRT. Since fine pitch can be achieved, the resolution can be increased.

The structure shown in Figure 11.35 has a shadow mask for color separation in a vacuum container surrounded by glass. Through the holes in the shadow mask, the three primary colors of RGB emitted from the three electron guns emit the three primary colors by separately stimulating the phosphors of the three colors applied on the screen. The intensity of luminescence of the three primary color phosphor differs in proportion to the amount of electron beam from the electron gun. Any gray-scale display can be made in any color by the principle of color addition. There is one electron gun corresponding to the RGB phosphors on the screen, and three electron guns, in general, are arranged inline in a row. To this electron gun the brightness and color signals are provided. The focused beam at the focus electrode of the electron gun is deflected horizontally and vertically with the deflection magnetic field of the deflection yoke attached to the outside of the neck portion of the CRT.

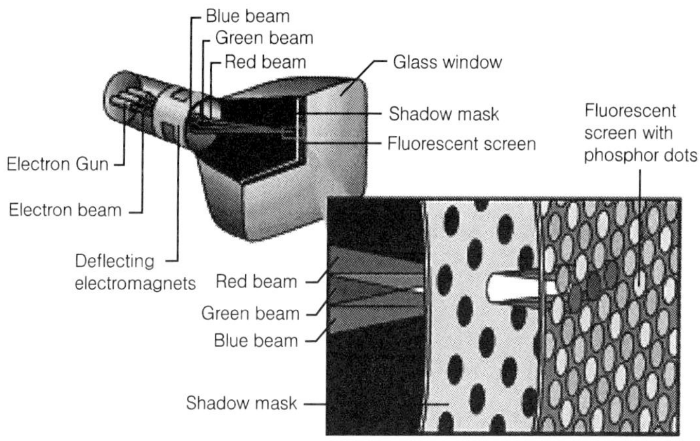

Figure 11.35

Shadow Mask type used in Flatron

The emitted electron beam from the electron gun is affected by the external magnetic field outside the CRT. For example, due to the external magnetic field the pre-determined trajectory of electron beam could be influenced, and a color change may occur. In order to block such unnecessary external magnetic fields, a magnetic shield is placed on the inner wall of the CRT. In order to efficiently emit the light of the phosphor excited by the electron beam in the front direction, the metal back made of aluminum is placed on the back surface (electron gun side) of the phosphor. This reflects light back to the back to increase the brightness of the light emitted to the front. Since the metal back is about 0.4 mm thick, the electrons accelerated by the high voltage to reach the phosphors through the aluminum shadow-mask to stimulate the phosphors. Some of the light emitted from the phosphors comes backward towards the mirror-like aluminum-shadow-mask. This light will be reflected by the shadow-mask and sent towards the screen. This process brightens the phosphors in the forward direction. The used electrons are finally grounded through the anode electrode.

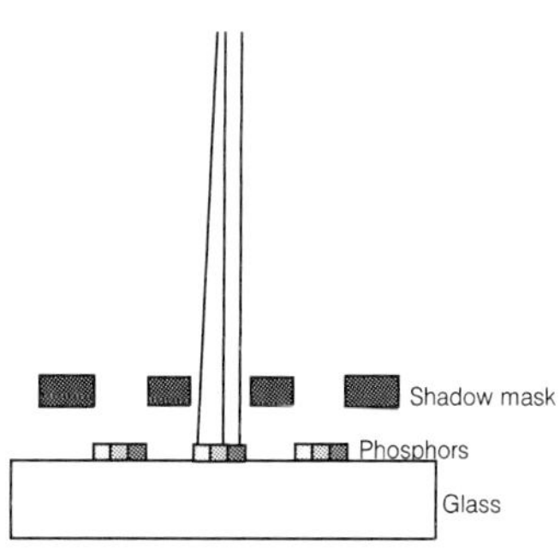

Figure 11.36

The process of electron beam reaching the phosphor in Flatron

2) Projection CRT

It reflects a small high-resolution CRT (Braun tube) on a mirror to form an image on the screen. It is the oldest method and has been improved in brightness and focus with continuous technology development. There are 1-tube CRT and 3-tube CRTs. In particular, CRTs used in 3-tube type use three high-performance CRTs separated by red, green, and blue, unlike CRTs used in conventional home TVs.

It has a relatively high resolution and excellent color reproducibility, fine tuning is possible, and smooth video production is possible. However, the high resolution needs high voltages, which generate a lot of heat. This is called Burn-in. To solve this problem, shield coating is needed at high voltages or lower voltage is required than the existing method. Since the image is reflected by the mirror in accordance to the to the optical design, system installation is hard to build and maintain.

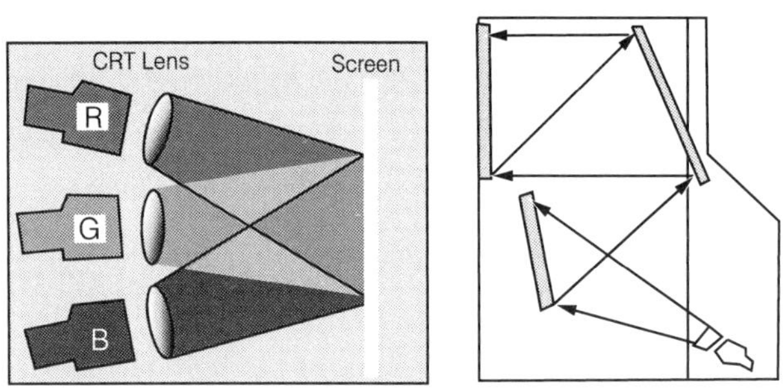

Figure 11.37
Principle of Projection TV

5.2 FED

Another type of FED is CNT FED (Carbon Nano Tube FED). The CNT FED consists of carbon nanotubes and emitters, which were once developed by Motorola, and prototypes were developed in May 2005. At one time, CNT FED was expected to be superior to LCD and PDP because it has no size limitation and high resolution. The reason for this is that the structural characteristics of the CNT-FED as shown in Figure 11.40 (a) are already very simple compared to the microtip FED (MT-FED). The ease of implementation is expected to be as high as in Figure 11.40 (b).

Emitters representing the FED are the most important factors in the performance of the FED, and emitter materials and structures have been developed variously. Among them, the micro tip emitter based on Mo material shows excellent performance in terms of image quality and power consumption in a display area of 20 inches or less. CNT–FED based on a carbon nanotube (CNT) shows image quality and mass productivity that can be applied to displays up to 40 inches.

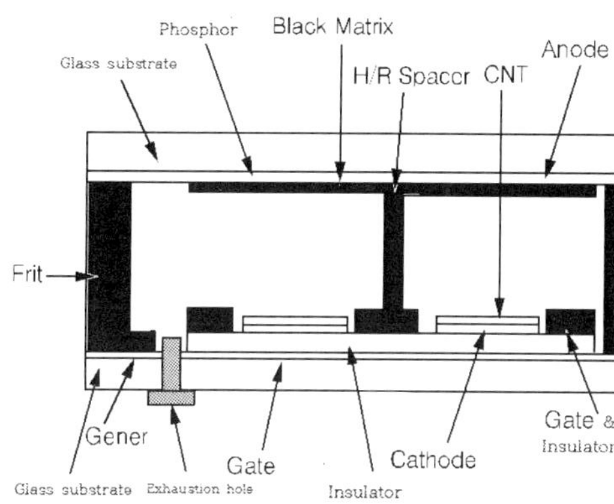

(a) CNT FED panel structure

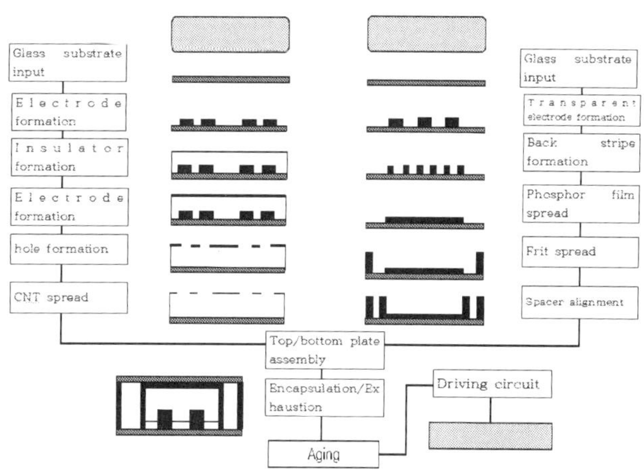

(b) Order of the process by screen printing

Figure 11.38

carbon nanotube FED panel structure and process order

6 Status of Electron beam Display

6.1 Pros and Cons of Electron Beam Display

The electron–beam–based display has a principle of emitting phosphors directly with electron beams. Therefore, it has many advantages over other displays. LCDs do not directly display black, whereas CRT and FED have a higher contrast ratio than LCDs because they are black when the pixel light is removed. Since it emits light directly from pixels, unlike LCD, there is no color distortion that changes color depending on the field of view. LCD has a disadvantage in the molecular dynamics, causing slow response of the display, resulting in a image retention and distortion. Since the electron–emitting display is a mode in which electrons strike, the response speed is very fast. The electron beam display has a very good color reproduction. However, there are many disadvantages. In the electron–beam display, when the phosphor is emitted by the electron gun and the screen is continuously scanned, that is, when the same pixel is continuously turned on, a burn–in phenomenon occurs in which the pixel is burned and the color is yellowed. In addition, due to the volume of CRTs, large–area displays are bulky and heavy. To solve this problem, the FED has been produced. However, the FED has a life expectancy of only 2,600 hours due to the phenomenon of Coulomb aging, in which electrons accumulate in the phosphor. To solve this problem, a high–voltage display must be produced. This leads to an increase in power consumption and also leads to a drawback of severe heat generation.

6.2 Current Status of Electron Beam Display

Since the invention of CRT in 1897, CRT has dominated the display market for about 100 years. CRT has many disadvantages, but as a development to solve this problem, a display using an electron beam different from the CRT of FED has also been made. However, due to the limitations of the method of utilizing the electron beam and the steady progress of solving the problem of the color reproduction rate, the contrast ratio, the viewing angle, etc., of the LCD and the like, the research and production of FED have been stopped.

3D Display

Chapter 12

1 Concept of 3D Display

1.1 3D Technology: Generation of Knowledge and Information in 3D

Along with the start of human civilization, the development of various communication methods has evolved with human culture. The industrial systems of the information society in the 21st century is expected to be transformed into a knowledge-intensive industrial structure based on human creativity, and the information and communication industry is expected to become the mainstream.

Since radio was developed in the past, the information has penetrated our lives through human ears. TV culture, which was first developed and commercialized in 1938, has evolved into HDTV (High Definition TV) from black and white to color TV, as shown in Figure 12.1 and has been a means of transmitting the most and image information to us so far. The emergence of such a TV can be said to be the opening of a two-dimensional visual information in our society.

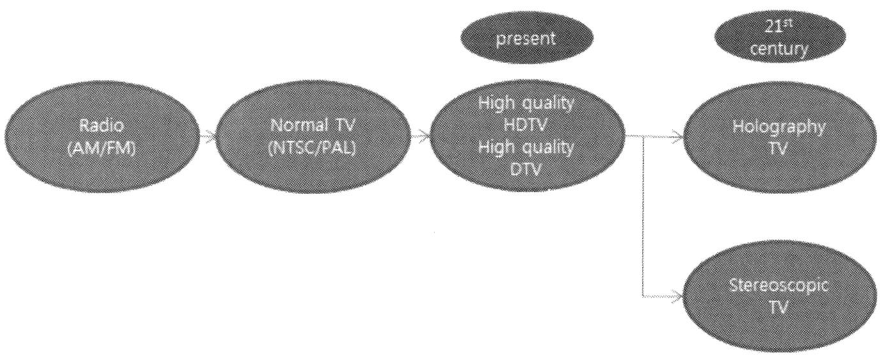

Figure 12.1
Direction of Broadcast Technology Development.

However, the information that human actually through eyes is 3D video. So there were needs that people pursue more nature-friendly and video that is closer to the reality. As a result, human need for more information grows continuously, That is, not only information from eyes and ears but also from feelings can be translated through the display. Research on a broadcast system like 3D TV has been on going worldwide actively.

The, Information Super Highway (ISH) announced from all advanced countries including Korea, enabled next generation multimedia communication service that has realism and a sense of nature through high speed, high capacity, three–dimensional, intelligence and humanized. ISH was built Broad band International Digital Network(B–ISDN) which will be generated in the future. VIP(Visual Intelligent and Personal) that is insisted by NTT in Japan, means added visibility intelligence and individualism, which give human reality and friendship. That service will be realized soon. For the realization of VIP, services needed to be developed are shown in Figure 12.2. It means that the present multimedia of watching and listening from the digital terminal will change to multimedia of "enjoying more natural looking" through 3D information terminals.

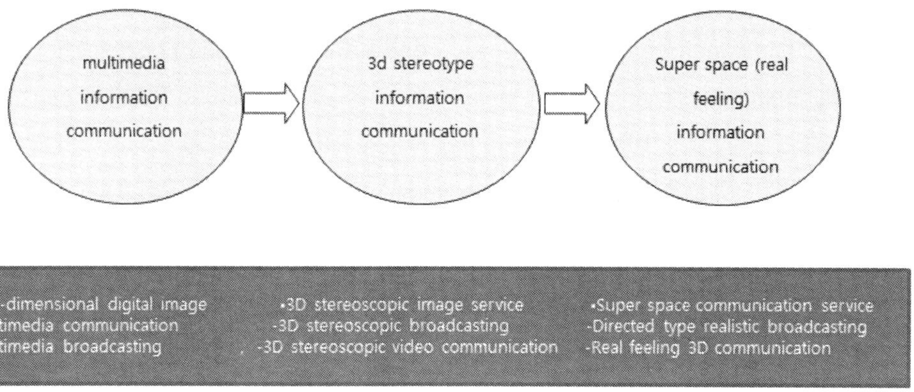

Figure 12.2
Direction of Communication and Information.

Therefore, with the development of broadcasting and multimedia technologies, 3D technology has been recognized as a high–tech technology that can create high value–added video industry in the next–generation around the world. Currently, 3D technology is being actively developed for practical use in advanced countries such as Japan, USA and Europe. This is due to the remarkable development of 3D display technology in the background. Especially, the developed technology of hardware like recent micro electronic industry, the appearance of brand new display like LCD, PDP, developed technology of processing digital video, and high image quality to movie level and 3–dimensional technology is remarkable results of those researches.

Therefore, this 3D stereoscopic image technology will lead the next–generation digital image culture as a new conceptual image media that elevates the quality level of visual information to almost similar to the real–world image that human beings see and feel, unlike the conventional 2D image.

1.2 Definition of 3D Display and Characteristics

(1) Definition of 3D Display

"3D stereoscopic image" refers to a more realistic image that simultaneously provides depth and spatial shape information, unlike 2D plane information. In other words, it is the image of the space recognition concept that is integrated into the process of recognition of learned environment by human binocular (left and right eyes) and then organizing in the brain. These 3D stereoscopic images provide a sense of presence, realism, and virtual reality as if looking at the scene in the field.

The 3D display is a stereoscopic display technology capable of expressing images having a depth that was impossible in a conventional two-dimensional flat display.

(2) Characteristics of 3D Display

3D video information is already well-known through EXPO exhibition or IMAX movie. The stereoscopic effect appears due to the difference between the left and right images of the object projected on the two eyes, and is recognized as a three-dimensional image through the synthesis process of the brain. Two eyes are located at a distance of about 65 mm apart so that they see images of two directions slightly different from each other, where binocular disparity occurs. Conventional CRTs or flat panel displays only two-dimensional images on the screen, making it difficult to view the image information on the front, rear, or side of the object, while 3D display technology simultaneously displays image information in various directions. Not only the depth information of the image but also the realistic feeling of being in the image space is important.

In order to realize a perfect 3D video communication that has the same realism and naturalness as the person feels in reality, it is necessary to be able to maximize the sense of presence. The maximization of the sense of presence is ultimately the three-dimensional display of the image. Further, it is required to realize a stereoscopic moving image capable of vividly expressing the motion of the object.

1.3 History of 3D Display

(1) History of 3D display

From ancient times, people tried to display the natural scenery that is visible in everyday life, the beauty of ever-changing nature, and realistic stereoscopic images. The desire to see in three-dimensional photos, movies and videos has accelerated the research and development of the principle of the 3D display. Based on various studies including stereoscopic photographs in the middle of the 19th century, it began to spread widely in Europe. In recent years, research and development on 3D display technology, which is emerging as a new next-generation high value-added video industry, has been actively developed, and it is not uncommon to further develop the 3D stereoscopic principle already designed in the late 19th century to early 20th century. Therefore, looking at the history of 3D research will greatly contribute to understanding the latest 3D research and development.

The simple question, "What is the structure that causes stereoscopic feeling?" is a cause of a long history of 3D research. Most of the painting techniques have been used to effectively create a three-dimensional effect on the flat canvas by the size of the object to be drawn, how to attach the shade and the adoption of the perspective method. In this way, the method of giving binocular disparity effect to the three-dimensional feeling that can be easily obtained is slightly out of the right-and-left image of the middle of the 19th century. The perception of the difference by these binomials greatly improves the stereoscopic effect, so that the so-called "pop-out" effect can be obtained. In 1839, a silver salt photograph was invented by Daruerre and binocular stereoscopic photography became active. A stereoscopic lens for dividing the two photographs into two left and right eyes was also created. These stereoscopes were installed in many entertainment venues at the end of the 19th century and became popular. The stereoscopic movies and stereoscopic hi-vision of polarized spectacles, which are currently popular at exhibition fairs and theme parks at the end of the 20th century, are based on the same principles as those of the end of the 19th century in the sense that they use stereoscopic effect by binocular disparity. The stereoscopic effect are divided into two major factors: the unification of the lens focusing, the size of the image, motion parallax, and binocular factors such as congestion and binocular parallax. Studies on their effects and structure are being pursued. Furthermore, it is interesting to note that the use of parallax barrier or lenticular sheet, which has been actively pursued as a principle of non-spectacles 3D TV system, is also derived from stereoscopic photographs of the early 20th century.

(2) Principle of parallax barrier and lenticular sheet

The principle of parallax barriers is to print or photograph the images to be seen in each of the left and right eyes alternately in a vertical pattern, which is regarded as an extremely thin vertical lattice barriers (see Figure 12.3). In this way, the vertical pattern image to be entered into the left eye and the vertical pattern image to be entered into the right eye are distributed by the barrier, so that another image slightly shifted into the left and right can be seen. The principle of the lenticular sheet is also realized by the refraction of "the distribution of the left and right images" by the row of the lens. These 3D stereoscopic display ideas were hardly verified by the material technology and processing technology of the early 20th century, but they were put to practical use with the progress of recent microelectronics and optical technology. In 1997, stereoscopic cameras were also commercially available.

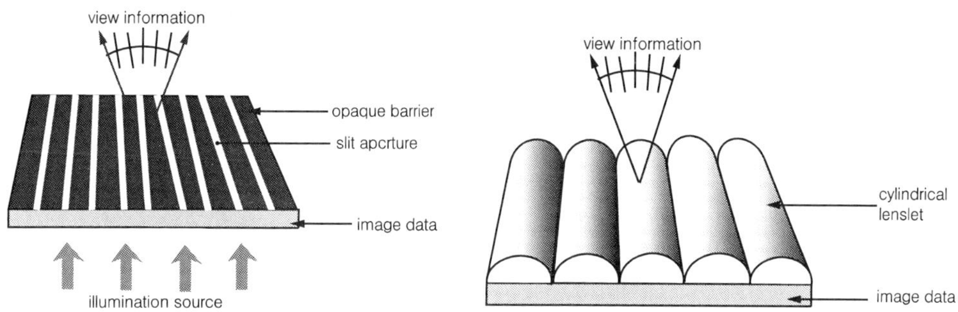

(a) parallax barrier (b) lenticular sheet

Figure 12.3
3D display mimetic diagram[1).

Display Engineering

* 2 Category and driving principle of 3d display

2.1 structure and principle of 3D display

Most of the 3D display portrays a stereoscopic image by using binocular parallax.

1) Computer Graphics, ACM SIGGRAPH, 31(2), May 1997, pp. 58–62. (http://web.media.mit.edu/~halazar/autostereo/disptech97.pdf)

However, binocular parallax is merely one of the factors that people can recognize a 3D space. Actually, people can recognize 3D space by more information such as congestion (rotation angle of the eye), control (alignment of eye focus), motion parallax (change due to the relative motion between the viewer and the object), and psychological factors (perspective, shading).

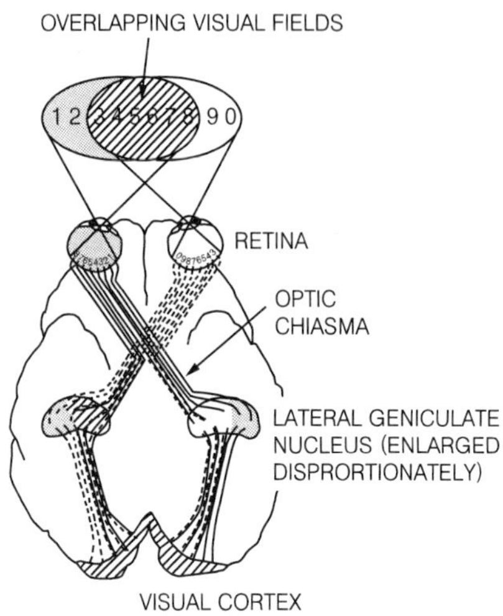

Figure 12.4
Model of human visual system

People have this binocular form of eyes anatomically, and how the human eye works systematically is shown as follows. The visual system of a pair of eyes has black pupils in the left and right eyes that are connected through a pathway through the optic nerve fibers from the visual cortex to a body part called the lateral geniculate body. As shown in Figure 12.4, the optic nerve cell fibers are separated at the optic chiasma and crossed to the opposite side. Therefore, the fibers from the right pupil pass through the left lateral fracture of the brain, while the fibers from the left pupil pass through the right lateral fracture of the brain. Here, a certain degree of information processing is performed. Visual information from the two lateral fractures reaches the visual area in the visual cerebral cortex. It provides a function for visual detection in this shell processing. That is, when light reflected from an external object enters the eye, it is focused through the cornea and the pupil lens. A number of light reflections depends on the thickness of the lens, which is controlled by the muscle of the eye as well as the cornea. And, unconsciously, the eyes move so that the image is formed on the retina. The retina consists of photodetectors that detect light, called rods and cones, which generate electrical pulses through the optic nerves, which absorb the light by initiating electrical reactions. By this method, the images generated from the two pupil senses are formed on the upper side. Therefore, important information processing in the visual information system occurs behind the pupil. In general, the final information processing results in one correct image, the left eye image is processed by the right half of the brain, and the right eye image is processed by the left half of the brain. The current 3D stereoscopic display is disregarded among the above information, which is a fundamental factor of the inconvenience of existing 3D viewing.

2.2 Classification of 3D display

Three-dimensional image display can be classified according to the 3D display system, the viewpoint, whether wearing glasses or not, the composition of the system, and the observation condition as shown in Table 12.1.

Table 12.1 3D Display type.

Section	Type	Glasses	Color	Motion Picture	Face to Face	Multiple Viewing	Viewing Angle	Moving Viewpoint
Stereoscopic Display	Box	×	O	O	×	×	×	×
	Anaglyph	O	×	O	O	O	O	×
	Polarizing glasses	O	O	O	O	O	O	×
	Time-sharing	O	O	O	O	Δ-O	O	×
	2 eyes Lenticular	×	O	O	O	×	×	×
Volumetric Display	multi eyes Lenticular	×	O	O	O	×-Δ	Δ	Δ
	Integral Photography	×	O	O	O	×-Δ	O	O
	sampling depth	×	O	O	Δ	Δ	O	O
	Holography	×	×-O	×-O	Δ	O	O	O

(o: necessary/possible ; x: unnecessary/impossible)

Recognition of the three-dimensional image is mainly caused by binocular parallax. If the observed image at different angles is input to both eyes, people can realize space sense by brain activity. Three-dimensional image display is classified into two types depending on the type of image recognition, the binocular system (stereoscopic display) and a three-dimensional system (volumetric display). The binocular method uses binocular disparity as shown in Figure 12.5. Depending on whether the observer wears glasses or not, there are glasses type polarizing method and time division method, and glasses-free type Parallax Barrier method and Lenticular method. The former is possible by using existing displays, but separate polarized or LCD shutter glasses should be worn. The volumetric display combines the existing display with an image splitter and cylindrical lens array, and it can observe the fixed range so it is limited to a small number of people. However, it is more practical since separate glasses are not needed. The glasses-free method is planned to expand the stereoscopic area by the multocular lenticular method, a point of view-following method corresponding to the observer's movement, and stereo TV system using a type of backlight division. The development of glasses-free stereo TV is active, but still in the research phase. The volumetric display method is divided into multivariate, holography, and depth sampling. The multocular method has a similar structure with a stereoscopic method. By displaying multiple images at multiple angles, it is possible to observe at a wider angle than binocular with only two images.

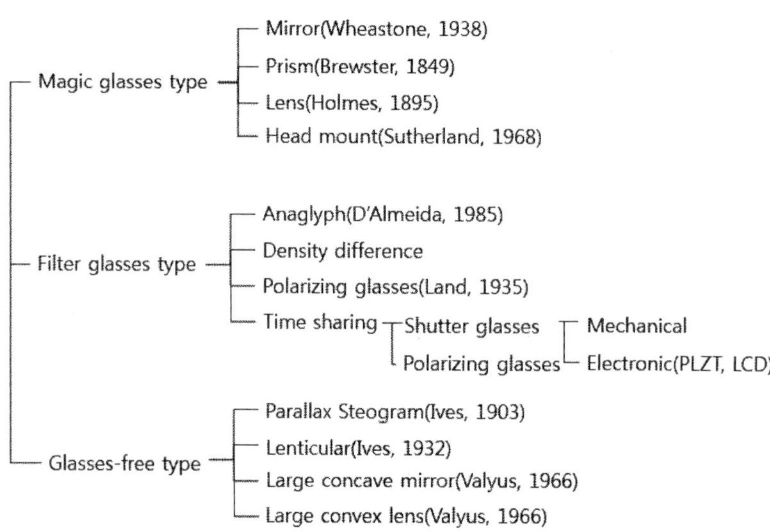

Figure 12.5
Binocular stereoscopic display classification

Holography is divided into optical holography and electro-holography method. The former is a method using the laser or white light source to the light-sensitive substrate as recording and reproducing the amplitude and the phase of the light waves in the interference pattern of the object, as shown in Figure 12.6. The latter is a method of expressing three-dimensional stereoscopic information by performing a large amount of image information processing by performing electrical signal processing in the form of an interference pattern. The depth sampling method is a method of displaying a stereoscopic image by superimposing various images in the depth direction of an object, such as a varifocal mirror method, a rotating cylinder method, and a display surface vibration method. This method expresses the depth sense so that it can be observed at any point in a predetermined range so that there is no restriction on the observation position and natural image display is possible, but the development of a large amount of information processing technology is the key.

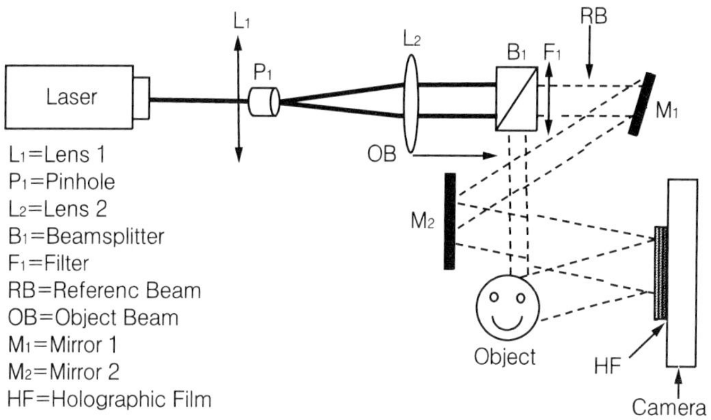

Figure 12.6
Optical holographic 3D display method (Israel Institute of technology).

2.3 Motion concept in accordance with the 3D display type

(1) Glasses type

1) Polarization type

A polarizing plate is invented by W. B. Herapath in 1852 and Anderton proposed how to use the principle of the polarization in the 3D projection in 1891. After that, a 3D film method using a polarizing plate was released by E. H. Land in 1935. This is the 3D display system using polarization filter, which is widely used today. The polarization filter method using a light-blocking effect by the combination of orthogonal polarization elements is to separate the left and right images. Figure 12.7 shows the structure of a projection type three-dimensional display by the polarizing filter. As shown in the figure, the left and right image were projected to screen by a video projector equipped with the orthogonal polarization filter and the observer can observe with an orthogonal polarizing filter attached spectacles.

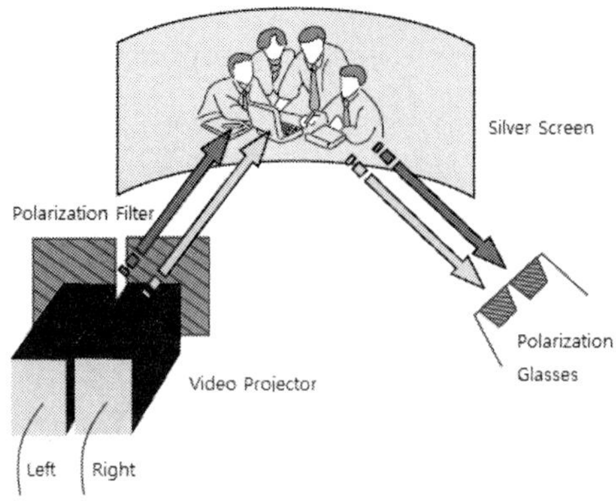

Figure 12.7
Polarized filter type projection type stereoscopic display.

Generally, a linear polarizing filter is often used for a polarizing filter system, but a circular polarizing filter is sometimes used. The circularly polarized light filter can convert the circularly polarized light into the right–handed or left–handed circularly polarized light by combining the linearly polarized light filter and the quarter–wave plate. Stereoscopic display using a circular polarizing film has an advantage that stereoscopic viewing can be performed even when a viewer tilts his/her face largely to the left and right, as compared with a case where a linear polarizing film is used. On the other hand, there is a problem that color crosstalk tends to occur in a stereoscopic image.

The advantages of the polarized glasses method are ① good color reproduction, ② a high resolution, ③ it is possible to display a full–color moving picture, ④ it can be seen by many audiences concurrently, ⑤ good sense of reality, ⑥ it is easier to realize. On the other hand, the disadvantages are ① a dark image is obtained because a polarizing filter having a low transmittance is used, ② the viewer has to wear glasses, ③ A special screen without polarization interference is needed for half mirror or screen due to reflection.

Polarized glasses method is widely used in events such as stereoscopic movies and Exposition since ancient times, and its typical example is the IMAX 3D exhibit at Suntory Hall of IMAX Corporation of Canada EXPO'90 (Osaka). The 3D high–vision system which was developed by NHK has taken attention as the application of the stereoscopic TV system. Because this system has the advantages of excellent stereoscopic effect and high image quality of 3D high vision and high–definition, various stereoscopic high–vision program software is produced as described in the edit processing technique, and practical use is progressing in various aspects.

It has been used in brain and ophthalmic surgery by surgical medical applications established in The Kobe meyitek Techno Center and a Tokai University Marine Science Museum.

2) Time–share system

Time–share system is a method in which a right and left side image with a time difference are alternately switched in sequence and presented in both eyes and a stereoscopic vision is performed using shutter glasses which are opened and closed synchronously. In the early days, mechanical shutter glasses were used, and electronic shutter glasses with electro–optic effects of PLZT, a transparent ceramic, were used. However, PLZT is rarely used because the transmittance of light is low by several % and the driving voltage is high. In recent years, it is possible to drive low voltage and low current, so small size, lightweight, and cheap liquid crystal shutter glasses are widely used. Figure 12.8 shows the basic principle of the liquid crystal shutter type 3D TV. In the time–division stereoscopic TV system, left and right images are alternately displayed for each field of view of the TV, and shutter glasses are opened and closed in synchronism with stereoscopic viewing. As a result, when the screen of the display is bright, flicker[2)] at 30 Hz is disturbed, and the vertical resolution is reduced to half (1/2F).

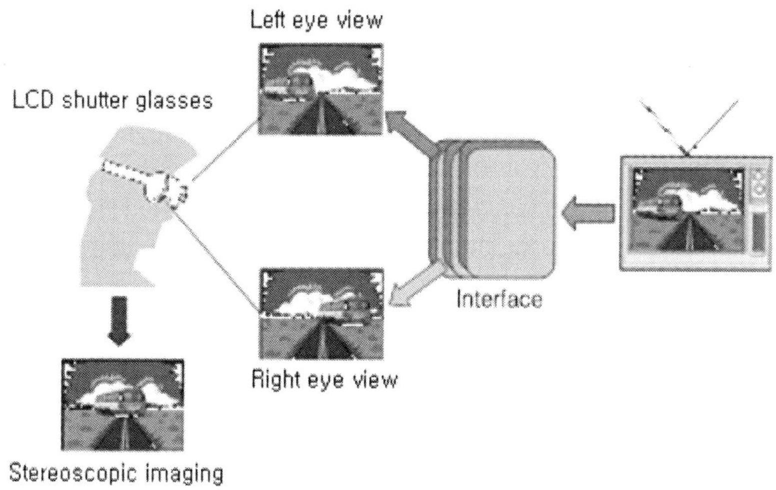

Figure 12.8
3D TV system with liquid crystal shutter.

2) flicker is a visible fading between cycles displayed

As a method for solving these problems, a double speed scanning type time division stereoscopic TV system has been devised. The principle is that the left and right images are first written into the field memory with a normal 60 Hz, 2:1 interlaced signal, and read out from the memory at a double speed of 120 Hz, 4:1 interlace scanning. Thereafter, the left and right images are alternately switched every 1/120sec. and displayed on the screen. The liquid crystal shutter is opened and closed synchronously with the left and right images.

In this case, a receiver operating at a double speed of the current TV system (horizontal 31.5 kHz, vertical 120 Hz) is required, and a multi-scan type display capable of coping with this is also commercialized. The liquid crystal shutter glasses need to operate at a speed of 120 Hz to have a high light shielding ratio (contrast ratio) at the time of shutter ON and OFF. In this regard, a TN type liquid crystal in which a response speed and a light shielding ratio should be improved or a shutter eyeglass using a ferroelectric liquid crystal, and the like are realized.

On the other hand, as another form of the time-division stereoscopic TV system, a system in which a high-speed liquid crystal shutter panel is mounted on the front surface of a CRT displaying an image and light circular polarized glasses are written on the observer side is also commercialized. In the time-division stereoscopic TV system, ① a stereoscopic image is displayed on a single display, ② High-quality full-color display is possible, ③ It is highly compatible with current TV system, ④ It is easy to realize the system at low cost, ⑤ It can be observed by many people at the same time.

Since the time division stereoscopic television system is a practical stereoscopic television system, it is developed not only for business purposes but also as a game machine for enjoying at home. For example, in 1986, Victor Corporation of Japan developed a VHD stereoscopic video disc recording a stereoscopic image of an NTSC (National Television System Committee) system. At the time of stereoscopic playback, left and right images are alternately reproduced by 1/60 sec, and a flicker of 30 Hz occurs at the time of reproduction by a liquid crystal stereoscopic scope which is opened and closed in 1/60 sec. Meanwhile, a flicker-free stereoscopic video disc was released in 1994 from Sony.

The images captured by the Hi-Vision camera separately are compressed in half in the longitudinal direction, synthesized, and converted into high-vision MUSE (multiple sub-Nyquist sampling encoding) signals to be recorded on a laser disk (LD). The image reproduced by the LD player of the MUSE system is doubled to be alternately displayed on the Hi-Vision TV by 1/120 sec, and solidified by using the liquid crystal shutter glasses which are opened and closed by 1/120 sec.

Although this system will have the half of 1125 scanning lines, you can enjoy high definition stereoscopic images without flicker by a combination of home high-definition TV and LD player for MUSE.

In addition, as a large-scale system for the fair, there is a space-shaped color 3D three-dimensional projection method disclosed by EXPO '90 (Fujitsu Pavilion) of Canadian IMAX. The dome with a diameter of 24 meter displays alternately time-division images of left and right images, and a wireless liquid-crystal shutter eyeglasses that opens and closes at 90Hz. In the huge space enough to cover the entire field of view of the audience, 3D CG (Computer Graphics) images with overwhelming power and impression were screened.

3) Other types of 3D display

After the principle of the two-color three-dimensional display system was proposed in 1853 by Rollman, anaglyphic (Anaglyph) scheme was announced in 1858 by d'Almeida. In this method, a stereo control image drawn in a complementary color (for example, red and blue) is selected by separating the left and right images with a color filter having no common transmission wavelength region and solidifying the image. When the color difference of the color filter is too large or the contrast condition is high, the binocular vision struggle (a phenomenon in which the left image and the allegorical image appear alternately appear at the outline part of the figure) appears and becomes unstable. Observation of bright screen for a long time causes a difference in color adaptation and causes fatigue.

The concentration difference method uses a phenomenon referred to Pulfrich effect. When a planar image with a motion is observed with filters having a different transmittance mounted on the left and right, an image with a sense of perception is displayed corresponding to the perceived time difference according to the difference in transmittance. This method has no effect in the still image because it obtains a sense of the depth from the motion component in the horizontal direction, and there is a problem that the forward/backward direction is reversed or the depth feeling changes depending on the direction or speed of the movement. Although the effect is small in a general image, a stereoscopic effect is apt to occur in a CG image.

(2) Glasses-free type

1) Parallax barrier type

It is annoying wearing special glasses to view the stereoscopic images. In 1903, F. E. Ives of the United States proposed a stereoscopic method called parallax stereogram. In this method, as shown in the principle of Fig. 12.9, two images of left and right eyes are arranged alternately at an appropriate interval behind the opening of a thin slit called parallax barriers to correctly separate them when viewed through this opening at a certain point. However, since only two left and right images are displayed, the viewpoint cannot be moved freely.

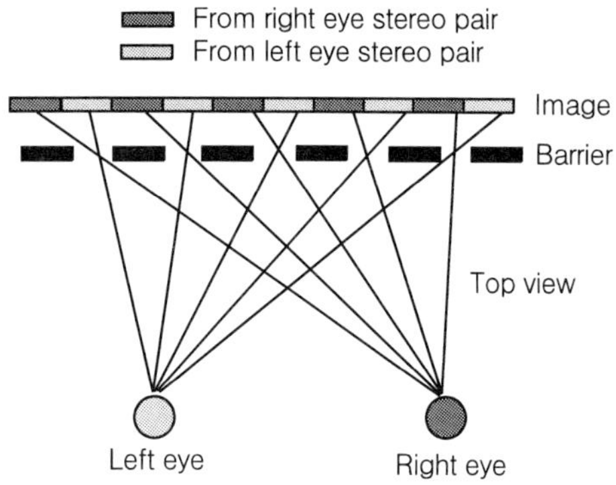

Figure 12.9
Parallax barrier method.

In 1918, C. W. Kanolt of the United States proposed a Parallax panoramagram to view three-dimensional images continuously even when changing the viewpoint. In this method, the slit spacing of the parallax barrier is changed to lower the aperture ratio, and instead, the multi-directional image is photographed on the image display surface in multiple directions. The first parallax panoramogram experiment was conducted by H. E. Ives in 1928. However, the parallax barrier method has hardly been popularized because of the reasons that the barrier is uncomfortable or the brightness of the stereoscopic image is impaired.

Recently, however, parallax barriers have come to be recognized again. A group at the University of Illinois in the United States created a 13-view image of a CG image and put it on a color slide film with a scanner, which then presented a bright three-dimensional still image that was illuminated by a backlight.

In addition, parallax barrier system for the personal computer stereoscopic color liquid crystal display is commercially available by US DIT. The three-dimensional image display has a special structure having a slender slit-shaped light source of the stripes on the back of the liquid crystal panel which displays the left and right images and it displays a bright and high contrast and three-dimensional color image. In addition, stereoscopic liquid crystal color TV which improved this principle was announced in Sanyo Electric in 1994. By using a bright light source for the backlight of a liquid crystal display, it is a fixed type that cannot change the position and shape of a parallax barrier. The liquid crystal parallax barrier method which can change the shape and the position of the barrier dynamically was published in NHK in 1992. In this method, two liquid crystal panels are stacked, and a slit barrier is displayed on one liquid crystal panel. Through this slit barrier, the stereoscopic image of the multi-points can be seen without glasses. This method can freely change the shape of the slit barrier displayed on the liquid crystal panel, the opening, the position, and the like in accordance with the number of viewpoints of the three-dimensional image. As a result, arbitrary three-dimensional image display from the binocular to the multi-view can be handled. Also, since the two-dimensional image can be displayed without lowering the resolution without displaying the slit barrier, it is also used as a "wall-hanging" non-eyeglass 3D TV device compatible with the current TV system. In addition, there is a feature that a two-dimensional image and a three-dimensional image can be mixedly displayed on the same screen and is used as a three-dimensional image display for a computer corresponding to a multimedia. By further developing this method, it is possible to display not only a one-dimensional slit barrier of the longitudinal pattern but also a slit barrier of a two-dimensional lattice shape on a liquid crystal panel.

By displaying the multi-viewpoint image having a parallax in the horizontal and vertical directions behind the slit barrier (pinhole) of a two-dimensional grid pattern, a three-dimensional image corresponding to observer's left and right, vertical direction movement can be seen.

2) Lenticular type

Many people have already seen postcards of stereoscopic photographs. Actually, a lenticular method is used for a postcard of this stereoscopic photograph. The history of the lenticular system has long been established, and in 1932 H. E. Ives has patented Lenticular Stereo. However, the processing technology and material technology were immature. Precise machining technology, plastics industries, photographs, printing technology, etc was progressed in the 1960s. Overall lenticular imaging techniques were advanced remarkably. In recent years, cheap 3D camera equipped 3 eyes lens which can shoot 3D lenticular photos and 3D pictures can be enjoyed easily. In addition, the lenticular method makes it possible to enjoy full–color 3D moving images without wearing glasses and it has been used in studies such as stereo TV system for work or artificial realism and position sensitive communications. From autumn in 1994, it has started practical work in the field of video games.

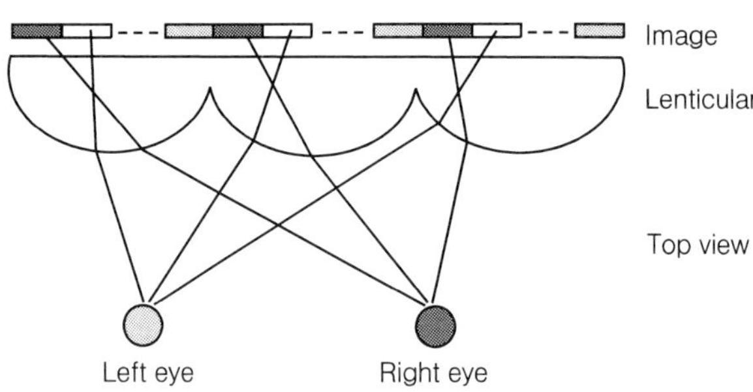

Figure 12.10
Lenticular display principle.

The principles of the lenticular method are to place the right and left images in the focal plane of the lens, called a lenticular screen in a stripe state and observe it through lens plate, then according to the directivity of the lens plate, a three–dimensional image is separated without glasses, as shown in Figure 12.10. The lenticular method has been thought for multi–eyes method which can make 3D images by viewing multi–directions, not just 2 eyes type. There is 8 eyes lenticular method as an example of the multi–eyes method. The stripe image in each direction is separated according to the lens action of the lenticular screen, and when the observer moves in the horizontal direction, the stereoscopic image can be seen in the other direction corresponding to the oblique view. In addition to optimizing optical parameters such as the pitch, thickness, and curvature of the lens plate to prevent moir (irregular grain patterns on the surface of the fabric), it is necessary to align the lens plate with the display image precisely.

2.4 Motion concept in accordance with the Volumetric Display 3D display type

(1) Full 3D type

Full 3D type is display technology including IP, volumetric and holography.

1) Integral Photography (IP)

IP is the method proposed by M. G. Lippmann in French in 1908, but after the proposal, it has been almost forgotten for 60 years because it requires a high degree of precision machine technology and high-resolution photo technology. Figure 12.11 shows the principle. Fly's eye lens is placed in front of the photographic plate. When the emulsion side of the dry plate is aligned with the focal plane of the lens, numerous small images are formed there. These images are the inverted images of the same object viewed from different directions.

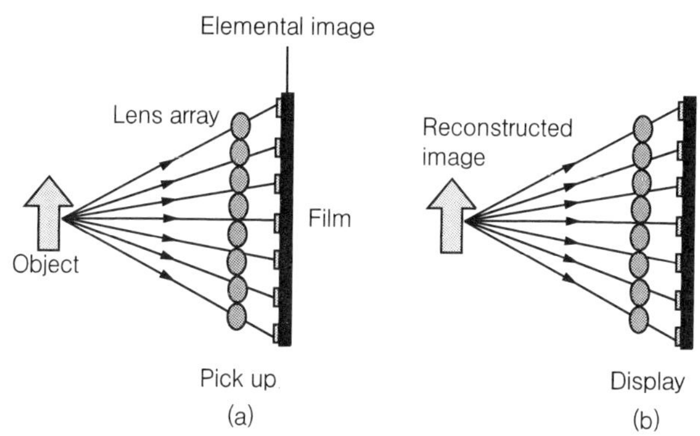

Figure 12.11
(a) shooting of IP (b) playback of IP.

Next, the dry plate is exposed and developed, and then the positive plate is inverted and copied at exactly the same distance. If you place it exactly in its original position and illuminate it with white diffused light from the back, it will reproduce the exact same path as when shooting, as shown in Fig. (b), resulting in a three-dimensional real image at the position of the original object.

When you look at this 3D object in the left direction, it looks as though the original object is in that position. But here, in fact, there is a problem. The image originally photographed becomes a concave inverted image when viewed from the right side of the subject. For example, if the subject is a person's face, the image to be reproduced becomes an unpleasant appearance on the inside of the mask. This inverted image is called a pseudoscopic image in optics. This is a very frequent problem with regard to 3D imaging technology.

Interestingly, Lippmann himself, who invented IP, has overlooked this problem. Thinking for a moment, this problem seems to have something eccentric and simple. But the reality is not so simple. The reason is that the problem is an inherent problem due to the fact that the image given by the IP is real. The solution came from H. E. Ives in 1931. It is a way of once again receiving the inverted reality from the IP and finally looking at it as shown in Figure 12.11 (b).

This is of course not a simple solution. It may seem like a double effort, but it is not that much. This is because in this process it is possible to simultaneously invert from the dark to the bright phase. It is easy to see that in the explanation of principle IP, it is convenient to make an inverted copy of a dry plate and put it exactly in its original position, but there will be considerable difficulties in doing so. In that respect, if Ives's two-stage IP method that is a fly-eye lens with a thickness that allows the focal plane and the back surface to coincide with the beginning, is used, a photoresist (or a photosensitive film) and dry plate can be integrated, then the difficulty of imaging is solved at once.

(2) Volumetric display

The principle of the five elements sampling method is that an image (a single layer of 1, 2, 3, ..., 10) of a two-dimensional cut surface with respect to each position of an object is moved to an imaging system and is synchronized and then switched at high speed to display an image, and an image is reproduced in a space using a residual image phenomenon of the eye.

In this method, since the complete three-dimensional image having the omnidirectional information is reproduced in the space, the stereoscopic image viewed from the position can be seen when the eye is moved vertically and horizontally.

This display method is suitable for synthesizing and displaying three-dimensional images on a plurality of single layers obtained by CT scanning, or for displaying three-dimensional images by CG. Other methods such as a method of vibrating or rotating a display surface at a high speed, a display area layering method, and the like have been proposed. The other five elements sampling methods are such as vibrating or rotating the display surface at a high rate and the display area platform.

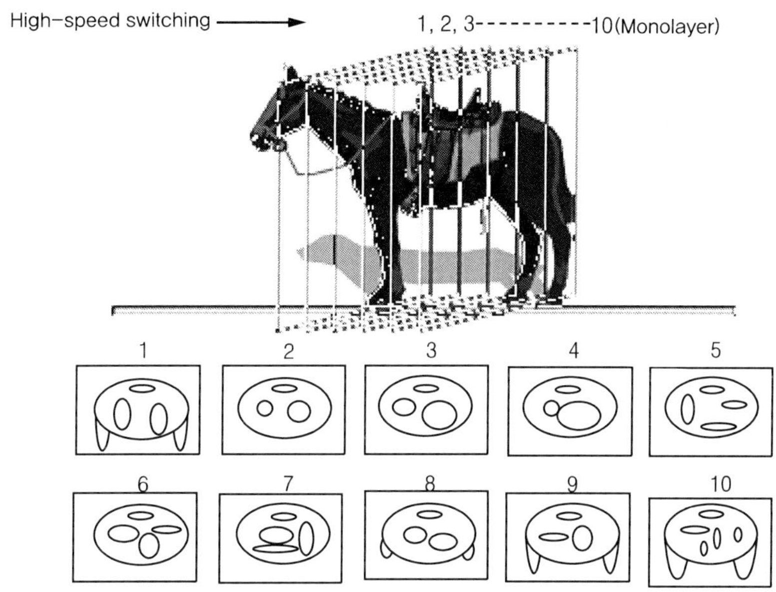

Figure 12.12

The principle of the five elements sampling method.

(3) Vibration, movement and rotation of the display surface method

A typical example of the surface vibration method is a variable focus mirror method which shows a 3D image by reflecting the image on the oscillating mirror as shown in Figure 12.13.

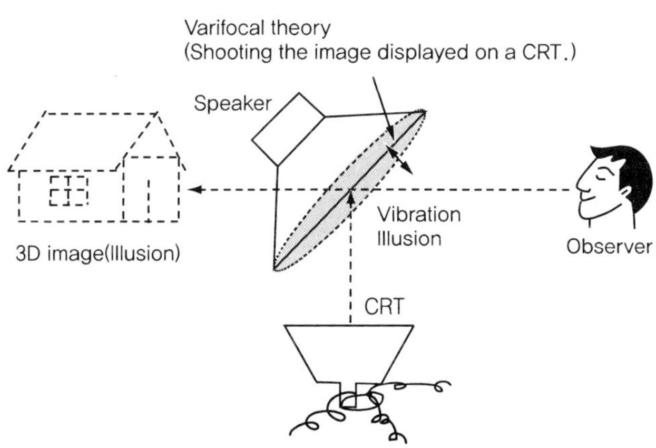

Figure 12.13

principle of Variable Focus Mirror Method.

In this system, an image reflected on a mirror can be observed in a three-dimensional image by vibrating a reflecting mirror with an electric signal and displaying an image synchronized with the rcflecting mirror on a CRT screen. The flexible polyester film is used for the mirror, and the outer surface is coated with a metal such as aluminum. The vibration of the mirror is about 30 to 60 Hz in terms of utilizing the afterimage characteristic of the eye. The three-dimensional image according to the variable focus mirror method has a problem that the phase distortion or magnification changes due to a change in the curved surface of the movable mirror although a plane depth feeling of more than 100 times the vibration amplitude of the mirror is obtained.

In 1981, Zenith Kosa solved this problem and released [Space Graph System] for CG. There is also a method called a moving screen method as shown in Figure 12.14. In this method, laser light is scanned and projected on a moving surface of an object to be displayed on a moving screen. The projected image corresponds to each position of the screen. If the speed of the screen is fast enough and synchronized, a three-dimensional image is recognized in a space constituted (the area of the screen x the stroke of the motion) by the afterimage of the eye.

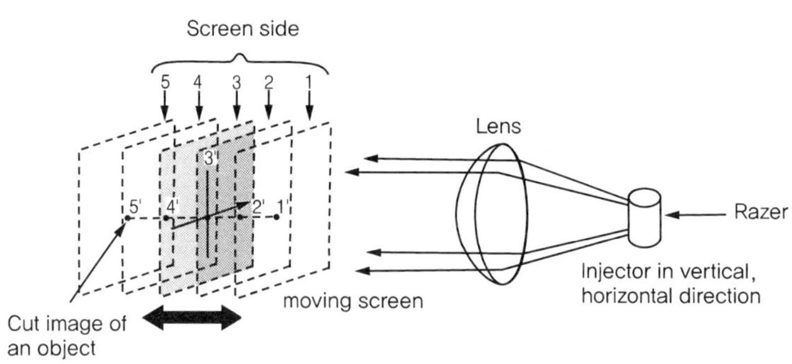

Figure 12.14
Principle of moving screen method

The group of Shibaura Institute of Technology uses this principle to present a three-dimensional display using a rotary moving screen and display a color three-dimensional CG image. In a recent example, a display surface in which 96 x 96 mm light emitting diodes (LEDs) array is moved in the vertical direction at a speed of 20 Hz, and 100 CG images are displayed at intervals of 1 mm in synchronization with this, Toshiba has announced a system that allows viewing of three-dimensional images in a 360° direction. On the other hand, as an example of the rotating screen system, a rotating cylindrical three-dimensional display system has been proposed in which laser light modulated in luminance with a video signal to be displayed is irradiated from above and scanned in synchronism with a rotating spiral screen.

In the USA, TI Corporation introduced Multiplanar 3–D display, a revolutionary type of rotation screen in 1990. The basic principle is to install a translucent wing which is a screen on the rotation axis of the motor and rotate it at high speed of 600 rpm. When a laser beam modulated in amplitude in synchronization with the rotation of the motor is irradiated, a three–dimensional moving image can be displayed on a cylindrical space having a diameter of 36 inches on which a wing moves as shown in Figure 12.15.

In the experiment, a cylinder having a diameter of 4 inches and a height of 4 inches was scanned by a laser beam within a range of a cube having a size of 4 inches on one side to realize the display of a pixel having a resolution of 750 x 750 x 750. This method can see stereoscopic moving images everywhere around 360°. When displaying a color image, the intensity of the three primary color laser light of red, blue, and green may be adjusted.

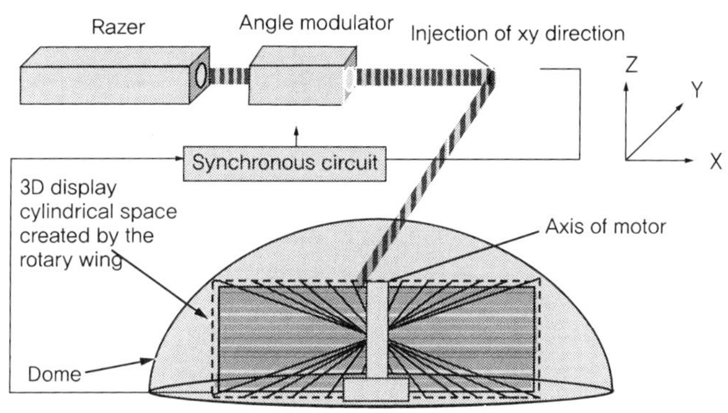

Figure 12.15
3–dimensional moving picture display with rotating screen.

(4) Laminated display screen type

An example of a three–dimensional image display system is a display panel laminating method proposed by T. S. Bugak. The method as shown in Figure 12.16, is to configure a combination of cells and chiral liquid crystal cell. A plurality of liquid crystal panels are laid out, and the image displayed on the CRT is reflected by a half mirror to be projected on a liquid crystal panel, and the cells in front of the liquid crystal cell corresponding to the image of each cut surface are controlled to reflect the image in the original direction.

In this method, there is no mechanical moving part, but there is a drawback that the brightness of the image darkens in proportion to the number of the passing of the cells.

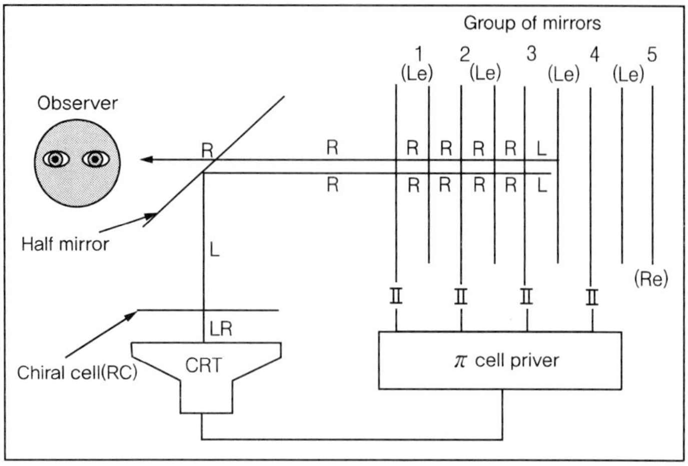

Figure 12.16
Display surface laminated 3D display

In the three-dimensional space reproduced by the above-described five-line sampling method, if the display of all the cut images is not completed within the after glow time of the eyes, not only the flicker is detected but an image of the back or the inside hidden in the front reproduction image is visually seen. A so-called phantom image is created. Therefore, it is suitable for three-dimensional display such as a CT image, an air traffic control, and a computer-processed image in which phantom images are allowed.

(5) Holography

Holography is wave memory technique using interference and diffraction of the light wave. Holography is invented by D. Gabor in 1948. Practical holography using a laser as a light source is suggested by E. Leith and J. Upatnieks. Up to date, efforts to apply holography to 3D image display has been carried out. Techniques of playing 3D image without a laser have been invented. The hologram can be synthesized at the normal picture. In recent years, electronic moving image holography technology capable of reconstructing a hologram using an electric stimulus in real time has appeared, and research and development toward a holographic TV have become active.

A narrow beam from the laser is made to be spherical wave using a lens. Two lights are overlapped to touch the photographic film: one is scattering light which is formed by spherical wave bumping into the subject, and the other one is parallel light which is reflected by the mirror. In overlapped region, there is interference fringe respecting to the angle of two lights. Exposure of interference fringe into photographic film and developing is called hologram.

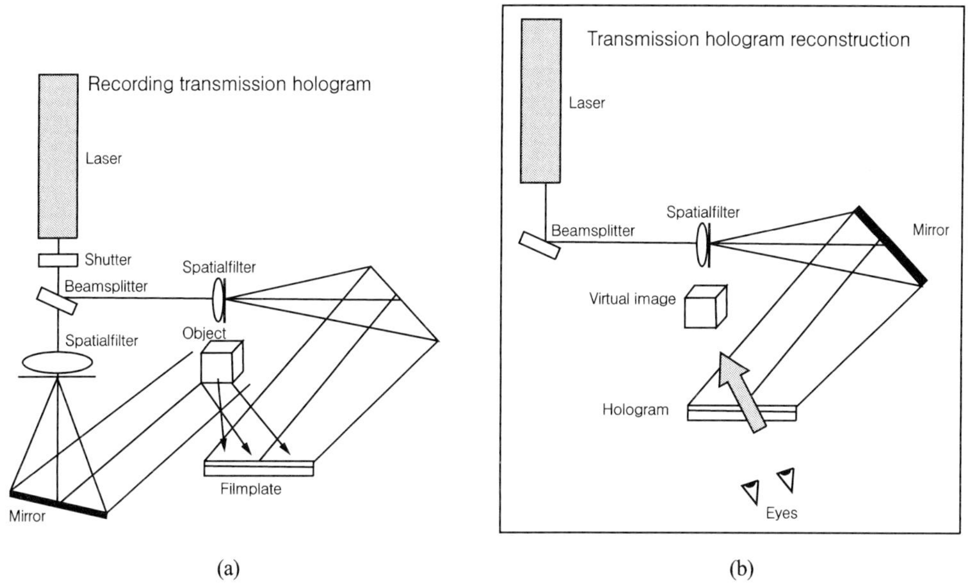

Figure 12.17
Recording (a) and reproduction (b) of an object by holography

Interference fringe in hologram becomes scattered stripes, and the average value of fringe spacing is as follows.

$$\delta = \lambda / \sin\theta$$

λ is the wavelength of light, θ is the angle between scattering wave and parallel wave. Also, the spatial frequency of interference fringe is $1/\delta$. If the source of light is He–Ne laser red light (λ=0.63 μm) and θ is 30°, δ is 1.3 μm and spatial frequency is 1260 Hz/mm. Therefore, very high–resolution photo sensitive material is needed to record this interference fringe. When this hologram is recorded with parallel light in the same direction, an image reappears like the original image. Holography is characterized in that a three–dimensional structure or position of an object can be directly recorded and displayed

(6) Interactive Technology

In 1993, a three–dimensional image display system without glasses called a backlight division method for selecting left and right images by having a directivity of backlight on a liquid crystal display was announced. As shown in the principle of Figure 12.18, this method first examines the observer's face using two types of near-infrared light diodes with different wavelengths on both sides.

Two single-color CCD cameras equipped with filters corresponding to each wavelength were used to photograph the left and right faces of the observer separately, and the images are placed in CRT which is backlight source of color liquid crystal panel. After the characteristic of self luminous image of CRT is exaggerated by Fresnel lens, it has directivity and can be used as light source of liquid crystal panel. It has characteristic of the self tracking 3D display.

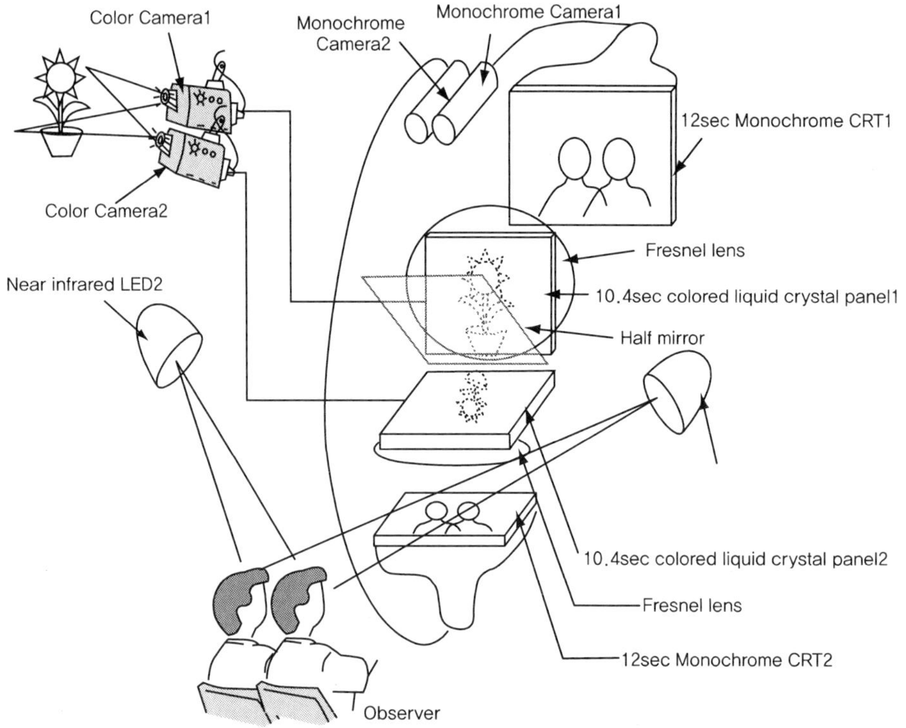

Figure 12.18
Backlight split 3D display

(7) System Technology

Head-Mount Display (HMD) is goggle-type image display device that provides three-dimensional images to the observer equipped with binocular CRT or liquid display. The first patent on the HMD was applied by Albert B. Pratt in 1916.

HMD is designed so that a display screen in front of eyes seems to be a big screen by the optical system. There is a sensor which detects the position of an object at the head of the helmet. According to position information based on the sensor, contents could be changed dynamically in response to the movement of the helmet. In other words, when observer turns his head to left, the image of left side will be shown and vice versa. As a result, the observer can see a video of entire circumference. That makes the observer enter in the virtual space that computer has created and has an experience like being in that space. Such a new interactive video experience of technology is referred to 'Virtual Reality.' Pioneering research about using virtual reality to computer and human interface has been promoted by Ames Research Center of NASA since the 1980s. HMD is a device for displaying virtual space. It is used not only for research of virtual reality but also for games and industries in amusement facilities and enjoying large-screen images in combination with TV or VTR at home.

Relationship of the area of visual angle and resolution (number of pixels in a minute view) which is needed for HMD is as follows. In order to view the video which has 35° of visual angle with enough definition, resolution of 4 million pixels (2000x2000 pixels in a minute view) which is more than double that of high-vision is needed.

For HMD of high resolution type, HMD of 1280x960 pixels which use micro color CRT of 1 inch is released at the n-vision company. In order to realize high resolution color image display, they use liquid crystal shutter that is possible to electronically express RGB color at monochrome CRT tube side, not CRT of shadow mask type. They equip liquid crystal shutter that is possible to express RGB color at 180 Hz which is three times higher than normal scan speed of CRT. The RGB liquid crystal shutter is switched in synchronism with this, so that the image presented in a time-division manner appears as full color due to the afterimage phenomenon of the eye.

In constituting HMD, lightweight and development of eyepiece with wide view angle are a second priority task. In order to gain enough realism and illusion of being absorbed in the video space, at least close to 90° viewing angle is needed. For HMD of wide visual range type, LEEP System company in America is realizing viewing angle that is more than 90° by zoom in LCD display unit by a special optical system that is similar to the eyepiece. However, if the viewing angle is to be widened, the correction of the lens aberration and a complicated optical system are required, so that the weight of the display device as a whole is increased even if the display device is reduced in size and weight.

Other hardware design considerations include adding a so–called "See Through" function that allows the user to check the external situation while using the HMD or making vision correction tool or attaching tool which could control the distance between the lens of HMD to binocular width by the user themselves considering eye strain.

2.5 Purpose and function of three dimensions display

Applications of the 3D display are restricted to medical display, game console and CAD at present. The technology that can express stereoscopic image with normal TV image is being developed. Table 12.2 shows purpose and function of three dimensions stereoscopic image.

Table 12.2 Use and function of 3D stereoscopic display.

Usage \ Function	Display function								
	Big screen	Flat/ Small	Multiple viewing	Non Glasses	Fixed	protruding Image	Floating Image	animation	Interactive
TV	△	◎	○	○	○			◎	
CAD		○			○				◎
Textbook			○	○	○	○	◎	○	
Decoration	△		○	○	◎	○	◎	○	
Measuring equipment		◎		◎			△		
Surveillance		◎						○	△
Art		○		○	○		○		
Books	△			○	○	○			
Theater	◎		◎	○	○	○		◎	
Game		○		△	△	○		◎	◎
Virtual reality	○			△	○	○	◎	◎	◎

Note (◎ : Great, ○ : Good, △ : Normal)

Currently, there is almost no domestic 3D information terminal market, and it is used in some companies, research institutes, and universities for experimental purposes. As the demand of the related industries has increased, the demand market has been formed with the participation of the companies in the medical, security, advertisement, broadcasting, etc. mainly in the software market including the 3D game.

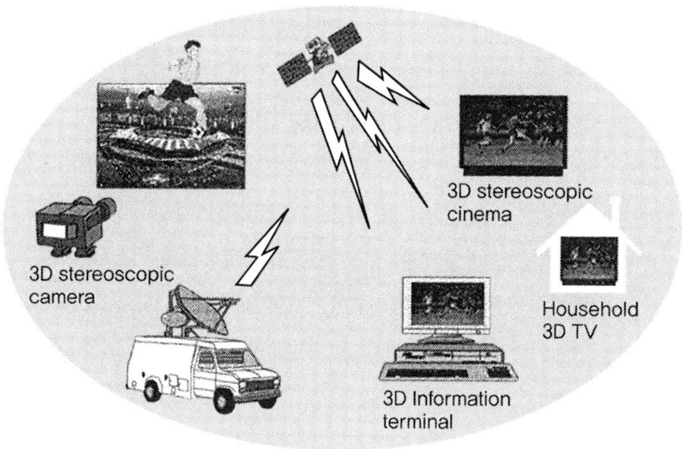

Figure 12.19
3D stereoscopic broadcasting concept

3 Technology development trend of three-dimensional display

3.1 Applications and market prospect of 3D technology

(1) Applications of 3D technology

The stereoscopic technology was applied to polarized stereoscopic movies at the beginning of development, and from the late 1990s due to the expansion of personal terminals, game machines, industrial instruments, and virtual reality, application studies in this field are actively proceeding. Especially, it is expected to be applied in the household, recreational, communication, cultural and other special industry fields related to the rapid expansion of the information.

Major application fields include 3D measurement, 3D game, 3D advertisement, and 3D display. The field of measurement includes real-time 3D measurement, structural processing, and costume design. In the game field, 3D graphics are applied to existing two-dimensional displays, and new markets are created in entertainment such as arcade, 3D game machines, and theme parks. In the field of advertising, large-scale

3D projection technology is already applied in exhibition hall, movie theater, and event. In the field of 3D display, there are active developments in the field of a CAD for 3D structural design, architectural design, 3D image surgeries using endoscopes, and techniques for application to aircraft simulators.

As a personal information terminal, it is expected to be utilized for a 3D game, Internet shopping, and 3D image data. There is a wide range of applications ranging from deep sea remote sensing which is difficult to reach directly by humans, robot vision, and virtual reality, VR glove in the future. Therefore, the 3D display is expected to have more industrial effects compared to when it changed from monochrome TV to color TV broadcasting in the past, and it is expected to develop 3D TV as next-generation display. The global market research firm, "NPD DisplaySearch" predicts that by the end of 2011, global 3D display market will reach the US $13.2 billion and sales volume will reach 50.8 million units through the recently announced "3D Display Technology and Market Forecast Report." Also, by 2019, the 3D display market will grow to $67 billion and sales volume will increase to 226 million units.

As shown in Figure 12.20, according to each display method and display ability, the application field of the 3D information terminal technology demand field is being applied to all industrial fields.

(2) Market Forecast of 3D Technology

The 3D display began to be applied to the field of exhibition and the promotion of stereoscopic polarized glasses since the early 80s. In the mid-90s, Sanyo and Toppan of Japan and VRex of USA commercialized shutter glasses, lenticular and polarizing systems. In 2000, with the development of LCD technology, the development of personal 3D monitors of 15 inch to 18 inches has been actively developed, and it is attempting commercialization in Sanyo of Japan, DTI of USA, and Dresden of Germany. In the future, the demand for stereoscopic images is expected to increase rapidly in the mid-2000s due to an increase in demand in special fields such as personal information terminals, game machines, medical, industrial use, and virtual reality. Looking at the market size, it has soared to $ 13.2 billion in 2012 and is expected to reach $ 67 billion by 2019. Table 12.3 shows the market outlook for 3D stereoscopic displays. Table 12.3 summarizes 3D display's market prospect.

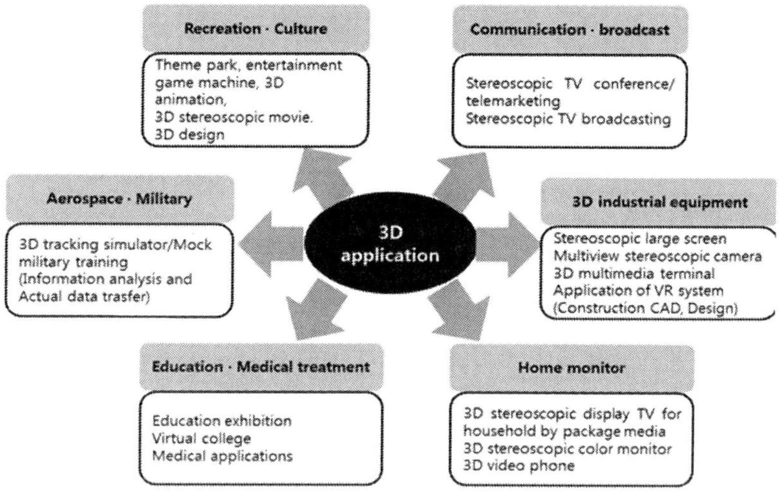

Figure 12.20

Applications of 3D technology.

Table 12.3 3D display's market prospect

Division	Early-stage	Market formation		Market Growth		
	Before 90s	1990~1995	1995~2000	2000~2005	2006~2010	2011~
Contents	·Initial stereoscopic technology ·Polarized spectacular stereoscopic movie ·Started NHK research	·Sanyo 3D LCD monitor development ·Next-generation 3D research after HDTV ·Increasing research trends(USA, Japan, Europe)	·Sanyo 2D, 3D image conversion technology development ·3D monitor development ·VR technology development	·3D Monitor Commercialization ·Development of multiview, 3D method	·Commercialization for personal and industrial use ·Development of 3D TV	·VR development progress ·Introducing various applications using AR by enhancing the webcam performance and linking with general mobile phone camera

http://artistchai.tistory.com/869

3.2 The develop trends of 3D technology in domestic and foreign

The development of technologies and devices related to 3D displays, which have a wide range of applications such as next-generation video media, broadcasting, communication, games, and medical care, has been investigated and developed in recent years in developed countries through national policy development. Technology studies are being actively conducted independently in order to preempt technology, mainly in the United States, Japan, and Europe, as shown in Table 12.4.

Table 12.4 3D display-related foreign research and development trends.

Features	Features	Research and Development Authority	Technology Overview	3D display main research and development trends
UAS	actively linked with other technology development by core technology related to 3D display rather than comprehensive form	CMU(Univ.)	• 3-D camera, super-viewpoint video synthesis	• HMD • Holovideo system • DMD(Volumetric) • Super viewpoint Image Synthesis • 3D CG Applications • Related research-based background excellent environment for 3D display
		MIT Media Lab.	• Holographic video	
		York Univ.(Canada)	• Three-dimensional image recognition process computer modeling	
		Smith-Kettlewell Lab	• Description of the physiological mechanisms stereopsis	
		DTI	• desktop Lenticular terminal(17 inch)	
		DMA	• Promotional terminal(Real expansion and spatial display	
		StanfordUniv. UAH, NASA	• Laser scanning 3D, HOE using 3D systems	
		TI	• DMD Development	
		HIT Lab	• Developing the display technology by using retina	
JAPAN	TAO Project	NHK lab	• Human factor/3D HDTV technology	• 40 ″ Lenticular prototyping • LCD/PDP technology • Point conversion technology • Tactile sensor and sound • Direction to pursue physical and psychological comfort
		ART lab	• Human Visual Perception mechanism	
		TAO	• 3D TV/FLA, Anamorphic method, Sharing space	
		Sony	• HMD, 3D image compression, Educational stereo TV	
		Sanyo	• Lenticular TV, Parallax barrier TV	
		Toko University	• Holographic 3D TV, HDTV	
		Toppan Printing	• 3D Image synthesis, Lenticular plate	

EUROPE	PANORAMA Project	HHI(Germany)	• Head tracking stereoscopic glasses-free display • The three-dimensional reconstruction in 3eyes camera	• video meeting 60inch terminal • Liquid crystal shutter method multi view 3D system • DISTIMA & PANORAMA large scale projects
		SIMENS (Germany)	• 3D image Digitization of compression / recording control system in 2eyes three-dimensional image	
		IRT(Germany)	• Polarized Glasses 2 eyes stereo TV	
		Phillips (Netherlands)	• Lenticular plate production	
		Cambridge (England)	• FLCD multi view projection 3D system	
		Hannover University (Greece)	• Stereoscopic terminal, Codecs	
		CNET(France)	• Stereoscopic Video Coding	

In the United States, a research project, "3D stereoscopic and graphic display technology" by AVDS of ARPS is being carried out. NASA, AT&T, MIT University are developing 「Feeling 3D multi media」 for aerospace, broadcasting, military, and medical. In Japan, NHK, NTT, ATR made prototype of 3D TV(next-generation) and they are researching about its practical use in national project that is TAO (Telecommunication Advanced Organization) of 「High 3D moving image Communication」

In Europe, PANORAMA (PAckage for New OpeRational Autostereoscopic Multiview system and Applications) project is promoted, following DISITIMA (Digital STereoscopic IMaging & Application) project, for development of 3D image transmission and display of the video-conference system. Besides, Rainbow group of the University of Cambridge is developing timeshare 50-inch 3D display systems as shown in Figure 12.21.

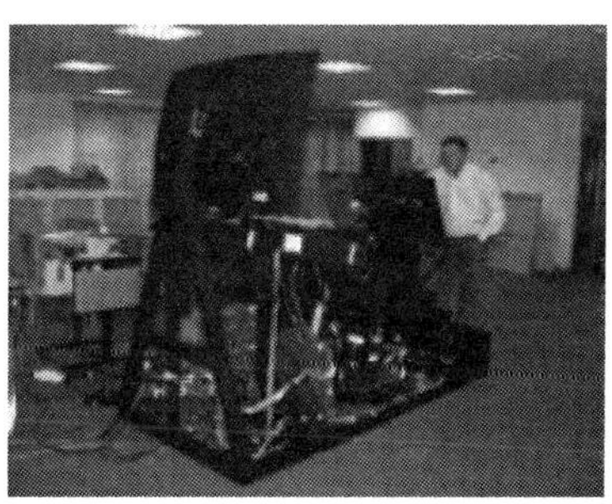

Figure 12.21
Timeshare 50-inch 3D display systems of University of Cambridge.

The domestic display industry occupies the top position in the world in the middle of 2000 and is leading the globalization of the global market. However, since the base of the backward industry is weak, it is time to reinforce the backward industry to secure competitiveness.

Recently, by the increase of interest in 3D image display technology emerged as the next-generation of advanced technology. As shown in Table 12.5, some companies, research institutes, and universities are actively pursuing research and development.

Table 12.5 3D display related domestic research and development trends.

Characteristic	R & D organization		Technical Development Overview	Major research and development trend of 3D display
Several universities, research institutes, and companies have independently conducted 3D technology research	Laboratory	Korea Game Support Center	3D Animation/3D Game/3D VR S/W	•Development of a 10-inch class stereoscopic monitor system •Research and development of 15-inch stereo stereoscopic monitor •Developing high-definition holographic screen technology •Need to develop multi-point 3D display technology
		Korea Standards Research Institute	Emotional Engineering (3D viewing environment)	
		KIST	Holographic Color Screen Multipoint System	
		KAERI	Polarization type stereoscopic image monitor / stereoscopic camera	
		ETRI	Broadcasting of 3D TV demonstration service(2002 World cup relay)	
	University	Kwangwoon University	•multi-point 3d display system •Holographic 3d moving picture video	
		Seoul National University	Three-dimensional display using integral photograph	
		Yonsei University	Multi-point stereo image processing	
		Kangwon National University	2D / 3D conversion technology, compression, and restoration of stereo image	
		Hannam University	Intermediate view image synthesis technology	

Characteristic	R & D organization		Technical Development Overview	Major research and development trend of 3D display
Several universities, research institutes, and companies have independently conducted 3D technology research	Company	Samsung Advanced Institute of Technology	LCD 3D monitor testing using HOE	
		Samsung Electronics	Developing a 15-inch glasses stereoscopic image monitor	
		Hyundai Electronics	3D flat panel display monitor	
		AIS(Co)	Developing 3D stereoscopic shooting technology, producing 3D content	
		3D Korea	Developing HMD, and 3D content	
		Forbes Corporation	Lenticular photo	
		Lentech World(Co)	Lenticular photo	
		Asol(Co)	3d adapter for camcorder	
		Wobo Electronics(Co)	H/W and S/W for Virtual reality	
		Infotech	Glasses free 3d image play back display device	
		Songshan(Co)	Hologram production (advertising, art)	
		HUHU(Co)	Polarization type stereoscopic monitor and stereoscopic camera	

3.3 Tree-Dimensional View of the Future Display

(1) 3D stereoscopic technology research and development direction

Up to now, more than a dozen types of 3D stereoscopic display technology have been developed, because display methods vary depending on the purpose and display type used. The direction of research and development of 3D stereoscopic image technology is going from stereoscopic to multi-resolution. In terms of size, it is bounded between 15 inches and 30 inches, and it can be divided into application fields such as information terminal and 3D TV.

Prior to 2000, large-sized projection systems of more than 100 inches using CRT or LCD project were mainly developed for promotional displays. Recently, as the LCD has become bigger, CAD or medical monitors have been developed for personal information terminals and game machines.

In stereoscopic technology, it is possible to display images in various directions. In the case of using two images on the left and right sides, the stereoscopic method is used. However, since the resolution of the display has to be increased as the display direction is increased, the stereoscopic type is mostly used at present. It is also referred to as 2.5D in order to distinguish it from the full parallax 3D technology. Complete three-dimensional techniques include real-life images floating in the air that can be seen in fancy films, and ways of viewing images in all directions using hologram technology. In the future, it will be possible to display a large amount of image information together with the development of a large-screen display and a high-resolution display, so that it is expected to develop a natural image display and a 3D TV capable of viewing by a large number of people.

A major challenge of the 3D display is the development of an autostereoscopic display method that automatically tracks the position of an observer and displays a three-dimensional image. Image recognition technologies such as head tracking, pupil tracking, and face tracking are being applied. In addition, dizziness, headache, eye fatigue, etc occur as bio-effects by three-dimensional imaging. Visual fatigue is known to appear about 30 minutes after the 3D display begins to appear. Human factor technology is needed as a research on human psychological effect to be able to display more natural feeling.

(2) Technology Development Trend of next-generation Display

Future society is called the age of information and communication. With the miniaturization of information devices, hand-held or wearable devices capable of being held in the hands or body are being developed. Imagine a human life in the future that wears a spectacular display and a computer or video device and sends and receives information and watch movies.

3D simulator technology that can stimulate all human senses, which can see, hear, touch and smell, is being developed, and virtual reality technology that can reproduce or experience the actual situation in virtual space is expected in the future.

However, without the development of display technology that can see and feel the actual image, the future information age is impossible. Already, next-generation displays have three technological development trends, along with technological development from CRT to flat panel displays such as LCD, PDP, and EL. These technological trends are interrelated and ultimately linked to the technology of 3D stereoscopic display.

The first is a compact miniaturization technology. In addition to the portability of information devices, eyeglasses with a palm-top or ultra-small micro-display of less than 1 inch are being developed. This is a hand-held or wearable display that can be applied to information communication devices, personal game devices, and stereoscopic glasses.

The second is large-screen technology. With the development of PDP and projection displays, wall-mounted TV has become possible, and the home theater market is also rapidly expanding. In the future, it is expected that it will be possible to display a small three-dimensional movie theater that only displays stereoscopic movies or a three-dimensional image for events in the field of architecture and entertainment.

The third is solidification. Here, there is a three-dimensional information terminal using an HMD system using a very small display and a flat display such as an LCD. LCD is mainly used as a display suitable for a three-dimensional display, and it is already increasingly used as a display device for various monitors in personal monitors and industrial fields. Therefore, if 2D/3D compatible technology is applied to existing two-dimensional display, the application field of the 3D display is cxpected to increase rapidly. That is, the 3D display is expected to be promising as a next-generation display after flat panel display such as LCD.

PRACTICE

1. Whether a target having a specific area of (X), a height (Y), and a depth (Z) is close to or remote from the observer, the observer perceives the width, height and depth of the subject stably and permanently. Perceiving the shape and depth of the object in a stable and persistent manner is called a three-dimensional shape and depth homeostasis. The importance of perceptual stability is underestimated because it is found in anybody and it is achieved in very short time by our visual system. Let's consider if the 3D shape and depth homeostasis is realized in the current 3D display.

2. The optical three-dimensional information about the target is collected by the video camera and delivered to the viewer, the physiological three-dimensional information which contains the information of distance to the target is not transmitted to the observer at the current level of technology. Let's consider why physiological three-dimensional information is not delivered to an observer and a realistic way to solve it (In here, the optical three-dimensional information is having the properties of light, and can be obtained easily by using the camera in a three-dimensional image display system. In addition, the physiological three-dimensional information is a three-dimensional information is collected by the observer's eye structure).

3. Johnston (1991) shows empirically that the human visual information processing system perceives the depth of the target by scaling the binocular parallax information formed on the retina through the actual psychophysical experiments. Then, let's consider how three-dimensional shape and depth homeostasis can be implemented in human information processing system and how to implement.

4. Various methods have been researched and developed for 3D display technology, and the standard has not yet been established in the world. One of the main evaluation factors of 3D display technology is that autostereoscopy can be performed without wearing special secondary glasses. Integral photography (IP) technology is one of the most remarkable autostereoscopy display technologies. In particular, this technology has evolved into an integral imaging (II) technology that is real-time IP with the development of imaging devices and display devices. Let's discuss the advantages of this technology.

5. Compare and discuss 2D and 3D visual display.

6. Explain the principle of the parallax barrier and a lenticular sheet.

7. Explain the term binocular disparity and presence.

8. Discuss about current 3D display's problem and improvement direction.

Display Application System

Chapter 13

1 What is display application system?
2 Existing Display application
3 Future display application system
4 Summary

1 What is display application system?

1.1 Display application system

The display device displays visual-information on a screen in the form of text or graphics. It is an output device that converts electrical signal from machine and delivers it as human readable visual information. The display is largely classified into an emissive and a non-emissive type. The emissive displays are CRT, ELD, FED, PDP, VFD and LED. The non-emissive displays are LCD, ECD, EPID, TBD and SPD. Most commonly used displays that can be found in home or office are CRT, LCD, PDP and OLED (organic EL). CRT (Cathode Ray Tube) is the earliest and most widely used device that works by electron emission, its control and hitting a fluorescent surface where the fluorescence creates the visual information. Whereas the LCD (Liquid Crystal Display) uses liquid crystal to display the information. Orientation of liquid crystal molecules can be changed by suitably applying an electric field across it. This change in liquid crystal molecules can change the polarization direction of light that are suitably controlled to create a visual information. A colored optical image or information is displayed by combining color filter and backlight. Depending on the driving method, it is divided into passive matrix type and active matrix type LCD. A passive matrix type LCD works by using scanning electrode and the signal electrode, that are placed in the XY type planar arrangement, where the crossing points of the electrodes are used as pixels. In an active matrix type LCD thin film transistors (TFT)s are used to control each pixel. LED (Light Emitting Diode) is a semiconductor device which can emit light when a suitable electric bias is applied. The LEDs can have a faster response time, lower power consumption and longer lifetime. In a PDP (Plasma Display Panel), Xe/Ne or Xe/Ne/He mixed gas is used to create plasma with the help of two electrodes, placed between two glass plates in the display device. UV radiation (of wavelength 147 nm) generated in the Xe gas, stimulates fluorescent substance in one of the glass plates. It is a self-luminous device. Since the generated light is emitted through top glass, the reflective fluorescent substance is used. Demand for PDP was expected to surge since PDP was light emitting having a clear display, high display quality with a large panel, fast response time, high reliability and long lifetime. However, it lost market competitiveness to LCD as the LCD consumes less power and it is less expensive. OLED stands for Organic Light Emitting Diode (OLED), which uses a phenomenon in which electrons and holes separated in a specific semiconductor material and then recombined to generate light. By using semiconductors of different band gap, the OLEDs can display different colors. Since OLED emits light directly, the backlight is not needed (as the backlight is necessary in LCD) and the range of light expression is wider than that of LCD. Due to high brightness, low power consumption and long lifetime, the OLED is getting much more attention for next-generation flat display technology than that of the LCD now.

Various electronic products such as TV, camera, autopilot and gaming devices have become part of human activity. Therefore, delivery of visual information between human and the electronic device is getting more important. Presently, the display technology is continuously evolving and is one of the most common ways of information exchange between human and machine. The application of the display to a particular field or product is called a display application.

2 Existing Display application

2.1 CRT (Cathode-ray Tube)

CRT was invented by German physicist Karl Ferdinand Braun in 1897. It is a special vacuum tube that converts an electrical signal to display images of shapes or letters by the action of an electron beam. Since CRT has gone through the research and development as a display for over 100 years, it has an excellent overall picture quality such as brightness, viewing angle, contrast, etc. and is a low cost device. However, it has disadvantages because it is heavier, thicker and larger in volume in comparison to flat displays. There are mono-chrome CRT and color CRT. Depending on the application, it is divided into CPT (Color Picture Tube) for TV and CDT (Color Display Tube) for a computer monitor.

2.2 Application of CRT

(1) CPT for TV

CPT for TV was developed with the application of cathode ray tube by Joseph John Thomson (1856~1940). Various cathode ray tubes developed in the 19th century. When a current flows between two electrodes, the one with the lower potential is called cathode and the other as anode. Cathode ray tube is high vacuum glass tube where emission and flow of electron is studied. Due to the circular structure of electron gun, the length of the CRT in TV are longer and corner of its screen is rounded. In contrast, the recently developed flat display technology such as LCD are thiner and rectangular display screen is possible with the LCD.

Figure 13.1
CPT for TV

(2) CDT for computer monitor

CDT for the computer monitor is a device that displays output of the computer. Monitors are integral part of popular computers became popular. In the early days the computer monitors were made from the CRT, therefore its size was also huge. The demand for CRT-based computer monitor is rapidly decreasing due to the advent of LCD monitor. However, unlike the CPT for TV, computer game enthusiasts mainly use CDT for computers. The reason is that there is no residual image processing as that can be found in LCD monitors. For CDT, eye strain is less when flipping the screen due to image residual. This is the main reason why computer game users are using CDT based CRT monitors.

Figure 13.2
CDT for computers

2.3 VFD (Vacuum Fluorescent Display)

VFD is a display element applying triode cathode consisting of a positive electrode (anode), a negative electrode (cathode) and a grid. The electrons that are emitted from the cathode excites the fluorescent substance on the anode, thereby fluorescent light is emitted. The basic principle is that the thermal electrons emitted from the cathode are controlled and accelerated by the grid voltage between the anode and cathode. These electrons collide with fluorescent substance on the anode emitting light. VFD has good randomness of the display pattern, high visual recognizability, low operating voltage and it can be directly driven by MOS IC.

Figure 13.3
applications of VFD

Since VFD is self-emitting, it is easy to view and displays vivid colors compared to LCD. Its operating temperature is wider as $-40 \sim 85°C$ and it is reliable having 10,000 hours of lifetime.

2.4 Applications of VFD

Due to its excellent visibility and a wide range of the green light emitting wavelength characteristic, a vacuum fluorescent display is widely used as a display device for audio and video players, vehicle-mounted watches, measuring instruments, car audios, etc.

2.5 PDP (Plasma Display Panel)

PDP (Plasma Display Panel) is an electronic display that uses neon gas for generating UV light. A gas mixture, Ne+Ar, Ne+Xe is used between the front glass and rear glass. By applying a voltage to the anode and cathode the neon UV light is emitted. Since it is a display device with a gas discharge phenomenon, it is also known as a gas discharge display device. When a voltage is applied to the anode and cathode, the confined gas becomes active and returns to its original state as time goes by, it emits a strong and beautiful light like aurora. This plasma phenomenon is used for PDP.

A small gap of 0.1 mm is maintained between two sheets of glass plates. A gas such as neon and xenon is used to fill in the discharge space configured by a number of partition walls and three R, G and B are coated phosphors. When a voltage is applied to electrodes, electric discharge occurs and UV light is gnerated. The generated ultraviolet light hits the phosphors emitting visible lights. This is used as the basic principle of PDP.

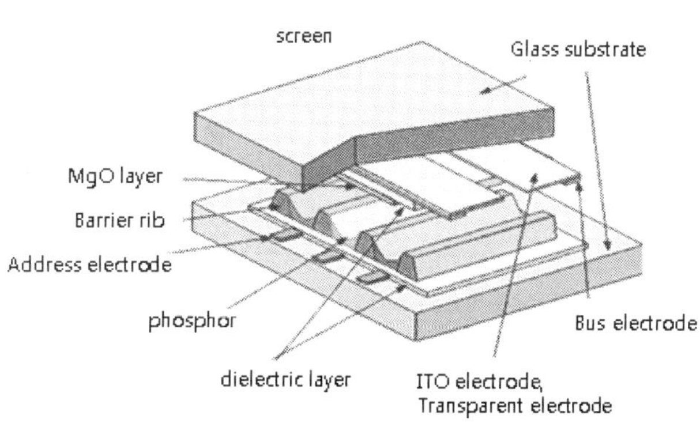

Figure 13.4
Structure of PDP

In the case of DV type PDP, electrodes are exposed to the discharge gas. As discharge is repeated, sputtering occurs at the electrodes and the electrodes degrade due to the metal evaporation. To compensate this, the electrodes are coated with dielectric films in the AC type PDP. However, there are other problems. Discharge occurs due to movement of ions and electrons. During discharge, ions hit phosphors and the phosphors can be damaged. Phosphor materials are very weak against ion bombardment, this results in short lifetime for both AC and DC type PDP.

PDP TV has better color capability compared to the LCD. While LCD dependents on the backlight, the gas-discharge tube in PDP emits light. In the LCD, very fast moving images, which are faster than the molecular dynamics, are not properly displayed, making an afterimage effect. In the PDP, the screen response speed is faster and the afterimage effect is lower. The viewing angle of the PDP screen is also wider.

2.6 Application of PDP

One of the display products using PDP technology is PDP TV. It was invented by Donald L. Bitzer and H. Gene Slottow of University of Illinois Urbana-Champaign for PLATO Computer System in 1964. The early monochrome PDP (mostly orange or green) did not use memory or electrical circuits to update image and this approach was very popular. However, its commercial sales began to decline in the late 1970s as the cost of semiconductor memory used for CRT TV became cheaper than that in the plasma or the PDP. In 1997, Pioneer Co., Japan first placed commercial plasma TV on the market. The screen size was 21 inch in 1992. In Consumer Electronic Show (CES) held in Las Vegas in 2006, Panasonic introduced 103 inch PDP TV.

Figure 13.5
103 inch PDP TV of Panasonic

PDP is a light-emitting device like the CRT but has the advantage of thinness. Later, the PDP TV started to replace the CRT TV.

The PDP displays were used for commercial, public display boards and display of game devices in the game room. The PDP technology used in these products is the same as that of PDP TV.

Figure 13.6

Applications of PDP

2.7 LCD (Liquid Crystal Display)

LCD is a thin display device consisting of millions of little cells in the form of a matrix. Liquid crystal is confined between 2 sheets of glass. Inside the glasses, electrodes are formed and the electrodes are electrically connected to the external terminal. The liquid crystal looks as a liquid but is at a specific state of optical anisotropy. It becomes liquid crystal at certain temperature range. It is organic compounds called 'Thermotropic Liquid Crystal'.

There are two types of LCD: reflection and transmission types. In the reflection type LCD, light is incident through the front glass, that is then reflected by the reflector at the back panel to display the information. For transmission type LCD, light that is generated at the back of the LCD, that is transmitted in a controlled manner.

2.8 Application of LCD

(1) Passive matrix LCD

Passive matrix LCD is usually used for calculators or watches. A voltage is applied to the columns and rows of grid-lines of the panel with arranged liquid crystals and the liquid crystal at the intersection is driven to constitute a screen. In passive matrix LCD, STN(Super Twisted Nematic) method which can rotate up to 270 degrees for each element is generally used. Due to its simple structure, the productivity of LCD is high and its production cost is low. However, the image quality is poor with slow response time and image retention occurs heavily. It is used in small products such as calculators or watches.

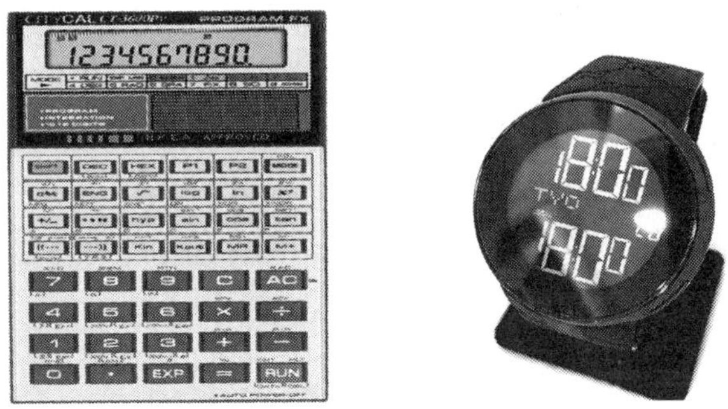

Figure 13.7
Passive LCD display

(2) Active matrix LCD

Active matrix LCD uses a TFT (Thin Film Transistor) for controlling each pixel. It is possible to implement a higher picture quality than the passive LCD and there is little after-image when displaying moving images due to the fast response time. However, its production cost is high because of the complicated structure. The majority of LCD TV or monitors use active-matrix LCD, especially the TFT method. There are several types of TFT LCD depending on the arrangement of liquid crystals. IPS or VA panel has less image distortion compared to TN panel when viewed from sides.

1) TN (Twisted Nematic) type TFT-LCD panel

Normally crystal liquid molecules are in the horizontal direction. When a voltage is applied, they can turn to the vertical direction. This principle is used in the display screen. It is the oldest and most used TFT LCD type. Since its response time is slow, after-image occurs. The production cost and power consumption is low. However, color display ability is poor with narrow viewing angle. The image gets distorted when viewed from sides. It is mainly used for low-end TVs or monitors.

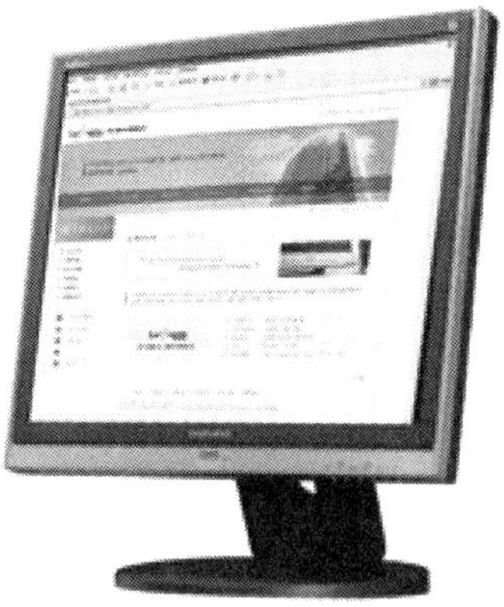

Figure 13.8
TN type TFT-LCD panel

2) IPS (In-Plane Switching) type TFT LCD panel

In IPS type TFT LCD, the display occurs by rotating the liquid crystal molecules. Its ability to display color is good and there is lower image distortion when viewed from various angles. It is also called wide viewing angle panel. However, in comparison to the TN type panel, its power consumption is higher, response is slower and production cost is higher. It is mainly used by professionals such as graphic designers in display machines.

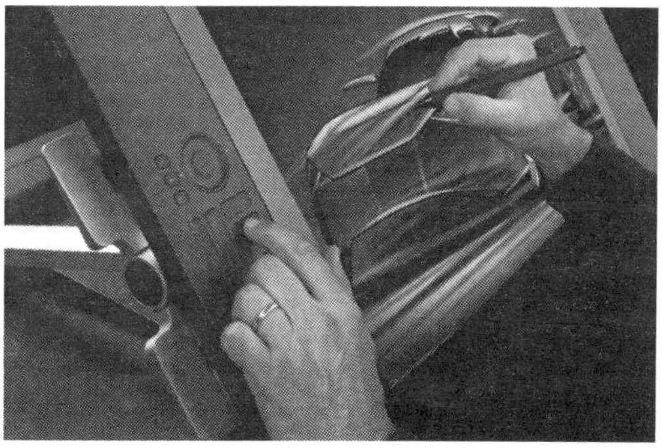

Figure 13.9
IPS type TFT-LCD panel

3) VA (Vertical Alignment) type TFT LCD panel

Normally liquid crystal molecules are oriented normal to the surface they remain attached to. With an applied electric field, this orientation can be changed and the molecules can align parallel to the applied electric field. This change in orientation of the liquid crystal molecules leads to a change in polarization of transmitted light. With the help of a pair of cross-optical-polarizers a desired display can be obtained. Such a display also has a wide viewing angle. Its ability to display color is in between that of the TN and IPS. Its response is slower but it has excellent contrast and good for viewing images. Its price is between that of the TN and IPS.

Figure 13.10
VA type TFT-LCD panel

2.9 LED (Light Emitting Diode)

LED is a GaAs-based semiconductor device (or diode) that emits light when current flows through it. In this p-n junction semiconductor diodes, minority carriers (electron or hole) are injected and light is emitted by recombination of these carriers. Since the conversion efficiency from electrical energy to light energy in LED is high, it is possible to reduce energy consumption for generating light by up to 90%. It has drawn a significant attention as a next-generation light source. It can replace incandescent lamps which have only about 5% energy efficiency, or fluorescent lamps. LEDs consist of a structure in which the oppositely charged particles, electrons and holes, combine near the p-n junction, and emits light. The color of light depends on the characteristics of this conductive material, such as its optical gap, doping density etc.

2.10 Application of LED

(1) Illumination

One of the applications of LED is light source. In these applications visible light emitting LEDs are used. LED emits 〉90% of electrical energy as light and its efficiency is higher than the other light sources. In comparison to incandescent bulbs, since it does not use filament to emit light it generation of heat is lower, operates at a lower temperature and more stable under mechanical vibration or shock. Many lighting products are rapidly being replaced by LED lights because they have a longer life and can produce light of desired color temperature and higher brightness.

(2) LCD Display Backlight

LED TV uses the LED as a source of illumination for the transmission type LCD display. Strictly speaking, LED TV is an advanced form of LCD TV. Earlier, the CCFL (Cold Cathode Fluorescent Lamp) was used to provide the backlight of LCD TV, that is replaced by LED BLU (Backlight Unit). The advantages of LED TV are: (a) The display devicc can be made thinner than the LCD TV. (b) The lifetime of LED TV is nearly twice than that of the LCD TV. (c) Since LED itself emits light, as against the mercury vapor lamp used in earlier LCD display, the power consumption is reduced to 50~60%. In the LED TV, spectra of the transmitted light can change depending on the lifetime of LED and temperature. The optical efficiency of LED is only 20~30 Lm/W while that of CCFL is 50~60 Lm/W.

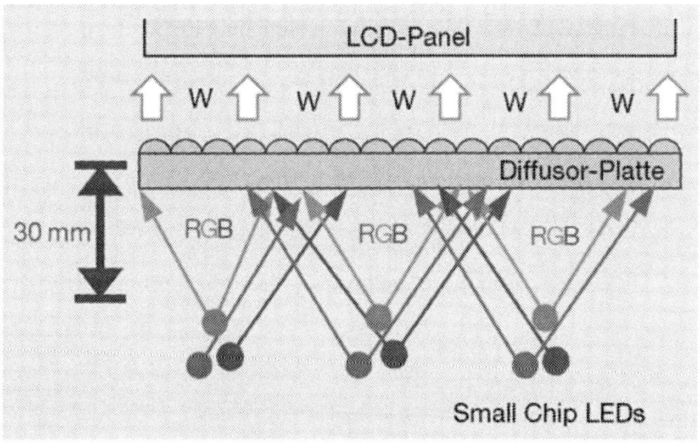

Figure 13.11
LED Backlight for LCD

(3) Large Display

Display that is installed in a large outdoor advertising and sports stadium is the large display and displays an image by having the same pixel arrangement as the existing flat-panel. The basic element for the LED display is the LED (Light-Emitting Diode). Since LED emits light through completely different ways, the light can be clear and bright.

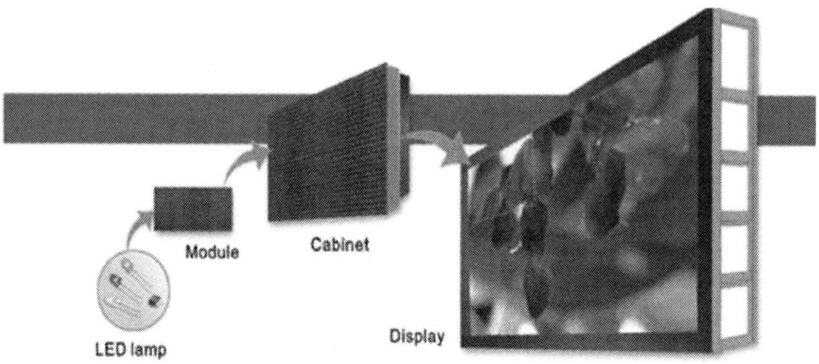

Figure 13.12 Structure of LED Display

LED consists of several modules composed of a number of pixels. A pixel is made of 2~5 LEDs. For large modules, more than 9 diodes could be used. Generally, a module contains 16 pixels arranged in X-Y plane in a square shaped area. In special cases, the module contains 32 pixels. In a module, two letters or two numbers can be displayed. Various images and characters can be displayed by a display consisting of a number of modules.

The largest LED display is in Suzhou, China, and its length is 500 m. The largest LED TV is Center Hung Video Display with a horizontal length of 49 m and the vertical height is of 22 m installed in Cowboys Stadium, USA. Advertising and entertainment-related markets of the large LED display are expected to grow, but recently the price of these displays is declining 20~30% every year due to fierce competition. Nevertheless, LED displays are expected to grow as a state-of-the-art multi-function LED electronic display, such as neon signs, high-definition, 3D and smart screen.

Figure 13.13
Large LED display

Display Engineering

3 Future display application system

3.1 Transparent display

The transparent displays have transparent screen where the back surface of the screen can be seen. The transparent displays are divided into see–through types like HUD (Head–Up Display) and HMD (Head–Mount Display) and direct viewing type like transparent OLED.

3.2 Application of transparent display

(1) Transport goods

HUD (Head–Up Display) is generally used is a transparent display that presents data without requiring users to look away from their usual viewpoints. HUD is developed primarily to visually assist pilot of an aircraft by delivering useful information. In the HUD, the user sees the information on a transparent screen. Initially, it was used in aircraft but recently, HUD is used as navigation and to display information. When HUD is used, the driver can see the outside along with additional information without gazing back and forth. If cameras are installed in the front and rear, left and right, the images can be seen through HUD installed in the windshield. The internal mirror or the outside mirrors would be unnecessary.

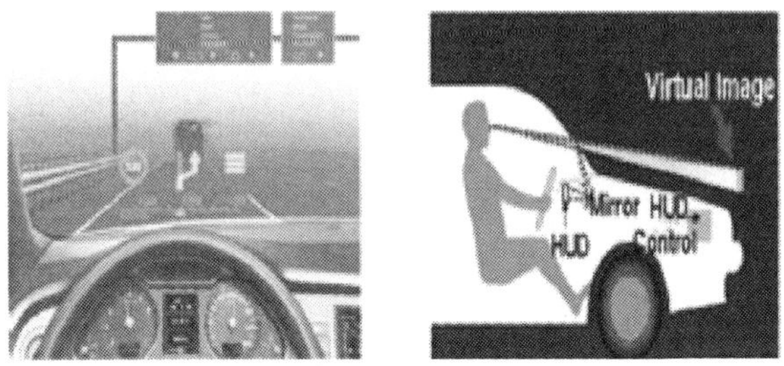

Figure 13.14
HUD principles for automobiles

Figure 13.15
HUD for aircraft

(2) HMD (Head-Mount Display)

HMD can be used in mobile devices as an accessory. It is a see-through type display device, similar to that of the HUD but instead of creating the image to a transparent screen, in HMD the image is reflected from a screen, thereby the observer can see the image. Well known HMD products are Google Glass developed by Google and Oculus Rift by Palmer Lucky. In general, the optical device is placed in front of user's eyes. The HMD can be used as a transparent display too. In that case the image to eyes by a transparent reflecting plate. However, the existing HMD is large and heavy, because of which it causes discomfort and inconvenience to users. It is also very expensive. By using a self-emitting small transparent display, few of the disadvantages of the HMD can be addressed. In that case, separate optical component engine or reflecting plate will be unnecessary and it will be easier to use.

(3) Smart Window Products

Smart window product is used as a glass window but it can pass or block the light as well as display information. Smart window display can be divided into LCD type and OLED type. The LCD display requires a very bright light source in order to obtain sufficient brightness, but it requires large space for the installation and power needed for adjusting the brightness. However, as the LCD is produced widely for display applications, the existing plants can be used for producing the LCD type smart window. In OLED type smart window, the polarizers and color filters are not used. Therefore, it has a transmissivity of ~60% while LCD has a transmissivity of 10~20%. It means that much thinner, lighter and less power consuming display can be fabricated with the OLED. It is expected that the future transparent display will be of OLED type.

Smart window display can be divided into 3 categories depending on primary application. (1) When the interest is on the objects, it can be used to in the windows of department stores, office walls, appliances used in the home such as refrigerators and microwaves. In this case the interest is on the control of optical transparency of the screen rather than information displayed on the screen. The transparency can be controlled by turning the current ON/OFF. (2) The screen information and objects behind the screen need to be checked. The examples can be a bus station and subway platform. Since displays used in such a situation are in a fixed place for a long time, the stability of the products is very important. In case damages occur by the external environment, its design should be so that repair and maintenance are possible. (3) In case of virtual reality for those who are interested in 3D images. So far, 3D desktop technology is being developed. For example, the computer can recognize the 3D movements of hands, and then aligns the objects on the screen, creating an environment similar to that of reality. Using this type of displays are expected to improve accuracy in any activity. It is expected to have wider application in medical, design, architecture, etc. Once the resolution and response speed are improved, it is expected to have a larger market than that exists now.

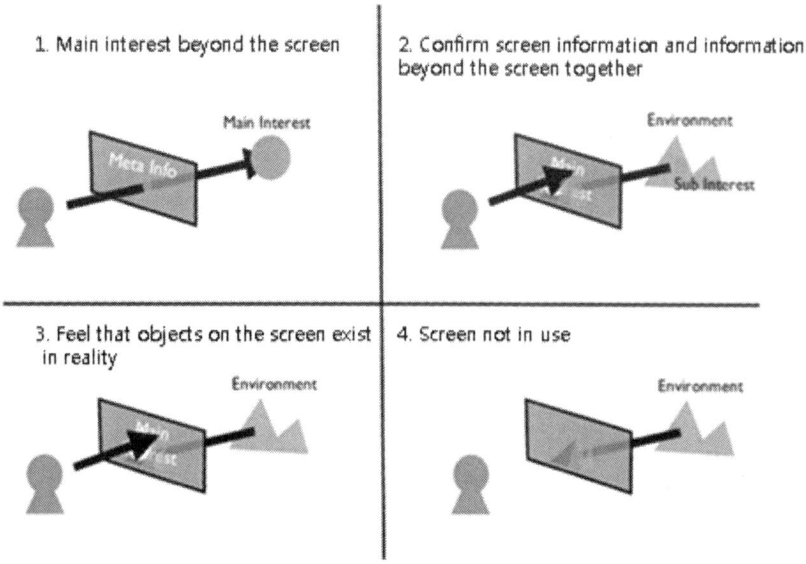

Figure 13.16
4 types of smart window

3.3 Flexible Display

The flexible display uses a flexible and thin substrate so that it can be bent or rolled without damage. It is lightweight, thin and unbreakable. The image quality of the flexible display is presupposed to be at least equivalent to LCD or OLED. The development steps of the flexible display are Thin & Light–weight, Bendable, Rollable and Foldable (depending on the degree of bending). Recently, Stretchable display tends to be included in the development stage of the flexible display.

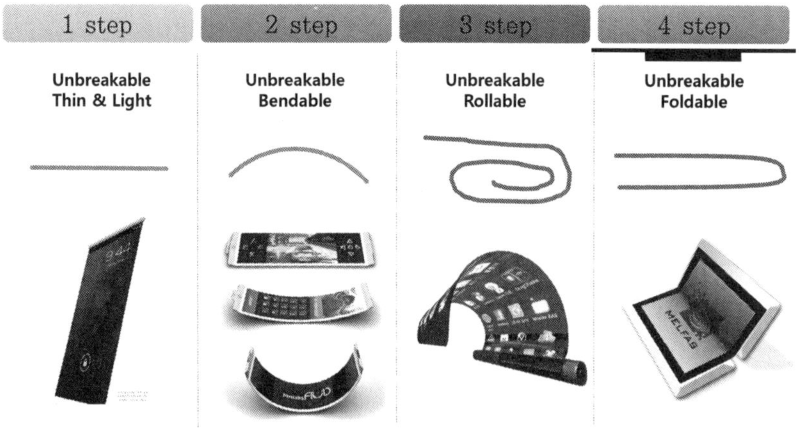

Figure 13.17
Development step of flexible display

3.4 Application of flexible display

(1) Mobile Display

The currently available commercialized product in mobile display uses flexible OLED. The research and development for the flexible LCD have been carried out for some time. However, full implementation of the flexible LCD has been difficult due to the need of the backlight. Therefore, OLED based flexible display, that do not require the backlight, first appeared in the market. Few of these products include the G Flex of LG with a radius of curvature of 700 mm and Samsung Galaxy Round with a fixed radius of curvature of 400 mm. This is a very rudimentary form of flexible display products with a large radius of curvature and there is not a big difference with the existing flat-panel displays. In other words, presently most of these cannot be bent freely, so these are far from the ideal flexible display. Since research and development on flexible battery, flexible materials and devices are being actively carried out by a number of companies and research institutes, it is expected that the flexible products of various designs can be found in the display market.

(2) e-paper Display

In main display products such as TV, tablet and mobile phones, the improvement of characteristics such as brightness, responsiveness and color reproducibility is important. However, still image display with memory which can change the display once in a minute or once in a day is taking much attention as well. In recent years, predictions for technical challenges of e-paper and new fields of application are actively carried out, as evidenced from presentation in technical seminars and conferences.

In addition to the application as an information terminal, it is highly likely that new market utilizing the features of e-paper will arise. For IC cards, it is likely that the display for a one-time password will be mounted instead of showing points or balance. The demand is expected to increase with an emphasis on convenience and security. Since the IC card requires mechanical flexibility to meet the ISO bending standard, environment friendliness considering the frequency of use of the wallet, and reduction of cost to 20% of the current cost, there is a high possibility of adopting electronic paper capable of cost reduction through mass production. The first interesting application for commercialization was e–book is to eliminate heavy backpacks of students. The release of 5~6 inch e–book started in the bookstore has received the attention of many reading enthusiasts. Presently, some of the e–books are launched by the Amazon as "Kindle" and by the Barnes & Noble as "Nook".

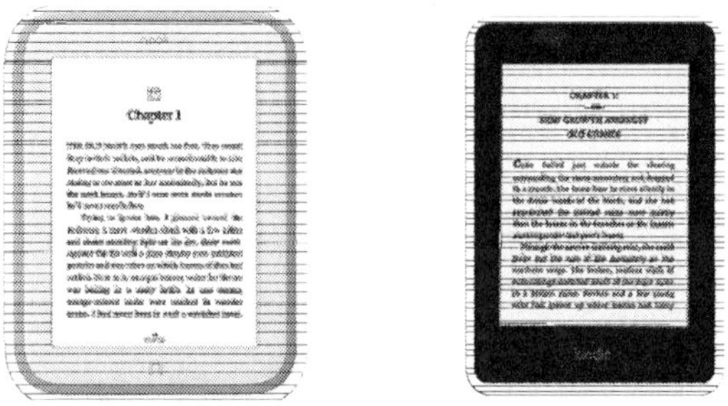

(a) (b)

Figure 13.18

e–book reader (a) Kindle (b) Nook

The mobile phones are likely to be developed into a new device according to the coloration of e–paper. Not just as the main display or the second display, there has been a trial to employ e–paper to the keyboard. In addition, the market has attracted attention as a possible expansion of e–paper as displaying available memory to an USB disk, digital photographic film, watches, etc.

Recently, prototypes of each company participating in the development of electronic paper were displayed in the US International Information Display Society held in Long Beach. LG Display presented 14.1 inches VGA class e–paper and 6–inch monochrome e–paper with 800x600 pixels. Samsung Electronics modified LCD panel manufacturing line for color e–paper production. Normal LCD manufacturing line does not allow the plastic to be used as the substrate for e–paper. Samsung made the LCD panel manufacturing line to produce e–paper by attaching the plastic substrate to support the substrate and newly developed TFT and color filter manufacturing line. Samsung presented 14.3–inch color e–paper and 8.1–inch color e–paper. Japan's Bridgestone also exhibited e–paper of A4 size with its brightness twice in comparison to the conventional one. Bridgestone had used a color filter for LCD for conventional e–papers. However, they used their own color filter technology for an e–book to ensure color purity, that significantly increased brightness of display.

Figure 13.19
e-paper based future examples

Expansion of information technology is taking place steadily and it is expected to penetrate into a variety of markets in future.

(3) Wearable Display

Wearable display literally means that a light weight and portable display device that can comfortably worn by any person for its portable applications. It is not wearable just as an accessory but it is a type of electronic device that can communicate and display information to help the user in a certain manner. The advantage of the wearable display based electronic devices is that they can monitor the detailed information of the surrounding environment or changes of individual's biological signals such as body temperature, heartbeat etc and display them to the users.

In future, the wearable display based electronic devices, smart glasses, as well as watch type wearable devices is expected to receive interest as important electronic devices. Samsung Electronics and Apple were leaders in launching Galaxy Gear and Eyewatch respectively as wearable devices. Since then a number of other electronics companies around the world are producing and releasing various types of watch-like or band-type wearable display devices. These wearable displays have built-in sensors (for acceleration, light intensity, touch-pressure, etc.) to monitor the user's physical movement. These devices can also check text messages, e-mail and telephone call. Various healthcare related applications are also evolving, specifically personal health monitoring is expected to be one of the most important application of these wearable display devices. In the health-care sector the wearable devices can become a very useful and informative electronic device.

Figure 13.20 Watch-like wearable displays

3.5 UHD (Ultra High Definition) Display

UHD Display has 4 to 16 times better resolution compared to the HD display. It can have more than 10 audio channels. The aim is to satisfy the user's need through highly realistic high-quality AV broadcasting service. In the public space, the size of the display is expected to be larger and larger. Since the human visual resolution characteristic is related to the display size, the image quality becomes poorer with the existing HD resolution when display size is larger than 60-inch. Therefore a higher resolution UHD technology is required for larger display devices. The technology for the TV is also developing into high quality, three-dimensional, widescreen, reality based and immersive, trying to satisfy the five human senses and the sense of space. This kind of virtual reality-based media is presently providing a three-dimensional experience in a limited scale, that is expected to expand to various industries, education, entertainment industry, healthcare, sports and other activities. In this respect the UHD Display can provide a vivid realization of high-resolution. Development of UHD is focused on virtual reality, convenience to perform a task and a connection to human emotion. To achieve all these, a further development and integration of OLED, IPS, 3D, big screen, immersive effect, e-Gallery, web OS etc. is necessary.

3.6 Application of UHD Display

(1) TV

UHD TV is the next-generation broadcast standard, where delicate and natural expression is possible due to its higher image quality than the HD TV. TV image quality is usually expressed as 4K, 8K. It means that UHD TV can represent 4 times (4K), 16 times (8K) more clear images compared to HD TV. Higher image quality also represents its closeness to reality.

The smart TV is a combination of UHD display and UI, and the UHD TV is a combination of broadcasting technology and UHD display. A 3D effect can be obtained in a UHD TV as its display screen is large and has a large number of pixels.

Figure 13.21
Resolution of UHD TV

1) Korean company Samsung Electronics, is releasing new types of TVs. Curved UHD TV is one of them. One of the advantages of the curved screen TV is impressive picture quality. In a smaller screen distortion may be jarring to the eyes rather than the impression of the observed images. A similar discomfort may exist for a screen that is too big. For this reason curvature of the curved screen is made under the assumption that there is no distortion of the image. As an example, assume the average distance for watching TV is 3.5~4 m at home, then the screen curvature of 4200R can create an impressive image in a Curved UHD TV, without significant distortion or loss of detail. Screen made in this way gives a panoramic effect. Therefore, it is expected to see more UHD TV in the market.

(a) (b)

Figure 13.22
Curved UHD TV (a)Samsung 105 inch LCD TV (b)LG 55 inch OLED TV

(2) Monitor

UHD monitor is a product from a niche market of UHD TV but it seems to be gradually expanding to various computer monitors and of various sizes. Due to the rapid cost reduction of UHD TV, low priced high–resolution monitors are expected to be available in near future. A further expansion of specialized market including monitors for professionals with premium–quality is also expected.

Figure 13.23
Curved UHD monitor

3.7 OLED (Organic Light Emitting Diode)

OLED is an organic light emitting diode, that emits light when electric current flows through it. It can emit light at a voltage lower in comparison to that of the

LED. Its main advantage is that it can easily be made as a thin film, therefore light can be emitted over an wider angular range. It also responds fast. In an OLED TV, due to its wide viewing angle there is little distortion in image quality with viewing angle. Application of OLED includes mobile phones, tablets and TV. It is expected to be part of next-generation flat screen display. With the continuous development of materials, processes technology and cost reduction, various OLED displays are expected to be found in the market.

3.8 Application of OLED

(1) Illumination

The efficiency of light emission of OLED is not sufficient enough until now to use it as a source of illumination. Considering that the light efficiency of the LED products is at least 50m/W, the potential commercialization of OLED illumination is at its initial step. However, industries are paying attention to use OLED as a source of illumination because of certain advantages. It is possible to use the OLED in large sized panel of thickness of only 2~3 mm. Experts suggest that it is suitable for indoor illumination which requires faint light. A variety of designs can also be implemented because the OLED is thin and flexible. The OLED consumes less power, so OLED as well as LED can be energy saving light source for next-generation lighting.

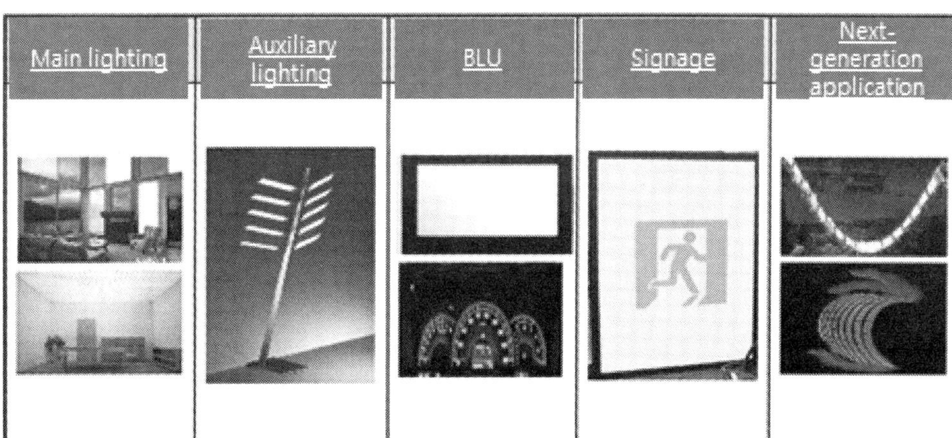

Figure 13.24

Types of OLED lighting

OLED illumination can be applied in various areas such as main lighting, auxiliary lighting, display BLU and display of various electronic devices. Development of surface lighting using fluorescent lamps which is a linear shaped light source, and LED, which is point light source, is actively progressing. OLED does not need auxiliary aids such as a light guide plate since it itself is the surface light type and a large area OLED fabrication is possible. If the brightness per unit area is adjusted to reduce glare, human-friendly lighting is possible, where OLED can have significant application. OLED can be applied in various other forms such as auxiliary lighting stands in the form of a surface light source, guide light, etc.

One of the high marketable applications of the OLED surface light source is display BLU and extreme slim display. Various display indicators in electronic products can be replaced by OLED surface light source. Few of the feasible surface light sources are surface illumination using flexible substrate and multiple functional windows using large transparent (or semi-transparent) substrate.

Canvas first introduced 99×99 mm panel with a light efficiency of 45 lm/W as a prototype. Hakunetu-sha was established in 1890 and has changed its name to Toshiba Lighting in 1989. The Toshiba Lighting has started the next-generation lighting business. Various applications of LED and OLED without shadow, are acclaimed. Panasonic Corp. achieved 142 lm/W using the high-refractive-index hemisphere lens in the laboratory. PIOL was founded for the sales of OLED illumination panels. Since 2012 it is in charge of research and development as well as sales. Illumination panels of brand name as "Eleaf", having a light efficiency of 30 lm/W and brightness of 3,000 cd/m^2 are being sold. Lighting Japan company unveiled illumination panels of size 150×50 mm and 200×50 mm in 2013. Osram produces both panels and modules. It introduced OLED prototype named 'ORBEOS' in 2009. Since then new type of products are are introduced every year. In April 2013 FC Bayern Erlebniswelt museum was newly decorated using LED and OLED devices. In 2014, transparent OLED illumination panel of size 180×65 mm, light efficiency of 20 lm/W and transparency of 57% was introduced.

OLED illumination market is just in the infancy in comparison to that of the LED. OLED lighting was 4 to 5 years behind the LED but it is expected that it will reach a level that can compete with the LED soon. The OLED may take time for commercialization due to its relatively low light efficiency and high prices. Implementation of various designs and development of technology to compensate the poor lifetime in comparison to that of the LED, price reduction and standardization are thought to be the key. Present day applications of light source are not just limited to the basic dome light but it also include walls being illuminated, automobiles, agricultural fields, apparel and medical applications.

Figure 13.25
Applications of OLED light

(2) Traffic Lights

The general function of the future traffic light is to provide an updatable traffic and lane information in real-time using the OLED display. Some of the signals are independent of the color, as it simply displays certain information. Unlike existing traffic light systems, installation of innovative suspended system is simple and electricity can be saved since the efficiency of OLED is higher than that of conventional systems.

The rear displays in cars with OLED immediately reacts to the movement of the vehicle. For example, if the wheel is to turn to left, the rear indicator can indicate that accordingly. The trailing driver can predict the movement of the front driver by the light indicator. However, a confusing flow of light can cause distraction and confusion to others.

OLED is a very thin and lightweight and it can have a vast set of applications. It's organic materials are sensitive and prone to damage.

Figure 13.26

Car rear light

(3) OLED Display

OLED displays are lighter, thinner, less power consuming and have higher purity of color in comparison to that of the widely used LCD displays. OLED displays do not need a backlight, unlike LCD. They have many advantages and can be used in transparent and flexible displays.

The first commercialized Active Matrix OLED (AMOLED) displays were 300 OLED displays for mobile phones, introduced by SK Display, Japan in 2002. In 2003, Full color 2.16 inch AMOLED display "ALE251" was mounted on Kodak digital camera. The full-fledged mass production of AMOLED display began with 260,000 pixels by Samsung Display from 2007.

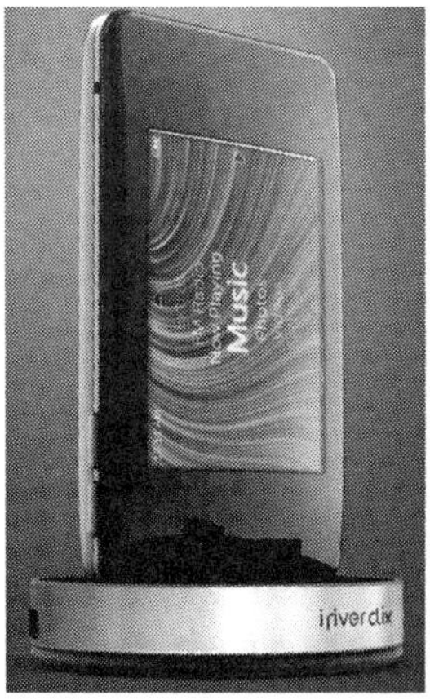

Figure 13.27

Clix of Iriver that opened full-fledged era of AMOLED display

Since 2010, AMOLED displays are used for various mobile electronic devices, high-resolution monitor, TV, wearable display, etc.

Figure 13.28

Commercialized products of AMOLED display

Figure 13.29

Future products applying AMOLED display

AMOLED has great possibility of making devices with lightweight, thin and flexible characteristics and can implement various shapes of display elements.

Display Engineering

4 Summary

The various existing display systems, like CRT, PDP, LCD, LED, OLED, transparent display, flexible display and UHD have been described.

The initial CRT displays are large in size and protrudes from the rear for stable screen output. The screen used to be curved and then became flat as the technology developed further. The production process and driving method for the CRT display are simple. It has wide viewing angle and rich in colors. However, to prevent image distortion due to the deflection of the electron beam and to improve the focus, it had to be thick. For this, the back of screen has to be sticking out causing it to be heavy and making it unfavourably large. Due to these disadvantages, it is not used much in these days.

PDP is a display that uses plasma effect (than the electron gun in CRT system) and its thickness was much reduced compared to the CRT. The color, brightness and sharpness of image do not change anywhere left or right since self-emitting display elements are used. However, the power consumption in PDP is high and a lot of heat is generated due to the discharge of plasma. The resolution of PDP is lower compared to the LCD or LED. Thus, PDP is disappearing from the flat display market that is now taken over mostly by LCD or LED based displays.

For LCD, the required thickness for the display can be below 10 cm when the screen is over 30 inches. LCD is divided into many types depending on the backlight unit. It can be classified as passive and active LCDs. For passive LCD, the productivity is high and production cost is low due to its simple structure. However, the image quality is poor and the response time is low. When displaying fast-moving images, a serious residual image occurs. Active LCD is divided into TN (Twisted Nematic), IPS (In-Plane Switching) and VA (Vertical Alignment) TFT panels depending on the liquid crystal arrangement of the panel. Unlike TN panel, the IPS or VA panel has lower image distortion when seen from sides. Since backlight is used in LCD displays, the displayed image can be very bright and sharp.

For LED, the conversion efficiency of light energy from electrical energy is very high. Since the energy consumption can be reduced up to 90%, it is getting attention as next-generation light source, replacing incandescent lamps and fluorescent light that have a lower conversion efficiency of ~5%. Depending on the wavelength of emitted light, there are various technologies due to which the LEDs can be used as visible, infrared and ultraviolet light source. An example for visible LED is LED TV. For this LED BLU is used instead of CCFL for the backlight of LCD. The advantages are its image quality is better than that of LCD and the power consumption is low. However, one of its disadvantages is that a separate cooling system is needed to remove a lot of heat generated in these displays.

OLED is an organic light emitting diode. Its image is 1000 times faster than that of LCD. OLED is divide into AMOLED and PMOLED. Since OLED uses elements which are self-emitting, it itself can be used as a display. Thus, it can be made very thin, has no afterimage due to fast response time and delivers high-quality clear images.

The transparent displays have transparent screen through which the back of the screen can be seen. Its technological implementation is diverse. It is divided into see-through types such as HUD, HMD and direct-view types, such as transparent LCD and transparent OLED displays.

In the HUD, the images are formed by the optical engine, then the images are projected on a transparent screen, where it is reflected to the user's eye. For HMD, the images are projected into user's eyes directly, without using the transparent screen. Transparent LCD and transparent OLED can be used for display devices such as smart windows. Since the OLED light source is better, it is more likely that transparent OLED would be used more in future.

The flexible display uses very thin and flexible substrate. It can be bent, folded or rolled without damaging the device. Its form factors such as light-weight, thin and unbreakable, the picture quality of the display are presupposed to be at least equivalent to that of LCD or OLED. The applications are a flexible OLED smartphone, E-paper, wearable devices, etc.

The resolution of the HUD display is 4 to 16 times higher compared to that of the HD and it has a multiple audio channels that is more than 10, providing a highly realistic ultrahigh quality AV broadcasting service. For UHD TV, a very fine screen display is possible due to its increased number-density of pixels. In a large screen, this technology can give a 3D effect. Such a TV can be watched at a close range due to the high number-density of pixels. UHD monitors are expected to create a specialized market such as monitors for professionals with a premium high-quality .

PRACTICE

1. Describe the pros and cons of CRT (Cathode–ray Tube)

2. Show an example of self–emitting display and what are excellent characteristics compared to non–emitting display.

3. Explain the difference between VA and IPS liquid crystal mode used for LCD products.

4. What are the applications of Flexible Display?

5. Explain the features and benefits of LCD and OLED.

6. What are the applications of transparent display?

7. Give an example and explain the expected change in our real life through various displays in future.

References

[1] Optical World 147-22 / Korean Optical Equipment Association (2013)

[2] Optical World 147 28-39 / Korea Optics Association (2013)

[3] Analysis of Electronic Communication Trends Volume 23, Issue 5, Volume 113 (2008.03) pp.153-163 /

[4] http://nge.itfind.or.kr

[5] http://www.ebn.co.kr

[6] Hyung-Cheol Lee, "Research Methodology of 3D Hummock Factor: Focusing on 3D Visual Fatigue"

[7] Seung Hyun Lee, "Understanding of 3D video" JinSam Media (2010.2)

[8] "3D display technology and market forecast report"

[9] "Autostereoscopic 3D Display in Signage and Professional Application" Insight Media (2009)

[10] "Human-Device Interaction Technology", "CT Insight", Volume 27, Korea Creative Content Agency (2012)

[11] The basic structure and the structure of the latest display technology, (2003)

[12] Introduction to KIDS Display School Display, Korea Information Display Society (2014)

Index

|D|

|E|

|F|

| M |

| N |

| O |

| P |

| Q |

| R |

| S |

| T |

| U |

| V |

| W |